Study Guide

Chemical Principles

EIGHTH EDITION

Steven Zumdahl
University of Illinois

Donald DeCoste
University of Illinois

Prepared by

Paul Kelter
Northern Illinois University

CENGAGE Learning

Australia • Brazil • Mexico • Singapore • United Kingdom • United States

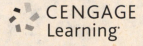

CENGAGE
Learning

WCN: 01-100-101

For product information and technology assistance, contact us at **Cengage Learning Customer & Sales Support, 1-800-354-9706**.

For permission to use material from this text or product, submit all requests online at **www.cengage.com/permissions**
Further permissions questions can be emailed to **permissionrequest@cengage.com**.

ISBN: 978-1-305-86712-3

Cengage Learning
20 Channel Center Street
Boston, MA 02210
USA

Cengage Learning is a leading provider of customized learning solutions with office locations around the globe, including Singapore, the United Kingdom, Australia, Mexico, Brazil, and Japan. Locate your local office at: **www.cengage.com/global**.

Cengage Learning products are represented in Canada by Nelson Education, Ltd.

To learn more about Cengage Learning Solutions, visit **www.cengage.com**.

Purchase any of our products at your local college store or at our preferred online store **www.cengagebrain.com**.

Printed in the United States of America
Print Number: 01 Print Year: 2015

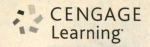

Table of Contents

To the Student

My thirty years of experience in general chemistry have shown me that there are several characteristics that successful students have in common:

- They keep up with the work on a daily basis;
- They study and, often, work with others, until they understand concepts deeply;
- They look for different ways to learn;
- They raise questions and seek answers to gain clarity.

This *Study Guide* can't help address the first item, but it can support you as you work through the other three parts of the list. The guide is written for the eighth edition of Steve Zumdahl's *Chemical Principles* textbook; my section descriptions and table and figure references match his. However there are a few instances in the guide when my approach will differ from that used by Dr. Zumdahl. Both ways have proven successful and we want you to use the method that works best for you.

There are over 1300 problems, many of them worked out, in the guide. Ideally, you will read the textbook, do the problems, then read this *Study Guide* and do its problems. The guide is not a substitute for the text. The textbook has a richness of information that the guide cannot approach. However, the guide can enrich and deepen your understanding. Please use it with that in mind.

Acknowledgment

Thanks to Brendan for his support.

Thanks to Barb for 24/7 for 35, and to Seth and Aaron for 27 and 25.

CHAPTER 1

Chemists and Chemistry

The Bottom Line: Chapter 1

The goal in this chapter is to introduce you to the incredible, though often unrecognized, impact of chemistry on our world. Key discussions include the *scientific method* as a framework for rational thinking, the role of chemistry in *industrial manufacturing*, and the experiences of people whose *chosen careers* are in the chemical sciences.

Note in Figure 1.1 in your textbook the wide variety of professions that value the expertise of chemists! Zumdahl makes the key point that the chemists in these fields may not be doing chemistry, but use their understanding of chemistry to work as lawyers, salespeople, etc. Chemists are problem solvers, and that is the theme of the next section.

1.1 Thinking Like a Chemist

The salient point here is that solving problems in chemistry is not a lock-step process. Although you try to establish a reasonable general direction, the work often proceeds in fits and starts. The notion that every problem can have unexpected twists gives chemistry its flavor. Creative problem solving involves understanding and persistence.

1.2 A Real-World Chemistry Problem

In this section, we dive right into a chemistry problem of significance—saving books that are rotting away due to the acidification of the book pages. Some universal ways of expressing chemical processes are introduced. The acid nature of aluminum is displayed in a reversible equation, and justified by considering the Al^{+3}–O bond strength. The polymer *cellulose* is displayed in Figure 1.2 in your textbook. The structure is written using one form of the chemist's shorthand, the ball-and-stick structure, which shows each type of atom as a different color ball and shows the connections between them—the "bonds" as thin lines or cylinders—"sticks". (Which are the oxygen atoms? Which are the hydrogen atoms? The carbon atoms?) Other representations of molecules depict the chemical connection of two atoms—the bonds—as lines. Still others use the "space-filling" model, in which the atoms are shown as spheres and there are no bonds shown. Look for each of these, and other representations of molecules, throughout the text. Which picture of molecules is best? That depends upon what you want to show. Here, Zumdahl wants to show a representation of the connections between the atoms that help to depict the shape of the molecule.

In order to fully understand the problem, you also need to know that acids can break cellulose apart in a process called hydrolysis and that bases can react to neutralize acids. Note as well that your book is really dealing with *two aspects* of "thinking chemically." The first is the *content background*. The second, and the real theme here, is the *process* of assessing and solving a problem. The schematic in Figure 1.3 in your textbook summarizes this. This thinking strategy is also the focus of Section 1.3.

1.3 The Scientific Method

As this chapter goes into the various parts of the *scientific method* and summarizes the approach in Figure 1.4 in your textbook, it is made clear that there is no single approach to rational problem solving. Flexibility and creativity give rise to the special insights that lead to new knowledge.

Among the various terms that are introduced (theory, law, etc.) is an important subtext—that we are affected by human biases and prejudices, and we must be as careful as possible not to anticipate specific outcomes merely because we want them to be so.

Note the example of the Law of Conservation of Mass, in which Zumdahl stresses the difference between a law (what happens) and a theory (why it happens).

1.4 Industrial Chemistry

This section broadens the scope of the discussion to include large-scale chemical processing and manufacturing. It cites the differences between small-scale syntheses done in academic settings and large-scale ones characteristic of industrial plants that produce billions of pounds of product.

One of the recurring themes of this book is that you approach new problems in imaginative ways. With that in mind, Example 1.4, below, makes the connection between industrial chemistry and economics.

Example 1.4 - Chemistry in the Marketplace

One popular food company currently produces five styles of margarine—regular, lite, extra lite, ultra, and ultra fat-free. They differ primarily in the fraction of water to fat in the mix, with the ultra (i.e., lower fat and calorie) types having much more water than the regular margarine. The cost to the consumer for all five margarine types is about the same. Is this reasonable? In thinking through the problem, please list the major kinds of costs to the company in getting the product from "the drawing board" to the store shelf.

Solution

It would seem, at first, that pricing all types of margarine the same would put consumers in the position of paying a stiff price for the water in the ultra brands. The list of drawing board to store shelf expenses might include research and development, safety testing of the product, the actual cost of the substances in the product, labor, advertising, packaging, refrigeration, shipping, environmental controls in the manufacturing process, and a profit margin for the company. Given these other expenses, *should the ultra styles be less expensive than the others*?

1.5 Polyvinyl Chloride (PVC): Real-World Chemistry

Although polyvinyl chloride is the material focus here, the real message in this section is concerned with the intricacy and applications of chemicals. Chemists don't have all the answers, but through systematic work uses can be found for substances. Please test your understanding of this section by answering the following questions:

1. Why is pure PVC almost useless?
2. Your book says that "The development of PVC illustrates the interplay of logic and serendipity." Summarize how this interplay was important in the development of PVC.
3. State the effect of each common PVC additive.

4. There were several unknowns regarding PVC. What were they? In general, does it make sense to proceed with a process even though you don't really know why it works? Explain your answer.
5. Historically, why were tin and antimony compounds preferred as stabilizers in spite of their relatively high cost?
6. What is the importance of antimony(III) oxide in PVC manufacture? How does it work? Why is it occasionally detrimental to the product?
7. At the end of the section, your book cites several factors to consider in product manufacture. Please list five *additional* factors.

CHAPTER 2

Atoms, Molecules, and Ions

The Bottom Line: Chapter 2

This chapter gives a thumbnail sketch of how the scientific method was used to shed light on chemical reactions that have been used for thousands of years and on the fundamental structure of chemical particles. Special emphasis is placed on the work of Dalton, Cannizzaro, and Rutherford. The last two sections of the chapter comprise most of the written text and deal with how we systematically organize the elements in the periodic table and how we name compounds formed from the elements.

2.1 The Early History of Chemistry

The first part of the section considers the work of the ancient Greeks. The book points out that the Greeks did not use experiments as a way to establish the veracity of their ideas. It can be added that according to experts on the culture, there was, in effect, a pecking order in which the aristocracy of society, including Plato, and later Socrates and Aristotle (as in *aristo*cracy), felt that experimentation was beneath them. Rather, such activities were the purview of the "non-intellectuals" in the society.

This section discusses *alchemy*. In addition to comments on the positive aspects of this **pseudo-science**, we can note that the opening of trade routes throughout Europe and Asia in the reign of Alexander the Great led to an important intercultural sharing of alchemical ideas between the Greeks, Egyptians, Indians, and Chinese. The word **alchemy** itself is probably derived from a combination of the Arabic definite article **al** with the Chinese **chin-i**, meaning **gold-making juice**. Then again, the Greek word for casting or pouring metals is **cheo**. However, the Egyptian word for black (the color of metals in preparation for alchemical treatment) is **khem**. Chemistry truly had international origins!

The section ends with the advent of experimental chemistry at the hands of Robert Boyle. The quantitative aspects of chemistry were especially refined with the work of Antoine Lavoisier, the focus of the first part of Section 2.2.

2.2 Fundamental Chemical Laws

This section describes **the law of conservation of mass** and **the law of definite proportion**. Based on the discussion, try the following exercises.

Example 2.2 A - The Law of Definite Proportion

A sample of chloroform is found to contain 12.0 g of carbon, 106.4 g of chlorine, and 1.01 g of hydrogen. If a second sample of chloroform is found to contain 30.0 g of carbon, how many grams of chlorine and how many grams of hydrogen does it contain?

Solution

Assuming that the law of definite proportion is true, the ratios of the masses of the elements in chloroform are constant. Therefore, if the amount of carbon is increased by a factor of 2.5 (30.0 g/12.0 g, as above) then the same must hold true for chlorine and hydrogen.

$$\text{g chlorine} = 106.4 \text{ g} \times 2.5 = 266. \text{ g chlorine}$$
$$\text{g hydrogen} = 1.01 \text{ g} \times 2.5 = 2.53 \text{ g hydrogen}$$

The law of multiple proportions is discussed in your textbook. The salient point is that, given a compound such as H_2O, if H_2O_2 exists, the ratio of the masses of oxygen atoms that combine with one gram of hydrogen will be a small whole number. The following exercise illustrates this.

Example 2.2 B - The Law of Multiple Proportions

Water, H_2O, contains 2.02 g of hydrogen and 16.0 g of oxygen. Hydrogen peroxide, H_2O_2, contains 2.02 g of hydrogen and 32.0 g of oxygen. Show how these data illustrate the law of multiple proportions.

Solution

In H_2O, 7.92 g (16.0/2.02) of oxygen combines with each 1.0 g of hydrogen. In H_2O_2, 15.84 g (32.0/2.02) of oxygen combines with each 1.0 g of hydrogen. The **ratio of the masses** of oxygen in the two compounds is:

$$\frac{15.84\text{ g}}{7.92\text{ g}} = 2.00$$

This is a **small whole number** that illustrates the law of multiple proportions.

2.3 Dalton's Atomic Theory

Read the four statements relating to Dalton's atomic theory given in your textbook. Use the following exercise to test your understanding of the theory.

Example 2.3 - Dalton's Atomic Theory

Match the chemical statement on the left side with the atomic theory statement on the right side.

	Chemistry		Atomic Theory
a.	Although graphite and diamond have different properties (due to the nature of interatomic bonding), they are both composed solely of carbon. The carbon atoms are identical.	1.	Each element is made up of tiny particles called atoms.
b.	$2H_2 + O_2 \rightarrow 2H_2O$, not CS_2 or NaCl	2.	The atoms of a given element are identical. The atoms of different elements are different in some fundamental way or ways.
c.	There are 6.02×10^{23} atoms in 55.85 g of iron.	3.	Chemical compounds are formed when atoms combine with one another other. A given compound always has the same relative numbers and types of atoms.
d.	$C + O_2 \rightarrow CO_2$. CO_2 is not CO, CO_3 or Fe_2O_3.	4.	Chemical reactions involve reorganization of the atoms—changes in the way they are bound together. The atoms themselves are not changed in a chemical reaction.

Solution

a. 2
b. 4
c. 1
d. 3

The statements in Dalton's atomic theory are **not necessarily mutually exclusive**. As a result, statement b can fit No. 3, and statement d can fit No. 4.

In reading this section in your text, don't neglect the discussion of **Avogadro's hypothesis**! The stated relationship between volumes and numbers of particles of gases sets the stage for realizing the *diatomic* nature of gases such as O_2, H_2, and Cl_2.

2.4 Cannizzaro's Interpretation

This section shows how the combination of elegant assumptions and many experiments to verify the work led to a common (and correct!) understanding of many atomic masses. The next exercise is a twist on Example 2.1 in your textbook.

Example 2.4 - Establishing Relative Atomic Masses

Chemistry students at an archeological dig site find a red metal piece of unknown composition. After an analysis of the material on the surface of the piece, they find that the mass ratio of metal to oxygen is 2.327 grams to 1.000 gram. If the formula of the compound is M_2O_3, what is the unknown metal, M?

Solution

There are all kinds of reasonable approaches to the problem. For example, if 2 atoms of M to 3 atoms of O yields a ratio of 2.327 g to 1.000 g, then 3 atoms of M to 3 atoms of O yields a mass ratio of

$$^{3}/_{2}\,(2.327 \text{ g}) = 3.490 \text{ g of the metal to } 1.000 \text{ g of oxygen.}$$

The actual mass of the metal is therefore 3.490 times the mass of oxygen

$$3.490 \times 16.00 = \mathbf{55.85\ grams}.$$

The unknown metal is **iron**, and the formula of the compound is $\mathbf{Fe_2O_3}$.

2.5 Early Experiments to Characterize the Atom

The focus here is the structure of the atom itself. The experiments of Thomson that determined the existence of electrons are described. The magnitude of the charge on an electron, as well as its mass, was determined by Millikan. The surprising results of Rutherford, described in "The Nuclear Atom" subsection in your textbook, elucidated the nature of the atom as having a small, dense, positively-charged nucleus surrounded by electrons moving around it at a relatively large distance.

2.6 The Modern View of Atomic Structure: An Introduction

Zumdahl sets the stage for this discussion with the start of the fourth paragraph, where he writes, "As with any theory in science, while it provides answers to questions it also brings about more questions." He then asks, "“If all atoms are composed of these same components, why do different atoms have different chemical properties?” The answer to this question lies in the number and the arrangement of the electrons."

The names, sizes and masses of the parts of the atom are discussed in your textbook. Pay special attention to Table 2.2 in your textbook. The key point is that although the mass of an atom is concentrated in its nucleus, most of the volume of the atom is taken up by electron movement. Since electrons are involved in bonding, we need a way to symbolize the number of protons, neutrons, and electrons in an atom. Given

$$^{A}_{Z}X$$

X = The element symbol, as read from the periodic table.
Z = The number of protons.
A = The number of protons plus neutrons. (The number of neutrons = A − Z.)

The number of **protons** determines the element. **Potassium** has **19 protons**. If an element has **20 protons**, it is **calcium**. The number of neutrons and electrons does not determine what element you have. Only the number of protons is important. For example,

$$^{31}_{15}P$$

contains **15 protons, 15 electrons** (in an atom, which is electrically neutral, the number of protons equals the number of electrons), and **16 neutrons** (31−15). The atomic mass is 31. If instead, we had

$$^{30}_{15}P$$

everything would be the same as in the previous example, except for the number of neutrons, which would be **15 neutrons** (30−15). This phosphorus atom is therefore an **isotope** of the previous phosphorus atom.

Example 2.6 - Protons, Neutrons, Electrons, and Symbols

Fill in the missing information in the following table:

Symbol	Element	Protons	Neutrons	Electrons
$^{14}_{6}C$	______	______	______	______
$^{235}_{92}$___	______	______	______	______
$^{_}_{_}$___	uranium	______	146	______
$^{55}_{_}$___	______	______	______	25

Solution

Symbol	Element	Protons	Neutrons	Electrons
${}^{14}_{6}C$	carbon	6	8	6
${}^{235}_{92}U$	uranium	92	143	92
${}^{238}_{92}U$	uranium	92	146	92
${}^{55}_{25}Mn$	manganese	25	30	25

Determining the number of each kind of particle becomes more challenging when **ions** are introduced. Both your textbook and this study guide deal with these ideas in our next sections.

2.7 Molecules and Ions

The key point of this section is that molecules contain atoms rather than ions. Certain atoms do tend to ionize. For example, CaF_2 contains three ions, a Ca^{2+} and two F^-. Compounds such as CaF_2 that contain ions are called ionic solids or salts. The positive and negative charges in salts exactly balance. In CaF_2, there are two positive and two negative charges, yielding a neutral salt. In future chapters you will learn how to predict *whether* molecules or salts will form. At this point it is only expected that you know one when you see one.

For the next exercise please keep in mind that the total charge of an ion is determined by whether you have an excess of protons (net positive charge) or electrons (net negative charge).

Fe^{3+} means that iron has **3 more protons than electrons** (26 protons, 23 electrons)
O^{2-} means that oxygen has **2 more electrons than protons** (8 protons, 10 electrons)

Example 2.7 - Protons, Neutrons, and Electrons in Ions

Fill in the missing numbers in the following table:

Symbol	Protons	Neutrons	Electrons	Charge
${}^{32}_{16}S^{2-}$	________	________	________	________
${}^{\_}_{\_}\_\_\_^{\_}$	56	81	54	________
${}^{\_}_{\_}Cl^-$	________	20	________	–1
${}^{52}_{24}\_\_\_^{7+}$	________	________	________	________

Solution

Symbol	Protons	Neutrons	Electrons	Charge
${}^{32}_{16}S^{2-}$	16	16	18	–2
${}^{137}_{56}Ba^{2+}$	56	81	54	+2
${}^{37}_{17}Cl^-$	17	20	18	–1
${}^{52}_{24}Cr^{7+}$	24	28	17	+7

2.8 An Introduction to the Periodic Table

This section points out that the periodic table is organized based on the properties that elements have in common with one another. You can combine the section reading with your own experience to identify the essential properties of metals vs. nonmetals.

Look at the periodic table, Figure 2.19 in your textbook. Please note:

1. The bold line from boron to polonium that separates metals from nonmetals.
2. Group names. You will be using these designations in your future science work.
3. The various properties of each group as described in the text.
4. Groups go down; periods go across.

Example 2.8 - Identify the Element

Given the following information, identify each element:

a. This is the only metal in group 6A.

b. This alkali metal is in the same period as iodine.

c. Two atoms of this element, which is in the same period as magnesium, combine with a calcium ion to form a salt.

Solution

a. The only metal in group 6A is **polonium.** It is to the left of the bold line on the periodic table. Elements that are next to that line are sometimes called metalloids because they have some properties of both metals and nonmetals.

b. The only alkali metal in the same period as iodine is **rubidium.**

c. Calcium forms Ca^{2+}. Therefore, if two atoms combine with it to make an electrically neutral salt, they each must have a −1 charge. Halogens readily form ions with a −1 charge. The halogen that is in the same period as magnesium is **chlorine.**

2.9 Naming Simple Compounds

There are really two keys to naming or assigning formulas to compounds. The first key is to know the names of the elements (typically, knowing numbers 1 through 54 is sufficient). The second key is to correctly assign charges to ions. IUPAC (International Union of Pure and Applied Chemistry) nomenclature standards are used in your textbook and in this study guide. That way, any chemist (or chemistry student) should know exactly what you are talking about when you name or give the formula of a compound.

Naming Compounds

Zumdahl discusses three types of binary compounds:

- **Type I**, which contains a metal that forms only one type of cation, such as Ca^{2+} or Li^{+};
- **Type II**, which contain metals that have more than one type of cation, such as Cr^{3+} and Cr^{6+}, and can form more than one compound with an anion, and;
- **Type III**, which are binary covalent compounds containing two nonmetals.

As we consider the rules for naming compounds, practice assigning each to a type, I, II, or III.

1. Binary Salts: A binary salt contains only one kind of cation and one kind of anion.

a. **Cations:** Ions with positive charges fall into two general classes. The first contains elements that ionize to only one oxidation state, including **all group 1A, 2A, and 3A elements**. Such cations form what your book calls **Type I binary ionic compounds**. To name the cation, simply give the name of the element, and add **ion**. For example, K^+ is called potassium ion, and Be^{2+} is called beryllium ion. The reverse process is also true. Calcium ion is labeled Ca^{2+}.

Elements that Ionize to a Single Oxidation State (Cation)	Oxidation State	Examples
All group 1A (alkaline)	+1	Na^+, Rb^+
All group 2A (alkaline earths)	+2	Mg^{2+}, Ba^{2+}
All group 3A	+3	Al^{3+}, Ga^{3+}

The second class of cations, which includes most transition elements, some halogens, and heavier noble gases, form **Type II binary ionic compounds**. The cations that form these compounds can have several oxidation states, such as Fe^{3+}, Fe^{2+}, and Fe^{6+}. We use Roman numerals in parentheses after the element to name such cations. For example, try the following example regarding cation nomenclature:

Example 2.9 A - Naming Cations

Fill in the following table:

Element	Oxidation State	Symbol	Name
phosphorus	+5	P^{5+}	________
________	________	Mn^{7+}	________
________	________	______	rubidium ion
________	________	______	bromine(V) ion
vanadium	+2	______	________
strontium	+2	______	________

Solution

Element	Oxidation State	Symbol	Name
phosphorus	+5	P^{5+}	phosphorus(V) ion
manganese	+7	Mn^{7+}	manganese(VII) ion
rubidium	+1	Rb^+	rubidium ion
bromine	+5	Br^{5+}	bromine(V) ion
vanadium	+2	V^{2+}	vanadium(II) ion
strontium	+2	Sr^{2+}	strontium ion

b. **Anions:** Negative ions (anions) that are made up of a single element are all named in the same fashion. You simply strip off the ending and add **ide**. For example,

Symbol	Element	Anion Name
Cl^-	chlorine	chlor**ide**
H^-	hydrogen	hydr**ide**
S^{2-}	sulfur	sulf**ide**
N^{3-}	nitrogen	nitr**ide**

Some anions can have several oxidation states. For purposes of naming the anion part of compounds, the oxidation state is unimportant. The *–ide* rule always applies. Similarly, when you see an *–ide* in a name, it indicates an anion.

THE OXIDATION STATE OF AN ANION IS EQUAL TO THE GROUP NUMBER MINUS EIGHT

For example, if nitrogen (group 5A) acts as an anion (it can also be a cation in many cases),

$$\text{Oxidation State} = \underset{\substack{\uparrow \\ \text{group \#}}}{5} - 8 = -3, \text{ or } N^{3-}$$

The following exercise tests your ability to interconvert between the anion and its name.

Example 2.9 B - Naming Anions

Fill in the following table:

Element	Oxidation State	Symbol	Anion Name
oxygen	−2	_____	________
________	_____	Se^{2-}	________
iodine	_____	_____	________
________	_____	_____	phosphide

Solution

Element	Oxidation State	Symbol	Anion Name
oxygen	−2	O^{2-}	oxide
selenium	−2	Se^{2-}	selenide
iodine	−1	I^{-}	iodide
phosphorus	−3	P^{3-}	phosphide

c. **Naming binary salts:** The process of naming binary salts involves combining the names of the cation and anion parts. The only complication comes with determining whether or not the cation comes with a Roman numeral. As you name your compounds keep in mind that the cation part always comes first, and the total charge for the compound must equal zero.

Let's name $\mathbf{CaF_2}$ together, working with the anion side first. According to our rules for anions, fluorine becomes **fluoride**. The charge due to fluoride is −1 per fluorine × 2 fluorines = −2 total.

Let's consider the cation (left) side. Because the fluorine ions contribute a total charge of −2, the calcium ion must be +2 to make a neutral compound. The cation is Ca^{2+}. (Actually, we knew this already because calcium is an alkaline earth and can only oxidize to a +2 state!) Therefore, according to our Type I rules, the cation is **calcium** ion rather than calcium(II). The entire compound name is therefore

$$\underset{\substack{\uparrow \\ \text{cation}}}{\text{calcium}} + \underset{\substack{\uparrow \\ \text{anion}}}{\text{fluoride}} = \textbf{calcium fluoride}$$

For a different problem, let's name $\mathbf{PbS_2}$. Working on the anion side as before we have two **sulfides**. Because sulfur is in group 6A, its oxidation state is −2, though it *can* vary when combining with more electronegative compounds, such as in SO_3. In PbS_2, the two sulfides contribute −**4** to the compound. On the cation side, the **lead** can have more than one oxidation state. In this case, to form an electrically neutral salt its oxidation state must be +**4**. The cation part is **lead(IV)** ion, and the name of the compound is **lead(IV) sulfide**.

Example 2.9 C - Naming Binary Salts

Name the following binary compounds:

a. KCl b. $SnCl_4$ c. IF_5 d. Fe_2O_3

Solution

1. Name the anion.
2. Establish the total charge due to the anion.
3. Name the cation.
4. Determine the charge on each cation. (Remember that the compound must be neutral.)
5. If the cation is not in groups 1A, 2A, or 3A, its oxidation should be specified with Roman numerals.
6. Combine the cation and anion to get the name of the compound.

   a. **potassium chloride** (potassium is in group 1A)
   b. **tin(IV) chloride**
   c. **iodine(V) fluoride** (Iodine can act as a cation! We will see later that we can also call this iodine pentafluoride.)
   d. **iron(III) oxide**

2. Salts With Polyatomic Ions

   a. **Conventional polyatomic ions:** Table 2.5 in your textbook contains formulas for, and names of, the polyatomic ions you should know. The rules that apply for binary salt nomenclature work here as well. Keep in mind that you do not change the name of the polyatomic ions when you form the compound name. For example, $NaC_2H_3O_2$ combines the sodium cation with the acetate anion, giving **sodium acetate**.

   b. **Oxyanions:** Many polyatomic anions, including several in Table 2.5, contain different numbers of **oxygen atoms** combined with an atom of a different element. These are **oxyanions**. An oxyanion series normally contains either 2 or 4 members. In a two-member series, the anion with **more** oxygens gets the suffix –**ate**, and the one with **fewer** oxygens gets the suffix –**ite**.

   NO_2^- = nitr**ite** SO_3^{2-} = sulf**ite**
   NO_3^- = nitr**ate** SO_4^{2-} = sulf**ate**

   When there are four oxygen atoms in the series (as in many halogen-containing oxyanions) the sequence is:

   IO^- = **hypo**iod**ite**
   IO_2^- = iod**ite**
   IO_3^- = iod**ate**
   IO_4^- = **per**iod**ate**

Example 2.9 D - Naming Compounds With Polyatomic Ions

Name the following compounds:

a.	$KMnO_4$	d.	NaH_2PO_4	g.	KIO_3
b.	$Ba(OH)_2$	e.	$(NH_4)_2Cr_2O_7$	h.	$Ca(OCl)_2$
c.	$Fe(OH)_3$	f.	$NaBrO_4$	i.	$Cr(NO_3)_3$

Solution

a. potassium permanganate
b. barium hydroxide (Barium is in group 2A—it is not a transition element with more than one non-zero oxidation state and doesn't require a Roman numeral.)
c. iron(III) hydroxide (Iron is a transition metal. A Roman numeral is required.)
d. sodium dihydrogen phosphate
e. ammonium dichromate
f. sodium perbromate (BrO_4^- is part of an oxyanion series.)
g. potassium iodate (IO_3^- is part of an oxyanion series.)
h. calcium hypochlorite (OCl^- is also part of an oxyanion series.)
i. chromium(III) nitrate (Chromium is a transition element.)

3. Binary Covalent Compounds

Such compounds are formed between **two nonmetals**. Your book calls these **Type III Covalent Compounds**. We have already considered IUPAC naming of such compounds (See IF_5 in Example 2.9 C in this study guide.). However, your book proposes a more common, and equally acceptable, nomenclature system. In Table 2.6 in your textbook, prefixes for you to use in naming both the cation and anion parts of covalent compounds are presented. Note that *mono–* is *not* put in front of a single cation or anion—the prefix is assumed. The compound IF_5, which was previously called iodine(V) fluoride, can also be named **iodine pentafluoride**. Your book uses common systematic names, and we will use them in this study guide as well.

Zumdahl summarizes the entire process in Figures 2.20 and 2.21 in your textbook, which are a flowchart and a strategy for naming binary compounds.

Example 2.9 E - Naming Binary Covalent Compounds

Name the following compounds:

a.	CO_2	c.	N_2O_3	e.	SO_3
b.	P_2O_5	d.	Cl_2O_7	f.	BrF_3

Solution

a. carbon dioxide
b. diphosphorus pentoxide (drop the "a" in penta—the "o" or "a" in the prefix is normally dropped when the element begins with a vowel.)
c. dinitrogen trioxide
d. dichlorine heptoxide
e. sulfur trioxide
f. bromine trifluoride

4. Acids

Acids are considered in detail in Chapters 4, 7, and 8 of your textbook. At this early point, acids are defined after Interactive Example 2.4 in your textbook so you can at the very least name them. The chemical clarity will come later on. The names and formulas of some important acids are given in Tables 2.7 and 2.8 in your textbook. A flowchart for naming acids is given as Figure 2.22 in your textbook. To name an acid, first determine if the anion part is polyatomic.

If the anion part is monatomic (such as Cl^- or S^{2-}), change the –**ide** suffix to –**ic acid**, and add the prefix *hydro*– to the name. For example, the acid H_2S contains the anion S^{2-}. The sulf**ide** becomes sulfur**ic** **acid**. We add the prefix **hydro**–, giving H_2S the name **hydrosulfuric acid**. The one exception to the rule is CN^-. Even though the anion is polyatomic, the acid is known as hydrocyanic acid.

If the anion part is polyatomic (such as NO_3^- or $C_2H_3O_2^-$), change –**ite** suffixes on anions to –ous acid and change –**ate** suffixes to –**ic acid**. For example, HIO_4 contains the anion IO_4^-, periodate. We change the –**ate** to –**ic acid**, giving **periodic acid**.

Example 2.9 F - Naming Acids

Name the following acids:

a.	HF	c.	$HBrO_3$	e.	HI
b.	$HC_2H_3O_2$	d.	HBrO	f.	HNO_2

Solution

a. hydrofluoric acid (monatomic anion)
b. acetic acid (polyatomic anion)
c. bromic acid (polyatomic anion)
d. hypobromous acid (the anion is hypobromate)
e. hydroiodic acid (monatomic anion)
f. nitrous acid (the anion is nitrite)

Example 2.9 G - Tying It All Together

Name the following compounds:

a.	PCl_5	d.	Sb_2S_3	g.	$NaC_2H_3O_2$
b.	$HClO_2$	e.	XeF_4	h.	$NaHCO_3$
c.	$Ni(NO_3)_2$	f.	NH_4OH	i.	LiH

Solution

a. phosphorus pentachloride
b. chlorous acid (The anion is chlorite.)
c. nickel(II) nitrate (The nickel is a transition metal.)
d. antimony(III) sulfide (3 sulfurs × −2 each = −6; each antimony must have +6/2 = +3 oxidation state.)
e. xenon(IV) fluoride or xenon tetrafluoride
f. ammonium hydroxide (two polyatomic ions)
g. sodium acetate
h. sodium bicarbonate
i. lithium hydride (Hydrogen is an anion here.)

Getting Formulas From Names

The key to getting formulas from names is to remember that chemical names contain both a cation and anion part. In virtually every case, one of the two parts will unequivocally indicate its oxidation state. Remember that the compound is neutral, and the oxidation state on one part of the compound dictates the oxidation state on the other part. Let's try **magnesium nitride** together. Magnesium is in group 2A and, as such, can only form the Mg^{2+} ion. Nitrogen is found in many oxidation states, depending on the compound. However, when it exists as an **anion** in an ionic binary compound, we need to use our rule that the oxidation state equals the group number minus eight. For the nitride,

$$\text{oxidation state} = 5 - 8 = -3.$$

Nitrogen is present as the N^{3-} ion. The simplest way for a +2 ion to combine with a −3 ion to maintain charge neutrality is to have **3 Mg^{2+}** and **2 N^{3-}**, or **Mg_3N_2**.

The same thinking applies to deriving the formulas of acids, such as **nitric acid**. An **acid** has hydrogen for a cation. Recall the rules for **anions of acids**. The ending **–ic acid** came from an anion with an **–ate** ending. The anion was nitr**ate**, or NO_3^-. Therefore, the formula is **HNO_3**.

As a final exercise in this section, let's work out the formulas of compounds from their names.

Example 2.9 H - Formulas From Names

Write the formulas for each compound:

a. sodium chloride
b. calcium fluoride
c. iron(III) nitrate
d. copper(I) chloride
e. hypoiodous acid
f. tin(IV) oxide
g. dinitrogen tetroxide
h. ammonium acetate

Solution

a. $NaCl$
b. CaF_2
c. $Fe(NO_3)_3$
d. $CuCl$
e. HIO (IO^- is the hypoiodite ion)
f. SnO_2
g. N_2O_4
h. $NH_4C_2H_3O_2$

Exercises

Section 2.2

1. In an exothermic (heat producing) reaction, chlorine reacts with 2.0200 g of hydrogen to form 72.926 g of hydrogen chloride gas. How many grams of chlorine reacted with hydrogen?

2. Sulfur and oxygen can react to form both sulfur dioxide and sulfur trioxide. In sulfur dioxide, there are 32.06 g of sulfur and 32.00 g of oxygen. In sulfur trioxide, 32.06 g of sulfur are combined with 48.00 g of oxygen.
   a. What is the ratio of the weights of oxygen that combine with 32.06 g of sulfur?
   b. How do these data illustrate the law of multiple proportions?

3. By experiment it has been found that 2.18 g of zinc metal combines with oxygen to yield 2.71 g of zinc oxide. How many grams of oxygen reacted with zinc metal?

4. A sample of H_2SO_4 contains 2.02 g of hydrogen, 32.07 g of sulfur, and 64.0 g of oxygen. How many grams of sulfur and grams of oxygen are present in a second sample of H_2SO_4 containing 7.27 g of hydrogen?

5. In a reaction, 15.0 g of chromium oxide reacts with 34.0 g of aluminum to produce chromium and aluminum oxide. If 22.0 g of chromium is produced, how much aluminum oxide is produced?

6. Carbon monoxide contains 12.01 g of carbon and 16.0 g of oxygen. Carbon dioxide contains 12.01 g of carbon and 32.0 g of oxygen. What is the ratio of the masses of oxygen that combine with 12.01 g of carbon in these compounds? Is this consistent with the Law of Multiple Proportions?

7. Iron reacts with sulfur to form iron sulfide. If 56 g of iron are used, how many grams of sulfur are needed to produce 88 g of iron sulfide?

Section 2.3

8. Describe the part of Dalton's Atomic Theory to which each chemical statement relates.
   a. $H_2 + Cl_2 \rightarrow 2HCl$
   b. There are 3.01×10^{23} atoms in 20.04 g of calcium.
   c. Lead does not change to chromium when it forms lead hydroxide.

Section 2.6

9. Identify each of the following elements:
   a. $^{91}_{40}X$ c. $^{33}_{16}X$ e. $^{51}_{23}X$
   b. $^{108}_{47}X$ d. $^{85}_{36}X$ f. $^{133}_{55}X$

10. Identify each of the following elements:
   a. $^{98}_{43}X$ c. $^{75}_{33}X$ e. $^{40}_{19}X$
   b. $^{186}_{75}X$ d. $^{14}_{6}X$ f. $^{131}_{54}X$

11. Identify each of the following elements:
   a. $^{208}_{82}X$ b. $^{239}_{94}X$ c. $^{61}_{28}X$

12. How many protons and neutrons are in each of the following elements?
   a. ^{89}Y c. $^{24}Mg^{2+}$ e. $^{35}Cl^{-}$
   b. ^{73}Ge d. ^{238}U f. ^{65}Zn

13. How many protons and neutrons are in each of the following elements?
   a. ^{227}Ac c. ^{11}B e. ^{239}Pu
   b. ^{70}Ga d. ^{251}Cf f. ^{64}Cu

14. Each of these isotopes is used in seeding the prostate to cure cancer. Identify these:
   a. $^{125}_{53}X$ b. $^{103}_{46}X$ c. $^{131}_{55}X$

15. Three or four atoms of the element ununoctium were likely first observed in 2006. The formation was the result of the reaction below. Identify "X" and "Q".

$$^{249}_{98}X + ^{48}_{20}Q \rightarrow ^{294}_{118}Uuo + 3n$$

Section 2.7

16. How many protons, neutrons, and electrons are in each of the following ions?
   a. $^{56}Fe^{3+}$ c. $^{19}F^{-}$ e. $^{127}I^{-}$
   b. $^{40}Ca^{2+}$ d. $^{31}P^{3-}$ f. $^{127}I^{7+}$

17. How many protons, neutrons, and electrons are in each of the following?

a. ${}^{195}Pt^{+}$ d. ${}^{16}O^{2-}$ g. ${}^{184}W$

b. ${}^{93}Nb$ e. ${}^{122}Sb^{2+}$ h. ${}^{133}Cs^{+}$

c. ${}^{40}Ar^{-}$ f. ${}^{56}Fe^{2+}$ i. ${}^{28}Si^{3-}$

18. Fill in the missing information in the following table:

Symbol	Protons	Neutrons	Electrons	Charge
${}^{80}_{35}Br^{-}$	____	____	____	____
${}^{—}_{—}$____$^{5+}$	35	45	____	+5
${}^{137}_{—}$____$^{—}$	56	____	54	____
${}^{108}_{47}Ag^{+}$	____	____	____	____
${}^{51}_{23}$____$^{5+}$	____	____	____	____
${}^{—}_{—}Co^{—}$	____	32	____	+2

19. Fill in the missing information in the following table:

Symbol	Protons	Neutrons	Electrons	Charge
${}^{27}_{—}$____$^{—}$	13	____	10	____
${}^{88}_{38}$____$^{—}$	____	____	____	+1
${}^{—}_{—}$____$^{2+}$	30	35	____	____
${}^{35}_{—}$____$^{—}$	____	18	18	____
${}^{—}_{—}Te^{2-}$	____	76	____	____
${}^{85}_{—}Rb^{—}$	____	____	____	+1

Section 2.8

20. Name the family to which each of the following elements belongs:

a. Fe c. Ar e. Rb

b. Cl d. Sr f. Nd

21. Are the following elements metals or nonmetals?

a. Mg d. Br g. Co

b. P e. O h. Mo

c. Hg f. Bi i. Xe

22. Name the family to which each of the following elements belongs:

a. Es d. Yb f. Fr

b. I e. Kr g. Ca

c. Au

23. Where in a group will you probably find atoms having the largest atomic radius? Explain.

24. Given the position in the periodic table, what is the most likely oxidation state that each element will have when forming an ion?

a. Cs
b. N
c. Br
d. K
e. Al
f. S

25. Would you expect the following atoms to gain or lose electrons when forming an ion? If so, how many would be gained or lost?

a. Be
b. Cl
c. Al
d. O
e. F
f. Li
g. P

Section 2.9

26. Name each of the following compounds:

a. PbI_2
b. NH_4Cl
c. Fe_2O_3
d. LiH
e. CsCl
f. $NaC_2H_3O_2$
g. $Cr(OH)_3$
h. OsO_4
i. $K_2Cr_2O_7$
j. Na_2SO_4
k. KH_2PO_4

27. Name each of the following compounds:

a. $MgSO_4$
b. N_2O_3
c. Ce_2O_3
d. $KMnO_4$
e. NiO
f. $BaSO_4$
g. $Fe(IO_4)_3$
h. SO_3
i. $KClO_4$

28. Name each of the following compounds:

a. NI_3
b. PCl_5
c. CO
d. P_4O_{10}
e. N_2O_4
f. NH_3

29. Name each of the following compounds:

a. P_4O_6
b. KOH
c. N_2
d. BF_3
e. AgCl
f. $KHCO_3$
g. $AgNO_3$

30. Name each of the following compounds:

a. HIO_3
b. HBr
c. HNO_2
d. HCN
e. $NaNO_2$
f. K_2SO_3
g. $NaHSO_3$

31. Name each of the following compounds:

a. UF_6
b. $Cu(NO_3)_2$
c. H_3PO_4
d. SF_6
e. $Mg(OH)_2$
f. $SnCl_2$
g. Na_2CO_3

32. Write formulas for each of the following compounds:

a. sodium cyanide
b. tin(II) fluoride
c. sodium hydrogen sulfate
d. lead(II) nitrate
e. iron(III) oxide
f. calcium phosphate
g. sodium bromate

33. Write formulas for each of the following compounds:

a. sodium sulfate
b. manganese(IV) oxide
c. potassium chlorate
d. potassium hypochlorite
e. lithium aluminum hydride
f. barium chloride
g. magnesium oxide
h. copper(I) oxide

34. Write formulas for each of the following compounds:

a. potassium carbonate
b. magnesium hydroxide
c. dinitrogen tetroxide
d. hypoiodous acid
e. iron(III) chloride
f. tin(IV) oxide
g. rubidium nitrate
h. potassium chlorate
i. carbon tetrachloride
j. sodium iodate
k. potassium permanganate
l. sulfurous acid
m. potassium hydrogen phosphate
n. ammonium acetate
o. ammonium dichromate
p. hydrobromic acid

35. Give the names of the following acids:

a. H_2SO_3
b. HI
c. HBr
d. HNO_2
e. H_3PO_4
f. HCl

36. Give formulas for the following acids:

a. nitric acid
b. hydrofluoric acid
c. sulfuric acid
d. hydrocyanic acid
e. hydrosulfuric acid
f. acetic acid

37. Give the alternate or common name for each of the following compounds or cations:

a. sodium hydrogen carbonate ($NaHCO_3$)
b. dinitrogen monoxide (N_2O)
c. nitrogen monoxide (NO)
d. iron(II) (Fe^{2+})
e. tin(IV) (Sn^{4+})
f. lead(II) (Pb^{2+})

Multiple-Choice Self-Test

1. The oxides of CO and CO_2 must have the following carbon-to-oxygen mass ratio

A. 12:16, 12:32 B. 12:12, 12:16 C. 12:8, 12:4 D. 12:12, 12:24

2. How many electrons and protons, respectively, are there in Ra^{2+}?

A. 88, 88 B. 86, 88 C. 224, 226 D. 228, 224

3. How many total protons are found in two molecules of $C_{20}H_{30}O$?

A. 102 B. 316 C. 302 D. 600

4. How many electrons does an ion with mass number 210, 125 neutrons, and a charge of −2 have?

A. 85 B. 83 C. 87 D. 89

5. An ion has a charge of +3 and 55 electrons. Which of the following elements is that ion?

A. Th B. Ce C. Mn D. Co

6. Atom A loses 1 electron, and atom B gains 2 electrons. What formula results if these two ions combine to produce a neutral compound?

A. AB B. A_2B C. AB_2 D. A_2B_3

7. Which group of elements belongs to the Transition Metals family?

A. Ru, C, Hg, Ir B. Pd, Ir, Ac, Re C. Bi, Sc, Pu, Rn D. Ti, Sc, Au, Fr

8. The formula for iron(III) carbonate is

A. $FeCO_3$ B. $Fe_2(CO)_3$ C. $Fe_2(CO_3)_3$ D. $Fe_3(CO)_3$

9. The compound $Co_2(CO_3)_3$ is named

A. cobalt(III) carbonate
B. cobalt carbonate
C. cobalt(II) carbonate
D. cobalt carbon trioxide

10. The ion SCN^- is named
A. sulfocyano ion B. thiocyano ion C. cyano ion D. thiocyanate ion

11. A compound in which the nitrite-to-metal ion ratio is 2:1 has the following formula
A. $M(NO_2)_2$ B. $M(NO_2)_4$ C. M_2NO_2 D. MNO_2

12. The formula for hydrosulfuric acid is
A. H_2S B. H_2SO_3 C. H_2SO_4 D. HSO_4

13. The oxoacid HIO is named
A. hypoiodous B. iodic C. iodous D. periodic

14. What is the formula for ammonium perchlorate?
A. NH_3ClO_3 B. NH_4ClO_4 C. NH_3ClO D. NH_3ClO_2

15. When hydrogen ions in sulfuric acid are replaced by an iron (3+) ion, the resulting compound has the following formula
A. $Fe_2(SO_4)_3$ B. Fe_3SO_4 C. FeSO D. $Fe_3(SO_4)_2$

16. When a calcium ion is combined with a sulfate ion, the following neutral compound is produced
A. $CaSO_4$ B. CaS C. $Ca_2(SO_4)_3$ D. $Ca(SO_4)_2$

Answers to Exercises

1. 70.906 g
2. a. The ratio is 1.5 to 1, or 3 to 2.
   b. The ratios are whole numbers.
3. 0.53 g
4. 115 g of sulfur and 230 g of oxygen
5. 27.0 g of aluminum oxide
6. The ratio of oxygen in carbon dioxide to oxygen in carbon monoxide is 2.00:1. Yes, this is consistent with the Law of Multiple Proportions.
7. 32 g of sulfur
8. a. Chemical reactions involve reorganization of the atoms.
   b. Each element is made up of tiny particles called atoms.
   c. Chemical reactions involve reorganization of the atoms.
9. a. Zr c. S e. V
   b. Ag d. Kr f. Cs
10. a. Tc c. As e. K
    b. Re d. C f. Xe
11. a. Pb b. Pu c. Ni
12. a. Y = 39 p, 50 n c. Mg^{2+} = 12 p, 12 n e. Cl^- = 17 p, 18 n
    b. Ge = 32 p, 41 n d. U = 92 p, 146 n f. Zn = 30 p, 35 n
13. a. Ac = 89 p, 138 n c. B = 5 p, 6 n e. Pu = 94 p, 145 n
    b. Ga = 31 p, 39 n d. Cf = 98 p, 153 n f. Cu = 29 p, 35 n

14. a. I b. Pd c. Cs

15. "X" = Cf, "Q" = Ca

11.

	protons	neutrons	electrons
a.	26	30	23
b.	20	20	18
c.	9	10	10
d.	15	16	18
e.	53	74	54
f.	53	74	46

17.

	protons	neutrons	electrons
a.	78	117	77
b.	41	52	41
c.	18	22	19
d.	8	8	10
e.	51	71	49
f.	26	30	24
g.	74	110	74
h.	55	78	54
i.	14	14	17

18.

Symbol	Protons	Neutrons	Electrons	Charge
$^{80}_{35}Br^{-}$	35	45	36	−1
$^{80}_{35}Br^{5+}$	35	45	30	+5
$^{137}_{56}Ba^{2+}$	56	81	54	+2
$^{108}_{47}Ag^{+}$	47	61	46	+1
$^{51}_{23}V^{5+}$	23	28	18	+5
$^{59}_{27}Co^{2+}$	27	32	25	+2

19.

Symbol	Protons	Neutrons	Electrons	Charge
$^{27}_{13}Al^{3+}$	13	14	10	+3
$^{88}_{38}Sr^{+}$	38	50	37	+1
$^{65}_{30}Zn^{2+}$	30	35	28	+2
$^{35}_{17}Cl^{-}$	17	18	18	−1
$^{128}_{52}Te^{2-}$	52	76	54	−2
$^{85}_{37}Rb^{+}$	37	48	36	+1

20. a. transition metal
b. halogen
c. noble gas
d. alkaline earth
e. alkali metal
f. lanthanide

21. a. metal
b. nonmetal
c. metal
d. nonmetal
e. nonmetal
f. metal
g. metal
h. metal
i. nonmetal

22. a. actinides
 b. halogens
 c. transition metals
 d. lanthanide
 e. noble gases
 f. alkali metals
 g. alkaline earth metal

23. The atomic radius increases going down a group because electrons are being added to shells that are, on average, farther from the nucleus.

24. a. 1+
 b. 3−
 c. 1−
 d. 1+
 e. 3+
 f. 2−

25. a. lose 2
 b. gain 1
 c. lose 3
 d. gain 2
 e. gain 1
 f. lose 1
 g. gain 3

26. a. lead(II) iodide
 b. ammonium chloride
 c. iron(III) oxide
 d. lithium hydride
 e. cesium chloride
 f. sodium acetate
 g. chromium(III) hydroxide
 h. osmium tetroxide (or osmium(VIII) oxide)
 i. potassium dichromate
 j. sodium sulfate
 k. potassium dihydrogen phosphate

27. a. magnesium sulfate
 b. dinitrogen trioxide
 c. cerium(III) oxide
 d. potassium permanganate
 e. nickel(II) oxide
 f. barium sulfate
 g. iron(III) periodate
 h. sulfur trioxide
 i. potassium perchlorate

28. a. nitrogen triiodide
 b. phosphorus pentachloride
 c. carbon monoxide
 d. tetraphosphorus decoxide
 e. dinitrogen tetroxide
 f. ammonia

29. a. tetraphosphorus hexoxide
 b. potassium hydroxide
 c. molecular nitrogen
 d. boron trifluoride
 e. silver chloride
 f. potassium hydrogen carbonate
 g. silver nitrate

30. a. iodic acid
 b. hydrobromic acid
 c. nitrous acid
 d. hydrocyanic acid
 e. sodium nitrite
 f. potassium sulfite
 g. sodium bisulfite

31. a. uranium(VI) fluoride
 b. copper(II) nitrate
 c. phosphoric acid
 d. sulfur hexafluoride
 e. magnesium hydroxide
 f. tin(II) chloride
 g. sodium carbonate

32. a. $NaCN$
 b. SnF_2
 c. $NaHSO_4$
 d. $Pb(NO_3)_2$
 e. Fe_2O_3
 f. $Ca_3(PO_4)_2$
 g. $NaBrO_3$

33. a. Na_2SO_4
 b. MnO_2
 c. $KClO_3$
 d. $KClO$
 e. $LiAlH_4$
 f. $BaCl_2$
 g. MgO
 h. Cu_2O

34. a. K_2CO_3
 b. $Mg(OH)_2$
 c. N_2O_4
 d. HIO
 e. $FeCl_3$
 f. SnO_2
 g. $RbNO_3$
 h. $KClO_3$
 i. CCl_4
 j. $NaIO_3$
 k. $KMnO_4$
 l. H_2SO_3
 m. K_2HPO_4
 n. $NH_4C_2H_3O_2$
 o. $(NH_4)_2Cr_2O_7$
 p. HBr

35. a. sulfurous acid
 b. hydroiodic acid
 c. hydrobromic acid
 d. nitrous acid
 e. phosphoric acid
 f. hydrochloric acid

36. a. HNO_3
b. HF
c. H_2SO_4
d. HCN
e. H_2S
f. $HC_2H_3O_2$

37. a. sodium bicarbonate
b. nitrous oxide
c. nitric oxide
d. ferrous ion
e. stannic ion
f. plumbous ion

Answers to Multiple-Choice Self-Test

1. A	4. C	7. B	10. D	13. A	15. A
2. B	5. B	8. C	11. A	14. B	16. A
3. B	6. B	9. A	12. A		

CHAPTER 3

Stoichiometry

The Bottom Line: Chapter 3

Your book seeks to make you comfortable with the symbols and equations—the **chemist's shorthand** representations that are used to determine relationships between reactants and products. The chapter also introduces strategies for problem solving. By the end of this chapter you should be ready to tackle a wide variety of computational challenges.

3.1 Atomic Masses

The **Mass Spectrometer** is the best instrument for measuring the masses of atoms and ions (charged particles). The essential operation of the mass spectrometer is given in your textbook.

Example 3.1 A - Calculation of Average Atomic Masses From Isotopic Data

A sample of metal "M" is vaporized and injected into a mass spectrometer. The mass spectrum tells us that 60.10% of the metal is present as ^{69}M, and 39.90% is present as ^{71}M. The mass values for ^{69}M and ^{71}M are 68.93 u and 70.92 u, respectively.

- What is the **average atomic** mass of the element?
- What is the element?

Solution

The basic question here is, "How much does each isotope contribute to the overall atomic mass of the element?" ^{69}M contributes 60.10%. ^{71}M contributes 39.90%. For every 100 atoms of M, on average 60.10 are ^{69}M and 39.90 are ^{71}M.

Mass of 100 atoms of M = Mass of ^{69}M + Mass of ^{71}M

$$= 60.10 \text{ atoms} \times 68.93 \frac{\text{u}}{\text{atom}} + 39.90 \text{ atoms} \times 70.92 \frac{\text{u}}{\text{atom}}$$

$$= 6972 \text{ u}$$

Average atomic mass of M = Mass of 1 atom of M

$$= \frac{\text{Mass of 100 atoms of M}}{\text{100 atoms}}$$

$$= \frac{6972 \text{ u}}{100 \text{ atoms}}$$

$$= \mathbf{69.72\ u/atom}$$

Look at your periodic table. The element with an average atomic mass of 69.72 u/atom is **Ga**.

Does the Answer Make Sense?

Many more of the isotopes are present as ^{69}M than ^{71}M. We would therefore expect the average atomic mass to be closer to 69 than to 71.

Example 3.1 B - Calculation of Relative Isotope Abundances

The element indium exists naturally as two isotopes. ^{113}In has a mass of 112.9043 u, and ^{115}In has a mass of 114.9041 u. The average atomic mass of indium is 114.82 u. Calculate the **percent relative abundance** of the two isotopes of indium.

Solution

We have 2 unknowns, the relative abundance of ^{113}In (expressed as a fraction between 0 and 1), and the relative abundance of ^{115}In (also expressed as a fraction between 0 and 1). Because we have two unknowns, we need two equations to solve the problem. The first equation expresses the **weighted average** of the two isotopes, as in Example 3.1A.

Equation No. 1:

atomic mass of ^{113}In × relative abundance of ^{113}In + atomic mass of ^{115}In × relative abundance of ^{115}In = **average atomic mass of In**

The second equation indicates that the sum of the relative abundances of the isotopes of indium equals 1. (The total of the two isotopic abundances must be 100% because all the isotopes present must be accounted for. This total, expressed as a whole number rather than a percentage, is equal to 1.)

Equation No. 2:

relative abundance of ^{113}In + relative abundance of ^{115}In = 1

Let the relative abundance of $^{113}In = X$, and let the relative abundance of $^{115}In = Y$. The two simultaneous equations for two unknowns are:

$$\begin{aligned} &\text{Equation No. 1:} \quad 112.9043(X) + 114.9041(Y) = 114.82 \\ &\text{Equation No. 2:} \quad X + Y = 1 \end{aligned}$$

Make the coefficients in front of Y in the two equations equal by multiplying equation No. 2 by 114.9041, so we can solve for X. Our two simultaneous equations become:

$$\begin{aligned} &\text{Equation No. 1:} \quad 112.9043(X) + 114.9041(Y) = 114.82 \\ &\text{Equation No. 2:} \quad 114.9041(X) + 114.9041(Y) = 1 \end{aligned}$$

Subtract equation No. 2 from equation No. 1, and solve for X.

$$\begin{aligned} -1.9998(X) + 0(Y) &= -0.0841 \\ X &= \frac{0.0841}{1.9998} \\ &= \mathbf{0.042} \end{aligned}$$

To solve for Y, plug the answer for X back into the original equation No. 2.

$$\begin{aligned} 0.042 + Y &= 1 \\ Y &= 1 - 0.042 \\ &= \mathbf{0.958} \end{aligned}$$

To calculate the percent relative abundance of each isotope, multiply *X* and *Y* by 100%. Thus the answer to the problem is:

Percent relative abundance of ^{113}In = 4.2%
Percent relative abundance of ^{115}In = 95.8%

Does the Answer Make Sense?

Since the average atomic mass of indium is very close to the mass of ^{115}In, we would expect most of the indium to be in that form. Be careful not to accidentally put the wrong percentage with the wrong isotope.

3.2 The Mole

The **mole** is the key to many chemical calculations. A **mole** is defined as **the number equal to the number of carbon atoms in exactly 12 grams of pure ^{12}C.**

1 mole of anything = 6.022 × 10^{23} units of that thing
1 mole of klingons = 6.022 × 10^{23} klingons
1 mole of compact discs = 6.022 × 10^{23} compact discs

One mole of **ANY** substance contains **Avogadro's number** (6.022 × 10^{23}) of particles. However, one mole of klingons is heavier than 1 mole of compact discs. The masses of 1 mole of various elements are given in Table 3.1 in your textbook.

Key Problem Solving Relationship: The average mass of **one atom** of a substance expressed in **u** is the same number as the mass of **one mole** of a substance expressed in **grams**.

1 atom of ^{20}Ne = 20.18 **u**
1 mole of ^{20}Ne = 20.18 **grams**
1 mole of ^{20}Ne = 6.022 × 10^{23} **atoms** of Ne
1 gram (exactly) = 6.022 × 10^{23} **u**

Example 3.2 A - Conversion Among Atoms, Moles, And Mass

A sample of elemental silver (Ag) has a mass of 21.46 g.

- How many **moles** of silver are in the sample?
- How many **atoms** of silver are in the sample?

Solution

We can find the number of moles of silver by multiplying mass by molar mass. If you are not certain of this relationship, the **units** in the problem tell you that this must be so.

Once we calculate the number of moles of Ag, we can use **Avogadro's number** to find the number of atoms. (Make sure your units cancel properly!)

$$21.46 \text{ g Ag} \times \frac{1 \text{ mole Ag}}{107.9 \text{ g Ag}} = \mathbf{0.1989 \text{ moles Ag}}$$

$$0.1989 \text{ moles Ag} \times \frac{6.022 \times 10^{23} \text{ atoms Ag}}{1 \text{ mole Ag}} = \mathbf{1.198 \times 10^{23} \text{ atoms Ag}}$$

Does the Answer Make Sense?

We have nearly 0.2 moles of silver, so we would expect there to be on the order of 10^{23} atoms of silver. Be careful not to accidentally **invert** conversion factors so that your answer comes out to be unfortunately small!

Example 3.2 B - Practice With Conversions

What is the weight of 7.81×10^{22} atoms of calcium?

Solution

You want **grams**. You are given **atoms**. The central relationship between the mass and the number of particles is the **mole**. We will go through moles to get to grams.

$$\text{atoms Ca} \xrightarrow{\text{Avogadro's number}} \text{moles Ca} \xrightarrow{\text{molar mass}} \text{grams Ca}$$

CAUTION: MAKE SURE YOUR UNITS CANCEL PROPERLY!

$$\text{g Ca} \parallel 7.81 \times 10^{22} \text{ atoms Ca} \times \frac{1 \text{ mole Ca}}{6.022 \times 10^{23} \text{ atoms Ca}} \times \frac{40.08 \text{ g Ca}}{1 \text{ mole Ca}} = \mathbf{5.20 \text{ g Ca}}$$

Does the Answer Make Sense?

You have 7.81×10^{22} atoms of Ca, which is a little more than 10% of a mole. Therefore, we would expect a mass of a little more than 10% of the molar mass.

3.3 Molar Mass

Recall that a **molecule** is a **covalently bonded collection of atoms**. Carbon dioxide (CO_2), benzene (C_6H_6), and ethanol (C_2H_6O) are all examples of molecules.

MOLAR MASS OF A MOLECULE =

The sum of the masses of the individual atoms in the molecule

In most chemical calculations, we express molar masses in **grams** using prefixes as necessary (mg, μg, kg, etc.).

Example 3.3 A - Converting Among Molar Mass, Moles, Mass, And Number Of Particles

How many μg are there in 3.82×10^{-7} moles of pyrogallol, $C_6H_6O_3$?

Solution

This is a two-step conversion:

$$\text{moles pyrogallol} \xrightarrow[\text{mass}]{\text{molar}} \text{g pyrogallol} \xrightarrow[\text{conversion}]{\text{unit}} \mu\text{g pyrogallol}$$

(Be careful to use the proper unit conversion from g to μg!)

The molar mass of pyrogallol (pyrgl) is:

C: 6×12.01	=	72.06 g
H: 6×1.008	=	6.048 g
O: 3×16.00	=	48.00 g
molar mass	=	**126.11 g/mole pyrogallol**

$$\mu g\ pyrgl \ \| \ 3.82 \times 10^{-7}\ \text{moles pyrgl} \times \frac{126.11\ \text{g pyrgl}}{1\ \text{mole pyrgl}} \times \frac{\overset{\text{exactly}}{\downarrow}\ 1\times10^{6}\ \text{g pyrgl}}{1\ \text{g pyrgl}} = \mathbf{48.2\ \mu g\ pyrogallol}$$

Example 3.3 B Practice With Conversions

Freon-12, which has the formula CCl_2F_2, is used as a refrigerant in air conditioners and as a propellant in aerosol cans. Given a 5.56 mg sample of Freon-12:

a. Calculate the number of molecules of Freon-12 in that sample.
b. How many mg of chlorine are in the sample?

Solution

Use **moles** as your **bridge** between milligrams and molecules. Also, always keep in mind that a conversion factor is valid in two ways, for example:

$$\frac{6.022\times10^{23}\ \text{molecules}}{1\ \text{mole}} \text{ and } \frac{1\ \text{mole}}{6.022\times10^{23}\ \text{molecules}} \text{ are both correct.}$$

You must be careful to use the conversion factor in the correct fashion so that the units cancel.

$$\text{mg Freon-12} \xrightarrow[\text{molar mass}]{\text{mg to grams, and}} \text{moles Freon-12} \xrightarrow{\text{Avagadro's number}} \text{molecules Freon-12}$$

The mass of 2 chlorine atoms **divided by** the molar mass of Freon-12 gives the **fraction** of the total mass that is chlorine. Multiplying the **chlorine fraction** by the **total sample size** gives the mg chlorine in the sample.

The molar mass of Freon-12 (Fr-12) is:

C:	1 × 12.01	=	12.01 g
Cl:	2 × 35.45	=	70.90 g
F:	2 × 19.00	=	38.00 g
molar mass		=	**120.91 g/mole Freon-12**

$$5.56\ \text{mg Fr-12} \times \frac{1\ \text{g Fr-12}}{1000\ \text{mg Fr-12}} \times \frac{1\ \text{mole Fr-12}}{120.91\ \text{g Fr-12}} \times \frac{6.022\times10^{23}\ \text{molecules Fr-12}}{1\ \text{mole Fr-12}}$$

$$= \mathbf{2.77 \times 10^{19}\ molecules\ Fr\text{-}12}$$

From the molar mass calculation above, Cl = 70.90 g of the molar mass of Freon-12.

$$\text{fraction of Freon-12 that is Cl} = \frac{70.90\ \text{g Cl}}{120.91\ \text{g Fr-12}} = \mathbf{0.586}$$

mg Cl in 5.56 mg of Freon-12 = 0.586 × 5.56 mg = **3.26 mg Cl**

3.4 Conceptual Problem Solving

The goal of this section is to help you to develop the tools to, "become a good problem solver." with a focus on tackling problems that are new to you. Zumdahl wants you to focus on the *process* of problem solving. To do so, he introduces several questions (asking the right questions is a key!) for you to consider:

- Where are we going?
- What do we know?
- How do we get there?
- Does my answer make sense?

This is the structure that is used in the examples in this chapter. Try to apply it throughout the text as your ability to solve problems grows.

3.5 Percent Composition of Compounds

There are three steps to calculating the **mass percent** of each element in a compound.

a. Compute the **molar mass** of the compound.
b. Calculate how much of the molar mass comes from each element.
c. Divide each element's mass contribution by the total molar mass, and multiply by 100 to convert to percent.

Example 3.5 - Practice With Mass Percents

Calculate the mass percent of each element in potassium ferricyanide, $K_3Fe(CN)_6$.

Solution

We can follow the problem solving strategy in the text, noting, "*Where are we going?*" We seek the mass percent of each element.

Continuing with the key questions, "*What do we know?*" We know the formula of the compound and the number of atoms of each element in the compound. The mass percent is the ratio of the mass of the element to the mass of the total compound, expressed as a percentage.

How do we get there – to the mass percent of each element? Using the same strategy as in the previous example,

a.

K:	3 × 39.10	=	117.30 g
Fe:	1 × 55.85	=	55.85 g
C:	6 × 12.01	=	72.06 g
N:	6 × 14.01	=	84.06 g
molar mass		=	**329.27 g/mole $K_3Fe(CN)_6$**

b. Contribution of each element:

K = 117.30 g; Fe = 55.85 g; C = 72.06 g; N = 84.06 g

c. Mass percent of each element:

$$\text{Mass percent of K} = \frac{117.30\text{ g K}}{329.27\text{ g }K_3Fe(CN)_6} \times 100\% = \mathbf{35.62\%\ K}$$

$$\text{Mass percent of Fe} = \frac{55.85\text{ g Fe}}{329.27\text{ g }K_3Fe(CN)_6} \times 100\% = \mathbf{16.96\%\ Fe}$$

$$\text{Mass percent of C} = \frac{72.06\text{ g C}}{329.27\text{ g }K_3Fe(CN)_6} \times 100\% = \mathbf{21.88\%\ C}$$

$$\text{Mass percent of N} = \frac{84.06\text{ g N}}{329.27\text{ g }K_3Fe(CN)_6} \times 100\% = \mathbf{25.53\%\ N}$$

Does the Answer Make Sense? Checking the Figures

Adding up the percentages gives **99.99%**, not 100.00%. This is an example of the kind you will see from time to time, in which rounding off to 2 significant figures after the decimal point leads to a **loss** of 0.01%. In such cases, we say that the answers are correct **within round-off error**.

COMMENT: Although it would have been correct to calculate the mass percent of N by subtracting (%K + %Fe + %C) from 100% in the previous problem, this eliminates your ability to double check your answer because you have, in effect, already "guaranteed" that the mass percents must equal 100%. We therefore recommend against this procedure.

3.6 Determining the Formula of a Compound

The **empirical formula** is represented by the **simplest whole number ratio** of atoms in a compound. Examples of empirical formulas are:

CH, CH_4, CH_2O, $K_2Cr_2O_7$, $K_3Fe(CN)_6$.

The **molecular formula** is the **actual ratio** of atoms in a compound. Examples of molecular formulas are:

C_6H_6, CH_4, $C_6H_{12}O_6$, $K_2Cr_2O_7$, $K_3Fe(CN)_6$.

The empirical and molecular formulas **can** be the same. Review the methods for experimentally determining the empirical and molecular formulas in your text, and note especially the step summary at the end of Section 3.6 in your textbook. Then try the following problems.

Example 3.6 A - Determination Of The Empirical Formula Of A Compound

Determine the empirical and molecular formulas for a deadly nerve gas that gives the following mass percent analysis:

C = 39.10% H = 7.67% O = 26.11% P = 16.82% F = 10.30%

The molar mass is known to be 184.1 grams.

Solution

Where are we going? We are looking for both the empirical and molecular formula of the compound.

What do we know? The empirical formula is the simplest whole number ratio of the atoms in the compound and the molecular formula is the actual formula for the compound. In both cases the formula will look like P_xO_y and we are trying to solve for x and y.

To solve empirical formula problems that do not involve combustion (added oxygen):

1. Assume the compound has a mass of exactly 100 grams. You can therefore convert percentage to grams.
2. Calculate the moles of each kind of atom present.
3. Determine the simplest whole-number ratios by dividing the moles of each compound by the smallest calculated mole value.

To determine the molecular formula, divide the molar mass by the empirical formula mass. This will give the number of empirical formula units in the actual molecule. For example, if the

empirical formula is CH, and you determine that the **molar mass** is **6 times** the empirical formula mass, the **molecular formula** is C_6H_6.

C = 39.10%   H = 7.67%   O = 26.11%   P = 16.82%   F = 10.30%

$$\text{moles of C} \parallel 39.10 \text{ g C} \times \frac{1 \text{ mole C}}{12.01 \text{ g C}} = \mathbf{3.26 \text{ moles C}}$$

$$\text{moles of H} \parallel 7.67 \text{ g H} \times \frac{1 \text{ mole H}}{1.008 \text{ g H}} = \mathbf{7.61 \text{ moles H}}$$

$$\text{moles of O} \parallel 26.11 \text{ g O} \times \frac{1 \text{ mole O}}{16.00 \text{ g O}} = \mathbf{1.63 \text{ moles O}}$$

$$\text{moles of P} \parallel 16.82 \text{ g P} \times \frac{1 \text{ mole P}}{30.97 \text{ g P}} = \mathbf{0.543 \text{ moles P}}$$

$$\text{moles of F} \parallel 10.32 \text{ g F} \times \frac{1 \text{ mole F}}{19.00 \text{ g F}} = \mathbf{0.543 \text{ moles F}}$$

Dividing through by the smallest number:

$$C_{\frac{3.26}{0.543}} \quad H_{\frac{7.61}{0.543}} \quad O_{\frac{1.63}{0.543}} \quad P_{\frac{0.543}{0.543}} \quad F_{\frac{0.543}{0.543}}$$

Gives an empirical formula of

$$C_{6.00} \quad H_{14.01} \quad O_{3.00} \quad P_1 \quad F_1 = \mathbf{C_6H_{14}O_3PF}$$

Does the Answer Make Sense?

The empirical formula mass = 184.1 grams. Therefore, the molecular formula and the empirical formula are the same. This consistency suggests that the answer makes sense.

NOTE on precision: Empirical formula determinations are not among the most precise experiments in chemistry. You may expect variations of up to about 5% in your values. Be flexible!

Example 3.6 B - Determining The Empirical Formula From Combustion Data

A combustion device was used to determine the empirical formula of a compound containing **ONLY carbon, hydrogen, and oxygen**. A 0.6349-g sample of the unknown produced 1.603 g of CO_2 and 0.2810 g H_2O. Determine the empirical formula of the compound.

Solution

The primary assumption is that the **combustion was complete**. Therefore, **all** of the carbon was converted to CO_2. All of the hydrogen was converted to H_2O. Then if you calculate the **grams of carbon in 1.603 g CO_2** and the **grams of hydrogen in 0.2810 g H_2O**, you know C and H in your original compound. You can calculate grams of oxygen in the original compound as:

grams O = total grams − (grams C + grams H).

a. Molar mass of CO_2 = 44.01 grams.

$$\text{Mass fraction of C in } CO_2 = \frac{12.01}{44.01} = 0.2729 \text{ (0.2729 g C in every g of } CO_2\text{)}$$

g C in 1.603 g CO_2 = 0.2729 × 1.603 g = **0.4374 g C** in CO_2 (and therefore in our original compound).

b. Molar mass of H_2O = 18.016 grams.

Mass fraction of H in $H_2O = \frac{2.016}{18.016} = 0.1119$ (0.1119 g H in every g of H_2O)

g H in 0.2810 g H_2O = 0.1119 × 0.2810 g = **0.0314 g H** in H_2O (and therefore in our original compound).

c. grams O in original compound = 0.6349 g − (0.4374 g C + 0.0314 g H)

= **0.1661 g O** in original compound.

d. Converting each to moles,

moles C ‖ $0.4374\text{ g C} \times \frac{1\text{ mole C}}{12.01\text{ g C}}$ = **0.0364 mole C**

moles H ‖ $0.0314\text{ g H} \times \frac{1\text{ mole H}}{1.008\text{ g H}}$ = **0.0312 mole H**

moles O ‖ $0.1661\text{ g O} \times \frac{1\text{ mole O}}{16.00\text{ g O}}$ = **0.0104 mole O**

e. dividing by smallest number

$$C_{\frac{0.0364}{0.0104}}\ H_{\frac{0.0312}{0.0104}}\ O_{\frac{0.0104}{0.0104}} = C_{3.5}H_3O_1$$

The empirical formula (simplest ***whole*** number ratio) = **$C_7H_6O_2$**.

Two step-by-step methods for obtaining the molecular formula of a compound are summarized at the end of Section 3.6 in your textbook.

3.7 Chemical Equations

Chemical equations are all of the form

Reactants → Products

Chemical equations describe chemical changes that occur in a reaction. The physical states that can be present (with their symbols) in a reaction are described in your textbook. The general information that can be gotten from a chemical reaction is given in Table 3.2 in your textbook.

Example 3.7 - Relating Reactions And Products Of A Chemical Equation

Show, by means of molar masses, that matter is neither created nor destroyed in the equation given below.

$$C_2H_6O(l) + 3O_2(g) \rightarrow 2CO_2(g) + 3H_2O(g)$$

Solution

The law of conservation of matter says that matter is neither created nor destroyed in a chemical reaction. Therefore,

number of grams of reactant = number of grams of product

Remember to take the **number of moles** of each compound into account when doing your calculations!

Compound	Molar Mass	Number of moles in equation	Total grams in equation
C_2H_6O	46.068 g	1	46.068 g
O_2	32.00 g	3	96.00 g
CO_2	44.01 g	2	88.02 g
H_2O	18.016 g	3	54.048 g

So the **total grams of reactant** = 46.068 + 96.00 = **142.07 g**
and the **total grams of product** = 88.02 + 54.048 = **142.07 g**
grams of reactant = grams of product

This relationship must hold in every chemical reaction.

3.8 Balancing Chemical Equations

A **balanced chemical equation** means that the **number of atoms of each element on the reactants side equals the number of atoms of each element on the products side.**

Example 3.8 - Practice With Balancing

Balance the following combustion reaction:

$$C_4H_{10}(g) + O_2(g) \rightarrow CO_2(g) + H_2O(g)$$

Solution

Chemists have many different strategies for balancing chemical equations. There is no one best way. There is an excellent strategy given in this section in your textbook.

An alternate strategy is:

1. Never change a molecular structure. Only use coefficients.
2. Find the atoms that are in only one compound on the reactant side. Balance those first.
3. In general, leave oxygen and hydrogen until the very end. They usually appear many times, and balancing other atoms will often force O and H to become balanced.
4. Always double check **AFTER** balancing to make sure that the atoms of each element are equal on the reactants and products sides.

We will proceed by balancing carbon first and saving hydrogen and oxygen for later.

a. Multiply **CO_2 by 4** to balance carbons.
b. Multiply **H_2O by 5** to balance hydrogens.

This leaves us with 4 carbons and 10 hydrogens on both sides. We have **2 oxygens on the left side** and **13 oxygens on the right**.

c. Multiply **O_2 by** $^{13}/_2$ to balance oxygens.

Our balanced equation becomes:

$$C_4H_{10}(g) + {}^{13}/_2O_2(g) \rightarrow 4CO_2(g) + 5H_2O(g)$$

We have:

$$4C, 10H, 13O \rightarrow 4C, 10H, 13O.$$

Fractional coefficients **are** used from time to time in chemistry. For clarity, however, we prefer to use simple **whole number ratios**. Therefore:

d. Multiply both sides of the reaction by 2. This converts the $^{13}/_2$ to 13, a whole number. The number of atoms of each element on both sides of the equation are **still equal** because we have multiplied both sides by 2.

Final answer:

$$\mathbf{2C_4H_{10}(g) + 13O_2(g) \rightarrow 8CO_2(g) + 10H_2O(g)}$$

3.9 Stoichiometric Calculations: Amounts of Reactants and Products

The most important reason for learning to balance chemical equations (as in Section 3.8) is to establish reactant-product mole (and therefore mass) relationships. For instance, in Example 3.8, we found that:

$$\mathbf{2\ moles\ of\ C_4H_{10} + 13\ moles\ of\ O_2 \rightarrow 8\ moles\ of\ CO_2 + 10\ moles\ of\ H_2O}$$

In terms of the grams of O_2 to grams of CO_2 relationship,

$13\text{ moles } O_2 \times \dfrac{32.00\text{ g } O_2}{1\text{ mole } O_2} =$ **416 g O_2** reacts with 2 moles of C_4H_{10} to yield 10 moles of H_2O and

$8\text{ moles } CO_2 \times \dfrac{44.01\text{ g } CO_2}{1\text{ mole } CO_2} =$ **352 g CO_2**.

The **key relationship** in problems such as these is the **moles of reactant to moles of product** conversion factor. Note that

$$\frac{13\text{ moles } O_2}{8\text{ moles } CO_2} \text{ and } \frac{8\text{ moles } CO_2}{13\text{ moles } O_2} \text{ are both correct.}$$

Example 3.9 A - Amount Of Product From A Given Amount Of Reactant

Given the following reaction:

$$Na_2S(aq) + AgNO_3(aq) \rightarrow Ag_2S(s) + NaNO_3(aq)$$

How many grams of Ag_2S can be generated from the reaction of 3.94 g of $AgNO_3$ with excess Na_2S?

Solution

1. **Always balance the chemical equation!**
2. The **mole ratio conversion factor** acts as the **bridge** between $AgNO_3$ and Ag_2S. This leads to the following strategy:

$$\text{g } AgNO_3 \xrightarrow[\text{mass}]{\text{molar}} \text{moles } AgNO_3 \xrightarrow[\text{ratio}]{\text{mole}} \text{moles } Ag_2S \xrightarrow[\text{mass}]{\text{molar}} \text{g } Ag_2S$$

It is recommended that you solve such problems with one set of equations rather than splitting the calculation up into separate steps. Too many calculations may lead to confusion and errors.

The balanced chemical equation is:

$$Na_2S(aq) + 2AgNO_3(aq) \rightarrow Ag_2S(s) + 2NaNO_3(aq)$$

Then use one equation:

$$\text{g } Ag_2S \parallel 3.94 \text{ g } AgNO_3 \times \frac{1 \text{ mole } AgNO_3}{169.88 \text{ g } AgNO_3} \times \frac{1 \text{ mole } Ag_2S}{2 \text{ mole } AgNO_3} \times \frac{247.8 \text{ g } Ag_2S}{1 \text{ mole } Ag_2S}$$

$$= \mathbf{2.87 \text{ g } Ag_2S}$$

Does the Answer Make Sense?

Although the answer looks reasonable, the best way to check your answer is by double checking your use of units. If they cancel properly, then your answer is likely to be correct.

Example 3.9 B - Amount Of Reactant Needed To Produce A Product

Aspirin (acetylsalicylic acid) is prepared by the reaction of salicylic acid ($C_7H_6O_3$) and acetic anhydride ($C_4H_6O_3$). How many grams of salicylic acid (Sal) are needed to make 500 aspirin tablets weighing 1.00 g each (assuming 100% yield)?

$$\underset{\text{salicylic acid}}{C_7H_6O_3} + \underset{\text{acetic anhydride}}{C_4H_6O_3} \rightarrow \underset{\text{aspirin}}{C_9H_8O_4} + \underset{\text{acetic acid}}{HC_2H_3O_2}$$

Solution

We will work this problem in basically the same fashion as we did the previous example. The only difference is that we are going from **product** to **reactant**.

number of aspirin tablets (start) $\xrightarrow{\text{mass per tablet}}$ grams of aspirin $\xrightarrow{\text{molar mass}}$ moles of aspirin

grams of salicylic acid $\xleftarrow{\text{molar mass}}$ moles of salicylic acid $\xleftarrow{\text{mole ratio}}$

$$\text{g salicylic acid} \parallel 500 \text{ aspirin tablets} \times \frac{1.00 \text{ g aspirin}}{1 \text{ aspirin tablet}} \times \frac{1 \text{ mole aspirin}}{180.15 \text{ g aspirin}} \times \frac{1 \text{ mole Sal}}{1 \text{ mole aspirin}}$$

$$\times \frac{138.12 \text{ g Sal}}{1 \text{ mole Sal}} = \mathbf{383 \text{ g salicylic acid}}$$

Remember that the most important relationships in all of chemistry are between **moles of reactants** and **moles of products**.

3.10 Calculations Involving a Limiting Reactant

In the previous section, we were given the amount of a reactant and asked to find the amount of product formed. The **assumption** then was that **we had as much of the other reactants as we required to make the reactant go completely**. This will not always be the case. For example, given the following reaction,

$$2H_2(g) + O_2(g) \rightarrow 2H_2O(g),$$

we need **2 moles of H₂** to completely react with **1 mole of O₂**, forming **2 moles of H₂O**. If we have only **1.5 moles of H₂,** only **0.75 moles of O₂** will react, forming **1.5 moles of H₂O**. That means that (1 mole O₂ – 0.75 mole O₂) **0.25 moles of O₂** will be left over, or **in excess**.

H₂ is said to be the **limiting reactant** in this reaction because it **limits the amount of product that can form**.

Example 3.10 A - Calculations Involving A Limiting Reactant

Sodium hydroxide reacts with phosphoric acid to give sodium phosphate and water. If 17.80 g of NaOH is mixed with 15.40 g of H_3PO_4,

 a. How many grams of Na_3PO_4 can be formed?
 b. How many grams of the excess reactant remain unreacted?
 c. If the <u>actual yield</u> of Na_3PO_4 was 15.00 g, what is the percent yield of Na_3PO_4?

Solution

We must first write a balanced chemical equation for this reaction:

$$3NaOH(aq) + H_3PO_4(aq) \rightarrow Na_3PO_4(aq) + 3H_2O(l)$$

a. The basic question to be addressed in limiting reactant problems is **"Which compound limits the amount of product formed?"** It is therefore recommended that you calculate the amount of product formed by **each** reactant. **The reactant that forms less product is limiting**, and that total amount of product will be formed:

$$\text{g of reactant} \xrightarrow[\text{mass}]{\text{molar}} \text{mol of reactant} \xrightarrow[\text{ratio}]{\text{mole}} \text{mol of product} \xrightarrow[\text{mass}]{\text{molar}} \text{g of product.}$$

b. To determine grams of excess reactant, you must calculate **how many moles of excess reactant were actually used**.

moles excess = # moles original – # moles used

 Then convert moles to grams.

c. The percent yield $= \dfrac{\text{actual yield}}{\text{theorertical yield}} \times 100\%.$

Determination of limiting reactant

$$\begin{array}{c}\text{g Na}_3\text{PO}_4\\ \text{from NaOH}\end{array} \left\| \; 17.80 \text{ g NaOH} \times \frac{1 \text{ mole NaOH}}{40.00 \text{ g NaOH}} \times \frac{1 \text{ mole Na}_3\text{PO}_4}{3 \text{ mole NaOH}} \times \frac{163.94 \text{ g Na}_3\text{PO}_4}{1 \text{ mole Na}_3\text{PO}_4}\right.$$

$$= \textbf{24.32 g Na}_3\textbf{PO}_4 \textbf{ from NaOH}$$

$$\begin{array}{c}\text{g Na}_3\text{PO}_4\\ \text{from H}_3\text{PO}_4\end{array} \left\| \; 15.40 \text{ g H}_3\text{PO}_4 \times \frac{1 \text{ mole H}_3\text{PO}_4}{98.00 \text{ g H}_3\text{PO}_4} \times \frac{1 \text{mole Na}_3\text{PO}_4}{1 \text{mole H}_3\text{PO}_4} \times \frac{163.94 \text{ g Na}_3\text{PO}_4}{1 \text{ mole Na}_3\text{PO}_4}\right.$$

$$= \textbf{25.76 g Na}_3\textbf{PO}_4 \textbf{ from H}_3\textbf{PO}_4$$

Therefore, **NaOH is the limiting reactant,** and **24.32 g Na₃PO₄ are formed**.

Determination of grams of excess reactant

H_3PO_4 is in excess. If 24.32 g Na_3PO_4 is formed, the **moles of H_3PO_4 used** is given by:

$$\text{moles } H_3PO_4 \text{ used} \parallel 24.32 \text{ g } Na_3PO_4 \times \frac{1 \text{ mole } Na_3PO_4}{3 \text{ mole NaOH}} \times \frac{1 \text{ mole } H_3PO_4}{98.00 \text{ g } H_3PO_4}$$

$$= \mathbf{0.1483 \text{ moles } H_3PO_4 \text{ used}}$$

The number of **moles of H_3PO_4 originally present** is:

$$15.40 \text{ g } H_3PO_4 \times \frac{1 \text{ mole } H_3PO_4}{98.00 \text{ g } H_3PO_4} = \mathbf{0.1571 \text{ moles } H_3PO_4 \text{ originally present}}$$

$$\text{moles } H_3PO_4 \text{ excess} = \text{moles } H_3PO_4 \text{ originally present} - \text{moles } H_3PO_4 \text{ used}$$

$$= 0.1571 \text{ moles} - 0.1483 \text{ moles}$$

$$= \mathbf{0.0088 \text{ moles } H_3PO_4 \text{ excess}}$$

$$\text{grams } H_3PO_4 \text{ excess} \parallel 0.0088 \text{ moles } H_3PO_4 \times \frac{98.00 \text{ g } H_3PO_4}{1 \text{ mole } H_3PO_4}$$

$$= \mathbf{0.86 \text{ g } H_3PO_4 \text{ excess}}$$

Determination of percent yield

$$\text{percent yield} = \frac{\text{actual yield}}{\text{theoretical yield}} \times 100\% = \frac{15.00 \text{ g}}{24.32 \text{ g}} \times 100\% = \mathbf{61.68\% \text{ yield}}$$

Example 3.10 B - Tying It All Together

The Space Shuttle environmental control system handles excess CO_2 (which the astronauts breathe out—it is 4% by mass of exhaled air) by reacting it with lithium hydroxide, LiOH, pellets to form lithium carbonate, Li_2CO_3, and water. If there are 7 astronauts on board the shuttle and each exhales 20 liters of air per minute, how long could clean air be generated if there were 25,000 g of LiOH pellets available for each shuttle mission? Assume the density of air is 0.0010 g/mL.

Solution

$$\frac{\text{g } CO_2 \text{ generated}}{\text{minute}} \parallel \frac{0.0010 \text{ g air}}{\text{mL}} \times \frac{20{,}000 \text{ mL air}}{\text{min} \bullet \text{astronaut}} \times 7 \text{ astronauts} \times \frac{4 \text{ g } CO_2}{100 \text{ g air}}$$

$$= \mathbf{5.6 \text{ g } CO_2 / \text{minute}}$$

The reaction of CO_2 with LiOH is

$$CO_2 + 2LiOH \rightarrow Li_2O_3 + H_2O$$

Grams of CO_2 that can react with 25,000 g of LiOH:

$$\text{g } CO_2 \parallel 2.5 \times 10^4 \text{ g LiOH} \times \frac{1 \text{ mol LiOH}}{23.949 \text{ g LiOH}} \times \frac{1 \text{ mol } CO_2}{2 \text{ mol LiOH}} \times \frac{44.01 \text{ g } CO_2}{1 \text{ mol } CO_2} = \mathbf{22{,}971 \text{ g } CO_2}$$

Number of days clean air can be generated:

$$22{,}971 \text{ g } CO_2 \times \frac{1 \text{ minute}}{5.6 \text{ g } CO_2} = 4102 \text{ minutes} = \mathbf{2.8 \text{ days of clean air}}$$

3.11 Solving a Complex Problem

Your textbook raises two key points about complex problems:

1. They use understanding of many concepts.
2. They can be solved more than one way.

Let's try this example to reinforce these two key points.

Example 3.11 – A Descriptive Example

Your textbook will consider solutions reacting in the next chapter. To show how asking the right questions using the knowledge you already have can be put to use solving even (for now) unfamiliar problems, try this:

The amount of vitamin C, also called ascorbic acid, in a solution can be determined by its reaction with an iodine solution to which some starch has been added to give it a deep blue color.

$$C_6H_8O_6(aq) + I_2(aq) \rightarrow C_6H_6O_6(aq) + HI(aq)$$

We have a solution of iodine (I_2) containing 0.01371 moles I_2 in each liter. We have a one-cup, or 238 mL sample of fruit juice in which we want to determine the amount of vitamin C, in milligrams. We take 50.00 mL of the fruit juice and carefully add I_2 solution until the reaction is complete (the product solution will stay blue at this point). We add a total of 28.09 mL of the I_2 solution.

How many mg of vitamin C are in one cup (238 mL) of the fruit juice?

Solution

Where are we going? Our goal is to determine the quantity of the vitamin C in one cup of a fruit juice.

What do we know? The equation that describes the reaction is

$$C_6H_8O_6(aq) + I_2(aq) \rightarrow C_6H_6O_6(aq) + HI(aq)$$

However, it is unbalanced so we'll need to take care of that. The balanced equation is

$$C_6H_8O_6(aq) + I_2(aq) \rightarrow C_6H_6O_6(aq) + \mathbf{2}HI(aq)$$

We also know the process. We are not reacting all of the fruit juice. Rather, we are combining 50.00 mL of the original 238-mL sample with 28.09 mL of the I_2 solution, which contains $\frac{0.1371 \text{ moles } I_2}{1 \text{ liter of } I_2 \text{ solution}}$.

How do we get there?

Method 1

We know by the balanced equation that one mole of vitamin C reacts with one mole of I_2. How many moles of I_2 did we add?

$$\frac{0.01371 \text{ moles } I_2}{\text{L } I_2 \text{ solution}} \times 0.02809 \text{ L } I_2 \text{ solution} = 0.0003851 \text{ moles of } I_2$$

Therefore, how many moles of vitamin C are in our 50.00 mL sample?

Moles vitamin C = moles of I_2 = 0.0003851 moles of vitamin C in 50.00 mL

How many moles of vitamin C are in one cup, or 238 mL of fruit juice?

$$\frac{0.0003851 \text{ moles vitamin C}}{50.00 \text{ mL}} \times \frac{238 \text{ mL}}{1 \text{ cup}} = 0.001833 \text{ moles vitamin C per cup of fruit juice}$$

Knowing the moles of vitamin C in one cup of fruit juice, how do we find mg in one cup? We can convert moles to grams, then milligrams. This requires us to find the molar mass of vitamin C (176.1 g/mol).

$$\frac{0.001833 \text{ mol vitamin C}}{\text{one cup fruit juice}} \times \frac{176.1 \text{ g vitamin C}}{1 \text{ mol vitamin C}} \times \frac{1000 \text{ mg}}{1 \text{ g}} = \begin{array}{l}\text{323 mg vitamin C in}\\ \text{one cup of fruit juice}\end{array}$$

Method 2

We can do a similar calculation, but find the moles, and then grams and mg of vitamin C in 50.00 mL of sample, then convert to one cup of fruit juice in the final step. That is,

$$\frac{\text{mol } I_2}{\text{L } I_2 \text{ soln}} \rightarrow \frac{\text{mol } I_2}{0.02809 \text{ L soln}} \rightarrow \frac{\text{mol vitamin C}}{50.00 \text{ mL vit C soln}} \rightarrow \frac{\text{g in vit C}}{50.00 \text{ mL vit C soln}}$$

$$\rightarrow \frac{\text{mg vit C}}{50.00 \text{ mL vit C soln}} \rightarrow \frac{\text{mg vit C}}{1 \text{ cup fruit juice}}$$

$$\frac{0.01371 \text{ mol } I_2}{\text{L } I_2 \text{ soln}} \times 0.02809 \text{ L } I_2 \text{ soln} \times \frac{1 \text{ mol vit C in 50.00 mL soln}}{1 \text{ mol } I_2} \times \frac{176.1 \text{ g vit C}}{1 \text{ mol vit C}}$$

$$\times \frac{1000 \text{ mg}}{1 \text{ g}} = 67.78 \text{ mg vitamin C}$$

$$67.78 \text{ mg vitamin C} \times \frac{238}{50.00} = 323 \text{ mg vitamin C in one cup of fruit juice}$$

There are other strategies to arrive at the answer, either by step-wise reasoning or dimensional analysis. They have in common that none are inherently better than the others. Rather pick the process that makes the most sense to you that gives the correct answer.

Exercises

Section 3.1

1. An element "E" is present as ^{10}E with a mass value of 10.01 u, and as ^{11}E with a mass value of 11.01 u. The natural abundances of ^{10}E and ^{11}E are 19.78% and 80.22%, respectively. What is the average atomic mass of the element? What is the element?

2. Naturally occurring sulfur consists of four isotopes, ^{32}S (95.0%), ^{33}S (0.76%), ^{34}S (4.22%), and ^{36}S (0.014%). Using these data, calculate the molar mass of naturally occurring sulfur. The masses of the isotopes are given in the table below.

Isotope	Atomic mass (u)
^{32}S	31.97
^{33}S	32.97
^{34}S	33.97
^{36}S	35.97

3. An unknown sample of mystery element "T" is injected into the mass spectrometer. According to the mass spectrum, 7.42% of the element is present as ^{6}T and 92.58% is present as ^{7}T. The mass values are 6.02 u for ^{6}T and 7.02 u for ^{7}T. Calculate the average atomic mass, and identify the mystery element.

4. A noble gas consists of three isotopes of masses 19.99 u, 20.99 u, and 21.99 u. The relative abundance of these isotopes is 90.92%, 0.257%, and 8.82%, respectively. What is the average atomic mass of this noble gas? What noble gas is this?

5. Chlorine has two stable isotopes. The mass of one isotope is 34.97 u. Its relative abundance is 75.53%. What is the mass of the other stable isotope?

6. Complete the following table of isotopic information for the element neon (Ne).

Isotope	Mass (u)	Abundance
^{20}Ne	19.99	______
^{21}Ne	20.99	0.257%
^{22}Ne	21.99	______

7. Silicon has three stable isotopes in nature as shown in the table below. Fill in the missing information.

Isotope	Mass (u)	Abundance
^{28}Si	27.98	______
^{29}Si	______	4.70%
^{32}Si	29.97	3.09%

8. Gallium has two stable isotopes of masses 68.93 u (^{69}Ga) and 70.92 u (^{71}Ga). What are the relative abundances of the two isotopes?

9. Magnesium exists as three isotopes in nature. One isotope (^{25}Mg) has a mass of 24.99 u and a relative abundance of 10.13%. The other two isotopes have masses of 23.99 u (^{24}Mg) and 25.98 u (^{26}Mg). What are their relative abundances? (atomic mass Mg = 24.305 u)

Section 3.2

10. How many moles are in a sample of 300 atoms of nitrogen? How many grams?

11. How many atoms of gold does it take to make 1 gram of gold?

12. If you buy 38.9 moles of M&M's®, how many M&M's® do you have? (1 mole of M&M's® = 6.022×10^{23} M&M's®)

13. A sample of sulfur has a mass of 5.37 g. How many moles are in the sample? How many atoms?

14. Give the number of moles of each element present in 1.0 mole of each of the following substances:

a.	Hg_2I_2	c.	$PbCO_3$	e.	$RbOH{\cdot}2H_2O$
b.	LiH	d.	$Ba_3(AsO_4)_2$	f.	H_2SiF_6

15. How many grams of zinc are in 1.16×10^{22} atoms of zinc?

16. How many u are in 3.68 moles of iron?

17. Calculate the molar mass of

a. $Zn(CN)_4$ b. $Cu(NH_3)_4{\cdot}8H_2O$

18. How many K ions are present in each of the following:

a. 3.0 moles KCl
b. 6.2 g of KNO_3
c. a mixture containing 12.6 g of K_3PO_4 and 5.4 g of KCl

19. How many sodium ions are present in each of the following:

a. 2.0 moles of Na_3PO_4
b. 5.8 g of NaCl
c. a mixture containing 14.2 g of Na_2SO_4 and 2.9 g of NaCl

20. What is the weight in grams of

a. 0.40 moles of CH_4
b. 11 moles of SO_4^{2-} ions
c. 5 moles of $Mg(OH)_2$

Section 3.3

21. Calculate the molar masses of each of the following:

a. Cu_2SO_4
b. NH_4OH
c. $C_{10}H_{16}O$
d. $Zr(SeO_3)_2$
e. $Ca_2Fe(CN)_6 \cdot 12H_2O$
f. $Cr_4(P_2O_7)_3$

22. What is the mass of 4.28×10^{22} molecules of water?

23. How many milligrams of Br_2 are in 4.80×10^{20} molecules of Br_2?

24. Determine the molar mass of $KAl(SO_4)_2 \cdot 12H_2O$.

25. How many moles of cadmium bromide ($CdBr_2$) are in a 39.25-g sample?

26. A sample of calcium chloride ($CaCl_2$) has a mass of 23.8 g. How many moles of calcium chloride is this?

27. If 0.172 moles of baking soda ($NaHCO_3$) were used to bake a chocolate cherry cake, how many grams of baking soda would the recipe call for?

28. How many moles are there in a sample of barium sulfate ($BaSO_4$) weighing 9.90×10^{-7} g?

29. How many grams are there in 0.36 moles of cobalt(III) acetate ($Co(C_2H_3O_2)_3$)? How many grams of cobalt are in this sample? How many atoms of cobalt?

30. How many milligrams of chlorine are there in a sample of 3.9×10^{19} molecules of chlorine gas (Cl_2)? How many atoms of chlorine?

31. Bauxite, the principle ore used in the production of aluminum cans, has a molecular formula of $Al_2O_3 \cdot 2H_2O$.

a. Determine the molar mass of bauxite.
b. How many grams of Al are in 0.58 moles of bauxite?
c. How many atoms of Al are in 0.58 moles of bauxite?
d. What is the mass in grams of 2.1×10^{24} formula units of bauxite?

Section 3.5

32. Calculate the mass percent of Cl in each of the following compounds:

a. ClF
b. $HClO_2$
c. $CuCl_2$
d. PuOCl

33. Calculate the mass percent of each element in $C_5H_{10}O$.

34. Calculate the mass percent of each element in potassium ferricyanide, $K_3Fe(CN)_6$.

35. Calculate the mass percent of each element in barium sulfite ($BaSO_3$).

36. Calculate the mass percent of each element in natural lucite ($KAlSi_2O_6$).

37. Calculate the mass percent of silver in each of the following compounds:

a. $AgCl$ b. $AgCN$ c. $AgNO_3$

38. Chlorophyll a is essential for photosynthesis. It contains 2.72% magnesium by mass. What is the molar mass of chlorophyll a assuming there is one atom of magnesium in every molecule of chlorophyll a?

Section 3.6

39. Calculate the mass percent of each of the elements in nicotine ($C_{10}H_{14}N_2$).

40. Which of the following formulas can be empirical?

a. CH_4 d. N_2O_5 g. Sb_2S_3
b. CH_2 e. B_2H_6 h. N_2O_4
c. $KMnO_4$ f. NH_4Cl i. CH_2O

41. Determine the empirical and molecular formulas of a compound that has a mass of 31.04 g/mole and contains the following percentages of elements by mass:

$$C = 38.66\%, H = 16.24\%, N = 45.10\%$$

42. The analysis of a rocket fuel showed that it contained 87.4% nitrogen and 12.6% hydrogen by weight. Mass spectral analysis showed the fuel to have a molar mass of 32.05 grams. What are the empirical and molecular formulas of the fuel?

43. A compound is found by mass spectral analysis to contain the following percentages of elements by mass:

$$C = 49.67\%, Cl = 48.92\%, H = 1.39\%$$

The molar mass of the compound is 289.9 g/mole. Determine the empirical and molecular formulas of the compound.

44. Vanillin, the pleasant-smelling ingredient used to bake chocolate-chip cookies, is often used in the production of vanilla extract. Vanillin has a mass of 152.08 g/mole and contains the following percentages of elements by mass:

$$C = 63.18\%, H = 5.26\%, O = 31.56\%$$

Determine the empirical and molecular formulas of vanillin.

45. Determine the empirical formula of a compound that contains the following percentages of elements by mass:

$$Mo = 43.95\%, O = 7.33\%, Cl = 48.72\%$$

46. A molecule with a molecular weight of approximately 110 g/mol is analyzed. The results show that it contains 10.05% of carbon, 0.84% of hydrogen, and 89.10% of chloride. Determine the molecular formula for this compound.

47. Using data provided, calculate the empirical formulas for the compounds indicated:

a. an oxide of nitrogen, a sample that contains 6.35 g of nitrogen and 3.65 g of oxygen.
b. an oxide of copper, one gram of which contains 0.7989 g of copper.
c. an oxide of carbon that contains 42.85% carbon.
d. a compound of potassium, chloride, and oxygen containing K = 31.97%, O = 39.34%.
e. a compound of hydrogen, carbon, and nitrogen containing H = 3.70%, C = 44.44%, and N = 51.85%.

Section 3.7

48. How many grams of product are formed in each of the following reactions?

    a. Two moles of H_2 react with one mole of O_2.
    b. One mole of silver nitrate reacts with one mole of sodium chloride.
    c. Three moles of sodium hydroxide react with one mole of phosphoric acid.

49. The following reaction was performed:

$$Fe_2O_3(s) + 2X(s) \rightarrow 2Fe(s) + X_2O_3(s)$$

It was found that 79.847 g of Fe_2O_3 reacted with element X to form 55.847 g of Fe and 50.982 g of X_2O_3. Identify element X.

50. Do these equations follow the conservation of matter?

    a. $Na_2SiO_3 + 6HF \rightarrow SiF_4 + 2NaF + 3H_2O$
    b. $3N_2O_4 + 2H_2O \rightarrow 4HNO_3 + 2NO_2$

51. Explain in words the substances that are brought together on the left side of the arrow and those that are formed on the right.

$$3CaCl_2 + Fe_2(SO_4)_3 \rightarrow 3CaSO_4 + 2FeCl_3$$

Section 3.8

52. Balance the following equation:

$$NH_4OH(l) + KAl(SO_4)_2{\cdot}12H_2O(s) \rightarrow Al(OH)_3(s) + (NH_4)_2SO_4(aq) + KOH(aq) + H_2O(l)$$

53. Fill in the blanks to balance the following chemical equations:

    a. ____ AgI + ___ Na_2S → ___ Ag_2S + ___ NaI
    b. ____ $(NH_4)_2Cr_2O_7$ → ___ Cr_2O_3 + ___ N_2 + ___ H_2O
    c. ____ Na_3PO_4 + ___ HCl → ___ $NaCl$ + ___ H_3PO_4
    d. ____ $TiCl_4$ + ___ H_2O → ___ TiO_2 + ___ HCl
    e. ____ Ba_3N_2 + ___ H_2O → ___ $Ba(OH)_2$ + ___ NH_3
    f. ____ HNO_2 → ___ HNO_3 + ___ NO + ___ H_2O

54. Balance the following equation:

$$Fe(s) + HC_2H_3O_2(aq) \rightarrow Fe(C_2H_3O_2)_3(aq) + H_2(g)$$

55. Complete the following reactions (making sure they are balanced):

    a. HNO_3 + ________ → $H_2O + KNO_3$
    b. ________ + $Na_3PO_4 \rightarrow Ca_3(PO_4)_2 + NaCl$
    c. $Mg(OH)_2 + HCl \rightarrow MgCl_2$ + ________
    d. ________ + $Cl_2 \rightarrow NaCl + Br_2$

56. Balance the following equations:

    a. ___ Ca + ___ C + ___ O_2 → ___ $CaCO_3$
    b. ___ FeS + ___ O_2 → ___ Fe_2O_3 + ___ SO_2
    c. ___ HNO_2 → ___ NO_2 + ___ H_2O + ___ NO
    d. ___ PCl_5 + ___ H_2O → ___ H_3PO_4 + ___ HCl

Section 3.9

57. How many grams of water vapor can be generated from the combustion of 18.74 g of ethanol?

$$C_2H_6O(g) + O_2(g) \rightarrow CO_2(g) + H_2O(g)\ \textbf{(unbalanced)}$$

58. How many grams of sodium hydroxide are required to form 51.63 g of lead hydroxide?

$$Pb(NO_3)_2(aq) + NaOH(aq) \rightarrow Pb(OH)_2(s) + NaNO_3(aq)$$ **(unbalanced)**

59. How many grams of potassium iodide are necessary to completely react with 20.61 g of mercury(II) chloride?

$$HgCl_2(aq) + KI(aq) \rightarrow HgI_2(s) + KCl(aq)$$ **(unbalanced)**

60. What mass of calcium carbonate ($CaCO_3$) would be formed if 248.6 g of carbon dioxide (CO_2) were exhaled into limewater, $Ca(OH)_2$? How many grams of calcium would be needed to form that amount of calcium carbonate? Assume 100% yield in each reaction.

61. The following reaction is used to form lead iodide crystals. What mass of crystals (PbI_2) could be formed from 1.0×10^3 g of lead(II) acetate ($Pb(C_2H_3O_2)_2$)?

$$Pb(C_2H_3O_2)_2(aq) + 2KI(aq) \rightarrow PbI_2(s) + 2KC_2H_3O_2(aq)$$

62. How many grams of precipitate (Hg_2Cl_2) would be formed from a solution containing 102.9 g of mercury that is reacted with chloride ion as follows:

$$2Hg^+(aq) + 2Cl^-(aq) \rightarrow Hg_2Cl_2(s)$$

63. You were hired by a laboratory to recycle 6 moles of silver ions. You were given 150 g of copper. How many grams of silver can you recover? Is this enough copper to recycle 6 moles of silver ions?

$$2Ag^+ + Cu \rightarrow 2Ag + Cu^{2+}$$

64. Fermentation converts sugar into ethanol and carbon dioxide. If you were to ferment a bushel of apples containing 235 g of sugar, what is the maximum amount of ethanol in grams that would be produced?

$$C_6H_{12}O_6 \rightarrow 2C_2H_6O + 2CO_2$$

65. The reaction between potassium chlorate and red phosphorus is highly exothermic and takes place when you strike a match on a matchbox. If you were to react 52.9 g of potassium chlorate ($KClO_3$) with red phosphorus, how many grams of tetraphosphorus decaoxide (P_4O_{10}) would be produced?

$$KClO_3(s) + P_4(s) \rightarrow P_4O_{10}(s) + KCl(s)$$ **(unbalanced)**

66. Oxygen reacts with 33.1 g of NH_3. Only 30.2 g of water vapor were formed. How much NH_3 actually reacted with oxygen?

$$5O_2 + 4NH_3 \rightarrow 4NO + 6H_2O$$

Section 3.10

67. A reaction combines 133.484 g of lead(II) nitrate with 45.010 g of sodium hydroxide. (See problem 58.)

    a. How much lead(II) hydroxide is formed?
    b. Which reactant is limiting? Which is in excess?
    c. How much of the excess reactant is left over?
    d. If the actual yield of lead(II) hydroxide were 80.02 g, what was the percent yield?

68. A reaction combines 64.81 grams of silver nitrate with 92.67 grams of potassium bromide.

$$AgNO_3(aq) + KBr(aq) \rightarrow AgBr(s) + KNO_3(aq)$$

a. How much silver bromide is formed?
b. Which reactant is limiting? Which is in excess?
c. How much of the excess reactant is left over?
d. If the actual yield of silver bromide were 14.77 g, what was the percent yield?

69. A reaction proceeds between 94.6 g of $KClO_3$ and 65.3 g of P_4. (See problem 65.)

a. How much potassium chloride is formed?
b. Which reactant is limiting? Which is in excess?
c. How much of the excess reactant is left over?
d. If the actual yield of potassium chloride was 21.0 g, what was the percent yield?

70. DDT, an insecticide harmful to fish, birds, and humans, is produced by the following reaction:

$$\underset{\text{chlorobenzene}}{2C_6H_5Cl} + \underset{\text{chloral}}{C_2HOCl_3} \rightarrow \underset{\text{DDT}}{C_{14}H_9Cl_5} + H_2$$

In a government lab 1142 g of chlorobenzene was reacted with 485 g of chloral.

a. How much DDT is formed?
b. Which reactant is limiting? Which is in excess?
c. How much of the excess reactant is left over?
d. If the actual yield of DDT was 200.0 g, what was the percent yield?

Multiple-Choice Self-Test

1. Bromine is composed of two isotopes. One of the isotopes, Br-X, makes up 49.7% of the total while the other, Br-Y (atomic mass = 78.9 u), makes up 50.3% of the total. Calculate the atomic mass of Br-X.

A. 80.0 B. 80.9 C. 89.7 D. 78.9

2. 18.0 cm^3 of water (assume density 1.00 g/cm^3) contain 1.0 mole of water molecules. If 0.9 cm^3 of water in a beaker is allowed to evaporate over a period of 24 hours, how many molecules of water evaporate per second? Assume a constant rate of evaporation.

A. 0.7×10^{15} B. 1.5×10^{23} C. 3.0×10^{23} D. 3.5×10^{17}

3. Determine the molecular formula of a compound that contains 26.7% P, 12.1% N, and 61.2% Cl and has a molecular weight of 580.

A. $(PNCl)_3$ B. $(PNCl_2)_5$ C. $(P_2NCl_2)_5$ D. $(PNCl_2)$

4. A 20-mg sample of C_xH_y is burned in oxygen to produce 60 mg of carbon dioxide and 32 mg of water. Calculate x and y by using these data.

A. 2, 4 B. 3, 6 C. 1, 4 D. 3, 8

5. Find the identity of the element X in the following equation

$$1C_3H_6X_3 + 3X_2 \rightarrow 3CX_2 + 3H_2O$$

A. O B. H C. Cl D. Br

6. The proper set of coefficients for the following equation is

$$\_C_3H_6O_3 + \_O_2 \rightarrow \_CO_2 + \_H_2O$$

A. 1, 3, 3, 3 B. 2, 4, 3, 3 C. 1, 2, 3, 3 D. 1, 6, 6, 6

7. For every liter of sea water that evaporates, 3.7 g of magnesium hydroxide are produced. How many liters of sea water must evaporate to produce 5.00 moles of magnesium hydroxide?

A. 78.8 B. 50 C. 143 D. 18.5

8. A solution of copper sulfate is treated with zinc metal. How many grams of copper are produced if 2.9 g of zinc is consumed?

$$CuSO_4 + Zn \rightarrow ZnSO_4 + Cu$$

A. 2.9 g B. 2.8 g C. 5.7 g D. 3.7 g

9. How many grams of carbon dioxide are produced from the burning of 1368 g of sucrose according to the following equation?

$$C_{12}H_{22}O_{11} + 12O_2 \rightarrow 12CO_2 + 11H_2O$$

A. 342 g B. 176 g C. 1056 g D. 2111 g

10. How many grams of sulfur dioxide are produced when 90.0 g of thionyl chloride react with excess water according to the following equation?

$$SOCl_2 + H_2O \rightarrow 2HCl + SO_2$$

A. 96.8 B. 90.0 C. 24.2 D. 48.5

11. Calcium oxide is a basic oxide that is not very soluble in water solutions. Calcium oxide can react with carbon dioxide to form calcium carbonate (according to the equation below). Calcium carbonate is an insoluble salt that forms stalactites and stalagmites. How many moles of carbon dioxide are removed from water if a 400.0-lb stalagmite is formed?

$$CaO + CO_2 \rightarrow CaCO_3$$

A. 1817 B. 908 C. 4000 D. 2258

12. Calculate the number of grams of $TiOCl_2$ required to react with 134 g of carbon.

$$2TiOCl_2 + 2C \rightarrow 2Ti + CO_2 + CCl_4$$

A. 134 B. 1.51×10^3 C. 536 D. 67

13. Calculate the number of grams of methane (CH_4) required to react with 25.0 g of chlorine according to the following equation:

$$3CH_4 + 4Cl_2 \rightarrow 2CH_3Cl + CH_2Cl_2 + 4HCl$$

A. 33.3 B. 18.8 C. 2.11 D. 4.23

14. Identify the limiting reagent, and calculate the number of grams left over of the reagent in excess for the following equation.

$$Cr_2O_3 + 3CCl_4 \rightarrow 2CrCl_3 + 3COCl_2$$

The reaction began with 5.00 g of Cr_2O_3 and 12.0 g of CCl_4.

A. Limiting: CCl_4; 1.05 g Cr_2O_3 left over
B. Limiting: Cr_2O_3; 10.0 g CCl_4 left over
C. Limiting: CCl_4; 3.70 g Cr_2O_3 left over
D. Limiting: Cr_2O_3; 6.75 g CCl_4 left over

15. Identify the limiting reagent and the number of grams left over for the following equation if 1.90 grams each of phosgene and sodium hydroxide are combined.

$$COCl_2 + 2NaOH \rightarrow 2NaCl + H_2O + CO_2$$

A. Limiting: $COCl_2$; 0.588 g NaOH left over
B. Limiting: $COCl_2$; 0.4 g NaOH left over
C. Limiting: NaOH; 0.520 g $COCl_2$ left over
D. Limiting: NaOH; 0.355 g $COCl_2$ left over

Answers to Exercises

1. The average atomic mass is 10.81 u. The element is boron.
2. The molar mass of sulfur is 32.06 u.
3. The average atomic mass is 6.95 u. The element is lithium.
4. The average atomic mass is 20.17 u. The gas is neon.
5. The mass of the other isotope is 36.93 u.
6. ^{20}Neon = 90.37%; ^{22}Neon = 9.37%
7. The abundance of silicon-28 is 92.21%. The atomic mass of silicon-29 is 29.01.
8. gallium-69 = 60.3%; gallium-71 = 39.7%
9. magnesium-24 = 79.13%; magnesium-25 = 10.13%; magnesium-26 = 10.74%
10. 4.98×10^{-22} moles of nitrogen; mass = 6.98×10^{-21} g
11. 3.06×10^{21} atoms of gold
12. 2.34×10^{25} M&M's®
13. 0.167 moles of sulfur; 1.01×10^{23} atoms in the sample
14. a. 2Hg/2I c. 1Pb/1C/3O e. 1Rb/3O/5H
    b. 1Li/1H d. 3Ba/2As/8O f. 2H/1Si/6F
15. 1.26 g Zn
16. 1.24×10^{26} u of iron
17. a. 169.46 g/mol b. 275.81 g/mol
18. a. 1.8×10^{24} ions b. 3.7×10^{22} ions c. 1.5×10^{23} ions
19. a. 3.6×10^{24} ions b. 6.0×10^{22} ions c. 1.5×10^{23} ions
20. a. 6.4 g b. 1100 g c. 300 g
21. a. 223.17 g/mole c. 152.23 g/mole e. 508.32 g/mole
    b. 35.05 g/mole d. 345.14 g/mole f. 729.82 g/mole
22. mass = 1.28 g
23. 127 mg Br_2
24. 474.4 g/mole
25. 0.1442 moles of cadmium bromide
26. 0.214 moles of calcium chloride
27. 14.4 g of $NaHCO_3$

28. 4.24×10^{-9} moles

29. 85 g of cobalt(III) acetate. There are 21 g of cobalt. There are 2.2×10^{23} atoms.

30. There are 4.6 mg of chlorine. There are 7.8×10^{19} atoms of chlorine.

31. a. bauxite = 137.99 g/mole
 b. 31.3 g Al
 c. 6.99×10^{23} atoms of Al
 d. 4.8×10^{2} g of bauxite

32. a. 65.11% Cl
 b. 51.78% Cl
 c. 52.73% Cl
 d. 12.0% Cl

33. Mass percent of C = 69.72%, H = 11.70%, O = 18.58%

34. Mass percent of C = 21.89%, Fe = 16.96%, N = 25.53%, K = 35.62%

35. Mass percent of Ba = 63.17%, S = 14.75%, O = 22.08%

36. Mass percent of K = 17.91%, Al = 12.36%, Si = 25.74%, O = 43.99%

37. a. Ag = 75.27% b. Ag = 80.57% c. Ag = 63.50%

38. chlorophyll a = 894 g/mole

39. Percent composition of C = 74.03%, H = 8.70%, N = 17.27%

40. Empirical formulas can be: a, b, c, d, f, g, i

41. Empirical formula = molecular formula = CH_5N

42. Empirical formula = NH_2; molecular formula = N_2H_4

43. Empirical formula = C_3HCl; molecular formula = $C_{12}H_4Cl_4$

44. Empirical formula = molecular formula = $C_8H_8O_3$

45. Empirical formula = $MoOCl_3$

46. $CHCl_3$

47. a. N_2O
 b. CuO
 c. CO
 d. $KClO_3$
 e. HCN

48. a. 36 grams
 b. 143 g AgCl
 c. 164 grams of sodium phosphate, 54 grams of water

49. X = aluminum

50. a. yes b. no

51. Three parts calcium chloride plus one part iron(III) sulfate produces three parts calcium sulfate and two parts iron(III) chloride.

52. $4NH_4OH(l) + KAl(SO_4)_2 \cdot 12H_2O(s) \rightarrow Al(OH)_3(s) + 2(NH_4)_2SO_4(aq) + KOH(aq) + 12H_2O(l)$

53. a. 2, 1, 1, 2
 b. 1, 1, 1, 4
 c. 1, 3, 3, 1
 d. 1, 2, 1, 4
 e. 1, 6, 3, 2
 f. 3, 1, 2, 1

54. $2Fe(s) + 6HC_2H_3O_2(aq) \rightarrow 2Fe(C_2H_3O_2)_3(aq) + 3H_2(g)$

55. a. $HNO_3 + KOH \rightarrow H_2O + KNO_3$
 b. $3CaCl_2 + 2Na_3PO_4 \rightarrow Ca_3(PO_4)_2 + 6NaCl$
 c. $Mg(OH)_2 + 2HCl \rightarrow MgCl_2 + 2H_2O$
 d. $2NaBr + Cl_2 \rightarrow 2NaCl + Br_2$

56. a. $2Ca + 2C + 3O_2 \rightarrow 2CaCO_3$
    b. $4FeS + 7O_2 \rightarrow 2Fe_2O_3 + 4SO_2$
    c. $2HNO_2 \rightarrow NO_2 + H_2O + NO$
    d. $PCl_5 + 4H_2O \rightarrow H_3PO_4 + 5HCl$

57. 21.99 g of water vapor

58. 17.12 g of NaOH

59. 25.20 g of KI

60. 565.4 g of $CaCO_3$; 226.4 g of Ca^{2+}

61. 1.4×10^3 g of PbI_2

62. 121.1 g of Hg_2Cl_2

63. 509. g of Ag recovered; no, you could recycle only 4.72 moles of Ag^+

64. 120. g of ethanol

65. 36.8 g of P_4O_{10}; [balanced equation: $10KClO_3(s) + 3P_4(s) \rightarrow 3P_4O_{10}(s) + 10KCl(s)$]

66. 19.0 g NH_3

67. a. 97.214 g of lead(II) hydroxide is formed.
    b. Lead nitrate is limiting. Sodium hydroxide is in excess.
    c. 12.771 g of sodium hydroxide is left over.
    d. The yield was 82.31%.

68. a. 71.63 g of silver bromide is formed.
    b. Silver nitrate is limiting. Potassium bromide is in excess.
    c. 47.27 g of potassium bromide is left over.
    d. The yield was 20.62%.

69. a. 57.5 g of KCl is formed.
    b. Potassium chlorate is limiting. Red phosphorus is in excess.
    c. 36.6 g of P_4 is left over.
    d. The yield was 36.5%.

70. a. 1166. g of DDT is formed.
    b. Chloral is limiting. Chlorobenzene is in excess.
    c. 401 g of chlorobenzene is left over.
    d. The yield was 17.1%.

Answers to Multiple-Choice Self-Test

1. B
2. D
3. B
4. D
5. A
6. A
7. A
8. B
9. D
10. D
11. A
12. B
13. D
14. A
15. B

CHAPTER 4

Types of Chemical Reactions and Solution Stoichiometry

The Bottom Line: Chapter 4

A variety of methods are presented here for solving problems relating to aqueous solutions. Although there are many sections in this chapter, the consistent theme is how substances behave in water.

4.1 Water, the Common Solvent

This section introduces you to the nature of interactions between atoms and electrons in the water molecule. Please pay special attention to the following ideas:

a. Water is *not* a linear molecule. It is bent at an angle of about 105°.

b. Electrons are not evenly distributed around the atoms in water. Notice the position of the partial charges on the molecule shown in Figure 4.1 of your textbook. The molecule is *polar* because the charges are not distributed symmetrically.

c. Like dissolves like. The following classes of molecules, in general, are miscible:

- polar and ionic
- polar and polar
- nonpolar and nonpolar

Most ionic salts dissolve in water. Compounds that contain *only* carbon and hydrogen are nonpolar. Given that information, please try the following example.

Example 4.1 A - Will The Substances Mix?

Predict whether each pair of substances will mix. State why or why not.

a. $NaNO_3$ and H_2O
b. C_6H_{14} and H_2O
c. I_2 and C_6H_{14}
d. I_2 and H_2O

Solution

a. Will mix. Sodium nitrate is ionic, and water is polar.

b. Will not mix. Cyclohexane (C_6H_{14}) is nonpolar while water is polar.

c. Will mix. The iodine molecule is composed of two identical atoms. Therefore the electrons are distributed symmetrically. Iodine is therefore almost completely nonpolar. Cyclohexane is also nonpolar.

d. Will not mix. The iodine molecule is nonpolar, and water is polar.

The dissociation of simple ionic salts in water is often written as shown in the following equations:

$$NaI(s) \xrightarrow{H_2O(l)} Na^+(aq) + I^-(aq)$$

$$K_2Cr_2O_7(s) \xrightarrow{H_2O(l)} 2K^+(aq) + Cr_2O_7^{2-}(aq)$$

$$Ba(OH)_2(s) \xrightarrow{H_2O(l)} Ba^{2+}(aq) + 2OH^-(aq)$$

Example 4.1 B - Practice With Equations

Complete each of the following dissociation equations:

a. $CaCl_2(s) \xrightarrow{H_2O(l)}$
b. $Fe(NO_3)_3(s) \xrightarrow{H_2O(l)}$
c. $KBr(s) \xrightarrow{H_2O(l)}$
d. $(NH_4)_2Cr_2O_7(s) \xrightarrow{H_2O(l)}$

Solution

a. $CaCl_2(s) \xrightarrow{H_2O(l)} Ca^{2+}(aq) + 2Cl^-(aq)$

b. $Fe(NO_3)_3(s) \xrightarrow{H_2O(l)} Fe^{3+}(aq) + 3NO_3^-(aq)$

c. $KBr(s) \xrightarrow{H_2O(l)} K^+(aq) + Br^-(aq)$

d. $(NH_4)_2Cr_2O_7(s) \xrightarrow{H_2O(l)} 2NH_4^+(aq) + Cr_2O_7^{2-}(aq)$

4.2 The Nature of Aqueous Solutions: Strong and Weak Electrolytes

In this section your book introduces **solute** and **solvent**. These terms are discussed in more detail in Chapter 17, but for now we will just define solute as the substance being dissolved and solvent as the dissolving medium. An **aqueous solution** means that **water is the solvent**.

Figure 4.4 in your textbook shows the effect of strong, weak, and nonelectrolytes on the ability to pass a current (conductivity) in an aqueous solution.

Electrolyte	Conductivity	Degree of Dissociation	Examples
Strong	high	total	strong acids, such as HCl; many salts such as NaCl and $Sr(NO_3)_3$; strong bases, such as NaOH, $Ba(OH)_2$, and other group I and II hydroxides.
Weak	low to moderate	partial	weak organic acids such as HCO_2H and $HC_2H_3O_2$; weak bases such as NH_3
Non	none	close to zero	sugar, AgCl, Fe_2O_3

Example 4.2 - Strong, Weak, Or Nonelectrolyte

List whether each of the following is a strong, weak, or nonelectrolyte.

a. $HClO_4$
b. C_6H_{12}
c. LiOH
d. NH_3
e. $CaCl_2$
f. $HC_2H_3O_2$

Solution

a. strong ($HClO_4$ is a strong acid)
b. non (cyclohexane contains only carbon and hydrogen and is therefore not soluble in water!)

c. strong
d. weak
e. strong
f. weak (acetic acid)

4.3 The Composition of Solutions

We deal here with the *preparation of solutions*. **Molarity (*M*) is defined as moles of solute per liter of solution.**

$$M = \frac{\text{moles of solute}}{\text{liter of solution}}$$

Keep in mind that $\frac{\text{moles}}{\text{liter}} = \frac{\text{millimoles}}{\text{milliliter}} = \frac{\text{micromoles}}{\text{microliter}}$, but $\mathbf{\frac{moles}{liter}}$ **DOES NOT EQUAL** $\mathbf{\frac{millimoles}{liter}}$ **OR** $\mathbf{\frac{moles}{microliter}}$. Be very careful with your units!

Example 4.3 A - Calculating Molarity

Calculate the molarity of a solution prepared by dissolving 11.85 g of solid $KMnO_4$ in enough water to make 750. mL of solution.

Solution

What are the units of molarity? We are looking for **moles of solute per liter of solution**. Therefore, you must first convert *grams of $KMnO_4$* to *moles of $KMnO_4$*, then divide by the volume. (Remember to convert volume from mL to L!)

The molar mass of $KMnO_4$ = 158.04 g/mole.

Method A

In this method we can proceed in two steps:

1. Convert from grams of $KMnO_4$ to moles of $KMnO_4$.

$$\text{moles KMnO}_4 \,\Big\|\, \frac{1\text{ mol KMnO}_4}{158.04\text{ g KMnO}_4} \times 11.85\text{ g KMnO}_4 = 0.07498\text{ mol KMnO}_4$$

2. Divide moles by volume (in liters!) to get molarity.

$$M = \frac{\text{mol}}{\text{L}} = \frac{0.07498\text{ mol KMnO}_4}{0.750\text{ L solution}} = 0.09997\ M\text{ KMnO}_4 = \mathbf{0.100}\ \boldsymbol{M}\ \mathbf{KMnO_4}$$

Method B

We can use dimensional analysis to solve the problem in one longer step:

$$\frac{\text{mol KMnO}_4}{\text{L solution}} \,\Big\|\, \frac{11.85\text{ g KMnO}_4}{0.750\text{ L soln}} \times \frac{1\text{ mol KMnO}_4}{158.04\text{ g KMnO}_4} = \mathbf{0.100}\ \boldsymbol{M}\ \mathbf{KMnO_4}$$

We will often use dimensional analysis in this study guide. However, it is never the ONLY correct method. Let's change the problem around a bit. Instead of using the mass of the solute to calculate the solution molarity, let's begin with **molarity** to calculate **mass**.

Example 4.3 B - Mass From Molarity

Calculate the mass of NaCl needed to prepare 175. mL of a 0.500 *M* NaCl solution.

Solution

What are we trying to solve? From a *dimensional analysis* point of view, the question we have to answer is, "How can we go from molarity (mol/L) to mass (g)?" Note that this means that somewhere along the line we will have to *cancel volume.*

$$\text{g NaCl} \parallel \frac{0.500 \text{ mol NaCl}}{\text{L soln}} \times \underset{\uparrow \text{ convert to grams}}{\frac{58.44 \text{ g NaCl}}{1 \text{ mol NaCl}}} \times \underset{\uparrow \text{ cancel volume}}{0.175 \text{ L soln}} = \mathbf{5.11 \text{ g NaCl}}$$

Our values for molarity and volume had 3 significant figures, so we rounded off accordingly.

Example 4.3 C - Volume From Molarity

How many mL of solution are necessary if we are to have a 2.48 *M* NaOH solution that contains 31.52 g of the dissolved solid?

Solution

What are we trying to solve? We are after the volume of sodium hydroxide solution. The units must cancel to give mL. Starting off by inverting molarity will put volume in the numerator. Remember to convert L to mL!

$$\text{mL solution} \parallel \frac{1.00 \text{ L soln}}{2.48 \text{ mol NaOH}} \times \underset{\uparrow \text{ convert from L to mL}}{\frac{1000 \text{ mL}}{1 \text{ L}}} \times \underset{\uparrow \text{ cancel moles NaOH}}{\frac{1 \text{ mol NaOH}}{40.00 \text{ g NaOH}}} \times \underset{\uparrow \text{ cancel grams NaOH}}{31.52 \text{ g NaOH}}$$

$$= \mathbf{318 \text{ mL solution}}$$

Does the Answer Make Sense?

The amount of NaOH, 31.52 g, is about 3/4 of a mole. If the solution is 2.48 *M*, you have about 1/3 as much NaOH as you need to make 1 L, or enough to make about 300 mL, so the answer seems to make sense.

In the previous problems we calculated the molarity of the solute. However, we have neglected to take into account the fact that each of the solutes, $KMnO_4$, NaCl, and NaOH, are **strong electrolytes and completely dissociate** in aqueous solution. (Recall Section 4.2.) For example,

$$KMnO_4 \xrightarrow{H_2O(l)} K^+(aq) + MnO_4^-(aq)$$

This means that while it is generally acceptable to discuss your solution concentration as **molarity of $KMnO_4$**, it is more chemically correct to discuss **molarity of K^+ ions and molarity of MnO_4^- ions**. Keeping the stoichiometry of the dissociation equation in mind, try the next problem.

Example 4.3 D - Molarity Of Ions In Solution

Determine the molarity of Cl^- ion in a solution prepared by dissolving 9.82 g of $CuCl_2$ in enough water to make 600. mL of solution.

Solution

What are we trying to solve? The goal is molarity of the chloride ion in the solution. We have the mass of the $CuCl_2$ and the volume of the solution. We can convert from mass per liter to moles per liter, and then recognize that there are 2 moles of chloride ion per mole of $CuCl_2$.

1. **Solute concentration:**

$$\frac{\text{mol CuCl}_2}{\text{L soln}} \parallel \frac{1 \text{ mol CuCl}_2}{134.45 \text{ g CuCl}_2} \times 9.82 \text{ g CuCl}_2 \times \frac{1}{0.600 \text{ L soln}} = \mathbf{0.1217}\ \boldsymbol{M}\ \mathbf{CuCl_2}$$

We retain all our figures because this is an *intermediate* calculation.

2. **Ion-to-solute ratio:**

$$CuCl_2(s) \xrightarrow{H_2O(l)} Cu^{2+}(aq) + 2Cl^-(aq)$$

Therefore the ratio $\frac{\text{ion}}{\text{solute}} = \frac{2 \text{ moles Cl}^-}{1 \text{ mole CuCl}_2}$.

3. **The final molarity of Cl^- in solution (also expressed as $[Cl^-]$):**

$$[Cl^-] = \frac{0.1216 \text{ mol CuCl}_2}{\text{L soln}} \times \frac{2 \text{ mol Cl}^-}{1 \text{ mol CuCl}_2} = \mathbf{0.243}\ \boldsymbol{M}\ \mathbf{Cl^-}$$

We rounded off because we were at the end of the problem. Three significant figures is correct because both mass and volume were given to three figures. Note that if we had rounded off earlier, we would have calculated that $[Cl^-] = 0.122\ M \times 2 = 0.244\ M$, which is not strictly correct.

ALTERNATIVELY, we can use dimensional analysis to solve the entire problem with one equation.

$$\frac{\text{mol Cl}^-}{\text{L soln}} \parallel \underbrace{\frac{2 \text{ mol Cl}^-}{1 \text{ mol CuCl}_2}}_{\text{from dissociation equation}} \times \underbrace{\frac{1 \text{ mol CuCl}_2}{134.45 \text{ g CuCl}_2}}_{\text{molar mass}} \times \underbrace{9.82 \text{ g CuCl}_2}_{\text{given}} \times \underbrace{\frac{1}{0.600 \text{ L soln}}}_{\text{given}} = \mathbf{0.243}\ \boldsymbol{M}$$

Example 4.3 E - Practice With Ion Concentration

Determine the molarity of Fe^{3+} ions and SO_4^{2-} ions in a solution prepared by dissolving 48.05 g of $Fe_2(SO_4)_3$ in enough water to make 800. mL of solution.

Solution

What are we trying to solve? As in the last example, we are looking for the molarity of ions in the solution. Note the mole ratios of each ion to each formula unit.

The dissociation equation is:

$$Fe_2(SO_4)_3(s) \xrightarrow{H_2O(l)} 2Fe^{3+}(aq) + 3SO_4^{2-}(aq)$$

So the mole ratio of $\frac{\text{ion}}{\text{solute}} = \frac{2 \text{ mol Fe}^{3+}}{1 \text{ mol Fe}_2(\text{SO}_4)_3}$ and $\frac{3 \text{ mol SO}_4^{2-}}{1 \text{ mol Fe}_2(\text{SO}_4)_3}$

Let's work with $[Fe^{3+}]$.

$$\frac{\text{mol Fe}^{3+}}{\text{L soln}} \parallel \frac{2 \text{ mol Fe}^{3+}}{1 \text{ mol Fe}_2(\text{SO}_4)_3} \times \frac{1 \text{ mol Fe}_2(\text{SO}_4)_3}{399.9 \text{ g Fe}_2(\text{SO}_4)_3} \times 48.05 \text{ g Fe}_2(\text{SO}_4)_3 \times \frac{1}{0.800 \text{ L soln}}$$

$$= \mathbf{0.300}\ \boldsymbol{M} = [\text{Fe}^{3+}]$$

Looking at the **mole ratios**, can you tell that the ratio of $\frac{[\text{SO}_4^{\ 2-}]}{[\text{Fe}^{3+}]}$ must equal $\frac{3 \text{ moles}}{2 \text{ moles}}$?

Therefore, since the volume is constant, $[\text{SO}_4^{\ 2-}] = \frac{3}{2}[\text{Fe}^{3+}] = \frac{3}{2}(0.300\ M)$

$$\mathbf{[SO_4^{\ 2-}] = 0.450}\ \boldsymbol{M}$$

An important part of your chemistry experience is to be able to *prepare dilute solutions from more concentrated (stock) solutions*. Your textbook points out that the most important idea in diluting solutions is that

moles of solute after dilution = moles of solute before dilution

$$\text{If molarity} = \frac{\text{moles of solute}}{\text{liter of solution}}, \text{ then}$$

$$\text{moles of solute} = \frac{\text{moles of solute}}{\text{liter of solution}} \times \text{liters of solution} = M \times V.$$

If the **moles of solute remain identical** before and after dilution (only the amount of water changes), then

$$M_i V_i = M_f V_f$$

where M_i = molarity of concentrated (initial) solution
V_i = volume of concentrated solution that you add to water to dilute…this will often be your **unknown volume**.
M_f = molarity of dilute (final) solution
V_f = total volume of your dilute solution

Try the following introductory example.

Example 4.3 F - Preparation Of A Dilute Solution

What volume of 12 M hydrochloric acid must be used to prepare 600. mL of a 0.30 M HCl solution?

Solution

Be aware that no matter how much you dilute your acid, **the number of moles of acid in the solution will remain the same**. Your molarity will change, but not the total number of moles. $M_i V_i = M_f V_f$ is your best bet in solving these types of problems.

Let M_i = 12 M HCl
M_f = 0.30 M HCl
V_i = unknown
V_f = 600. mL (or 0.600 L) The units of molarity cancel, so you may use any volume unit you want.

$$12\ M \times V_i = 0.30\ M\ (600.\ \text{mL}) = \frac{0.30\ M\ (600.\ \text{mL})}{12\ M} = V_i = \mathbf{15\ mL\ of\ 12}\ \boldsymbol{M}\ \mathbf{HCl}$$

Double Check

$$\frac{12 \text{ mol}}{\text{L}}(0.015 \text{ L}) = \frac{0.30 \text{ mol}}{\text{L}}(0.600 \text{ L})$$

$$0.18 \text{ mol} = 0.18 \text{ mol}$$

The number of moles, 0.18 mol, is the same on each side.

Although not discussed in Chapter 4 of your textbook, some of the end of chapter exercises in your text use the term **parts per million**.

$$1 \text{ part per million of "X" (ppm)} = \frac{1 \text{ part X}}{1\times10^6 \text{ parts solution}},$$

$$= \frac{1 \text{ g X}}{1\times10^6 \text{ g soln}} = \frac{1 \text{ μg X}}{1 \text{ g soln}}$$

If the solution is water (whose density = 1.00 g/mL),

$$\mathbf{1 \text{ ppm X}} = \frac{\mathbf{1 \text{ μg X}}}{\mathbf{1 \text{ mL soln}}} = \frac{\mathbf{1 \text{ mg X}}}{\mathbf{1 \text{ L soln}}}$$

Note that this differs from molarity in that the units are *mass per volume*, not moles per volume. To set the stage for your work with the additional exercises, consider the following problem.

Example 4.3 G - Parts Per Million

An aqueous solution with a total volume of 750 mL contains 14.38 mg of Cu^{2+}. What is the concentration of Cu^{2+} in parts per million?

Solution

What is a part per million? How can you express that in terms of the mass of solute and volume of solution?

The use of ppm DOES NOT necessarily require you to calculate moles on the way to your answer. The dimensional analysis unit of choice is **mg/L**.

$$\frac{14.38 \text{ mg } Cu^{2+}}{0.750 \text{ L soln}} = \mathbf{19.2 \text{ ppm } Cu^{2+}}$$

The next problem requires that you deal with moles.

Example 4.3 H - Molarity To Parts Per Million

A solution is 3×10^{-7} *M* in manganese(VII) ion. What is the Mn^{7+} concentration in ppm?

Solution

You want $\frac{\text{mg } Mn^{7+}}{\text{L soln}}$, and you have $\frac{\text{moles } Mn^{7+}}{\text{L soln}}$. Dimensional analysis is a good approach.

$$\frac{\text{mg } Mn^{7+}}{\text{L soln}} \parallel \frac{3\times10^{-7} \text{ mol } Mn^{7+}}{\text{L soln}} \times \frac{54.94 \text{ g } Mn^{7+}}{1 \text{ mol } Mn^{7+}} \times \frac{1000 \text{ mg } Mn^{7+}}{1 \text{ g } Mn^{7+}}$$

$$= \frac{0.0164 \text{ mg } Mn^{7+}}{\text{L soln}} \text{ or } \mathbf{0.02 \text{ ppm } Mn^{7+}} \text{ (rounded off)}$$

4.4 Types of Chemical Reactions

This section points out that reactions are divided into precipitation reactions, acid-base reactions, and oxidation-reduction reactions.

4.5 Precipitation Reactions

The major ideas in this section are that:

1. Many salts dissociate into ions in aqueous solution.
2. If a solid forms from a combination of selected ions in solution, the solid must contain an anion part and cation part, and the net charge on the solid must be zero.
3. There are some simple solubility rules you can use to help you predict the products of reactions in aqueous solutions.

Table 4.1 in your textbook lists solubility rules for salts in water. The information is important enough to reprint below.

Simple Rules for the Solubility of Salts in Water:

1. Most nitrate (NO_3^-) salts are soluble.
2. Most salts of Na^+, K^+, and the ammonium ion (NH_4^+) are soluble.
3. Most chloride salts are soluble. Notable exceptions are AgCl, $PbCl_2$, and Hg_2Cl_2.
4. Most sulfate salts are soluble. Notable exceptions are $BaSO_4$, $PbSO_4$, and $CaSO_4$.
5. Most hydroxide salts are only slightly soluble. The important soluble hydroxides are NaOH, KOH, and $Ca(OH)_2$ (marginally soluble).
6. Most sulfide (S^{2-}), carbonate (CO_3^{2-}), and phosphate (PO_4^{3-}) salts are only slightly soluble.

If two soluble substances (call them AX and BZ) are combined, you can assume that the products will be AZ and BX.

$$AX(aq) + BZ(aq) \rightarrow AZ + BX$$

Your goal is to determine, based on your knowledge of solubility rules, whether AZ or BX will form a solid (precipitate). Let's look at the following reaction.

$$AgNO_3(aq) + Na_2S(aq) \rightarrow \text{products.}$$

The reactants are electrolytes that will dissociate to form the ions

$$Ag^+(aq) + NO_3^-(aq) + 2Na^+(aq) + S^{2-}(aq).$$

If a solid forms, and if it is to have zero charge, it can be either $\mathbf{Ag_2S}$ or $\mathbf{NaNO_3}$. According to **solubility rule No. 1**, $NaNO_3$ is soluble. According to **solubility rule No. 6**, Ag_2S is insoluble and will therefore precipitate. The correct (balanced) reaction therefore is

$$2AgNO_3(aq) + Na_2S(aq) \rightarrow Ag_2S(s) + 2NaNO_3(aq)$$

Example 4.5 - Predicting Precipitates

Complete and balance the following reactions, determining in each case if a precipitate is formed.

a. $KCl(aq) + Pb(NO_3)_2(aq) \rightarrow$
b. $AgNO_3(aq) + MgBr_2(aq) \rightarrow$
c. $Ca(OH)_2(aq) + FeCl_3(aq) \rightarrow$
d. $NaOH(aq) + HCl(aq) \rightarrow$

Solution

a. $2KCl(aq) + Pb(NO_3)_2(aq) \rightarrow \mathbf{PbCl_2(s)} + 2KNO_3(aq)$
b. $2AgNO_3(aq) + MgBr_2(aq) \rightarrow \mathbf{2AgBr(s)} + Mg(NO_3)_2(aq)$
c. $3Ca(OH)_2(aq) + 2FeCl_3(aq) \rightarrow \mathbf{2Fe(OH)_3(s)} + 3CaCl_2(aq)$
d. $NaOH(aq) + HCl(aq) \rightarrow H_2O(l) + NaCl(aq)$

↑ water is the solvent, hence "(*l*)"

4.6 Describing Reactions in Solution

This section discusses the three kinds of equations that are used to describe reactions in aqueous solution. Specific definitions for the **molecular**, **complete ionic**, and **net ionic equations** are given in your textbook. Let's look at how the aqueous reaction of *silver nitrate* with *sodium sulfide* can be expressed with each type of equation.

a. **Molecular**: This gives the overall reaction. While it does give information on stoichiometry, it gives no information on whether or not compounds really exist as ions in solution. Molecular form:

$$2AgNO_3(aq) + Na_2S(aq) \rightarrow Ag_2S(s) + 2NaNO_3(aq)$$

b. **Complete Ionic**: This gives the equation including all ions in solution. Because **all** compounds and ions are present, some information may be redundant. Complete ionic form:

$$2Ag^+(aq) + 2NO_3^-(aq) + 2Na^+(aq) + S^{2-}(aq) \rightarrow Ag_2S(s) + 2Na^+(aq) + 2NO_3^-(aq)$$

c. **Net Ionic**: This gives information on only those species that undergo a chemical change. Ions that appear in the same form on both sides of the complete ionic equation are called **spectator ions** and are not included in the net ionic equation. In our sample equation, Na^+ and NO_3^- are unaltered during the reactions. They are therefore omitted from the net ionic equation. Net ionic form:

$$2Ag^+(aq) + S^{2-}(aq) \rightarrow Ag_2S(s)$$

The **molecular** and **net ionic** forms of equations are the most commonly used. The **complete ionic** form helps us determine the net ionic form.

Example 4.6 - Molecular, Complete Ionic, And Net Ionic Equations

Write the molecular, complete ionic, and net ionic forms for each of the following equations.

a. Aqueous nickel(II) chloride reacts with aqueous sodium hydroxide to give a nickel(II) hydroxide precipitate and aqueous sodium chloride.
b. Solid potassium metal reacts with water to give aqueous potassium hydroxide and hydrogen gas.
c. Aqueous sodium hydroxide reacts with aqueous phosphoric acid to give water and aqueous sodium phosphate.

Solution

a. **molecular**: $NiCl_2(aq) + 2NaOH(aq) \rightarrow Ni(OH)_2(s) + 2NaCl(aq)$
complete ionic: $Ni^{2+}(aq) + 2Cl^-(aq) + 2Na^+(aq) + 2OH^-(aq) \rightarrow Ni(OH)_2(s) + 2Na^+(aq) + 2Cl^-(aq)$
net ionic: $Ni^{2+}(aq) + 2OH^-(aq) \rightarrow Ni(OH)_2(s)$

b. **molecular**: $2K(s) + 2H_2O(l) \rightarrow 2KOH(aq) + H_2(g)$
complete ionic: $2K(s) + 2H_2O(l) \rightarrow 2K^+(aq) + 2OH^-(aq) + H_2(g)$
net ionic: $2K(s) + 2H_2O(l) \rightarrow 2K^+(aq) + 2OH^-(aq) + H_2(g)$

Note that in part b **every** reactant undergoes some chemical change. Therefore the complete and net ionic equations are the same.

c. **molecular**: $3NaOH(aq) + H_3PO_4(aq) \rightarrow 3H_2O(l) + Na_3PO_4(aq)$
complete ionic: $3Na^+(aq) + 3OH^-(aq) + 3H^+(aq) + PO_4^{3-}(aq) \rightarrow$
$3H_2O(l) + 3Na^+(aq) + PO_4^{3-}(aq)$
net ionic: $3OH^-(aq) + 3H^+(aq) \rightarrow 3H_2O(l)$

4.7 Selective Precipitation

Precipitation reactions allow us to target specific substances and separate and recover them from a solution. Note the solubility rules in Table 4.1 in your textbook, and please try the next examples.

Example 4.7 A - Recovering The Silver

A solution contains Ag^+, Mg^{2+}, Na^+, Al^{3+}, and K^+. What anion might you add to precipitate the silver ion?

Solution

The chloride ion would precipitate the silver while the other cations would remain in solution.

Example 4.7 B - Selective Precipitation

A solution contains Ca^{2+}, Cu^{2+}, and Pb^{2+}. What anions can we add, and in what order, to separate and recover each cation?

Solution

The order of precipitation is important here. For example, if sulfate is added, Pb^{2+} and possibly Ca^{2+} will precipitate together. Similarly, Cu^{2+} and Pb^{2+} will precipitate if sulfide, S^{2-}, is added. Taking the solubility rules into account, one strategy (of several) would be:

1. **Add Cl^-**. The Cu^{2+} and Ca^{2+} stay in solution. **$PbCl_2$ precipitates out.**
2. **Add F^-**. The Cu^{2+} stays in solution. **CaF_2 precipitates out.**
3. **Add S^{2-}. CuS precipitates out.**

4.8 Stoichiometry of Precipitation Reactions

The material in this section is very important because it combines most of the previous chemical ideas you have learned up until now. Solving problems involving precipitates from solution makes use of *molarity, solubility rules, balancing equations, and limiting reactant calculations.*

Your textbook suggests **SIX STEPS** to solving a stoichiometry problem involving reactions in solution. Let's use these ideas in the following example.

Example 4.8 A - An Introduction To Problems Based On Precipitation Reactions

Calculate the mass of Ag_2S produced when 125. mL of 0.200 *M* $AgNO_3$ is added to excess Na_2S solution. ("Excess" Na_2S means that you have more than you need; $AgNO_3$ is the limiting reactant.)

Solution

Where are we going? The goal is to find the mass of silver sulfide. This is produced when the silver nitrate and sodium sulfide reactants are combined.

Before we do any calculations we first have to determine what is happening in the solution. This is best done by writing a **balanced molecular equation**.

The problem says that $AgNO_3(aq)$ is being added to $Na_2S(aq)$. According to our solubility rules (Section 4.5), insoluble silver sulfide will be formed.

$$\mathbf{2AgNO_3(\mathit{s}) + Na_2S(\mathit{aq}) \rightarrow Ag_2S(\mathit{s}) + 2NaNO_3(\mathit{aq})}$$

The critical point here is that the *mole ratio* of Ag_2S to $AgNO_3$ is

$$\mathbf{\frac{1\ mol\ Ag_2S}{2\ mol\ AgNO_3}}$$

Now that we know what is going on in solution, we need to know **how many moles of reactant ($AgNO_3$)** we have. You learned how to do such calculations in Section 4.3 in your textbook and this study guide.

$$\text{moles } AgNO_3 \parallel \frac{0.200\text{ mol } AgNO_3}{\text{L solution}} \times 0.125\text{ L solution} = \mathbf{0.0250\ mol\ AgNO_3}$$

How many moles of product (Ag_2S) can be formed from this many moles of reactant? Given the mole ratio of 2:1 that we determined earlier,

$$\text{moles } Ag_2S \parallel 0.0250\text{ mol } AgNO_3 \times \frac{1\text{ mol } Ag_2S}{2\text{ mol } AgNO_3} = \mathbf{0.0125\ mol\ Ag_2S\ formed}$$

To finish up, we need to convert **moles of Ag_2S to grams of Ag_2S**.

$$\text{g } Ag_2S \parallel 0.0125\text{ mol } Ag_2S \times \frac{247.9\text{ g } Ag_2S}{1\text{ mol } Ag_2S} = \mathbf{3.10\ g\ Ag_2S\ formed}$$

Note that early on, once we wrote down the correct equation, we could have solved the problem with one long dimensional analysis equation:

$$\text{g } Ag_2S \parallel \frac{247.9\text{ g } Ag_2S}{1\text{ mol } Ag_2S} \times \frac{1\text{ mol } Ag_2S}{2\text{ mol } AgNO_3} \times \frac{0.200\text{ mol } AgNO_3}{\text{L solution}} \times 0.125\text{ L solution}$$

$$= \mathbf{3.10\ g\ Ag_2S\ formed}$$

However you choose to solve such problems, a stepwise approach is helpful.

Example 4.8 B - Practice With Precipitation Problems

What mass of $Fe(OH)_3$ is produced when 35. mL of a 0.250 M $Fe(NO_3)_3$ solution is mixed with 55 mL of a 0.180 M KOH solution?

Solution

Where are we going? We want the mass of iron(III) hydroxide that results from the reaction of iron(III) nitrate and potassium hydroxide.

Remember your systematic procedure for solving precipitation problems. The complication here is that, by combining different amounts of reactant, you have a **limiting reactant** problem. Also, remember to **properly balance your chemical equation**!

The reaction of interest is:

$$\mathbf{Fe(NO_3)_3(\mathit{aq}) + 3KOH(\mathit{aq}) \rightarrow Fe(OH)_3(\mathit{s}) + 3KNO_3(\mathit{aq})}$$

Using a stepwise approach,

$$\text{moles } Fe(NO_3)_3 \text{ before reaction} \parallel \frac{0.250 \text{ mol } Fe(NO_3)_3}{\text{L solution}} \times 0.035 \text{ L soln}$$

$$= \mathbf{0.00875 \text{ moles } Fe(NO_3)_3}$$

$$\text{moles KOH before reaction} \parallel \frac{0.180 \text{ mol KOH}}{\text{L solution}} \times 0.055 \text{ L soln} = \mathbf{0.00990 \text{ moles KOH}}$$

According to the balanced equation, 0.00875 moles of **$Fe(NO_3)_3$** can yield **0.00875 moles of $Fe(OH)_3$** (1:1 mole ratio), and 0.00990 moles of **KOH** can yield **0.00330 moles of $Fe(OH)_3$** (3:1 mole ratio).

Therefore, **KOH** is the **limiting reactant** (See Section 3.10) and **0.00330 moles of $Fe(OH)_3(s)$ will be produced**. Converting from moles of $Fe(OH)_3$ to grams of $Fe(OH)_3$ (F.W. = 106.9 g/mol) gives **0.35 g $Fe(OH_3)$ produced**.

Example 4.8 C - Gravimetric Analysis

An ore sample is to be analyzed for sulfur. As part of the procedure, the ore is dissolved, and the sulfur is converted to sulfate ion, SO_4^{2-}. Barium nitrate is added, which causes the sulfate to precipitate out as $BaSO_4$.

The original sample had a mass of 3.187 g. The dried $BaSO_4$ has a mass of 2.005 g. What is the percent of sulfur in the original ore?

Solution

All the sulfur in the original ore sample is (we assume) still present, but is now in $BaSO_4$ instead of the ore. Therefore, the key question here is, "How much sulfur is in 2.005 g of dried $BaSO_4$?"

$$\text{g S in } BaSO_4 \parallel 2.005 \text{ g } BaSO_4 \times \frac{1 \text{ mol } BaSO_4}{233.36 \text{ g } BaSO_4} \times \frac{1 \text{ mol S}}{1 \text{ mol } BaSO_4} \times \frac{32.07 \text{ g S}}{1 \text{ mol S}}$$

$$= \mathbf{0.276 \text{ g S}}$$

$$\%\ S = \frac{0.276 \text{ g S}}{3.187 \text{ g in the ore}} \times 100\% = \mathbf{8.65\%\ S \text{ in the ore}}$$

4.9 *Acid-Base Reactions*

The key to solving acid-base problems is to know that they require the same strategy as most of the other types of problems in this chapter. **Writing down a balanced chemical equation is always your first, and most important, step.**

AN ACID IS A PROTON DONOR.
A BASE IS A PROTON ACCEPTOR.

Some STRONG ACIDS	Some STRONG BASES
HCl	$NaOH$
HNO_3	KOH
H_2SO_4	NH_2^-
$HClO_4$	

You may assume that the acid-base reactions used in this section go to completion.

The steps for solving acid-base problems are given (in the "Steps: Calculations for Acid-Base Reactions" and accompanying flow chart) in this section in your textbook. Let's solve the following problem together to show how these steps are used.

Example 4.9 A - Neutralization Of A Strong Acid

How many mL of a 0.800 *M* NaOH solution is needed to just neutralize 40.00 mL of a 0.600 *M* HCl solution?

Solution

Neutralization is often used in acid-base chemistry. Neutralization of an acid implies stoichiometric addition (i.e., "just enough") of a strong base so that no acid remains. The analogous definition applies to the neutralization of a base by a strong acid.

1. **List the species** present in solution before reaction.

   $\underset{\text{from HCl}(aq)}{H^+, Cl^-}$ $\quad$ $\underset{\text{from NaOH}(aq)}{Na^+, OH^-}$ $\quad$ and H_2O

2. **Write the balanced net ionic reaction.**

   $$H^+(aq) + OH^-(aq) \rightarrow H_2O(l)$$

3. Find out the **number of moles of acid** we need **to neutralize.**

   $$\text{moles } H^+ \;\Big\|\; \frac{0.600 \text{ moles } H^+}{1 \text{ L soln}} \times \underset{\text{change mL to L}}{0.4000 \text{ L soln}} = \mathbf{0.0240 \text{ moles } H^+}$$

4. Because the stoichiometry of the reaction is 1 mole of base to 1 mole of acid, **0.0240 moles OH^- are required to neutralize the acid.**

5. **Determine the volume of OH^-** (from NaOH) **needed** to give that many moles.

   $$\text{L NaOH soln} \;\Big\|\; \frac{1 \text{ L NaOH soln}}{0.800 \text{ moles NaOH}} \times 0.0240 \text{ moles NaOH} = 0.0300 \text{ L soln}$$

   $$= \mathbf{30.0 \text{ mL NaOH solution}}$$

You should double check to make sure the **moles of acid** equals the **moles of base**.

Interactive Example 4.11 in your textbook is really a limiting-reactant problem such as we did in Section 3.10 of this study guide and your textbook. Review that material, and remember when doing limiting reactant problems to **convert everything to moles**.

The remainder of this section deals with **VOLUMETRIC ANALYSIS**. This kind of analysis uses **precisely measured amounts of liquid** to carry out an analysis. There are several new terms introduced such as **titration, buret, equivalence point, indicator**, and **endpoint**. When solving volumetric analysis problems, the same chemical rules apply as with most other acid-base problems:

- Write down the reaction.

- Convert to moles, and relate moles of acid to moles of base.
- Don't be frightened by "wordy" railroad problems.
- Ignore superfluous information.

See if you can do the next example on your own.

Example 4.9 B - Acid-Base Titration

You wish to determine the molarity of a solution of sodium hydroxide. To do this, you titrate a 25.00-mL aliquot of your sample, which has had 3 drops of phenolphthalein indicator added so that it is pink, with 0.1067 *M* HCl. The sample turns clear (indicating that the NaOH(*aq*) has been precisely neutralized by the HCl solution) after the addition of 42.95 mL of the HCl. Calculate the molarity of your NaOH solution.

Solution

1. The net ionic reaction is:

$$\mathbf{H^+(\mathit{aq}) + OH^-(\mathit{aq}) \rightarrow H_2O(\mathit{l})}$$

2. The total number of moles of H^+ added to neutralize the NaOH equals:

$$\text{moles } H^+ \parallel \frac{0.1067 \text{ mol } H^+}{\text{L soln}} \times 0.04295 \text{ L soln} = \mathbf{4.583 \times 10^{-3} \text{ mol } H^+}$$

3. The net ionic reaction tells us that **the number of moles of H^+ equals the number of moles of OH^-**. Therefore our OH^- sample of unknown molarity contains **4.583×10^{-3} mol OH^-**.

4. We had a 25.00-mL (0.02500 L) aliquot of our NaOH solution. The molarity of the solution is:

$$\frac{\text{mol OH}^-}{\text{L soln}} = \frac{4.583\times10^{-3} \text{ mol OH}^-}{0.02500 \text{ L soln}} = \mathbf{0.1833\ \mathit{M}\ NaOH}$$

Does the Answer Make Sense?

You had 25.00 mL of NaOH solution. It took almost twice the volume of 0.1067 *M* HCl to neutralize it. Because you have the **same number of moles in about half the solution volume,** it makes sense that the NaOH solution would be almost twice the concentration.

Example 4.9 C - Molar Mass Of An Acid

You want to determine the molar mass of an acid. The acid contains one acidic hydrogen per molecule. You weigh out a 2.879-g sample of the pure acid and dissolve it, along with 3 drops of phenolphthalein indicator, in distilled water. You titrate the sample with 0.1704 *M* NaOH. The pink endpoint is reached after addition of 42.55 mL of the base. Calculate the molar mass of the acid.

Solution

As always, we write the equation first. Let "HA" be your acid. It has one acidic hydrogen. Therefore it will react with base in a 1:1 ratio.

$$HA + NaOH \rightarrow H_2O + NaA,$$

or

$$\mathbf{H^+(\mathit{aq}) + OH^-(\mathit{aq}) \rightarrow H_2O(\mathit{l})}$$

We have the number of grams of our acid, **2.879 g**. To calculate the molar mass (g/mol) of HA, we need to know **how many moles** of HA we have.

moles NaOH = moles HA

Our task, then, is to find moles of NaOH (present as OH^- in solution).

$$\text{moles } OH^- \,\|\, \frac{0.1704 \text{ moles } OH^-}{\text{L soln}} \times 0.04255 \text{ L soln} = \mathbf{7.251 \times 10^{-3}\ mol\ OH^-}$$

The reaction stoichiometry is 1 mole of HA per mole of NaOH. Therefore there are 7.251×10^{-3} moles HA in 2.879 g HA.

$$\textbf{molar mass of the acid} = \frac{2.879 \text{ g}}{7.251\times10^{-3} \text{ mol}} = \mathbf{397.1\ g/mole}$$

4.10 Oxidation-Reduction Reactions

Both this section and Section 4.11 seem, at first reading, to be unrelated to the general topic of solution chemistry. This section deals with definitions and assignments relating to **electron exchange**, and the next deals with balancing such equations. These skills are keys to dealing with electron exchange.

Your textbook defines **oxidation-reduction reactions** as being those in which **one or more electrons are transferred**.

$$H_2(g) + Cl_2(g) \rightarrow 2HCl(g)$$

In this case, electrons are transferred from the hydrogen to the chlorine.

In the last section we saw that hydrogen can form H^+, or hydrogen with a "+1" oxidation state. It has one more proton than electron. Chlorine can form Cl^-, or chlorine with a "−1" oxidation state. Many elements are more energetically stable when they gain or lose electrons so that they have a **noble gas electronic configuration**. Sodium forms Na^+, calcium forms Ca^{2+}, and sulfur can form S^{2-}. All these are common **oxidation states**.

A set of six rules for **assigning oxidation states** is given in Table 4.3 in your textbook. You must, for now, memorize these rules. They will become second nature as you do more and more "redox" problems. Use these rules in the following problem.

Remember that the **sum of the oxidation states in a neutral compound must equal zero** and must be equal to the **overall charge** in an **ionic compound**.

Example 4.10 A - Practice With Oxidation States

Assign oxidation states to each atom in the equation (treat each compound separately).

$$Fe_2O_3 + 2Al \rightarrow Al_2O_3 + 2Fe$$

Solution

$Fe_2O_3(s)$	+ $2Al(s)$	→ $Al_2O_3(s)$	+ $2Fe(l)$
+3 per Fe	0 per Al	+3 per Al	0 per Fe
−2 per O	(rule #1)	−2 per O	(rule #1)

Let's look at what happened to each element as a result of the reaction.

- Fe went from Fe^{3+} to Fe^0 (gained 3 electrons).
- Al Went from Al^0 to Al^{3+} (lost 3 electrons).
- O remained the same.

THE IRON(III) ION GAINED ELECTRONS. IT HAS BEEN REDUCED.
THE ALUMINUM LOST ELECTRONS. IT HAS BEEN OXIDIZED.

Remember **OIL RIG**: **Oxidation Involves Loss** (of electrons).
Reduction Involves Gain (of electrons).

This next idea is tricky: Something that is **reduced** is called an **oxidizing agent** (it causes something else to be oxidized). Something that is **oxidized** is called a **reducing agent** (it causes something else to be reduced).

In the previous example:

- Iron(III) was reduced.
- Aluminum was oxidized.
- Iron(III) oxide was the oxidizing agent.
- Aluminum was the reducing agent.

Example 4.10 B - Which Atoms Undergo Redox?

For each reaction, identify the atoms that undergo reduction or oxidation. Also, list the oxidizing and reducing agents.

a. $2H_2(g) + O_2(g) \rightarrow 2H_2O(g)$
b. $Zn(s) + Cu^{2+}(aq) \rightarrow Zn^{2+}(aq) + Cu(s)$
c. $2AgCl(s) + H_2(g) \rightarrow 2H^+(aq) + 2Ag(s) + 2Cl^-(aq)$
d. $2MnO_4^-(aq) + 16H^+(aq) + 5C_2O_4^{2-}(aq) \rightarrow 2Mn^{2+}(aq) + 10CO_2(g) + 8H_2O(l)$

Solution

For purposes of assigning oxidation states, treat each compound or ion by itself.

a. $\underset{0}{2H_2(g)} + \underset{0}{O_2(g)} \rightarrow \underset{+1\text{ per H; }-2\text{ per O}}{2H_2O(g)}$

oxidized: hydrogen (0 to +1) | oxidizing agent: molecular oxygen
reduced: oxygen (0 to −2) | reducing agent: molecular hydrogen

b. $\underset{0}{Zn(s)} + \underset{+2}{Cu^{2+}(aq)} \rightarrow \underset{+2}{Zn^{2+}(aq)} + \underset{0}{Cu(s)}$

oxidized: zinc (0 to +2) | oxidizing agent: copper
reduced: copper (+2 to 0) | reducing agent: zinc

c. $\underset{+1\ -1}{2AgCl(s)} + \underset{0}{H_2(g)} \rightarrow \underset{+1}{2H^+(aq)} + \underset{0}{2Ag(s)} + \underset{-1}{2Cl^-(aq)}$

oxidized: hydrogen (0 to +1) | oxidizing agent: silver chloride
reduced: silver (+1 to 0) | reducing agent: molecular hydrogen

d. $\underset{+7\ -2}{2MnO_4^-(aq)} + \underset{+1}{16H^+(aq)} + \underset{+3\ -2}{5C_2O_4^{2-}(aq)} \rightarrow \underset{+2}{2Mn^{2+}(aq)} + \underset{+4\ -2}{10CO_2(g)} + \underset{+1\ -2}{8H_2O(l)}$

oxidized: carbon (+3 to +4) | oxidizing agent: permanganate ion (MnO_4^-)
reduced: manganese (+7 to +2) | reducing agent: oxalate ion ($C_2O_4^{2-}$)

4.11 Balancing Oxidation-Reduction Equations

In Section 4.10 you learned how to assign oxidation states to atoms in a compound. You also learned that in redox reactions oxidation states on some atoms change. We will use this information now to learn how to **balance redox equations using the oxidation states method and the half-reaction method.**

When you learn about electrochemistry later in the year, you will find that splitting redox reactions into oxidation and reduction halves is very convenient for describing what happens at different electrodes. The **half-reaction method** is therefore especially important to master.

The steps involved in balancing redox reactions by this method are given in your textbook. When you work redox problems, always remember that **you can add to your equation only substances that are already in solution** (such as H_2O, H^+ in acid solution and OH^- in basic solution).

Let's try to balance the following equation together.

Example 4.11 A - Balancing By The Oxidation States Method

Please balance the reaction for respiration.

$$C_6H_{12}O_6 + O_2 \rightarrow CO_2 + H_2O$$

Solution

Following the procedure in your textbook, the oxidation states are,

$$\underset{0\ +1\ -2}{C_6H_{12}O_6} + \underset{0}{O_2} \rightarrow \underset{+4\ -2}{CO_2} + \underset{+1\ -2}{H_2O}$$

Each carbon is oxidized from 0 to +4, while each oxygen (in O_2) is reduced from 0 to −2. To balance the electron exchange, we will therefore need *twice as many* oxygen atoms (from O_2) as carbon atoms. There are 6 carbons so we will need 12 oxygens from O_2, or $6O_2$.

$$\underset{\substack{\uparrow \\ 24 \\ \text{electrons} \\ \text{lost} \\ (6C \times 4)}}{C_6H_{12}O_6} + \underset{\substack{\uparrow \\ 24 \\ \text{electrons} \\ \text{gained} \\ (12O \times 2)}}{6O_2} \rightarrow \text{products}$$

Balancing the rest of the equation by inspection,

$$\mathbf{C_6H_{12}O_6 + 6O_2 \rightarrow 6CO_2 + 6H_2O}$$

Example 4.11 B - Practice With Oxidation States Method Balancing

Please balance the equation for the combustion of hydrazine, N_2H_4, with dinitrogen tetroxide, N_2O_4. This reaction helps keep the Space Shuttle in Earth orbit.

$$N_2H_4 + N_2O_4 \rightarrow N_2 + H_2O$$

Solution

$$\underset{-2\ +1}{N_2H_4} + \underset{+4\ -2}{N_2O_4} \rightarrow \underset{0}{N_2} + \underset{+1\ -2}{H_2O}$$

The nitrogen in hydrazine is oxidized from −2 to 0.
The nitrogen in dinitrogen tetroxide is reduced from +4 to 0.

The ratio of N_2H_4 to N_2O_4 must therefore be 2 to 1.

$$2N_2H_4 + N_2O_4 \rightarrow \text{products}$$

Balancing the remainder of the equation by inspection,

$$\mathbf{2N_2H_4 + N_2O_4 \rightarrow 3N_2 + 4H_2O}$$

Example 4.11 C - Balancing Redox Equations By The Half-Reaction Method

Balance the following equation **in acid solution** using the half-reaction method.

$$Cu(s) + HNO_3(aq) \rightarrow Cu^{2+}(aq) + NO(g)$$

Solution

Step 1: Identify and write equations for the half-reactions.

$$\underset{0}{\mathbf{Cu}}(s) + H\underset{+5}{\mathbf{N}}O_3(aq) \rightarrow \underset{+2}{\mathbf{Cu}^{2+}}(aq) + \underset{+2}{\mathbf{N}}O(g)$$

Copper is being oxidized: $Cu \rightarrow Cu^{2+}$
Nitrogen is being reduced: $HNO_3 \rightarrow NO$ ($N^{5+} \rightarrow N^{2+}$)

Step 2: Balance each half-reaction.

The oxidation is balanced atomically. We need to add two electrons to the right-hand side to balance electronically.

i. (oxidation) $Cu \rightarrow Cu^{2+} + \mathbf{2e^-}$
ii. (reduction) $HNO_3 \rightarrow NO$

a. **Balance all atoms that are neither oxygen nor hydrogen**. (Nitrogens are already balanced.)

b. **Balance oxygens** by adding water to the side that needs oxygen. (The left-hand side has 3 oxygens, and the right-hand side has one, so **2** waters must be added to the right-hand side.)

$$HNO_3 \rightarrow NO + \mathbf{2H_2O}$$

c. **Balance hydrogens** by adding H^+ to the side that needs hydrogen. (The left-hand side has 1 hydrogen, and the right-hand side has 4, so **3** hydrogen ions must be added to the left-hand side.)

$$HNO_3 + \mathbf{3H^+} \rightarrow NO + 2H_2O$$

The half-reaction is now balanced atomically, but not electronically.

d. **Balance charges** by adding **electrons** to the side that is more positive. (The left-hand side has 3 positives; the right-hand side is neutral. Therefore we need to add **3** electrons to the left-hand side.)

$$HNO_3 + 3H^+ + \mathbf{3e^-} \rightarrow NO + 2H_2O$$

Both half-reactions are now balanced.

Step 3: Equalize electron transfer.

The same number of electrons must be gained as are lost in the reaction. Therefore we must multiply each reaction by numbers that will allow both reactions to have **the same** number of electrons exchanged.

With our reactions, the lowest common denominator of electrons is **6**. Therefore we would multiply the oxidation by **3**, and multiply the reduction by **2**.

$$3Cu \rightarrow 3Cu^{2+} + 6e^-$$

$$2HNO_3 + 6H^+ + 6e^- \rightarrow 2NO + 4H_2O$$

Step 4: Add the half-reactions, and cancel appropriately to get a complete redox reaction.

$$3Cu \rightarrow 3Cu^{2+} + 6e^-$$
$$2HNO_3 + 6H^+ + 6e^- \rightarrow 2NO + 4H_2O$$
$$3Cu + 2HNO_3 + 6H^+ + 6e^- \rightarrow 3Cu^{2+} + 2NO + 4H_2O + 6e^-$$

Canceling electrons, we get our final result:

$$\mathbf{3Cu + 2HNO_3 + 6H^+ \rightarrow 3Cu^{2+} + 2NO + 4H_2O}$$

Double Check

Do we have the same number of each kind of atom on both sides? Yes, there are 3 coppers, 2 nitrogens, 6 oxygens, and 8 hydrogens on each side. Are the charges the same on both sides? Yes, +6. The equation is balanced.

Example 4.11 D - Practice With The Half-Reaction Method

Balance the following equation in acidic solution:

$$Cr_2O_7^{2-}(aq) + NO(g) \rightarrow Cr^{3+}(aq) + NO_3^-(aq)$$

Solution

Step 1: $\mathbf{Cr_2O_7^{2-} + NO \rightarrow Cr^{3+} + NO_3^-}$

Nitrogen is being oxidized: N^{2+} to N^{5+}
Chromium is being reduced: Cr^{6+} to Cr^{3+}

Step 2: $Cr_2O_7^{2-} \rightarrow Cr^{3+}$
$Cr_2O_7^{2-} \rightarrow \mathbf{2}Cr^{3+}$
$Cr_2O_7^{2-} \rightarrow 2Cr^{3+} + \mathbf{7H_2O}$
$Cr_2O_7^{2-} + \mathbf{14H^+} \rightarrow 2Cr^{3+} + 7H_2O$
$Cr_2O_7^{2-} + 14H^+ + \mathbf{6e^-} \rightarrow 2Cr^{3+} + 7H_2O$ (balanced)

$NO \rightarrow NO_3^-$
$NO + \mathbf{2H_2O} \rightarrow NO_3^-$
$NO + 2H_2O \rightarrow NO_3^- + \mathbf{4H^+}$
$NO + 2H_2O \rightarrow NO_3^- + 4H^+ + \mathbf{3e^-}$ (balanced)

Step 3: $2[NO + 2H_2O \rightarrow NO_3^- + 4H^+ + 3e^-]$

$2NO + \mathbf{4H_2O}$	$\rightarrow$	$2NO_3^- + \mathbf{8H^+} + 6e^-$	(oxidation)
$Cr_2O_7^{2-} + \mathbf{14H^+} + 6e^-$	$\rightarrow$	$2Cr^{3+} + \mathbf{7H_2O}$	(reduction)
$\mathbf{Cr_2O_7^{2-}(aq) + 2NO(g) + 6H^+(aq)}$	$\rightarrow$	$\mathbf{2Cr^{3+}(aq) + 2NO_3^-(aq) + 3H_2O(l)}$	

Double Check

There are 2 chromium, 2 nitrogen, 9 oxygen, and 6 hydrogen atoms on each side. Also, each side has a total charge of +4. The equation is balanced.

One of the conditions of balancing equations is that you can add to equations only what is actually in the solution. In **acid** solutions, you have a lot of H^+ ions. In **basic** solutions, however, you have a lot of OH^- ions. **We need to add OH^- ions to balance for hydrogen when balancing a reaction that takes place in basic solution**. Your textbook summarizes the four steps to use.

Example 4.11 E - Balancing Redox Equations In Basic Solution

Balance the following equation (IT IS ALREADY balanced in acid) assuming it is in basic solution.

$$Cr_2O_7^{2-}(aq) + 2NO(g) + 6H^+(aq) \rightarrow 2Cr^{3+}(aq) + 2NO_3^-(aq) + 3H_2O(l)$$

Solution

We need to get rid of the excess H^+ because OH^- is the dominant acid-base related species. Therefore, **add $6OH^-$ to both sides**. (Whatever is done to the left side must be done to the right if the equation is already balanced.)

$$Cr_2O_7^{2-} + 2NO + 6H^+ + \mathbf{6OH^-} \rightarrow 2Cr^{3+} + 2NO_3^- + 3H_2O + \mathbf{6OH^-}$$

The H^+ will combine with the OH^-, giving H_2O.

$$Cr_2O_7^{2-} + 2NO + \mathbf{6H_2O} \rightarrow 2Cr^{3+} + 2NO_3^- + 3H_2O + 6OH^-$$

The 3 H_2O's on the right-hand side can be cancelled (with 3 of the waters on the left-hand side), which gives the final balanced equation.

$$\mathbf{Cr_2O_7^{2-}(aq) + 2NO(g) + 3H_2O(l) \rightarrow 2Cr^{3+}(aq) + 2NO_3^-(aq) + 6OH^-(aq)}$$

4.12 Simple Oxidation-Reduction Titrations

Solving problems involving redox titrations involves the same general strategy as any other titration problem. That is, you must:

a. Balance the redox equation.
b. Determine the moles of titrant.
c. Use the balanced redox equation to determine the moles of unknown.
d. Convert from moles of unknown to grams, percent, molarity, or whatever.

Though specific strategies may vary, **balancing the redox equation** and converting data to **moles** will always be the most important steps.

The following problem illustrates redox titration problem-solving strategy.

Example 4.12 - Redox Titration

The use of potassium permanganate ($KMnO_4$) as an oxidizing agent is described in your textbook. A 0.0483 *M* $KMnO_4$ solution was used to titrate a solution containing 0.8329 g of impure calcium oxalate, CaC_2O_4. (The ionic form of the equation is given below.) If 30.25 mL of the $KMnO_4$ solution was required to reach the titration endpoint, calculate the **percent purity** of the CaC_2O_4.

$$MnO_4^-(aq) + C_2O_4^{2-}(aq) \rightarrow Mn^{2+}(aq) + CO_2(g) \quad \text{(unbalanced)}$$

Solution

You must learn to separate the wheat from the chaff in such problems. List only the important information. Also, draw a "mole map" outlining your strategy.

Balance equation ⟶ Determine moles of MnO_4^- used $\xrightarrow[\text{relationship}]{\text{critical}}$ Determine moles of $C_2O_4^{2-}$ used

Calculate percent $CaC_2O_4^{2-}$ in sample ⟵ Determine grams of $CaC_2O_4^{2-}$ in sample ⟵

The balanced equation is:

$$2MnO_4^-(aq) + 5C_2O_4^{2-}(aq) + 16H^+(aq) \rightarrow 2Mn^{2+}(aq) + 10CO_2(g) + 8H_2O(l)$$

$$\text{moles of } MnO_4^- \text{ used } \| \frac{0.0483 \text{ moles } MnO_4^-}{\text{L solution}} \times 0.03025 \text{ L soln}$$

$$= \mathbf{1.46_1 \times 10^{-3} \text{ moles } MnO_4^- \text{ used}}$$

$$\text{moles of } C_2O_4^{2-} \text{ used } \| 1.46_1 \times 10^{-3} \text{ moles of } MnO_4^- \text{ used} \times \frac{5 \text{ moles } C_2O_4^{2-}}{2 \text{ moles } MnO_4^-}$$

$$= \mathbf{0.00365_3 \text{ moles of } C_2O_4^{2-} \text{ used}}$$

$$\text{g of } CaC_2O_4 \text{ in the sample } \| 0.00365_3 \text{ moles } C_2O_4^{2-} \times \frac{1 \text{ mol } CaC_2O_4}{1 \text{ mol } C_2O_4^{2-}} \times \frac{128.10 \text{ g } CaC_2O_4}{1 \text{ mol } CaC_2O_4}$$

$$= \mathbf{0.467_9 \text{ g } CaC_2O_4 \text{ in the sample}}$$

$$\% \, CaC_2O_4 \text{ in the sample} = \frac{0.467_9 \text{ g } CaC_2O_4}{0.8329 \text{ g sample}} \times 100\% = \mathbf{56.2\% \; CaC_2O_4}$$

[Note that you should keep one extra (insignificant) figure all the way through. Drop it only at the very end.]

We could have done most of this problem using one equation that contains most of our mole map:

$$\text{g } CaC_2O_4 \| \frac{0.0483 \text{ moles } MnO_4^-}{\text{L solution}} \times 0.03025 \text{ L soln} \times \frac{5 \text{ moles } C_2O_4^{2-}}{2 \text{ moles } MnO_4^-} \times \frac{1 \text{ mol } CaC_2O_4}{1 \text{ mol } C_2O_4^{2-}}$$

$$\times \frac{128.10 \text{ g } CaC_2O_4}{1 \text{ mol } CaC_2O_4} = \mathbf{0.4679 \text{ g } CaC_2O_4}$$

Exercises

Section 4.1

1. Write dissociation equations for the following when they are dissolved in water:

   a. HF
   b. $SrBr_2$
   c. $MgBr_2$
   d. NH_4Cl
   e. $NaNO_3$
   f. $Al_2(SO_4)_3$

2. Write dissociation equations for the following when they are dissolved in water:

   a. $Na_2SO_4(s)$
   b. $KCl(s)$
   c. $NaOH(s)$
   d. $Na_2CrO_4(s)$
   e. $Mg(OH)_2(s)$
   f. $HCOOH(l)$

3. Which of the following pairs of substances are miscible, and which are not? Give a reason.

a. CH_3CH_2OH and H_2O c. C_6H_6 and H_2O
b. C_6H_6 and C_6H_{12} d. LiBr and H_2O

Section 4.2

4. Classify the following as strong, weak, or nonelectrolyte:

a. CH_3CH_2OH c. HCl e. C_6H_{12}
b. $C_{12}H_{22}O_{11}$ (sugar) d. NH_3

5. Determine whether each of the following is a strong, weak, or nonelectrolyte.

a. $MgCl_2$ c. $Be(OH)_2$ e. $NaCH_3COO$
b. C_4H_6 d. HNO_3

Section 4.3

6. Calculate the molarity of the following solutions:

a. 49.73 g H_2SO_4 in enough water to make 500 mL of solution
b. 4.739 g $RuCl_3$ in enough water to make 1.00 L of solution
c. 5.035 g $FeCl_3$ in enough water to make 250 mL of solution
d. 27.74 g $C_{12}H_{22}O_{11}$ in enough water to make 750 mL solution
e. 218.7 g HCl in enough water to make 500 mL of solution

7. Calculate the molarity of the following solutions:

a. 18.92 g of HNO_3 in enough water to make 500 mL of solution
b. 5.761 g of KOH in enough water to make 350 mL of solution
c. 21.18 g of $Fe(NO_3)_3$ in enough water to make 1.000 L of solution
d. 72.06 g of $BaCl_2$ in enough water to make 800 mL of solution

8. Calculate the concentrations of each of the ions in the following solutions:

a. 0.25 *M* Na_3PO_4 b. 0.15 *M* $Al_2(SO_4)_3$ c. 0.87 *M* Na_2CO_3

9. Calculate the concentrations of each of the ions in the following solutions:

a. 0.62 *M* $K_2Cr_2O_7$ c. 0.14 *M* $Co(NO_3)_2$ e. 0.23 *M* $(NH_4)_2Cr_2O_7$
b. 0.35 *M* NaOH d. 0.07 *M* Na_3PO_4 f. 0.49 *M* $Al_2(SO_3)_3$

10. Describe how you would prepare the following solutions:

a. 100. mL of 1.00 *M* NaCl
b. 250. mL of 1.00 *M* Na_2SO_4
c. 1.50 L of 0.500 *M* $K_2Cr_2O_7$

11. Describe how you would prepare the following solutions:

a. 400. mL of 0.100 *M* HCl c. 1.00 L of 1.50 *M* $KMnO_4$
b. 750. mL of 0.350 *M* $Ba(NO_3)_2$ d. 250 mL of 0.20 *M* $AgNO_3$

12. Describe how you would prepare the following solutions:

a. 500 mL of 1.0 *M* H_2SO_4 from 17.8 *M* H_2SO_4
b. 1.5 L of 0.25 *M* $KMnO_4$ from 1.0 *M* stock solution
c. 1.0 L of 0.15 *M* $KBrO_3$ from solid $KBrO_3$
d. 100 mL of 0.01 *M* $AgNO_3$ from 0.5 *M* stock solution
e. 1 L of 0.5 *M* $AgNO_3$ from solid $AgNO_3$

13. Describe how you would prepare the following solutions:

 a. 250 mL of 0.1 *M* HCl from 12.5 *M* HCl
 b. 500 mL of 1.5 *M* NaCl from 7.3 *M* NaCl
 c. 800 mL of 0.2 *M* $NiCl_2$ from 4.6 *M* $NiCl_2$
 d. 750 mL of 0.05 *M* $FeSO_4$ from 0.1 *M* $FeSO_4$

14. A standard solution of KHP ($C_8H_5O_4K$) was made by dissolving 3.697 g of KHP in enough water to make 100.0 mL of solution. Calculate the KHP concentration.

15. How many moles of KHP are contained in 30.00 mL of the solution in problem 14?

16. A solution of ammonium acetate ($NH_4C_2H_3O_2$) was made by dissolving 3.85 g of ammonium acetate in enough water to make 500 mL of solution. Calculate the solute concentration.

17. How many moles of ammonium acetate are in 17 mL of the solution in Problem 16?

18. How many milliliters of 0.136 *M* NaOH is required to react with the H_2SO_4 in 10 mL of a 0.202 *M* solution? The reaction of the two is:

$$2NaOH + H_2SO_4 \rightarrow Na_2SO_4 + 2H_2O$$

19. How many milliliters of 0.50 *M* $Ca(OH)_2$ are required to react with the HCl in 30 mL of a 0.12 *M* solution? The reaction of interest is:

$$2HCl + Ca(OH)_2 \rightarrow CaCl_2 + 2H_2O$$

20. What is the concentration of the following in ppm?

 a. 1.0×10^{-2} g Cu^{2+} in 2.0 L of solution.
 b. Pb^{2+} in 2.1×10^{-5} *M* $Pb(NO_3)_2$.

21. What is the concentration of the following in ppm?

 a. 6.2×10^{-3} g Be^{+2} in 750 mL of solution.
 b. 255 mg $NaIO_3$ in 1.5 L of solution.

Section 4.5

22. Balance the following reactions:

 a. $C_3H_8 + O_2 \rightarrow CO_2 + H_2O$
 b. $K_2CO_3 + Al_2Cl_6 \rightarrow KCl + Al_2(CO_3)_3$
 c. $Mg_3N_2 + H_2O \rightarrow MgO + NH_3$
 d. $Ca_3(PO_4)_2 + H_2SO_4 \rightarrow CaSO_4 + H_3PO_4$
 e. $KOH + H_3PO_4 \rightarrow K_3PO_4 + H_2O$
 f. $KClO_3 + C_{12}H_{22}O_{11} \rightarrow KCl + CO_2 + H_2O$

23. Complete and balance the following reactions:

 a. $NaCl(aq) + Hg_2(NO_3)_2(aq) \rightarrow$
 b. $Ca(OH)_2(aq) + Na_2CO_3(aq) \rightarrow$
 c. $Na_2S(aq) + FeCl_3(aq) \rightarrow$

Section 4.6

24. Write molecular, complete ionic, and net ionic equations for the following reactions:

 a. aqueous sodium sulfide reacts with aqueous copper(II) nitrate
 b. aqueous hydrogen fluoride reacts with aqueous potassium hydroxide to give water and aqueous potassium fluoride

Section 4.7

25. A solution contains Ag^+, Pb^{2+}, and Fe^{3+}. If you want to precipitate the Pb^{2+} selectively, what anion would you choose?

Section 4.8

26. What mass of $Mg(OH)_2$ is produced when 100. mL of 0.42 *M* $Mg(NO_3)_2$ is added to excess NaOH solution?

27. What mass of $BaSO_4$ is produced when 15.0 mL of 3.00 *M* H_2SO_4 is added to 20.0 mL of 0.100 *M* $BaCl_2$?

28. Calculate the mass of $CaSO_4$ produced when 10 mL of 6.0 *M* H_2SO_4 is added to 100 mL of 0.52 *M* $Ca(NO_3)_2$.

29. What mass of $CaCO_3$ is produced when 250 mL of 6.0 *M* Na_2CO_3 is added to 750 mL of 1.0 *M* CaF_2?

30. Calculate the mass of Al_2S_3 produced when 100 mL of 0.50 *M* $AlCl_3$ is added to 100 mL of 0.50 *M* Na_2S.

31. You are given the equation, $AgBr + 2S_2O_3^{2-} \rightarrow Ag(S_2O_3)_2^{3-} + Br^-$. What mass of AgBr can be dissolved by 750. mL of 0.300 *M* $Na_2S_2O_3$?

32. Given the following chemical equation, determine the theoretical yield of B_2H_6 if exactly 100.0 g of $LiAlH_4$ was allowed to react with 225 g of BF_3.

$$3LiAlH_4 + 4BF_3 \rightarrow 3LiF + 3AlF_3 + 2B_2H_6$$

Section 4.9

33. Which is the acid and which is the base?

    a. $H{-}C{\equiv}C{:}^-$ and HF
    b. H_2N^- and H_2O

34. How many mL of 2.3 *M* HNO_3 is needed to neutralize 0.92 L of 0.5 *M* KOH?

35. What volume of 0.1379 *M* HCl is required to neutralize 10.0 mL of 0.2789 *M* NaOH solution?

36. How many mL of 1.50 *M* NaOH is required to neutralize 275 mL of 0.5 *M* H_2SO_4?

37. What is the molarity of a solution of HCl if it requires 29.31 mL of a 0.0923 *M* NaOH solution to reach a phenolphthalein endpoint for the titration of a 10.0-mL aliquot of the HCl solution?

38. Given the following two equations, determine the number of grams of manganese dioxide required to prepare enough chlorine gas to produce 25.0 g of potassium hypochlorite.

$$MnO_2 + 4HCl \rightarrow MnCl_2 + Cl_2 + 2H_2O$$
$$Cl_2 + 2KOH \rightarrow KCl + KClO + H_2O$$

39. A titration is done using 0.1302 *M* NaOH to determine the molar mass of an acid. The acid contains one acidic hydrogen per molecule. If 1.863 g of the acid require 70.11 mL of the NaOH solution, what is the molar mass of the acid?

40. A 2.000-g sample of silver alloy was dissolved in nitric acid and then precipitated as AgBr. After drying, the sample of silver bromide weighed 2.000 g. Calculate the percentage of silver in the alloy.

41. Complete and balance each acid-base equation (assume complete neutralization):

a. $H_2SO_4 + NaOH \rightarrow$
b. $H_3PO_4 + Mg(OH)_2 \rightarrow$
c. $H_2SO_3 + NaOH \rightarrow$
d. $HC_2H_3O_2 + Ba(OH)_2 \rightarrow$

42. What volume of 0.2 *M* NaOH is required to neutralize 50 mL of 0.1 *M* H_2SO_3?

43. The neutralization of a 25.00-mL sample of an unknown base requires 18.34 mL of 0.100 *M* HCl. Assuming the acid-base stoichiometry is 1:1, what is the concentration of the unknown base?

44. Assuming that the stoichiometry is 1:1, what is the concentration of an unknown acid if a 20.0-mL sample of it is neutralized with precisely 33.4 mL of 0.250 *M* base?

45. A 25.0-mL sample of an ammonia solution is analyzed by titration with HCl. The reaction is given below. It took 18.96 mL of 0.150 *M* HCl to titrate the ammonia. What is the concentration of the original ammonia solution?

$$NH_3 + H^+ \rightarrow NH_4^+$$

Section 4.10

46. Determine the oxidation number for Mn in each of the following:

a. $KMnO_4$
b. $LiMnO_2$
c. MnO_2
d. K_2MnCl_4
e. Mn_2O_7

47. Determine the oxidation number for each atom in the following compounds or ions:

a. H_3O^+
b. P_4O_{10}
c. S_8
d. H_2CO
e. NH_4ClO_4

48. Determine the oxidation number for each atom in the following compounds:

a. $MgBr_2$
b. Na_2SO_4
c. $Cr_2O_7^{2-}$
d. $CaCO_3$
e. $NaClF_4$
f. HNO_3

Section 4.11

49. Balance the following oxidation-reduction reactions. Which species in each is the oxidizing agent, the reducing agent?

a. $P + Cl_2 \rightarrow PCl_5$
b. $Sn^{2+} + Cu^{2+} \rightarrow Sn^{4+} + Cu^+$
c. $Cu + H^+ + NO_3^- \rightarrow Cu^{2+} + NO_2 + H_2O$
d. $Br_2 + SO_2 + H_2O \rightarrow H^+ + Br^- + SO_4^{2-}$
e. $H_2SO_4 + HBr \rightarrow SO_2 + Br_2 + H_2O$

50. Balance the following redox reactions. Identify the oxidizing agent and the reducing agent.

a. $ClO^- + H^+ + Cu \rightarrow Cl^- + H_2O + Cu^{2+}$
b. $H_2O + Cr^{3+} + XeF_6 \rightarrow Cr_2O_7^{2-} + H^+ + Xe + F^-$
c. $H_2O + SO_3^{2-} + Fe^{3+} \rightarrow SO_4^{2-} + Fe^{2+} + H^+$

51. Balance the following in basic solution:

a. $P_4 \rightarrow PH_3 + HPO_3^{2-}$
b. $Cl_2 + OH^- \rightarrow Cl^- + ClO_3^-$
c. $Zn + NO_3^- \rightarrow Zn^{2+} + NH_3$

52. Balance the following redox reactions that take place in an acid solution:

   a. $H_3AsO_4 + Zn \rightarrow AsH_3 + Zn^{2+}$
   b. $HS_2O_3^- \rightarrow S + HSO_4^-$
   c. $Cr_2O_7^{2-} + Cl^- \rightarrow Cr^{3+} + Cl_2$
   d. $MnO_2 + Hg + Cl^- \rightarrow Mn^{2+} + Hg_2Cl_2$

53. Balance the following redox reactions that take place in a basic solution:

   a. $HXeO_4^- + Pb \rightarrow Xe + HPbO_2^-$
   b. $ClO_4^- + I^- \rightarrow ClO_3^- + IO_3^-$
   c. $Co(OH)_3 + Sn \rightarrow Co(OH)_2 + HSnO_2^-$

Section 4.12

54. Iron(II) concentration can be determined by redox titration with a cerium(IV) solution. The oxidation-reduction reaction is

$$Fe^{2+} + Ce^{4+} \rightarrow Fe^{3+} + Ce^{3+}$$

What is the concentration of Fe^{2+} if it requires 21.35 mL of 0.3136 *M* Ce^{4+} to titrate a 10.00-mL aliquot to the endpoint?

55. A solution containing an unknown concentration of Sn^{2+} ions was titrated with a solution containing Ce^{4+} ions. One liter of the unknown solution required 22.9 mL of a 0.270 *M* Ce^{4+} solution to reach the titration endpoint. Balance the following equation, and calculate the concentration of Sn^{2+} ions in the unknown solution.

$$Ce^{4+} + Sn^{2+} \rightarrow Ce^{3+} + Sn^{4+}$$

56. A 0.0621 *M* $KMnO_4$ solution was used to titrate a solution containing CaC_2O_4. If 22.76 mL of $KMnO_4$ solution was required to reach the titration endpoint, calculate the number of grams of CaC_2O_4 in the sample.

$$MnO_4^- + C_2O_4^{2-} \rightarrow Mn^{2+} + CO_2$$

Multiple-Choice Self-Test

1. Which of the following solvents would probably be the best one to dissolve NaBr?

   A. CH_4
   B. CH_3CH_3
   C. H_2O
   D. $CH_3CH_2CH_2CH_2CH_2CH_2CH_2OH$

2. Calculate the molarity when 18.5 g of nitric acid is dissolved in enough water to prepare 100.0 mL of solution.

   A. 2.94 *M* B. 5.78 *M* C. 3.51 *M* D. 0.287 *M*

3. Calculate the molarity of a solution when 0.500 pounds of silver nitrate is dissolved in enough water to prepare 16.75 L of solution.

   A. 1.34 *M* B. 0.0797 *M* C. 2.67 *M* D. 0.615 *M*

4. How many grams of nitric acid are present in 250.0 mL of 6.70 *M* acid solution?

   A. 16.8 g B. 106 g C. 335 g D. 211 g

5. How many liters of water are required to prepare a 0.1590 *M* silver nitrate solution if 1.00 pound of silver nitrate is used?

A. 8.38 L B. 16.8 L C. 7.61 L D. 36.9 L

6. Considering that calcium chloride is a strong electrolyte, what is the molarity of the chloride ions you would find in a solution prepared by mixing 8.99 g of calcium chloride with enough water to prepare 150.0 mL of solution?

A. 0.540 *M* B. 0.270 *M* C. 66.00 *M* D. 1.08 *M*

7. When 25.0 mL of a 10.6 *M* HCl solution is diluted to 200.0 mL, what is the final molarity of this acidic solution?

A. 1.33 *M* B. 2.65 *M* C. 3.98 *M* D. 3.53 *M*

8. Which of the following compounds do you expect to precipitate in an aqueous solution?

A. $AgNO_3$ B. $PbSO_4$ C. LiBr D. KI

9. With which of the following solutions would you mix a silver nitrate solution to precipitate a silver salt?

A. Lead sulfate solution
B. Potassium nitrate solution
C. Sodium chloride solution
D. None of the above

10. Which of the following solutions would form a precipitate when mixed with $Ba(NO_3)_2$?

A. KCl B. $Pb(NO_3)_2$ C. KNO_3 D. None of these

11. How many grams of $K_4Fe(CN)_6$ are required to precipitate all the Cd^{2+} ions as $Cd_2Fe(CN)_6$ from 4.00 mL of 0.15 *M* cadmium chloride solution according to the following equation?

$$K_4Fe(CN)_6 + 2CdCl_2 \rightarrow 4KCl + Cd_2Fe(CN)_6$$

A. 0.11 g B. 7.4 g C. 2.4 g D. 0.78 g

12. How many mL of 1.00 *M* sulfuric acid solution are required to precipitate out all the barium ions from 40.0 mL of 0.250 *M* barium chloride solution?

$$H_2SO_4 + BaCl_2 \rightarrow BaSO_4 + 2HCl$$

A. 10.0 mL B. 8.25 mL C. 20.4 mL D. 5.25 mL

13. How many grams of calcium phosphate are precipitated when 25.0 mL of 0.220 *M* calcium chloride solution is allowed to react with 15.0 mL of 0.880 *M* phosphoric acid solution?

A. 8.32 g B. 1.13 g C. 4.08 g D. 0.568 g

14. How many mL of 1.00 *M* KOH are required to just neutralize 600.0 mL of a 1.5 *M* HCl solution?

$$KOH + HCl \rightarrow H_2O + KCl$$

A. 900 mL B. 400 mL C. 2.50 L D. 40.0 mL

15. 1.122 g of an unknown monoprotic base dissolved in 50.0 mL of water is titrated with 20.0 mL of a 1.00 *M* HCl solution. Identify the base.

A. NaOH B. NH_3 C. KOH D. LiOH

16. Determine the oxidation state of oxygen in KO_2.

A. −½ B. +½ C. −2 D. −1

17. Determine the average oxidation number of carbon in $C_6H_{12}O_6$ (glucose).

A. −2 B. +4 C. +2 D. 0

18. In the following reaction, select the element that is the oxidizing agent.

$$2MnO_2 + KClO_3 + 2KOH \rightarrow 2KMnO_4 + KCl + H_2O$$

A. Mn B. K C. Cl D. H

19. Balance the following equation

$$\_\_\_H_2S + \_\_\_H^+ + \_\_\_MnO_4^- \rightarrow \_\_\_Mn^{2+} + \_\_\_S + \_\_\_H_2O$$

A. 5, 6, 2, 2, 5, 8 B. 5, 8, 2, 3, 6, 5 C. 1, 1, 1, 1, 1, 2 D. 1, 1, 1, 1, 1, 1

20. What is the proper set of coefficients for the following reaction?

$$\_\_\_CuS + \_\_\_H^+ + \_\_\_SO_4^{2-} \rightarrow \_\_\_Cu^{2+} + \_\_\_SO_2 + \_\_\_H_2O + \_\_\_S$$

A. 1, 2, 1, 3, 3, 2, 1 B. 1, 4, 1, 1, 1, 2, 1 C. 3, 2, 1, 1, 1, 3, 3 D. 4, 1, 1, 1, 2, 4, 4

Answers to Exercises

1.
a. $HF(g) \xrightarrow{H_2O(l)} H^+(aq) + F^-(aq)$
b. $SrBr_2(s) \xrightarrow{H_2O(l)} Sr^{2+}(aq) + 2Br^-(aq)$
c. $MgBr_2(s) \xrightarrow{H_2O(l)} Mg^{2+}(aq) + 2Br^-(aq)$
d. $NH_4Cl(s) \xrightarrow{H_2O(l)} NH_4^+(aq) + Cl^-(aq)$
e. $NaNO_3(s) \xrightarrow{H_2O(l)} Na^+(aq) + NO_3^-(aq)$
f. $Al_2(SO_4)_3(s) \xrightarrow{H_2O(l)} 2Al^{3+}(aq) + 3SO_4^{2-}(aq)$

2.
a. $Na_2SO_4(s) \xrightarrow{H_2O(l)} 2Na^+(aq) + SO_4^{2-}(aq)$
b. $KCl(s) \xrightarrow{H_2O(l)} K^+(aq) + Cl^-(aq)$
c. $NaOH(s) \xrightarrow{H_2O(l)} Na^+(aq) + OH^-(aq)$
d. $Na_2CrO_4(s) \xrightarrow{H_2O(l)} 2Na^+(aq) + CrO_4^{2-}(aq)$
e. $Mg(OH)_2(s) \xrightarrow{H_2O(l)} Mg^{2+}(aq) + 2OH^-(aq)$
f. $HCOOH(l) \xrightarrow{H_2O(l)} H^+(aq) + COOH^-(aq)$

3.
a. miscible; polar O-H bonds in both
b. miscible; both nonpolar
c. immiscible; C_6H_6 is nonpolar, water is polar
d. miscible; LiBr is ionic and H_2O is polar

4.
a. nonelectrolyte
b. nonelectrolyte
c. strong
d. weak
e. nonelectrolyte

5.
a. strong
b. nonelectrolyte
c. strong
d. strong
e. strong

6.
a. 1.01 *M*
b. 0.0228 *M*
c. 0.124 *M*
d. 0.108 *M*
e. 12.0 *M*

7.
a. 0.601 *M*
b. 0.293 *M*
c. 0.08756 *M*
d. 0.433 *M*

8. a. $0.75\ M\ Na^+$, $0.25\ M\ PO_4^{3-}$
   b. $0.3\ M\ Al^{3+}$, $0.45\ M\ SO_4^{2-}$
   c. $1.74\ M\ Na^+$, $0.87\ M\ CO_3^{2-}$

9. a. $1.24\ M\ K^+$, $0.62\ M\ Cr_2O_7^{2-}$
   b. $0.35\ M\ Na^+$, $0.35\ M\ OH^-$
   c. $0.14\ M\ Co^{2+}$, $0.28\ M\ NO_3^{2-}$
   d. $0.21\ M\ Na^+$, $0.07\ M\ PO_4^{3-}$
   e. $0.46\ M\ NH_4^+$, $0.23\ M\ Cr_2O_7^{2-}$
   f. $0.98\ M\ Al^{3+}$, $1.47\ M\ SO_3^{2-}$

10. a. 5.84 g NaCl in enough water to make 100 mL
    b. 35.5 g of Na_2SO_4 in enough water to make 250 mL
    c. 221 g of $K_2Cr_2O_7$ in enough water to make 1.5 L

11. a. 1.46 g of HCl in enough water to make 400 mL
    b. 68.6 g of $Ba(NO_3)_2$ in enough water to make 750 mL
    c. 237 g of $KMnO_4$ in enough water to make 1 L
    d. 8.5 g of $AgNO_3$ in enough water to make 250 mL

12. a. 28 mL of 17.8 M H_2SO_4 diluted to 500 mL
    b. 375 mL of 1 M $KMnO_4$ diluted to 1.5 L
    c. 25 g of $KBrO_3$ in enough water to make 1.0 L
    d. 2 mL of 0.5 M $AgNO_3$ diluted to 100 mL
    e. 85 g $AgNO_3$ in enough water to make 1 L

13. a. 2 mL of 12.5 M HCl diluted to 250 mL
    b. 103 mL of 7.3 M NaCl diluted to 500 mL
    c. 35 mL of 4.6 M $NiCl_2$ diluted to 800 mL
    d. 375 mL of 0.1 M $FeSO_4$ diluted to 750 mL

14. 0.1810 M KHP

15. 5.430×10^{-3} mole KHP

16. 0.0999 M $NH_4C_2H_3O_2$

17. 1.7×10^{-3} mole $NH_4C_2H_3O_2$

18. 29.7 mL NaOH

19. 3.6 mL $Ca(OH)_2$

20. a. 5 ppm Cu^{2+}
    b. 4.4 ppm Pb^{2+}

21. a. 8.3 ppm Be^{2+}
    b. 170 ppm $NaIO_3$

22. a. $C_3H_8 + 5O_2 \rightarrow 3CO_2 + 4H_2O$
    b. $3K_2CO_3 + Al_2Cl_6 \rightarrow 6KCl + Al_2(CO_3)_3$
    c. $Mg_3N_2 + 3H_2O \rightarrow 3MgO + 2NH_3$
    d. $Ca_3(PO_4)_2 + 3H_2SO_4 \rightarrow 3CaSO_4 + 2H_3PO_4$
    e. $3KOH + H_3PO_4 \rightarrow K_3PO_4 + 3H_2O$
    f. $8KClO_3 + C_{12}H_{22}O_{11} \rightarrow 8KCl + 12CO_2 + 11H_2O$

23. a. $2NaCl(aq) + Hg_2(NO_3)_2(aq) \rightarrow Hg_2Cl_2(s) + 2NaNO_3(aq)$
    b. $Ca(OH)_2(aq) + Na_2CO_3(aq) \rightarrow CaCO_3(s) + 2NaOH(aq)$
    c. $3Na_2S(aq) + 2FeCl_3(aq) \rightarrow Fe_2S_3(s) + 6NaCl(aq)$

24. a. molecular: $Na_2S(aq) + Cu(NO_3)_2(aq) \rightarrow CuS(s) + 2NaNO_3(aq)$
    complete ionic: $2Na^+(aq) + S^{2-}(aq) + Cu^{2+}(aq) + 2NO_3^-(aq) \rightarrow CuS(s) + 2Na^+(aq) + 2NO_3^-(aq)$
    net ionic: $S^{2-}(aq) + Cu^{2+}(aq) \rightarrow CuS(s)$

24. b. molecular: $HF(aq) + KOH(aq) \rightarrow H_2O(l) + KF(aq)$
complete ionic: $H^+(aq) + F^-(aq) + K^+(aq) + OH^-(aq) \rightarrow H_2O(l) + K^+(aq) + F^-(aq)$
net ionic: $H^+(aq) + OH^-(aq) \rightarrow H_2O(l)$

25. SO_4^{2-}

26. 2.4 g $Mg(OH)_2$

27. 0.467 g $BaSO_4$

28. 7.1 g $CaSO_4$

29. 75 g $CaCO_3$

30. 2.5 g Al_2S_3

31. 21.1 g AgBr

32. 45.9 g B_2H_6

33. a. base, acid b. base, acid

34. 200 mL of 2.3 *M* HNO_3

35. 20.2 mL HCl

36. 183 mL NaOH

37. 0.271 *M* HCl

38. 24.0 g MnO_2

39. 204.1 g/mole

40. 57.45% Ag in the alloy

41. a. $H_2SO_4 + 2NaOH \rightarrow 2H_2O + Na_2SO_4$
b. $2H_3PO_4 + 3Mg(OH)_2 \rightarrow 6H_2O + Mg_3(PO_4)_2$
c. $H_2SO_3 + 2NaOH \rightarrow 2H_2O + Na_2SO_3$
d. $2HC_2H_3O_2 + Ba(OH)_2 \rightarrow 2H_2O + Ba(C_2H_3O_2)_2$

42. 50 mL NaOH

43. 7.34×10^{-2} *M* base

44. 0.418 *M* acid

45. 0.114 *M* NH_3

46. a. +7 c. +4 e. +7
b. +3 d. +2

47. a. H = +1; O = −2 d. H = +1; C = 0; O = −2
b. P = +5; O = −2 e. N = −3; H = +1; Cl = +7; O = −2
c. S = 0

48. a. Mg = +2; Br = −1 d. Ca = +2; C = +4; O = −2
b. Na = +1; S = +6; O = −2 e. Na = +1; Cl = +3; F = −1
c. Cr = +6; O = −2 f. H = +1; N = +5; O = −2

49. a. $2P + 5Cl_2 \rightarrow 2PCl_5$; ox. agent = Cl_2; red. agent = P
b. $Sn^{2+} + 2Cu^{2+} \rightarrow Sn^{4+} + 2Cu^+$; ox. agent = Cu^{2+}; red. agent = Sn^{2+}
c. $Cu + 4H^+ + 2NO_3^- \rightarrow Cu^{2+} + 2NO_2 + 2H_2O$; ox. agent = NO_3^-; red. agent = Cu

49. d. $Br_2 + SO_2 + 2H_2O \rightarrow 4H^+ + 2Br^- + SO_4^{2-}$; ox. agent = Br_2; red. agent = SO_2
e. $H_2SO_4 + 2HBr \rightarrow SO_2 + Br_2 + 2H_2O$; ox. agent = H_2SO_4; red. agent = HBr

50. a. $ClO^- + 2H^+ + Cu \rightarrow Cl^- + H_2O + Cu^{2+}$; ox. agent = ClO^-; red. agent = Cu
b. $7H_2O + 2Cr^{3+} + XeF_6 \rightarrow Cr_2O_7^{2-} + 14H^+ + Xe + 6F^-$; ox. agent = XeF_6; red. agent = Cr^{3+}
c. $H_2O + SO_3^{2-} + 2Fe^{3+} \rightarrow SO_4^{2-} + 2Fe^{2+} + 2H^+$; ox. agent = Fe^{3+}; red. agent = SO_3^{2-}

51. a. $P_4 + 2H_2O + 4OH^- \rightarrow 2PH_3 + 2HPO_3^{2-}$
b. $3Cl_2 + 6OH^- \rightarrow 5Cl^- + ClO_3^- + 3H_2O$
c. $4Zn + NO_3^- + 6H_2O \rightarrow 4Zn^{2+} + NH_3 + 9OH^-$

52. a. $H_3AsO_4 + 4Zn + 8H^+ \rightarrow AsH_3 + 4Zn^{2+} + 4H_2O$
b. $3HS_2O_3^- + H^+ \rightarrow 4S + 2HSO_4^- + H_2O$
c. $Cr_2O_7^{2-} + 6Cl^- + 14H^+ \rightarrow 2Cr^{3+} + 3Cl_2 + 7H_2O$
d. $MnO_2 + 2Hg + 2Cl^- + 4H^+ \rightarrow Mn^{2+} + Hg_2Cl_2 + 2H_2O$

53. a. $HXeO_4^- + 3Pb + 2OH^- \rightarrow Xe + 3HPbO_2^-$
b. $3ClO_4^- + I^- \rightarrow 3ClO_3^- + IO_3^-$
c. $2Co(OH)_3 + Sn + OH^- \rightarrow 2Co(OH)_2 + HSnO_2^- + H_2O$

54. 0.6695 M Fe^{2+}

55. $2Ce^{4+} + Sn^{2+} \rightarrow 2Ce^{3+} + Sn^{4+}$; 3.09×10^{-3} M Sn^{2+}

56. 0.453 g CaC_2O_4

Answers to Multiple-Choice Self-Test

1. C
2. A
3. B
4. B
5. B
6. D
7. A
8. B
9. C
10. A
11. A
12. A
13. D
14. A
15. C
16. A
17. D
18. C
19. A
20. B

CHAPTER 5

Gases

The Bottom Line: Chapter 5

The goal in this chapter is to help you learn about the behavior of gases on both molecular and macroscopic levels. A critical subtext of the chapter is the importance of structured observation as a critical approach to understanding.

5.1 Early Experiments

Your textbook describes several 17th-century experiments that led to the development of the **manometer** and the **barometer**. In order to really know how they work, you must understand force, pressure, and the units of both.

Force = mass × acceleration

When you weigh yourself on a scale, you are measuring the **force** that your body exerts on the scale. Let's say that your body mass is 68 kilograms. Also, the **acceleration** due to **gravity** that the earth exerts on you is a constant 9.8 meters/second^2. The **total force** exerted by your body on the scale is

$$F = m \times a$$

$$F = 68 \text{ kg} \times \frac{9.8 \text{ m}}{\text{s}^2} = \frac{670 \text{ kg m}}{\text{s}^2}$$

The S.I. unit of force is the **Newton (N)**.

$$1 \text{ N} = \frac{1 \text{ kg m}}{\text{s}^2}$$

Therefore, your force on the scale is **670 N**. This corresponds to your weight of **150 pounds** (4.47 N/pound).

Example 5.1 A - The Units Of Force

Let's say that you (with your mass of 68 kg) are on the moon, where the acceleration due to gravity is about 1.6 m/s^2. What force would you exert on a scale ("How much would you weigh?") in Newtons and in pounds?

Solution

What are we trying to solve? We are asked for the force. *How is that related to the data we have?* Force is equal to the mass × the acceleration.

$$F = m \times a = 68 \text{ kg} \times \frac{1.6 \text{ m}}{\text{s}^2} = \frac{108.8 \text{ kg m}}{\text{s}^2} = \mathbf{110\ N}$$

$$\text{weight in pounds} = 108.8 \text{ N} \times \frac{1 \text{ pound}}{4.47 \text{ N}} = \mathbf{24\ pounds}$$

Does the Answer Make Sense?

The acceleration due to gravity on the moon is considerably less than that on the earth, so the force you exert on the scale should be less. The answer makes sense.

$$\textbf{Pressure} = \frac{\textbf{Force}}{\textbf{Area}}$$

If you weigh 670 N (670 kg m/s^2) on Earth, and you are standing on a 0.5 m × 0.5 m square, your **area** = **(0.5 m)^2 = 0.25 m^2**, and the pressure you exert is

$$P = \frac{F}{A} = \frac{670 \text{ kg m/s}^2}{0.25 \text{ m}^2} = \frac{\textbf{2700 kg}}{\textbf{m s}^2} \text{ or } \frac{\textbf{2700 N}}{\textbf{m}^2}$$

The S.I. unit of pressure is the **Pascal (Pa)**.

$$1 \text{ Pa} = \frac{1 \text{ kg}}{\text{m s}^2} = \frac{1 \text{ N}}{\text{m}^2}$$

Therefore, the pressure you exert is **2700 Pa**.

Example 5.1 B - The Units Of Pressure

Let's say that someone is wearing high heels with a total area (for both heels) of **1 × 10^{-4} m^2**. The force is **670 N**. Calculate the pressure that person exerts.

Solution

$$P = \frac{F}{A} = \frac{670 \text{ kg m/s}^2}{1\times10^{-4} \text{ m}^2} = \frac{6.7\times10^6 \text{ N}}{\text{m}^2} = \mathbf{6.7\times10^6 \text{ Pa}}$$

This pressure is quite high! That is why it hurts if someone wearing a stiletto heel stands on your foot.

Air exerts pressure as well. The **standard atmosphere** is equal to 101,325 Pa, but this is an awkwardly large number. We commonly express air pressure in units of **atmospheres (atm)**, **torr**, or **mm Hg**.

1 atm = 101,325 Pa
1 atm = 760 torr = 760 mm Hg

In a margin note your textbook notes that "standard pressure" equals 1 bar = 100,000 Pa = 0.987 standard atmospheres.

5.2 The Gas Laws of Boyle, Charles, and Avogadro

There are **three gas laws** you need to know about and be able to use. In the next section these laws will be combined to give a more general law governing the behavior of gases.

A. Boyle's Law

This law says **the pressure exerted by a gas is inversely proportional to the volume the gas occupies**. In other words, as you squeeze the gas, it exerts more pressure. As an example, fill a zip-lock freezer bag with air, and then seal it. Try to squeeze it (**reduce the volume**). It is very difficult to squeeze because the **pressure of the gas is increasing** as you reduce the volume of the bag.

Pressure × Volume = Constant

$$PV = k$$

If PV is a constant for a gas at constant temperature, then

$$\mathbf{P_1V_1 = P_2V_2} \textbf{ (at constant temperature)}$$

where P_1 = initial pressure
V_1 = initial volume
P_2 = final pressure
V_2 = final volume

Finally, when doing gas law problems it is especially important to *make sure your answer makes physical sense.*

Example 5.2 A - Boyle's Law

A gas that has a pressure of 1.3 atm occupies a volume of 27 L. What volume will the gas occupy if the pressure is increased to 3.9 atm at constant temperature?

Solution

What are we trying to solve? We are asked for the volume when the pressure is increased (at constant temperature).

What should happen to the volume when the pressure increases? If the pressure on the gas is **increased** by a factor of 3 (3.9 atm/1.3 atm—you are squeezing the gas), we would expect the volume to **decrease** proportionately. Therefore the volume should be 1/3 of its original value.

$P_1 = 1.3$ atm $P_2 = 3.9$ atm
$V_1 = 27$ L $V_2 = ?$

$$P_1V_1 = P_2V_2$$

$$1.3 \text{ atm } (27 \text{ L}) = 3.9 \text{ atm } (V_2)$$

$$V_2 = \frac{1.3 \text{ atm } (27 \text{ L})}{3.9 \text{ atm}} = \mathbf{9.0\ L}$$

The final volume is 9.0 L (1/3 of 27), the answer we predicted. Always consider, *"Does my answer make sense?"*

Example 5.2 B - The Units Of P × V

The PV constant in the previous example was $35._1$ L atm (1.3 atm × 27 L). Convert this constant to

a. L Pa
b. fundamental SI units.

Solution

a. $35._1 \text{ L atm} \times \dfrac{101{,}325 \text{ Pa}}{\text{atm}} = \mathbf{3.5_6 \times 10^6\ L\ Pa}$

b. We know from Section 5.1 that **1 Pascal = 1 kg m^{-1} s^{-2}**. 1 L = 1000 mL or 1000 cm^3 or (10 cm)^3. This equals (0.1 m)^3 or **1 L = 1 × 10^{-3} m^3**. Therefore

$$3.5_6 \times 10^6 \text{ L Pa} = 3.56 \times 10^6 \text{ L Pa} \times \frac{1 \times 10^{-3} \text{ m}^3}{1 \text{ L}} \times \frac{1 \text{ kg m}^{-1}\text{s}^{-2}}{1 \text{ Pa}} = \mathbf{3.5_6 \times 10^3\ kg\ m^2\ s^{-2}}$$

The unit kg m^2 s^{-2} is the SI unit of **energy** called the **Joule**. *PV, then, is really a unit of energy!* We will use this information in later chapters.

B. Charles's Law

Charles's law says that, **at constant pressure the volume of a gas is directly proportional to the temperature (in Kelvin) of the gas.**

$$\frac{\textbf{Volume}}{\textbf{Temperature}} = \textbf{Constant}$$

$$\frac{V}{T} = b$$

Using the same reasoning as we did in Boyle's law, if we change the temperature (at constant pressure), the volume will change so the ratio of volume to temperature will remain constant.

$$\frac{V_1}{T_1} = \frac{V_2}{T_2} \text{ (at constant pressure)}$$

V_1 = initial volume
T_1 = initial temperature
V_2 = final volume
T_2 = final temperature

Example 5.2 C - Charles's Law

A gas at 30.0°C and 1.00 atm occupies a volume of 0.842 L. What volume will the gas occupy at 60.0°C and 1.00 atm?

Solution

The temperature is increased; therefore, according to Charles's law the volume occupied should increase as well. We predict our **final volume will be larger**. (Remember to convert temperature to Kelvins!)

$V_1 = 0.842$ L
$T_1 = 30.0°\text{C} + 273 = 303$ K
$V_2 = ?$
$T_2 = 60.0°\text{C} + 273 = 333$ K

$$\frac{V_1}{T_1} = \frac{V_2}{T_2}$$

$$\frac{0.842 \text{ L}}{303 \text{ K}} = \frac{V_2}{333 \text{ K}}$$

$$V_2 = \frac{0.842 \text{ L } (333 \text{ K})}{303 \text{ K}} = \mathbf{0.925 \text{ L}}$$

The final volume agrees with our prediction, so it makes sense.

C. Avogadro's Law

Avogadro's law says that, **for a gas at constant temperature and pressure, the volume is directly proportional to the number of moles of gas.**

Volume = Constant × number of moles (at constant *T*, *P*)

$$V = an \text{ (at constant } T, P)$$

If you triple the number of moles of gas (at constant temperature and pressure), the volume will also triple.

$$\frac{V_1}{n_1} = \frac{V_2}{n_2} \text{ (at constant } T, P\text{)}$$

V_1 = initial volume
n_1 = initial number of moles
V_2 = final volume
n_2 = final number of moles

Example 5.2 D - Avogadro's Law

A 5.20-L sample at 18.0°C and 2.00 atm pressure contains 0.436 moles of a gas. If we add an **additional** 1.27 moles of the gas at the same temperature and pressure, what will be the **total volume** occupied by the gas?

Solution

According to Avogadro's law, as we increase the number of moles of a gas the volume should increase proportionately. Therefore we would predict that our volume would increase.

$V_1 = 5.20$ L $\qquad V_2 = ?$
$n_1 = 0.436$ moles $\qquad n_2 = 0.436 + 1.27 = 1.70_6$ moles

$$\frac{V_1}{n_1} = \frac{V_2}{n_2}$$

$$\frac{5.20 \text{ L}}{0.426 \text{ mol}} = \frac{V_2}{1.70_6 \text{ mol}}$$

$$V_2 = \frac{5.20 \text{ L} (1.70_6 \text{ mol})}{0.436 \text{ mol}} = \mathbf{20.3 \text{ L}}$$

The answer agrees with our prediction.

Keep in mind that we have assumed our gases behave **ideally**. That is not always a valid assumption as your textbook considers further in Section 5.10.

5.3 The Ideal Gas Law

The ideal gas law is a combination of Boyle's, Charles's, and Avogadro's laws. It relates pressure, temperature, volume and the number of moles of a gas. The derivation of the ideal gas law is given in your textbook. The equation of interest is

Pressure × Volume = # of moles of the gas × a constant × temperature

$$PV = nRT$$

P = pressure in **atm**
V = volume in **L**
n = number of **moles**
R = 0.08206 **L atm/K mol**
T = temperature in **Kelvins**

Please keep in mind that

1. This relationship assumes the gas behaves **ideally**. As we will see in Section 5.10, there are certain conditions under which a gas will not behave ideally, and correction factors must be added to the ideal gas law.
2. You need to keep track of your dimensions. Many ideal gas law problems are best solved using **dimensional analysis**.
3. Always **list** what you are given. You may be able to simplify the problem.

Let's try a few examples.

Example 5.3 A - Ideal Gas Law

A sample containing 0.614 moles of a gas at 12.0°C occupies a volume of 12.9 L. What pressure does the gas exert?

Solution

List those things that you are given, including constants. Once you see what you have and what you are being asked to find, you can decide how to manipulate your equation.

$P = ?$ $\quad R = 0.08206$ L atm/K mol

$V = 12.9$ L $\quad n = 0.614$ moles

$T = 285$ K

$$PV = nRT$$

or

$$P = \frac{nRT}{V}$$

$$P = \frac{0.614 \text{ mol } (0.08206 \text{ L atm/K mol})(285 \text{ K})}{12.9 \text{ L}} = \mathbf{1.11\ atm}$$

You can check your answer by using dimensional analysis as shown below.

$$\text{atm} \parallel \frac{0.08206 \text{ L atm}}{\text{K mol}} \times 0.614 \text{ mol} \times 285 \text{ K} \times \frac{1}{12.9 \text{ L}} = \mathbf{1.11\ atm}$$

Example 5.3 B - Practice With Gas Laws

A sample of methane gas (CH_4) at 0.848 atm and 4.0°C occupies a volume of 7.0 L. What volume will the gas occupy if the pressure is increased to 1.52 atm and the temperature increased to 11.0°C?

Solution

According to Section 5.2 in your textbook, if the **pressure is increased**, the **volume should decrease** (Boyle's law). If the **temperature is increased**, the **volume should also increase** (Charles's law). However, the pressure almost doubles while the temperature increase is relatively small. Therefore the pressure effect will be dominant. Overall then, we would expect the **volume to decrease**.

You are given two sets of data.

$P_1 = 0.848$ atm $\quad P_2 = 1.52$ atm

$V_1 = 7.0$ L $\quad V_2 = ?$

$T_1 = 277$ K $\quad T_2 = 284$ K

$$n_1 = n_2$$

$$R = R$$

$$P_1V_1 = n_1RT_1 \quad \text{and} \quad P_2V_2 = n_2RT_2$$

$$\frac{P_1V_1}{T_1} = n_1R \quad \text{and} \quad \frac{P_2V_2}{T_2} = n_2R$$

$$n_1R = n_2R \text{ (because } n_1 = n_2\text{), so}$$

$$\frac{P_1V_1}{T_1} = \frac{P_2V_2}{T_2}$$

This is often called the combined gas law. The keys to setting up this equation are to **list what you are given** and to **know what you are being asked to solve for**. By converting the combined gas law to solve for V_2, we find that

$$V_2 = \frac{P_1V_1T_2}{P_2T_1} = \frac{0.848 \text{ atm } (7.0 \text{ L})(284 \text{ K})}{1.52 \text{ atm } (277 \text{ K})} = \mathbf{4.0\ L}$$

This agrees with our prediction.

Example 5.3 C - More Practice With Gas Laws

How many moles of a gas at 104°C would occupy a volume of 6.8 L at a pressure of 270 mm Hg?

Solution

$$P = 270 \text{ mm Hg} \times \frac{1 \text{ atm}}{760 \text{ mm Hg}} = 0.355 \text{ atm}$$

$$V = 6.8 \text{ L} \qquad T = 377 \text{ K} \qquad n = ?$$

$$n = \frac{PV}{RT} = \frac{0.355 \text{ atm } (6.8 \text{ L})}{0.08206 \text{ L atm/K mol } (377 \text{ K})} = \mathbf{0.078\ moles}$$

Note the units are in moles. You can use dimensional analysis as an additional check.

$$\text{moles} \,\|\, \frac{1 \text{ K mol}}{0.08206 \text{ L atm}} \times 6.8 \text{ L} \times \frac{1}{377 \text{ K}} = \mathbf{0.078\ moles}$$

5.4 Gas Stoichiometry

This section uses the ideal gas law to perform a variety of calculations, including molar mass, density, and volume determination. Your textbook defines the volume occupied under **standard temperature and pressure (STP)**.

Technically, **STP MEANS 0°C and 1 bar = 100,000 Pa**. However, for our purposes, this is close enough to 1 atm that we will use 1.00 atm as the standard pressure.

Many gas law problems involve calculating the **volume** of a gas produced by the reaction of volumes of other gases. The problem-solving strategy we have used throughout your chemistry course remains the same. That is, you want to relate **moles of reactants to moles of products**. The ideal gas law will allow you to use the following strategy:

Volume of reactants $\xrightarrow{\text{ideal gas law}}$ moles of reactants $\xrightarrow{\text{stoichiometry}}$ moles of products

volume of products $\xleftarrow{\text{ideal gas law}}$

In the next problem, we are given moles of reactant and asked to find the volume of product.

Example 5.4 A - Reactions And The Ideal Gas Law

A sample containing 15.0 g of dry ice, $CO_2(s)$, is put into a balloon and allowed to sublime according to the following equation:

$$CO_2(s) \rightarrow CO_2(g)$$

How big will the balloon be (i.e., what will be the volume of the balloon) at 22.0°C and 1.04 atm after all the dry ice has sublimed?

Solution

What are you trying to solve? You are given *moles of* $CO_2(s)$. You want *volume of* $CO_2(g)$. The mole ratio of the solid to the gas is 1 to 1. Given the mass of CO_2, we can go from

$$\text{g } CO_2 \xrightarrow{\text{molar mass}} \text{moles } CO_2(g) \xrightarrow{\text{ideal gas law}} \text{volume } CO_2$$

$$\text{moles } CO_2 \; \| \; 15.0 \text{ g } CO_2 \times \frac{1 \text{ mol } CO_2}{44.0 \text{ g } CO_2} = \mathbf{0.341 \text{ mol } CO_2}$$

$$V = \frac{nRT}{P} = \frac{0.314 \text{ mol } (0.08206 \text{ L atm/K mol})(295 \text{ K})}{1.04 \text{ atm}} = \mathbf{7.94 \text{ L}}$$

Using dimensional analysis as a check:

$$\text{L } CO_2 \; \| \; \frac{0.08206 \text{ L atm}}{\text{K mol}} \times 295 \text{ K} \times \frac{1}{1.04 \text{ atm}} \times 0.341 \text{ mol} = \mathbf{7.94 \text{ L}}$$

↑ get the proper units on top

Example 5.4 B - Practice With The Ideal Gas Law

0.500 L of $H_2(g)$ are reacted with 0.600 L of $O_2(g)$ **at STP** according to the equation

$$2H_2(g) + O_2(g) \rightarrow 2H_2O(g)$$

What volume will the H_2O occupy at 1.00 atm and 350°C?

Solution

Because you have two reactants, this may be a **limiting reactant** problem. As with any such problem, you must find out how many moles of each reactant you have so you can determine the limiting reactant. Once you have done those calculations, you can determine the moles and then the volume the product will occupy.

reactant volume $\xrightarrow{\text{ideal gas law}}$ moles of reactants $\xrightarrow{\text{stoichiometry}}$ moles of products

volume of products $\xleftarrow{\text{ideal gas law}}$

Use the fact that the reactants are initially at STP to help you solve the problem.

$$\text{moles } H_2 \text{ at STP } \| \ \frac{1 \text{ mol}}{22.4 \text{ L}} \times 0.500 \text{ L} = \mathbf{0.0223 \text{ moles } H_2}$$

$$\text{moles } O_2 \text{ at STP } \| \ \frac{1 \text{ mol}}{22.4 \text{ L}} \times 0.600 \text{ L} = \mathbf{0.0268 \text{ moles } O_2}$$

To determine the limiting reactant,

$$\text{moles } H_2O \text{ from } H_2 \ \| \ 0.0223 \text{ mol } H_2 \times \frac{2 \text{ mol } H_2O}{2 \text{ mol } H_2} = 0.0223 \text{ mol } H_2O$$

$$\text{moles } H_2O \text{ from } O_2 \ \| \ 0.0268 \text{ mol } O_2 \times \frac{2 \text{ mol } H_2O}{1 \text{ mol } O_2} = 0.0536 \text{ mol } H_2O$$

The limiting reactant is $\mathbf{H_2}$, and **0.0223 mol H_2O** will be formed.

$$\text{Volume } H_2O = \frac{nRT}{P} = \frac{0.0223 \text{ mol } (0.08206 \text{ L atm/K mol}) (623 \text{ K})}{1.00 \text{ atm}} = \mathbf{1.14 \text{ L}}$$

Example 5.4 C - Density And Molar Mass

A gas at 34.0°C and 1.75 atm has a density of 3.40 g/L. Calculate the molar mass (M.M.) of the gas.

Solution

$$\text{M.M.} = \frac{dRT}{P}$$

$$\text{M.M.} = \frac{3.40 \text{ g/L } (0.08206 \text{ L atm/K mol}) (307 \text{ K})}{1.75 \text{ atm}} = \mathbf{48.9 \text{ g/mol}}$$

As a check, use dimensional analysis:

$$\frac{\text{grams}}{\text{mol}} \ \| \ \underset{\substack{\uparrow \\ \text{density}}}{\frac{3.40 \text{ g}}{\text{L}}} \times \underset{\substack{\uparrow \\ \text{R}}}{\frac{0.08206 \text{ L atm}}{\text{K mol}}} \times \frac{1}{1.75 \text{ atm}} \times 307 \text{ K} = \mathbf{48.9 \text{ g/mol}}$$

5.5 Dalton's Law of Partial Pressures

Dalton's law of partial pressures states that, for a mixture of gases in a container, the **total pressure** is the **sum** of the pressures each gas would exert if it were alone. Your textbook shows that, because the ideal gas law holds,

$$P_{\text{total}} = P_1 + P_2 + \ldots + P_n = (n_1 + n_2 + \ldots + n_n) \left[\frac{RT}{V}\right]$$

because $\frac{RT}{V}$ will be the same for each of the different gases in the same container.

The **key problem-solving strategy** with regard to partial pressure problems is to **use the ideal gas law to interconvert between pressure and moles of each gas**. Let's try the following example.

Example 5.5 A - Partial Pressures

A volume of 2.0 L of He at 46°C and 1.2 atm pressure was added to a vessel that contained 4.5 L of N_2 at STP. What are the **total pressure** and **partial pressure of each gas at STP** after the He is added?

Solution

What are we trying to solve? We are asked to find the partial pressure of the helium and nitrogen gases, as well as the total pressure exerted by the gases.

What data do we have, and what are we missing in order to use the ideal gas law? We have the partial pressure of nitrogen at STP, so we already have part of our answer. We also have the partial pressure of helium at non-STP conditions.

We can find the **number of moles** of He at the original conditions. This will ultimately lead to finding the partial pressure of He at STP.

$$n = \frac{PV}{RT}$$

$$n_{He} = \frac{(1.2 \text{ atm})\ (2.0 \text{ L})}{(0.08206 \text{ L atm/K mol})\ (319 \text{ K})} = \mathbf{0.091_7 \text{ mol He}}$$

When the gases are combined under STP conditions, the partial pressure of He will change. That of N_2 (already at STP) will remain the same.

$$P_{He} = \frac{nRT}{V} = \frac{(0.091_7 \text{ mol})\ (0.08206 \text{ L atm/K mol})\ (273 \text{ K})}{4.5 \text{ L}} = \mathbf{0.45_7 \text{ atm}}$$

$$P_{N_2} = \mathbf{1.00 \text{ atm}}$$

$$P_{total} = 1.00 \text{ atm} + 0.46 \text{ atm} = \mathbf{1.4_6 \text{ atm} = 1.5 \text{ atm}}$$

The equations for **MOLE FRACTION** are derived in your textbook. Recalling,

$$\chi_i = \frac{n_i}{n_{total}} = \frac{P_i}{P_{total}}$$

The key idea here is that, once you have either the number of moles **OR** the pressure of each component of your system, you can calculate the mole fraction. The next example has several parts.

Example 5.5 B - Mole Fraction And Partial Pressure

a. Calculate the number of moles of N_2 present in the previous example (5.5 A).
b. Calculate the mole fractions of N_2 and He given the following data from Example 5.5 A.
   i. mole data
   ii. pressure data

Solution

a. In calculating moles of N_2 that occupy 4.5 L (as given in 5.5 A), we can take advantage of STP conditions.

$$\textbf{True at STP:} \quad \frac{1 \text{ mol } N_2}{22.4 \text{ L}} = \frac{x \text{ mol } N_2}{4.5 \text{ L}}$$

$$x = \mathbf{0.20_1 \text{ mol } N_2}$$

b. i. The total number of moles = mol N_2 + mol He = $0.20_1 + 0.091_7 = 0.293$ mol

$$\chi_{N_2} = \frac{0.20_1 \text{ mol}}{0.293 \text{ mol}} = \mathbf{0.69}$$ dimensionless because the units cancel on top and bottom

$$\chi_{He} = \frac{0.091_7\text{mol}}{0.293 \text{ mol}} = \mathbf{0.31}$$

ii. Using partial-pressure data, the total pressure is 1.46 (1.45_7) atm.

$$\chi_{N_2} = \frac{1.00 \text{ atm}}{1.45_7 \text{ atm}} = \mathbf{0.69}$$

$$\chi_{He} = \frac{0.45_7 \text{ atm}}{1.45_7 \text{ atm}} = \mathbf{0.31}$$

5.6 The Kinetic Molecular Theory of Gases

At the beginning of this section, your textbook makes clear that the **kinetic molecular theory** is simply a **model** that attempts to explain the properties of an ideal gas. The postulates of the kinetic molecular theory are listed and discussed. Your textbook points out that a model is considered successful if it correctly **predicts** the behavior of the system. The postulates are:

1. The volume of the individual particles of a gas can be assumed to be negligible.
2. The particles are in constant motion. The collisions of the particles with the walls of the container are the cause of the pressure exerted by the gas.
3. The particles are assumed to exert no forces on each other.
4. The average kinetic energy of a collection of gas particles is assumed to be directly proportional to the Kelvin temperature of the gas.

Your textbook uses these postulates, along with definitions of force, momentum, and pressure and some geometry, to derive the ideal gas law.

The idea that **temperature is a measure of the average kinetic energy of a gas** is of critical importance.

$$(\text{KE})_{\text{average}} = \frac{3}{2}RT$$

Because KE = $\frac{1}{2}mv^2$ (where m = mass and v = velocity), $\frac{1}{2}mv^2 = \frac{3}{2}RT$, and velocity (as indicated by the random motions of the particles of a gas) increases with higher temperature. **Temperature is a measure of kinetic energy**.

Root Mean Square Velocity

The expression dealing with the average velocity of gas particles is called the **root mean square velocity** and is derived in your textbook.

$$\mu_{\text{rms}} = \sqrt{\frac{3RT}{M}}$$

where $\boldsymbol{R} = 8.3145$ J/K mol = $\dfrac{8.3145 \text{ kg m}^2/\text{s}^2}{\text{K mol}}$ (because 1 J = 1 kg m^2/s^2)

$\boldsymbol{T}$ = temperature in Kelvins

$\boldsymbol{M}$ = mass of a mole of the gas in **kilograms** ("kg/mol")

This is certainly a situation where the proper use of units will help you make sure that you have the correct answer.

Example 5.6 A - Root Mean Square Velocity

Calculate the root mean square velocity for the atoms in a sample of oxygen gas at

a. 273 K
b. 300°C

Solution

As the temperature is increased, we would expect the average kinetic energy of our particles to increase. The mass of the particles is constant; therefore, we would expect the rms velocity to increase when the temperature is raised to 300.°C from 273 K.

a. at 273 K: $R = 8.3145$ kg m^2/s^2 K mol
$T = 273$ K

$$M = \frac{\text{kg}}{\text{mol}} \,\|\, \frac{32.0\text{ g}}{1\text{ mol}} \times \frac{1\text{ kg}}{1000\text{ g}} = 0.3020\text{ kg/mol}$$

$$\mu_{rms} = \left[\frac{3\left(\frac{8.3145\text{ kg m}^2/\text{s}^2}{\text{K mol}}\right)(273\text{ K})}{0.0320\text{ kg/mol}}\right]^{1/2} = \left[\frac{2.128 \times 10^5\text{ m}^2}{\text{s}^2}\right]^{1/2} = \mathbf{461\text{ m/s}}$$

b. at 300.°C: R = as above
M = as above
$T = 573$ K

$$\mu_{rms} = \left[\frac{3\left(\frac{8.3145\text{ kg m}^2/\text{s}^2}{\text{K mol}}\right)(573\text{ K})}{0.0320\text{ kg/mol}}\right]^{1/2} = \left[\frac{4.467 \times 10^5\text{ m}^2}{\text{s}^2}\right]^{1/2} = \mathbf{668\text{ m/s}}$$

The increase in rms velocity agreed with our prediction.

Example 5.6 B - Other Velocity Calculations

Please determine the most probable and average velocities for a sample of oxygen gas at 300.°C. How do they compare with the root mean square velocity? Why does your textbook introduce three different measures of velocity?

Solution

The expressions for the velocities are given in your textbook. The most probable velocity, $\mu_{mp} = \sqrt{\frac{2RT}{M}}$. The average velocity is $\sqrt{\frac{3RT}{\pi M}}$. The ratios of the speeds are also given in this section of your textbook. The key to the different measures is that there is a *range* of velocities for any gas.

$$\mu_{mp} = \left[\frac{2(8.3145\text{ kg m}^2\text{ s}^{-2}\text{ K}^{-1}\text{ mol}^{-1})\ 573\text{ K}}{0.0320\text{ kg mol}^{-1}}\right]^{1/2} = \left[\frac{2.9776 \times 10^5\text{ m}^2}{\text{s}^2}\right]^{1/2} = \mathbf{546\text{ m/s}}$$

The ratios of $\mu_{mp} : \mu_{avg} : \mu_{rms} = 1.000 : 1.128 : 1.225$,

so $\mu_{avg} = 546 \text{ m/s} \times 1.128 = \mathbf{616 \text{ m/s}}$

$\mu_{rms} = 546 \text{ m/s} \times 1.225 = \mathbf{669 \text{ m/s}}$

which agrees (within round-off) with the answer in Example 5.6 A above.

Finally, your textbook points out two very important ideas. The first is that particles collide with one another and exchange energy after traveling a very short distance. This **distance between collisions** is called the **mean free path** and is typically very small. Because the exchange of energy happens at different times, particles are speeding up and slowing down. They have an average rms velocity but rarely have precisely that velocity. **Therefore particles have a large range of velocities, the average of which is the rms velocity**.

5.7 Effusion and Diffusion

Your textbook points out that **diffusion and effusion** are two different processes. It also gives a fun discussion relating to a former method for enrichment of uranium for use in nuclear reactors.

- **Diffusion** relates to the mixing of gases.
- **Effusion** relates to the passage of a gas through an orifice into an evacuated chamber.

Graham's Law of Effusion

If the kinetic energies of two gases, 1 and 2, in a system are the same at a given temperature, then for a mole of the particles

$$(\text{KE})_{avg} = N_A(\tfrac{1}{2}m_1\overline{u_1^2}) = N_A(\tfrac{1}{2}m_2\overline{u_2^2})$$

or, if M, molar mass, is equal to $N_A m$ *,

$$\tfrac{1}{2}M_1\overline{u_1^2} = \tfrac{1}{2}M_2\overline{u_2^2}$$

by cross multiplying and canceling, this becomes

$$\frac{M_2}{M_1} = \frac{\overline{u_1^2}}{\overline{u_2^2}}, \text{ or } \sqrt{\frac{M_2}{M_1}} = \frac{u_{\text{rms}_1}}{u_{\text{rms}_2}} = \frac{\text{rate of effusion}_1}{\text{rate of effusion}_2}$$

This is Graham's law of effusion. The **higher** the molar mass, the **slower** the rate of effusion through a small orifice.

Example 5.7 - Graham's Law Of Effusion

How many times faster than He would NO_2 gas effuse?

Solution

$$M_{NO_2} = 46.01 \text{ g/mol}$$

$$M_{He} = 4.003 \text{ g/mol}$$

$$\sqrt{\frac{M_{NO_2}}{M_{He}}} = \frac{\text{rate}_{He}}{\text{rate}_{NO_2}}$$

---

* *Note that "M" has changed meanings. In Chapter 4 it meant **molarity (mol/L)**. Here, it means **molar mass**. Such shifts in meaning are rare, but they do happen.*

$$\sqrt{\frac{46.01}{4.003}} = \frac{\text{rate}_{He}}{\text{rate}_{NO_2}}$$

$$3.390 = \frac{\text{rate}_{He}}{\text{rate}_{NO_2}}$$

He would effuse 3.39 times as fast as NO_2.

Note: You could have solved the problem as

$$\sqrt{\frac{M_{He}}{M_{NO_2}}} = \frac{\text{rate}_{NO_2}}{\text{rate}_{He}}$$

in this case; $0.295 = \frac{\text{rate}_{NO_2}}{\text{rate}_{He}}$. The conclusion is the same.

Does the Answer Make Sense?

NO_2 has a much higher mass than He. We would expect it to effuse more slowly.

With regard to **diffusion**, the important idea is that although gases travel very rapidly (hundreds of meters per second), their motions are in all directions, so mixing is relatively slow. In the case of diffusion, the basic structure of Graham's law holds.

$$\frac{\text{Distance traveled}_2}{\text{Distance traveled}_1} = \sqrt{\frac{M_1}{M_2}}$$

5.8 Collisions of Gas Particles with the Container Walls

This section is the first of three that develop the idea of what really happens with gases on a molecular level. This section deals with collisions with container walls. Section 5.9 considers collisions among the particles themselves. Finally, Section 5.10 considers the implications of these collisions. Your textbook rationalizes the equation relating collisions per second (a reflection of gas pressure) with the container walls to

- number of particles in the container $\left(\frac{N}{V}\right)$,
- average velocity of the particles (μ_{avg}),
- area of the wall section (A).

Note in Interactive Example 5.9 of your textbook that the equation is not used in a "plug and chug" fashion. In that example, the ideal gas law is used to calculate N/V!

Example 5.8 - Collisions Per Second

An inert gas has an impact rate of 5.81×10^{24} s^{-1} on a 5.00-cm^2 section of a container. The pressure and temperature of the gas are 1.00 atm and 57.0°C, respectively. Please identify the gas.

Solution

$$Z_A = A\,\frac{N}{V}\sqrt{\frac{RT}{2\pi M}}$$

We are given $A = 2.5 \times 10^{-3}\ \text{m}^2$
$R = 0.08206$ L atm K^{-1} mol^{-1} or 8.3145 J K^{-1} mol^{-1} (remember J = kg m^2 s^{-2})
$T = 330$ K
$P = 1.00$ atm
and $Z = 5.81 \times 10^{24}\ s^{-1}$

We need to find M in order to identify the inert gas. We can use the ideal gas law to find n/V and then N/V,

$$\frac{n}{V} = \frac{P}{RT} = \frac{1.00\ \text{atm}}{0.08206\ \text{L atm K}^{-1}\ \text{mol}^{-1}\ (330\ \text{K})} = 0.0369_3\ \text{mol/L}$$

$$\frac{N}{V} = \frac{0.0369_3\ \text{mol}}{\text{L}} \times \frac{6.022 \times 10^{23}\ \text{(molecules)}}{\text{mol}} \times \frac{1000\ \text{L}}{\text{m}^3} = 2.22_4 \times 10^{25}\ \text{(molecules)/m}^3$$

Rearranging the collision equation to solve for M,

$$Z_A^2 = A^2\left(\frac{N}{V}\right)^2 \frac{RT}{2\pi M},\ \text{or}\ \left[\frac{Z}{A\left(\frac{N}{V}\right)}\right]^2 = \frac{RT}{2\pi M},\ \text{or}$$

$$M = \frac{RT}{\left[\frac{Z}{A\left(\frac{N}{V}\right)}\right]^2 2\pi}$$

$$M = \frac{8.3145\ \text{kg m}^2\ \text{s}^{-2}\ \text{K}^{-1}\ \text{mol}^{-1}\ (330\ \text{K})}{\left[\frac{5.81 \times 10^{24}\ \text{s}^{-1}}{2.5 \times 10^{-3}\ \text{m}^2\ (2.22_4 \times 10^{25}\ \text{m}^{-3})}\right]^2 6.28} = \mathbf{0.0400\ kg/mol}$$

The molar mass is **40.0 g/mol**; therefore the gas is **Argon**.

5.9 Intermolecular Collisions

This section presents the next piece of the puzzle involving what really happens in gases on a molecular level. The goal here is to show the high number of collisions among particles and the **small distances between collisions (the mean free path)**. *Food for thought: Why does the collision frequency among molecules have less impact on the overall pressure than collision frequency between molecules and the walls of the container?*

Example 5.9 - Collision Frequency And Mean Free Path

What are the collision frequency and mean free path of the Ar sample under conditions given in this study guide's Example 5.8? The diameter of an Ar atom is 0.98×10^{-10} m.

Solution

From the previous example,

$N/V = 2.22_4 \times 10^{25}$ (molecules)/m^3
$T = 330$ K
and $M = 0.0400 \times 10^{-2}$ kg/mol

$$Z = 4\frac{N}{V}d^2\sqrt{\frac{\pi RT}{M}}$$

$$Z = 4\,(2.22_4 \times 10^{25}\ \text{m}^{-3})\,(0.98 \times 10^{-10}\ \text{m})^2 \left[\frac{(3.142)\ 8.3145\ \text{J K}^{-1}\ \text{mol}^{-1}\ (330\ \text{K})}{0.0400\ \text{kg mol}^{-1}}\right]^{1/2}$$

$$\boldsymbol{Z} = 3.97 \times 10^8 \approx \mathbf{4.0 \times 10^8\ s^{-1}}$$

$$\text{Mean Free Path} = \lambda = \frac{1}{\sqrt{2}(N/V)(\pi d^2)} = \frac{1}{1.414\,(2.22_4 \times 10^{25}\ \text{m}^{-3})(3.142)(0.98 \times 10^{-10}\ \text{m})^2}$$

$$\lambda = \frac{1}{9.49 \times 10^5\ \text{m}^{-1}} = \mathbf{1.1 \times 10^{-6}\ m}$$

5.10 Real Gases

We know that no gas behaves in a truly ideal fashion. This section is devoted to **determining the deviation of a real gas from ideality**. At the beginning of the section, the point is made that models are based on the best available data, are necessarily approximations, and when they fail, can help us learn more about our system. In this case, our "ideal gas" model fails under two important conditions: **high pressure** and **low temperature**. This forces corrections to the volume and pressure terms in the ideal gas equation.

1. ***Volume Correction***: Because gas molecules **do** take up space, the free volume of the container is not as large as it would be if it were empty.

$$\text{Volume}_{\text{available}} = V_{\text{container}} - \underset{\uparrow\ \text{\# of moles of gas}}{\overset{\text{correction factor}\ \downarrow}{nb}}$$

   Therefore pressure can be expressed

$$P = \frac{nRT}{(V - nb)}$$

   We are halfway home.

2. ***Pressure Correction***: Because gas molecules can interact with each other, they do not <u>collide with the walls of the container</u> ("exert pressure") to as great extent as when there is no intramolecular interaction (as with an ideal gas). Therefore,

$$P_{\text{observed}} = P_{\text{ideal}} - \text{a correction factor.}$$

As discussed in your textbook, the correction factor that accounts for the decrease in pressure due to intramolecular interactions equals

$$P_{obs} = P_{ideal} - a\left(\frac{n}{V}\right)^2$$

Where a = a constant which depends upon the gas, and
n = the number of moles of gas.

Combining P_{obs} (we will call this P) and $V_{corrected}$ into the ideal gas equation, the **CORRECTED** equation is called **van der Waals** equation and is given by

$$\underbrace{\left[P + a\left(\frac{n}{V}\right)^2\right]}_{\text{correction for pressure}} \underbrace{(V - nb)}_{\text{correction for volume}} = nRT$$

The constants a and b have been tabulated for different gases and are given in Table 5.3 in your textbook. Though the numbers get a bit messy, all you really are doing is determining corrected values for P and V.

Interactions among molecules are greatest at **low temperatures (low rms velocities), high pressure, and low volumes**. Thus deviations from ideality are expected to be greatest under these conditions.

Example 5.10 - Van Der Waals Equation

Calculate the pressure exerted by 0.3000 mol of He in a 0.2000-L container at −25.0°C

a. using the ideal gas law
b. using van der Waals equation

Solution

a. $PV = nRT$ for this ideal situation.

b. $P = \frac{nRT}{V}$. But correcting for non-ideality,

$$\left[P + a\left(\frac{n}{V}\right)^2\right](V - nb) = nRT$$

$$\text{or } P + a\left(\frac{n}{V}\right)^2 = \frac{nRT}{(V - nb)}$$

$$\text{or } P = \frac{nRT}{(V - nb)} - a\left(\frac{n}{V}\right)^2$$

This is one of the few equations you will be working with in general chemistry that you really cannot derive. You just have to know it.

From Table 5.3 in your textbook

a = 0.034 atm L²/mol² b = 0.0237 L mol

a. $P = \frac{nRT}{V} = \frac{0.3000\text{ mol }(0.08206\text{ L atm/K mol})(248.2\text{ K})}{0.2000\text{ L}}$

P_{ideal} **= 30.55 atm**

b. $P = \frac{0.3000\text{ mol }(0.08206\text{ L atm/K mol})(248.2\text{ K})}{0.2000\text{ L} - 0.3000\text{ mol }(0.0237\text{ L/mol})} - \frac{0.034\text{ atm L}^2}{\text{mol}^2}\left[\frac{0.3000\text{ mol}}{0.2000\text{ L}}\right]^2$

$P_{\text{real}} = 31.68\text{ atm} - 0.077\text{ atm} =$ **31.60 atm**

There is a pressure difference of one atmosphere between the ideal equation and van der Waals equation in this case. The error is about three percent.

In rereading this section in the text, you need to keep in mind the **conditions** under which gases deviate from ideality. Be able to discuss **why** certain conditions lead to nonideal behavior.

5.11 Characteristics of Several Real Gases

The key ideas in this section are:

- The van der Waals equation contains a term, "a", which is a pressure correction.
- H_2 has a low value of "a" because it exhibits very weak intermolecular forces.
- Real gases differ from ideal gases largely due to the intermolecular forces that make gases deviate from ideal behavior.

5.12 Chemistry in the Atmosphere

Composition of Our Atmosphere

In the first part of this section, your textbook makes some key points with regard to the **composition of our atmosphere**.

1. The atmosphere is composed of 78% N_2, 21% O_2, 0.9% Ar, and 0.03% CO_2 along with trace gases.
2. The composition of the atmosphere varies as a function of distance from the earth's surface. Heavier molecules tend to be near the surface due to gravity.
3. Upper atmospheric chemistry is largely affected by ultraviolet, x-ray, and cosmic radiation emanating from space. The ozone layer is especially reactive to ultraviolet radiation.
4. Manufacturing and other processes of our modern society affect the chemistry of our atmosphere. Air pollution is a direct result of such processes.

Air Pollution

The reactions that cause air pollution are extremely complex and only partially understood. Your textbook gives background reactions on **photochemical smog** and **sulfur-based pollution**. This serves as but a tiny introduction to the myriad of pollution-related problems and mechanisms.

The general reactions causing smog are:

1. $N_2(g) + O_2(g) \xrightarrow[\text{temperatures}]{\text{high}} 2NO(g)$
2. $2NO(g) + O_2(g) \rightarrow 2NO_2(g)$

3. $NO_2(g) + \xrightarrow[\text{energy}]{\text{radiant}} NO(g) + O(g)$

4. $O(g) + O_2(g) \rightarrow \underset{\text{ozone}}{O_3(g)}$

Higher ozone levels that are characteristic of smog cause lung and eye irritation and can be very dangerous for people with asthma, emphysema, and other respiratory conditions. Ozone can also react with other pollutants, as discussed in the text, to form the hydroxyl radical (•OH), a very reactive oxidizing agent that can react with nitrogen oxides and hydrocarbons to further increase levels of air pollutants.

Clearing the Air

Your textbook discusses "scrubbing" and its impact on air pollution. The next example reminds us of some the key reaction.

Example 5.12 – Sulphur Dioxide Pollution

List the reactions that result from the interaction of sulfur dioxide and oxygen, as well as sulfur trioxide and water. In addition, list the "scrubbing" reactions that remove sulfur dioxide from the exhaust of coal-burning power plants.

Solution

The reactions are:

1. $2SO_2(g) + O_2(g) \rightarrow 2SO_3(g)$
2. $SO_3(g) + H_2O(l) \rightarrow H_2SO_4(aq)$
3. a. $CaCO_3(s) \rightarrow CaO(s) + CO_2(g)$

   b. $CaO(s) + SO_2(g) \rightarrow CaSO_3(s)$

Exercises

Section 5.1

1. In a science demonstration, 4 to 6 plastic bags are arranged under a 2.0 m × 2.0 m piece of plywood. A volunteer stands on the plywood, and others blow into the bags to "levitate" the volunteer. If four bags are used and a 140-lb person stands on the plywood, what is the pressure that must be supplied by each of the four blowers (in Pa)? What if 6 bags and blowers were used? (Assume the plywood has no mass.)

2. How much force is required to inflate a high-pressure bicycle tire to 95 pounds per square inch (655 kPa) with a hand pump that has a plunger with an area of 5.0 cm^2?

3. An object exerts a force of 500. N and sits on an area of 4.5 m × 1.5 m. Calculate, in torr, the amount of pressure exerted by the object.

4. Calculate the density of mercury. (This can be done using the fact that 760 mm Hg = 101,325 Pa.) $F = ma$ where $a = 9.81$ m/s^2. Hint: Consider a column 76.0 cm high with a cross section of 1 cm^2.

5. During your travels through deep space, you discover a new solar system. You land on the outermost planet and determine that the acceleration due to gravity is 2.7 m/s^2. If your mass back on Earth is 72 kg, what force would you exert on a scale in pounds while standing on the planet's surface?

6. As you proceed on to the next planet, some of your unbreakable equipment breaks, including that top-of-the-line machine that determines acceleration due to gravity.

   a. How do you determine the acceleration due to gravity of this planet?
   b. Calculate the acceleration due to gravity if your 72-kg mass exerts a force of 18 pounds on the planet's surface.

Section 5.2

7. A diver at a depth of 100 ft (pressure approximately 3 atm) exhales a small bubble of air with a volume equal to 100 mL. What will be the volume of the bubble (assume the same amount of air) at the surface?

8. What would the volume of gas contained in an expandable 1.0-L cylinder at 15 MPa (1 MPa = 10^6 Pa) be at 1 atm (assuming constant temperature)?

9. A sample tube containing 103.6 mL of CO gas at 20.6 torr is connected to an evacuated 1.13 liter flask. (The new volume is the sum of those of the tube and flask.) What will be the pressure when the CO is allowed into the flask?

10. A gas has a pressure of 3.2 atm and occupies a volume of 45 L. What will be the pressure if the volume is compressed to 27 L at a constant temperature?

11. The volume of a gas (held at constant pressure) is to be used "as a thermometer." If the volume at 0.0°C is 75.0 cm^3, what is the temperature when the measured volume is 56.7 cm^3?

12. If a 16.6-L sample of a gas contains 9.2 moles of F_2, how many moles of gas would there be in a 750-mL sample at the same temperature and pressure?

13. An 11.2-L sample of gas is determined to contain 0.50 moles of N_2. At the same temperature and pressure, how many moles of gas would there be in a 20.-L sample?

14. Consider a 3.57-L sample of an unknown gas at a pressure of 4.3×10^3 Pa. If the pressure is changed to 2.1×10^4 Pa at a constant temperature, what will be the new volume of the gas?

15. Calculate the volume occupied at 87.0°C and 950. torr by a quantity of gas that originally occupied 20.0 L at 27.0°C and 570. torr.

16. What is the volume of 16.0 g of SO_2 at 20.0°C and 740. torr pressure?

17. A given weight of oxygen has a volume of 100. mL at 740. torr and 25.0°C. State whether the volume will be greater than 100. mL, less than 100. mL, or unchanged for each of the following new conditions (mass of oxygen remains constant):

    a. 700. torr and 25°C
    b. 740. torr and 50°C
    c. 2220. torr and 600. K

18. A quantity of gas at 27.0°C is heated in a closed vessel (constant volume) until the pressure is doubled. To what temperature is the gas heated?

Section 5.3

19. A weather balloon is filled with 0.295 m^3 of helium on the ground at 18°C and 756 torr. What will be the volume of the balloon at an altitude of 10 km where the temperature is −48°C and the pressure is 0.14 atm?

20. A sample of gas occupies 3.8 L at 15°C and 1.00 atm. What does the temperature need to be for the gas to occupy 8.3 L at 1.00 atm?

21. Calculate the volume of O_2 present in a sample containing 0.89 moles of O_2 at a temperature of 40°C and a pressure of 1.00 atm.

22. Water is decomposed to $H_2(g)$ and $O_2(g)$ by electrolysis. By measuring the current, it was determined that 0.365 moles of water decomposed. After the gases are dried and collected at 24.5°C and 757 torr, what are the volumes of each?

23. What pressure would be exerted by 50.0 g of He at 25.0°C in a volume of 350. L?

24. A vacuum line used in a research lab has a volume of 1.013 L. The temperature in the lab is 23.7°C, and the vacuum line is evacuated to a pressure of 1×10^{-6} torr. How many gas particles remain?

25. A 10.5-g sample of CO_2 gas occupies a volume of 7.00 L at a pressure of 1.5 atm. What must be the temperature of the gas?

26. A flask that can withstand an internal pressure of 2500. torr, but no more, is filled with a gas at 21.0°C and 758 torr and heated. At what temperature will it burst?

27. Calculate the number of moles present in a quantity of gas that occupies 26,880 mL at 564°C and 380. torr.

Section 5.4

28. The density of liquid nitrogen is 0.808 g/mL at −196°C. What volume of nitrogen gas at STP must be liquefied to make 10.0 L of liquid nitrogen?

29. Calculate the volume occupied by 2.5 mol of an ideal gas at STP.

30. A hydrocarbon (compound containing only hydrogen and carbon) was analyzed to be 85.7 mass percent carbon and 14.3 mass percent hydrogen. At 26°C and 745 torr pressure, a sample with a volume of 1.13 L had a mass of 1.904 g. Determine the molecular formula. (You may wish to review Chapter 3.)

31. An unknown gas has a density of 7.06 g/L at a pressure of 1.50 atm and 280 K. Calculate the molar mass of the gas.

32. $HCl(g)$ can be prepared by reaction of NaCl with H_2SO_4. What mass of NaCl is required to prepare enough HCl to fill a 340.-mL cylinder to a pressure of 151 atm at 20.0°C?

33. You are not sure whether to fill a balloon with He or hot air. To what temperature would the air have to be heated for a balloon to rise to the same height as a balloon filled with He at 25.0°C?

34. A 27.7-mL sample of $CO_2(g)$ was collected over water at 25.0°C and 1.00 atm. What is the pressure in torr due to $CO_2(g)$? (The vapor pressure of water at 25.0°C is 23.8 torr.) What will the volume of $CO_2(g)$ be at the same temperature and pressure after removing the water vapor?

Section 5.5

35. A gas-tight vessel has a volume of 1000 m^3. It is filled with air at 27°C and 1.00 atm. Assuming air to be 79% N_2 and 21% O_2 (by volume), calculate the following:

    a. partial pressure of N_2
    b. partial pressure of O_2

36. A gaseous mixture of O_2, H_2, and N_2 has a total pressure of 1.50 atm and contains 8.20 g of each gas. Find the partial pressure of each gas in the mixture.

37. The mole fraction of argon in dry air is 0.00934. How many liters of air at STP will contain enough argon to fill a 35.4-L cylinder to a pressure of 150. atm at 20°C?

38. Assume that the mole fraction of nitrogen in the air is 0.8902. Calculate the partial pressure of N_2 in the air when the atmospheric pressure is 820 torr.

39. The lower limit of flammability of $H_2(g)$ in air at room temperature and 1 atm is a partial pressure of 0.040 atm. Assuming a 4:1 ratio of N_2 to O_2 in air, what is the ratio of H_2 to O_2 at its flammability limit?

40. A flask with a volume of 1.20 L is filled with carbon dioxide at room temperature to a pressure of 650. torr. A second flask, with a volume of 900. mL, is filled with nitrogen at room temperature to a pressure of 800. torr. A stopcock connecting the two volumes is then opened, and the gases are allowed to mix at room temperature. What is the partial pressure of each gas in the final mixture, and what is the total pressure of the mixture?

Section 5.6

41. Calculate the temperature of a mole of oxygen molecules if the internal energy is 1.16×10^4J. Assume ideal gas behavior.

42. Calculate the root mean square speed of O_2 gas molecules at 300. K.

43. What happens to the average kinetic energy of a mole of an ideal gas if

    a. the volume is doubled resulting in a decrease in pressure at constant temperature?
    b. the temperature is increased at a constant pressure?
    c. absolute zero is obtained?

Section 5.7

44. Ammonia, $NH_3(g)$, and HCl(g) react to form a solid precipitate, NH_4Cl. Two cotton swabs, one moistened with ammonia and the other with hydrochloric acid, are inserted into opposite ends of a 1-meter-long glass tube. How far from the hydrochloric acid end of the tube would you expect to see the white NH_4Cl precipitate?

45. Calculate the rate of effusion of PH_3 molecules through a small opening if NH_3 molecules pass through the same opening at a rate of 8.02 cm^3/s. Assume the same temperature and equal partial pressures of the two gases.

46. What are the relative rates of diffusion for methane, CH_4, and oxygen, O_2? If $O_2(g)$ travels 1.00 m in a certain amount of time, how far will methane be able to travel under the same conditions?

47. Which gas would effuse faster, Ne or CO_2? How much faster?

Section 5.10

48. Arrange the following according to expected values for b (volume correction value) in van der Waals equation.

    He, CO_2, H_2O, HF, SF_6

49. Put the following gases in order from smallest to largest according to van der Waals constant "a."

    H_2, N_2, CH_4, Ne, H_2O

50. Put the following gases in order from smallest to largest according to van der Waals constant "b".

    Kr, Cl_2, NH_3, O_2, He

51. Calculate the pressure exerted by 1 mole of Xe(*g*) using the ideal gas law and van der Waals equation

    a. in a 100.0-L container at 23°C
    b. in a 1.000-L container at 23°C

52. Calculate the pressure exerted by 100. moles of Cl_2 gas in a 20.-L container at 25.0°C using van der Waals equation and the constants in Table 5.3 of your textbook.

53. Calculate the density of a mixture of 21% O_2 and 79% N_2 as well as the density of He at 760. torr and 0.0°C. Will the ratio of densities be different at 100. atm and 6.0°C?

54. Why are all gases not perfect gases?

Multiple-Choice Self-Test

1. If a barometer were built using water instead of Hg, how high would be the column of water if the pressure were 1 atm, knowing that the density of water is 13.6 times lower than that of mercury?

   A. 10.3 m B. 3.17 m C. 20.0 m D. 33.0 m

2. What is the pressure, in mm Hg, of a gas that has a pressure of 15.0 lb/in^2?

   A. 0.113 mm Hg B. 776 mm Hg C. 1.02 mm Hg D. 27.6 mm Hg

3. A 0.86 L sample of helium is heated from 68°F to 68°C. At constant pressure, what volume does this sample occupy at 68°C?

   A. 1.0 L B. 1.6 L C. 0.9 L D. 2.7 L

4. A 3.00-L sample of xenon is heated from 100°F to 200°F and an initial pressure 70.0 cm rising to 120 cm of Hg. What is the final volume, in L, of the gas?

   A. 1.80 L B. 2.06 L C. 3.00 L D. 6.00 L

5. How many moles of an ideal gas are present in a sample of 1.25 L at 311 K and a pressure of 25.0 lb/in^2?

   A. 0.0833 mol B. 0.0510 mol C. 0.0328 mol D. 0.0102 mol

6. A 3.25-L sample of a gas at 80.0°C is heated until a final volume of 32.5 L is reached. What is the final temperature of the gas in Kelvin at constant pressure?

   A. 3.53×10^3 K B. 151 K C. 1.08×10^3 K D. 1.34×10^3 K

7. Calculate the number of grams of acetylene (C_2H_2) in a 30.0-L cylinder at a temperature of 20.0°C and a pressure equal to 2500 lb/in^2.

   A. 8.47×10^3 g B. 1000 g C. 5.55×10^3 g D. 2.40×10^3 g

8. A 50.0-L cylinder at a temperature of 47°C and a pressure of 50.0 atm contains how many molecules per cm^3?

   A. 1.15×10^{21} B. 2.30×10^{22} C. 2.30×10^{19} D. 6.75×10^{18}

9. A 50.0-L cylinder of Cl_2 at 20.0°C and a pressure of 103,401 torr springs a leak. The following day the pressure is found to be 41,361 torr. How many moles of chlorine gas escaped during this time?

   A. 170 mol B. 280 mol C. 85.0 mol D. 113 mol

10. Tin reacts with hydrochloric acid to produce hydrogen gas and tin(II) chloride. How many liters of hydrogen gas are produced at 27.0°C and a pressure of 710. torr if 2.80 g of tin reacts with excess hydrochloric acid?

$$Sn(s) + 2HCl(aq) \rightarrow SnCl_2(aq) + H_2(g)$$

A. 0.620 L B. 0.320 L C. 2.00 L D. 1.25 L

11. How many cm^3 of carbon tetrachloride are produced when 8.0 L of chlorine are allowed to react with 0.75 L of methane at STP?

$$4Cl_2(g) + CH_4(g) \rightarrow 4HCl(g) + CCl_4(g)$$

A. 1500 cm^3 B. 750 cm^3 C. 360 cm^3 D. 1080 cm^3

12. Calculate the final pressure, in atm, after 9.06 g of krypton reacts with 10.0 g of fluorine at 300 K in a 10.0-L container.

$$Kr(g) + F_2(g) \rightarrow KrF_2(s)$$

A. 0.591 atm B. 0.382 atm C. 0.700 atm D. 1.90 atm

13. Calculate the density change, g/L, if 700 g of $C_2H_6(g)$ are removed from a 200.-L cylinder at 200. psi (lb/in^2) and a temperature of 20°C.

A. 3.5 g/L B. 15.0 g/L C. 1.7 g/L D. 16.2 g/L

14. Calculate P_T, in atm, for three different gases at partial pressures of 144.0 cm, 800.0 mm, and 1.3 m of Hg.

A. 1.90 atm B. 2.58 atm C. 1.06 atm D. 4.66 atm

15. 1.0 L of hydrogen gas is collected over water at 308 K at a pressure of 728 torr. How many grams of iron are required to react with excess HCl(*aq*) to produce this volume of hydrogen gas? The vapor pressure of water is 42.2 torr. The products of the reaction are iron(II) chloride and hydrogen gas.

A. 4.7 g B. 2.35 g C. 2.0 g D. 1.3 g

16. Gas A diffuses twice as fast as gas B. Gas B has a molecular weight = 60.0 g/mol. What is the molar mass of gas A?

A. 15.0 g/mol B. 120 g/mol C. 30 g/mol D. 90 g/mol

17. The rate of effusion of freon-12 to freon-11 is 1.07:1. The molar mass of freon-11 is 137.4 g/mol. Calculate the molar mass, in g/mol, of freon-12.

A. 100 g/mol B. 182 g/mol C. 120 g/mol D. 118 g/mol

18. Using the van der Waals equation, calculate the pressure exerted by 10. g of methane (CH_4) in a 2.1-L container at 330 K ($a = 2.253\ L^2\ atm/mol^2$, $b = 0.0458$ L/mol). Calculate using the ideal gas equation, and find the difference between the ideal gas pressure and van der Waals pressure.

A. 2.0 atm B. 0.5 atm C. 0.2 atm D. 1.5 atm

Answers to Exercises

1. 39 Pa for each of 4 blowers; 26 Pa for each of 6.
2. 329 N = 73 lbs
3. 0.56 torr

4. 13.6 g/cm^3 [Steps: 1. Force = $P \times A$; 2. mass = F/a; 3. density = mass/volume, so ultimately, density = (P × area)/(volume × acceleration)]

5. 44 lbs

6. a. Find out what force you exert on the surface, and work backwards.
   b. 1.1 m/s^2

7. 300 mL

8. 150 L

9. 1.73 torr

10. 5.3 atm

11. 206 K, −67°C

12. 0.42 moles

13. 0.89 moles

14. 0.73 L

15. 14.4 L

16. 6.17 L

17. a. more than b. more than c. less than

18. 600. K

19. 1.6 m^3

20. 356°C

21. 23 L

22. 8.95 L H_2, 4.47 L O_2

23. 0.873 atm

24. 3×10^{13} particles

25. 536 K or 263°C

26. 970.15 K or 697°C

27. 0.196 moles

28. 6.46×10^3 L

29. 56 L

30. C_3H_6

31. 108 g/mol

32. 125 g

33. 2140 K or about 1870°C

34. 736 torr, 26.8 mL

35. a. 0.79 atm b. 0.21 atm

36. O_2 = 0.0832 atm; H_2 = 1.32 atm; N_2 = 0.0951 atm

37. 5.30×10^5 L

38. 730 torr

39. 0.2 H_2:1 O_2; 0.04 H_2:0.2 O_2

40. $P_{CO_2} = 0.489$ atm, $P_{N_2} = 0.451$ atm, total = 0.940 atm

41. 930 K

42. 484 m/s

43. a. no change b. increase c. goes to zero

44. 40 cm

45. 5.67 cm^3/s

46. 1.41 CH_4:1 O_2; 1.41 m

47. Ne would effuse 1.48 times as fast as CO_2.

48. (smallest) He, HF, H_2O, CO_2, SF_6 (largest)

49. Ne, H_2, N_2, CH_4, H_2O

50. He, O_2, NH_3, Kr, Cl_2

51. a. ideal = 0.243 atm, van der Waals = 0.243 atm
 b. ideal = 24.3 atm, van der Waals = 21.4 atm

52. 7.9 atm

53. air = 1.29 g/L; He = 0.179 g/L; no

54. All gases are not perfect gases because intermolecular attractions can exist, albeit for a short time, even with nonpolar substances. This becomes important at low temperatures and high pressures.

Answers to Multiple-Choice Self-Test

1. A	5. A	8. A	11. B	14. D	17. C
2. B	6. A	9. A	12. B	15. C	18. C
3. A	7. C	10. A	13. A	16. A	
4. B					

CHAPTER 6

Chemical Equilibrium

The Bottom Line: Chapter 6

This chapter considers the implication of equilibrium to chemical systems. Your textbook introduces the concept of **equilibrium** by noting that **no reaction goes fully to completion**. Some reverse reaction, however small, always exists. The direction of equilibrium may lie "**far to the right**" (reaction almost complete), "**far to the left**" (negligible reaction), or somewhere in between. Whatever the extent of reaction, the key point is that equilibrium is reached when **the concentrations of reactants and products remain constant with time**.

6.1 The Equilibrium Condition

Try the following questions to test your general understanding of equilibrium.

Example 6.1 - Background Questions

1. What is dynamic equilibrium?
2. What is true about the initial rate of forward and reverse reactions in a system where only reactants are present?
3. What is true about the rates of forward and reverse reactions **at equilibrium?**
4. Why does equilibrium occur?
5. What are some of the factors that determine the equilibrium position of a reaction?

Solution

1. Dynamic equilibrium is a process in which the rates of the forward and reverse reaction are equal. As a result, the overall reaction appears static, but is not.
2. Initially there is only a forward reaction. After some product builds up, the reverse reaction begins.
3. The rates will be **equal** at equilibrium.
4. Equilibrium occurs because products, once they are produced, can combine to form reactants at a rate that will eventually **equal** that of product formation by reactants.
5. Among the factors listed in your textbook are initial concentrations, relative energies of the reactants and products, and the energy and organization of the products.

6.2 The Equilibrium Constant

Section 6.2 opens with the statement of the **law of mass action**. For a reaction

$$j\mathrm{A} + k\mathrm{B} \rightleftharpoons l\mathrm{C} + m\mathrm{D}$$

the equilibrium constant K is given by

$$K = \frac{[\mathrm{C}]^l[\mathrm{D}]^m}{[\mathrm{A}]^j[\mathrm{B}]^k},$$

One way to determine the significance of K is to assume that the forward reaction involves jA and kB in its rate-determining step, and the reverse reaction involves lC and mD in its rate-determining step.

$$\text{rate}_{\text{forward}} = k_f\,[A]^j[B]^k$$

$$\text{rate}_{\text{reverse}} = k_r\,[C]^l[D]^m$$

At equilibrium, $\text{rate}_{\text{forward}} = \text{rate}_{\text{reverse}}$, or $k_f\,[A]^j[B]^k = k_r\,[C]^l[D]^m$. Rearranging,

$$\frac{k_f}{k_r} = K = \frac{[C]^l[D]^m}{[A]^j[B]^k}$$

K is thus the **ratio** of the forward to reverse **rate constants** (not the rate)! As with the kinetic rate-determining step, reactant or product **coefficients become exponents** when put in the mass action expression (or **equilibrium expression**).

There is a significant margin note that discusses the use of K_{observed} when dealing with observed concentrations. These K_{obs} values **do** have units!

There are several types of equilibrium problems you will need to solve in this chapter:

- Solving for K (knowing equilibrium concentrations).
- Solving for equilibrium concentrations (knowing K).
- Solving for equilibrium concentrations (knowing K and initial concentrations).

We will work with the first two types of problems below.

Example 6.2 A - Calculating An Equilibrium Constant

Calculate the equilibrium constant, K, for the following reaction at 25°C,

$$H_2(g) + I_2(g) \rightleftharpoons 2HI(g)$$

if the equilibrium concentrations are $[H_2] = 0.106\ M$, $[I_2] = 0.022\ M$, and $[HI] = 1.29\ M$.

Solution

The mass action (or equilibrium) expression is

$$\boldsymbol{K = \frac{[HI]^2}{[H_2][I_2]}}$$

We are given the **equilibrium concentrations.** We can substitute to get

$$K = \frac{(1.29)^2}{(0.106)(0.022)} = \mathbf{7.1 \times 10^2} \text{ (dimensionless—the units cancel)}$$

Example 6.2 B - Practice With Equilibrium Expressions

Using the same reaction as in our previous problem (with $K = 7.1 \times 10^2$ at 25°C), if the **equilibrium concentrations** of H_2 and I_2 are 0.81 M and 0.035 M, respectively, calculate the equilibrium concentration of HI.

Solution

As before, the equilibrium expression is valid.

$$K = \frac{[HI]^2}{[H_2][I_2]}$$

Substituting,

$$7.1 \times 10^2 = \frac{[HI]^2}{(0.81)(0.035)}$$

$[HI]^2 = 20.13$, $\mathbf{[HI] = 4.49\ \mathit{M} = 4.5\ \mathit{M}}$ (2 sig figs)

Checking Your Answer

In the majority of equilibrium problems, you will be *given* K and asked to find the concentration of a substance. After finding the solution, the best way of checking your answer is to *substitute concentration values into the equilibrium expression to make sure you get the known value of K*. If you do (with a reasonable round-off error), your solution is correct. If you don't, there is probably an error. Checking by solving for K,

$$\mathbf{K} = \frac{(4.49)^2}{(0.81)(0.035)} = \mathbf{7.1 \times 10^2}$$

Our [HI] is therefore correct.

Example 6.2 C - Changing K When Changing The Reaction Coefficients Or Direction

Your textbook discusses certain conclusions regarding the value of K when you modify a balanced chemical equation. Reread the discussion after Example 6.1 in your textbook. Given our balanced chemical equation and value for K from the past two problems, calculate the value of K for the following reactions.

a. $2HI \rightleftharpoons H_2 + I_2$
b. $^1/_3H_2 + {}^1/_3I_2 \rightleftharpoons {}^2/_3HI$
c. $8HI \rightleftharpoons 4H_2 + 4I_2$

Solution

a. The equation is the **reverse** of the original.

$$K' = 1/K = 1/710 = \mathbf{1.4 \times 10^{-3}}$$

b. The equation is **1/3** of the original.

$$K' = K^{1/3} = (710)^{1/3} = \mathbf{8.9}$$

c. The equation is **reversed** and is **four times** the original.

$$K' = 1/K^4 = K^{-4} = (710)^{-4} = \mathbf{3.9 \times 10^{-12}}$$

6.3 Equilibrium Expressions Involving Pressures

Your textbook points out that equilibria for gases can be expressed in either **concentrations** or in **pressures terms.** We use $\mathbf{K}$ for the equilibrium constant gotten by using **concentrations**, and $\mathbf{K_p}$ for that using **pressures.**

K and K_p can be related. Before we explicitly show this relationship, let's solve a problem that uses pressures of gases rather than concentrations.

Example 6.3 A - Calculating K_p

Calculate K_p for the following reaction

$$CH_3OH \rightleftharpoons CO(g) + 2H_2(g)$$

given the equilibrium pressures as follows:

$P_{CH_3OH} = 6.10 \times 10^{-4}$ atm $\quad P_{CO} = 0.387$ atm $\quad P_{H_2} = 1.34$ atm

Solution

The form of the equilibrium expression is the same whether it involves concentration or pressure data. That is,

$$K = \frac{\textbf{products}}{\textbf{reactants}}$$

For this example,

$$K_p = \frac{P_{H_2}{}^2 P_{CO}}{P_{CH_3OH}} = \frac{(1.34\text{ atm})^2(0.387\text{ atm})}{(6.10\times10^{-4}\text{ atm})} = \mathbf{1.14 \times 10^3\ atm^2}$$

Example 6.3 B - Relating K To K_p

Derive an expression that relates K to K_p for the reaction given in the previous problem. Then **calculate K** for the reaction at 25°C using the value for K_p.

Solution

According to the ideal gas equation, $PV = nRT$. Concentration units can be expressed as moles/liter, or $\boldsymbol{n/V}$. Relating this to the ideal gas equation,

$$\frac{n}{V} = \frac{P}{RT}$$

$$K = \frac{[H_2]^2[CO]}{[CH_3OH]} = \frac{\left[\frac{n}{V}H_2\right]^2\left[\frac{n}{V}CO\right]}{\left[\frac{n}{V}CH_3OH\right]}$$

Substituting P/RT for n/V,

$$K = \frac{(P_{H_2}/RT)^2(P_{CO}/RT)}{P_{CH_3OH}/RT}$$

Factoring out P's,

$$K = \frac{P_{H_2}{}^2 P_{CO}}{P_{CH_3OH}}\left[\frac{(RT)^{-2}(RT)^{-1}}{(RT)^{-1}}\right]$$

Note how this agrees with the formula given in your textbook (given in terms of K_p).

$$\mathbf{K = K_p(RT)^{-\Delta n}}$$

Using $R = 0.08206$ L atm K^{-1} mol^{-1}
$T = 298.2$ K,

$$K = 1.14 \times 10^3 \text{ atm}^2 [(0.08206 \text{ L atm K}^{-1} \text{ mol}^{-1})(298.2 \text{ K})]^{-2}$$
$$= 1.14 \times 10^3 \text{ atm}^2 (1.67 \times 10^{-3} \text{ L}^{-2} \text{ mol}^2 \text{ atm}^{-2})$$
$$\mathbf{K = 1.90 \text{ mol}^2 \text{ L}^{-2}}$$

6.4 The Concept of Activity

Your textbook introduces **activity** as **the ratio of the equilibrium pressure (or concentration) for a given substance to its reference pressure**.

Food for Thought:

1. Why is activity introduced at this point in the text?
2. What does it tell us about the equilibrium constant?
3. Under what circumstance will $K = K_p$? Can you surmise *why* $K = K_p$ in that case?

6.5 Heterogeneous Equilibria

Equilibrium expressions involve concentrations (or pressures) of substances that **change** from initial to the equilibrium conditions. A pure substance, such as water, changes **amount** but not **concentration.**

The concentrations of pure solids and liquids remain constant. That means that pure solids and liquids **can be incorporated into the equilibrium constant.** For example, the equilibrium expression for

$$NH_4NO_2(s) \rightleftharpoons N_2(g) + 2H_2O(g)$$

is not $K = \dfrac{[H_2O]^2[N_2]}{[NH_4NO_2]}$ because the activity of NH_4NO_2, $a_{NH_4NO_2} = \dfrac{[NH_4NO_2]}{[NH_4NO_2]} = 1$.

Therefore the equilibrium expression becomes $K = [H_2O]^2[N_2]$ or $K_p = P_{H_2O}{}^2 P_{N_2}$.

Example 6.5 - Equilibrium Expressions Involving Pure Solids And Liquids

Please write equilibrium expressions for each of the following reactions:

a. $Ba(OH)_2(s) \rightleftharpoons Ba^{+2}(aq) + 2OH^-(aq)$
b. $HCl(g) + NH_3(g) \rightleftharpoons NH_4Cl(s)$
c. $Zn(OH)_2(s) + 2OH^-(aq) \rightleftharpoons Zn(OH)_4{}^{-2}(aq)$
d. $CH_3CO_2H(aq) \rightleftharpoons H^+(aq) + CH_3CO_2{}^-(aq)$
e. $Al(NO_3)_3(s) + 6H_2O(l) \rightleftharpoons [Al(H_2O)_6]^{3+}(aq) + 3NO_3{}^-(aq)$

Solution

Remember that concentrations of pure solids and liquids are **not** put into the equilibrium expression. They are incorporated into the equilibrium **constant**.

a. $K = [Ba^{2+}][OH^-]^2$

b. $K = \dfrac{1}{[HCl][NH_3]}$ or $K_p = \dfrac{1}{P_{HCl}P_{NH_3}}$

c. $K = \dfrac{[Zn(OH)_4^{2-}]}{[OH^-]^2}$

d. $K = \dfrac{[H^+][CH_3CO_2^-]}{[CH_3CO_2H]}$

e. $K = [NO_3^-]^3[Al(H_2O)_6^{3+}]$

6.6 Applications of the Equilibrium Constant

Your textbook begins this section by reviewing the things K can and cannot indicate:

- K **does not** reflect **how fast** a reaction goes.
- A **large** value of K means that mostly **products** will be present at equilibrium.
- A **small** value of K means that mostly **reactants** will be present at equilibrium.

The **theme** of this section is the use of the **reaction quotient, Q,** which predicts the direction that the reaction will go to reach equilibrium. You calculate Q by using the law of mass action on the **initial, not equilibrium,** concentrations (or pressures) of the reaction substances.

- If **Q is equal to K,** the system is **at equilibrium.**
- If **Q is greater than K** mathematically, there is too much product present. The system will **shift to the left** to reach equilibrium.
- If **Q is less than K,** there is too much reactant present. The system will **shift to the right** to reach equilibrium.

Example 6.6 A - Predicting The Direction Of Equilibrium

Let's reexamine the reaction between hydrogen gas and iodine gas that we used in Examples 6.2 A and B.

$$H_2(g) + I_2(g) \rightleftharpoons 2HI(g)$$
$$K = 7.1 \times 10^2 \text{ at } 25°C$$

Predict the direction the system will shift in order to reach equilibrium given each of the following initial conditions:

a. $Q = 427$
b. $Q = 1522$
c. $[H_2]_0 = 0.81\ M$ $[I_2]_0 = 0.44\ M$ $[HI]_0 = 0.58\ M$
d. $[H_2]_0 = 0.078\ M$ $[I_2]_0 = 0.033\ M$ $[HI]_0 = 1.35\ M$
e. $[H_2]_0 = 0.034\ M$ $[I_2]_0 = 0.035\ M$ $[HI]_0 = 1.50\ M$

Solution

a. **$Q < K$,** so the reaction will form more products and **shift to the right.**

b. **$Q > K$,** so the reaction will form more reactants and **shift to the left.**

c. $Q = \dfrac{[HI]_0^2}{[H_2]_0[I_2]_0} = \dfrac{(0.58)^2}{(0.81)(0.44)} = 0.94$

$Q < K$, so the reaction will **shift to the right**.

d. $Q = \dfrac{(1.35)^2}{(0.078)(0.033)} = 7.1 \times 10^2$

$\boldsymbol{Q = K}$, so the system is **at equilibrium**.

e. $Q = \dfrac{(1.50)^2}{(0.034)(0.035)} = 1.9 \times 10^3$

$\boldsymbol{Q > K}$, so the reaction will **shift to the left**.

Earlier in our discussion (See Section 6.2.) we mentioned three types of equilibrium problems. The third, and by far most interesting, is the problem in which we are *given initial concentrations and K* and are asked to *find equilibrium concentrations.* In the remainder of this section, we will introduce this type of problem. We will examine it in some detail in the next section.

Example 6.6 B - Equilibrium Concentrations: Large Value For K

Let's continue to use our data from our hydrogen and iodine reaction ($K = 7.1 \times 10^2$ at 25°C). Calculate the **equilibrium concentrations** if a 5.00-L vessel initially contains 15.7 g of H_2 and 294 g of I_2.

Solution

The key is to determine **the direction of the reaction**. Thus we must **compare *Q* to *K*.** (Note that our data must be converted from g/L to mol/L.)

$$[H_2]_0 = \frac{15.7\text{ g}}{5.00\text{ L}} \times \frac{1\text{ mol}}{2.016\text{ g}} = 1.56\ M$$

$$[I_2]_0 = \frac{2.94\text{ g}}{5.00\text{ L}} \times \frac{1\text{ mol}}{253.8\text{ g}} = 0.232\ M$$

$$[HI]_0 = 0.00\ M$$

$$\boldsymbol{Q} = \frac{[HI]_0^{\ 2}}{[H_2]_0[I_2]_0} = \frac{0^2}{1.56\ (0.232)} = \mathbf{0}$$

$\boldsymbol{Q < K,}$ so the reaction will **go to the right** (which makes sense, as there is no product!).

Our next step is critical—setting up the **table of initial and final conditions.** In order to do this well, we must **take *K* into account**. You **cannot** memorize the equilibrium table set-up. You must **evaluate** each problem separately.

The value of *K* is 710. We can **assume** the reaction will go **essentially to completion** although there will be some small amount that will be unreacted. If this is true, the **stoichiometry** of the reaction dictates that **I_2 is the limiting reactant** and that H_2 will be in excess.

	H_2	+	I_2	$\rightleftharpoons$	2HI
initial	1.56 *M*		0.232 *M*		0 *M*
final	1.328 *M*		0 *M*		0.464 *M*

However, the reaction does not go all the way. There will be some I_2 left over. Call the amount **$+x$**. An identical amount of H_2 will be left over ($+x$). Twice the amount of HI will **not be formed** (2 moles of HI to 1 mole of I_2 or H_2), so the amount that will not be formed is **$-2x$**. The "–" indicates that some small amount will remain on the reactant side. Summarizing in table form,

	H_2	+	I_2	$\rightleftharpoons$	2HI
initial	1.56 *M*		0.232 *M*		0 *M*
change	$-0.232 + x$		$-0.232 + x$		$+0.464 - 2x$
equilibrium	$1.328 + x$		$+x$		$0.464 - 2x$

Assumption

Since the reaction is fairly complete, neglect *x* **relative to** 1.328 and 0.464. If the assumption is valid, *x* will be <5% of 1.328 and 0.464. Therefore, $[H_2] \approx 1.328\ M$, $[I_2] = x$, and $[HI] \approx 0.464\ M$. Solving,

$$K = \frac{[HI]^2}{[H_2][I_2]} \quad \Rightarrow \quad 710 = \frac{(0.464)^2}{(1.328)(x)}$$

$$\mathbf{x = [I_2] = 2.3 \times 10^{-4}\ M}$$

$$\mathbf{[H_2] = 1.33\ M,\ [I_2] = 2.3 \times 10^{-4}\ M,\ [HI] = 0.464\ M}$$

Testing Our Assumption

Is 2.28×10^{-4} less than 5% of 1.328? Yes. Is it less than 5% of 0.464? Yes. If it were not, we would solve the equation by the quadratic formula. (See the next section.)

Double Check

Solve for *K*.

$$K = \frac{(0.464)^2}{1.328\,(2.28 \times 10^{-4})} = \mathbf{7.1 \times 10^2}$$

Remember, the key to solving equilibrium problems lies with **making and testing valid assumptions.**

6.7 Solving Equilibrium Problems

At the beginning of this section, your textbook has a box listing the procedure for solving equilibrium problems. When following that procedure you must always think about what assumptions can make your problem solving easier, and **test the validity** of your assumptions.

We have already dealt with **large values of *K*.** Now let's do an example that involves a **small value of *K*.** (What assumptions might you make?)

Example 6.7 A - Equilibrium Calculations: Small Value For K

The reaction between nitrogen and oxygen to form nitric oxide has a value for the equilibrium constant at 2000 K of $\mathbf{K = 4.1 \times 10^{-4}}$. If 0.50 moles of N_2 and 0.86 moles of O_2 are put into a 2.0-L container at 2000 K, what would be the equilibrium concentrations of all species?

$$N_2(g) + O_2(g) \rightleftharpoons 2NO(g)$$

Solution

The **most important question** you must raise in doing equilibrium problems is, "*What does the value of K tell me about the **extent of the reaction?***"

With **very small values of *K*** such as this one, the reaction will **stay far to the left.** Only a small amount of N_2 and O_2 (called the amount *x*) will react. The small amount of NO that will be formed will be **2*x*** because of the 2:1 stoichiometry between NO and N_2 or O_2. We may now set up our equilibrium table.

$$[N_2]_0 = \frac{0.50 \text{ moles}}{2.0 \text{ L}} = 0.25\ M \qquad [O_2]_0 = \frac{0.86 \text{ moles}}{2.0 \text{ L}} = 0.43\ M$$

	N_2	+	O_2	$\rightleftharpoons$	$2NO$
initial	$0.25\ M$		$0.43\ M$		$0\ M$
change	$-x$		$-x$		$+2x$
equilibrium	$0.25 - x$		$0.43 - x$		$2x$

Because K is small, **let us assume that x is negligible** compared to 0.25 and 0.43. That is, $0.25 - x \approx 0.25$ and $0.43 - x \approx 0.43$. We can now solve for x (and test the validity of our assumption).

$$K = \frac{[NO]^2}{[N_2][O_2]} \quad \Rightarrow \quad 4.1 \times 10^{-4} = \frac{(2x)^2}{(0.25)(0.43)}$$

$$4x^2 = 4.41 \times 10^{-5} \quad \Rightarrow \quad \mathbf{x = 3.3 \times 10^{-3}\ M}$$

$$\mathbf{[NO] = 2x = 6.6 \times 10^{-3}\ M} \qquad \mathbf{[N_2] = 0.25\ M} \qquad \mathbf{[O_2] = 0.43\ M}$$

We are not finished yet. We must **test our assumption** and **check the answer.** Is x less than 5% of 0.25 M? Yes. (It is about 1.3%.) Is x less than 5% of 0.43 M? Yes. (It is about 0.8%.) Our assumption was O.K.

We can check our math by **solving for K** using our equilibrium concentrations,

$$K = \frac{(6.6 \times 10^{-3})^2}{(0.25)(0.43)} = 4.0_5 \times 10^{-4}$$

Acceptably close, considering round-off.

We have solved problems with very large K's and very small K's. The majority of our problems will have these kinds of equilibrium constants. Occasionally you will have to solve a problem where the **value for K is intermediate** (roughly between 0.01 and 100). But remember, there are **no absolutes** in solving these kinds of problems. Assumptions that you make must **always be tested**.

In problems with intermediate K's, you **cannot (in general) make assumptions** based on the extent of reaction. It is too uncertain. The trade-off is that you must explicitly solve for x, often using the **quadratic formula**.

Example 6.7 B - Equilibrium Calculations: Intermediate Value For K

Sulfurous acid dissociates in water as follows:

$$H_2SO_3(aq) \rightleftharpoons H^+(aq) + HSO_3^-(aq)$$

If $[H_2SO_3]_0 = 1.50\ M$ and $[H^+] = [HSO_3^-] = 0\ M$, calculate the equilibrium concentrations of all species at 25°C if $K = 1.20 \times 10^{-2}$ for this reaction.

Solution with Assumptions

The value of K is small. Let us **assume** that we can **neglect dissociation of H_2SO_3** (i.e., x is negligible compared to 1.5 M).

	H_2SO_3	$\rightleftharpoons$	H^+	+	HSO_3
initial	$1.50\ M$		$0\ M$		$0\ M$
change	$-x$		$+x$		$+x$
equilibrium	$1.50 - x$ ($\approx$1.50??)		X		x

$$K = \frac{[H^+][HSO_3^-]}{[H_2SO_3]} \quad \Rightarrow \quad 1.20 \times 10^{-2} = \frac{x^2}{1.50}$$

$$\mathbf{x = 0.134\ M = [H^+] = [HSO_3^-]}$$

Let's test our assumption that x is negligible compared to 1.5 M. In fact, **0.134 is 9% of 1.5! Our assumption is not valid!** We must solve the problem explicitly.

Explicit Solution

$$1.20 \times 10^{-2} = \frac{x^2}{1.50 - x}$$

You may use the quadratic equation. (See Section 6.7 in your textbook.) Multiplying out,

$$0.0180 - 0.0120x = x^2$$

Setting our equation equal to zero,

$$x^2 + 0.0120x - 0.0180 = 0$$

The quadratic formula is

$$x = \frac{-b \pm \sqrt{b^2 - 4ac}}{2a}$$

Because "$-b$" is negative (−0.012), the "±" in the quadratic must be "+" or one of our solutions will be a negative concentration, a physical impossibility. That is,

$$x = \frac{-b + \sqrt{b^2 - 4ac}}{2a}$$

$a = 1$
$b = 0.0120$
$c = -0.0180$

$$x = \frac{-0.0120 + \sqrt{(0.0120)^2 - 4(1)(-0.0180)}}{2(1)}$$

$$x = \frac{-0.0120 + 0.2686}{2} = 0.128\ M$$

$[H_2SO_3] = 1.50 - 0.128 = \mathbf{1.37\ M}$ $\qquad$ $[H^+] = [HSO_3^-] = \mathbf{0.128\ M}$

$$\text{Checking, } K = \frac{(0.128)^2}{1.37} = 1.20 \times 10^{-2}$$

The key to doing equilibrium problems is to **make assumptions when you can** to simplify the math. But you must **test all assumptions that you make.**

6.8 Le Châtelier's Principle

Your textbook defines **Le Châtelier's Principle:** "If a change in conditions is imposed on a system **at equilibrium**, the equilibrium position will shift in a direction that tends to reduce that change in conditions."

Your textbook uses the **Haber process**,

$$N_2 + 3H_2 \rightleftharpoons 2NH_3$$

to conceptualize the following effects on the position of equilibrium:

- concentration
- pressure
- temperature

Look over the effects of each on the direction of equilibrium; then try the following problem.

Example 6.8 A - Le Châtelier's Principle

Nitrogen gas and oxygen gas combine at 25°C in a closed container to form nitric oxide as follows:

$$N_2(g) + O_2(g) \rightleftharpoons 2NO(g) \qquad \Delta H = +1.81 \text{ kJ}, K_p = 3.3 \times 10^{30}$$

What would be the effect on the direction of equilibrium (i.e., would it shift to the **left, right**, or **not at all**) if the following changes are made to the system?

a. N_2 is added.
b. He is added.
c. The container is made larger.
d. The system is cooled.

Solution

a. From a mathematical point of view, if P_{N_2} is larger, Q_p will be smaller than K_p. To compensate, more product must be formed. The reaction will shift **to the right.**

b. The addition of an **inert gas** such as He does not alter the partial pressure of any of the gases. Therefore Q_p is still equal to K_p, and the position of equilibrium **remains the same.**

c. **Enlarging** the container will favor the side with **more gas molecules** because they can exist with fewer collisions per unit time. In this case, both sides have the same number of molecules. Therefore there will be **no change** in the position of equilibrium.

d. This is an **endothermic reaction** (ΔH = "+"). It requires heat to go. Cooling the system will thus shift the equilibrium **to the left.**

Example 6.8 B - Practice With Le Châtelier's Principle

The combination of hydrogen gas and oxygen gas to give water vapor can be expressed by

$$2H_2(g) + O_2(g) \rightleftharpoons 2H_2O(g) \qquad \Delta H = -484 \text{ kJ}$$

Predict the effect on the direction of equilibrium of each of the following changes to the system.

a. H_2O is removed as it is being generated.
b. H_2 is added.
c. The system is cooled.

Solution

a. Moves **to the right** (removal of product forces Q to be less than K, and the system compensates by making more product).

b. Moves **to the right** (same as above, except Q is less than K because $[H_2]_0$ is larger than at previous equilibrium).

c. Moves **to the right** because an **exothermic** reaction gives off heat. Cooling the system increases the temperature gradient between the system and surroundings, thus allowing a continued flow of heat to the surroundings.

6.9 Equilibria Involving Real Gases

The key point in this section is that just as gases deviate from ideality under high pressure conditions, values of the equilibrium constant also deviate at high pressures. Try answering the following questions:

1. What is the relationship between K_p^{obs} at less than and greater than 1 atm? Why is this behavior observed?
2. What is the importance of Table 6.5 in your textbook?
3. How is the "true" value of K_p determined?

Exercises

Section 6.2

1. Write the equilibrium expression for each of the following reactions:
   a. $2H_2(g) + O_2(g) \rightleftharpoons 2H_2O(g)$
   b. $Cl_2(g) + 2Fe^{2+}(aq) \rightleftharpoons 2Fe^{3+}(aq) + 2Cl^-(aq)$
   c. $Cu^{2+}(aq) + 4NH_3(aq) \rightleftharpoons Cu(NH_3)_4^{2+}(aq)$

2. Write the equilibrium expression for each of the following reactions:
   a. $2NO(g) + O_2(g) \rightleftharpoons 2NO_2(g)$
   b. $Ag^+(aq) + I^-(aq) \rightleftharpoons AgI(s)$
   c. $Fe^{3+}(aq) + 3OH^-(aq) \rightleftharpoons Fe(OH)_3(s)$

3. Write the equilibrium expression for each of the following reactions:
   a. $Zn_2Fe(CN)_6(s) \rightleftharpoons 2Zn^{2+}(aq) + Fe(CN)_6^{-4}(aq)$
   b. $H^+(aq) + OH^-(aq) \rightleftharpoons H_2O(l)$

4. Calculate the equilibrium constant, K, at 25°C for the reaction

$$2NO(g) + O_2(g) \rightleftharpoons 2NO_2(g)$$

   if the equilibrium concentrations are $NO_2 = 0.55$ atm, $NO = 6.5 \times 10^{-5}$ atm, $O_2 = 4.5 \times 10^{-5}$ atm.

5. Calculate the equilibrium, K, at 25°C for the Haber process

$$3H_2(g) + N_2(g) \rightleftharpoons 2NH_3(g)$$

   if the equilibrium concentrations are $[H_2] = 0.85\ M$, $[N_2] = 1.33\ M$, $[NH_3] = 0.22\ M$.

6. Calculate the value of K for the reaction: $2NO(g) + Cl_2(g) \rightleftharpoons 2NOCl(g)$ if the equilibrium concentrations are:

$$[NOCl] = 4.2 \times 10^{-2}\ mol/L,\ [NO] = 6.7 \times 10^{-1}\ mol/L,\ [Cl] = 2.9 \times 10^{-3}\ mol/L$$

7. Write the equilibrium expression for each of the following reactions:
   a. $N_2(g) + 3H_2(g) \rightleftharpoons 2NH_3(g)$
   b. $I_2(s) + Cl_2(g) \rightleftharpoons 2ICl(g)$
   c. $2B(s) + 3F_2(g) \rightleftharpoons 2BF_3(g)$

8. Given your answer from Problem 4, calculate the value for *K* at 25°C for each of the following reactions:

   a. $^{1}/_{2}NO(g) + ^{1}/_{4}O_2(g) \rightleftharpoons ^{1}/_{2}NO_2(g)$
   b. $2NO_2(g) \rightleftharpoons 2NO(g) + O_2(g)$
   c. $NO_2(g) \rightleftharpoons NO(g) + ^{1}/_{2}O_2(g)$

9. Given your answer from Problem 5, calculate the value for *K* at 25°C for each of the following reactions:

   a. $2NH_3(g) \rightleftharpoons 3H_2(g) + N_2(g)$
   b. $NH_3(g) \rightleftharpoons ^{3}/_{2}H_2(g) + ^{1}/_{2}N_2(g)$
   c. $6NH_3(g) \rightleftharpoons 9H_2(g) + 3N_2(g)$

10. Write equilibrium expressions for each of the following reactions:

    a. $CaCO_3(s) \rightleftharpoons CaCO(s) + CO_2(g)$
    b. $2HCN(aq) + Zn(s) \rightleftharpoons H_2(g) + 2CN^-(aq) + Zn^{2+}(aq)$
    c. $2NaHCO_3(s) + 2CaHPO_4(s) \rightleftharpoons 2H_2O(g) + 2CO_2(g) + 2CaNaPO_4(s)$

11. The dissociation of acetic acid, CH_3COOH, has an equilibrium constant of 1.8×10^{-5} at 25°C. The reaction is

$$CH_3COOH(aq) \rightleftharpoons CH_3COO^-(aq) + H^+(aq)$$

If the equilibrium concentration of CH_3COOH is 0.46 moles in 0.500 L of water and that of CH_3COO^- is 8.1×10^{-3} moles in the same 0.500 L, calculate $[H^+]$ for the reaction.

12. Calculate the equilibrium constant, *K*, for the following reaction at 25°C if the equilibrium concentrations are $[Cl_2] = 0.371$ *M*, $[F_2] = 0.194$ *M*, and $[ClF] = 1.02$ *M*.

$$Cl_2(g) + F_2(g) \rightleftharpoons 2ClF(g)$$

13. Write the equation for the reaction in Problem 11 if the value of *K* is 5.6×10^4 $(1/K_{original})$.

14. Write the equilibrium expression for each of the following:

    a. $2H_2S(g) \rightleftharpoons 2H_2(g) + S_2(g)$
    b. $4NH_3(g) + 7O_2(g) \rightleftharpoons 4NO_2(g) + 6H_2O(l)$
    c. $2NO_2(g) + 7H_2(g) \rightleftharpoons 2NH_3(g) + 4H_2O(g)$
    d. $PCl_3(g) + Cl_2(g) \rightleftharpoons PCl_5(g)$

15. For the general reaction

$$A \rightleftharpoons B$$

in which K = x, what is the equilibrium constant if $^{1}/_{5}B \rightleftharpoons ^{1}/_{5}A$?

16. If the equilibrium constant at 444°C for the equilibrium

$$2HI(g) \rightleftharpoons H_2(g) + I_2(g)$$

is 1.39×10^{-2}, calculate the equilibrium constant at 444°C for

$$H_2(g) + I_2(g) \rightleftharpoons 2HI(g)$$

17. For which of the following cases does the reaction go farthest toward completion: $K = 1$, $K = 10^{10}$, or $K = 10^{-10}$?

18. For the system:

$$2HI(g) \rightleftharpoons H_2(g) + I_2(g)$$

the specific rate constant of the forward reaction, k, is 0.018 at 444°C. Calculate the specific rate constant of the reverse reaction, k', if $K = 1.39 \times 10^{-2}$.

19. Indicate the effect of the following on (a) the speed of a reaction and (b) the position of equilibrium:

    i. catalyst
    ii. pressure
    iii. temperature
    iv. concentration

Section 6.3

20. Derive an expression that relates K to K_p, and calculate the value of K at 25°C for the reaction given in Problem 4.

21. At 700 K, the measured values for the partial pressures of ammonia, hydrogen, and nitrogen are 0.400 atm, 7.20 atm, and 2.40 atm, respectively. Calculate K_p and K_c at 700 K for the ammonia synthesis, $N_2(g) + 3H_2(g) \rightleftharpoons 2NH_3(g)$.

22. For the following process at 700°C, what is the partial pressure of each gas at equilibrium if the total pressure is 0.750 atm?

$$C(s) + CO_2(g) \rightleftharpoons 2CO(g) \qquad K_p = 1.50$$

23. Calculate the value for K_p at 25°C if the value for K is 3.7×10^9 L mol^{-1} for the reaction

$$CO(g) + Cl_2(g) \rightleftharpoons COCl_2(g)$$

24. Given the initial partial pressures of PCl_5 = 0.0500 atm, PCl_3 = 0.150 atm, and Cl_2 = 0.250 atm at 250°C for the following reaction, what must each equilibrium partial pressure be?

$$PCl_5(g) \rightleftharpoons PCl_3(g) + Cl_2(g) \qquad K_p = 2.15$$

25. Consider the dimerization of nitrogen dioxide:

$$2NO_2(g) \rightleftharpoons N_2O_4(g) \qquad K_p = 8.8$$

If the temperature is 298 K and the total pressure is 0.220 atm, what are the equilibrium partial pressures?

Section 6.6

26. The reaction of methane with water is given by the following equation:

$$CH_4(g) + H_2O(l) \rightleftharpoons CO(l) + 3H_2(g) \qquad K = 5.67$$

Predict the direction the system will shift in order to reach equilibrium given the following initial values of Q.

    a. $Q = 11.85$
    b. $Q = 3.8 \times 10^{-4}$
    c. $Q = 5.67$

27. Determine what the system will do to reach equilibrium given the following values:

    a. $K = 2.9 \times 10^2$; $Q = 3.1 \times 10^1$
    b. $K = 0.621$; $Q = 6.21 \times 10^{-1}$
    c. $K = 7.3 \times 10^2$; $Q = 8.2 \times 10^2$

28. $K_p = 0.133$ atm at a particular temperature for the reaction:

$$N_2O_4(g) \rightleftharpoons 2NO_2(g)$$

Calculate the reaction quotient, Q, given the partial pressures of N_2O_4 = 0.048 atm and NO_2 = 0.056 atm.

29. Using the same reaction as in Problem 26, determine the direction the system will shift in order to reach equilibrium given the **initial concentrations.**

	$[CH_4]$	$[H_2O]$	[CO]	$[H_2]$
a.	0.500	0.300	0.620	0.100
b.	4.6×10^{-3}	0.800	0.200	1.00
c.	0.818	0.750	0.650	2.00

30. The reaction

$$2NO(g) \rightleftharpoons N_2(g) + O_2(g)$$

has a value of $K = 2.4 \times 10^3$ at 2000 K. If 0.61 g of NO are put in a previously empty 3.00-L vessel, calculate the equilibrium concentrations of NO, N_2, and O_2.

31. At 250°C the equilibrium constant for the reaction

$$PCl_5(g) \rightleftharpoons PCl_3(g) + Cl_2(g)$$

is 2.15. If PCl_5 was the only gas initially present in the reaction vessel at 0.012500 atm, calculate the partial pressures of all the gases after equilibrium has been reached.

32. Using the reaction in Problem 30 at 2000 K, calculate the equilibrium concentrations of NO, N_2, and O_2 if the initial concentrations of each species are [NO] = 0 M, $[N_2]$ = 0.850 M, and $[O_2]$ = 0.560 M.

33. Using the reaction in Problem 24, calculate the equilibrium partial pressures if the initial partial pressures are PCl_5 = 0.850 atm, PCl_3 = 0.440 atm, and Cl_2 = 0.935 atm.

Section 6.7

34. Hypobromous acid, HOBr, dissociates in water according to the following reaction:

$$HOBr(aq) \rightleftharpoons OBr^-(aq) + H^+(aq) \qquad K = 2.06 \times 10^{-9} \text{ at } 25°C$$

Calculate $[H^+]$ of a solution originally 1.25 M in HOBr.

35. The following reaction has an equilibrium constant of 6.2×10^2 at a certain temperature. Calculate the equilibrium concentrations of all species if 4.5 mol of each component was added to a 3.0-L flask.

$$H_2(g) + F_2(g) \rightleftharpoons 2HF(g)$$

36. Ammonia undergoes hydrolysis according to the following reaction:

$$NH_3(aq) + H_2O(l) \rightleftharpoons NH_4^+(aq) + OH^-(aq) \qquad K = 1.8 \times 10^{-5} \text{ at } 25°C$$

Calculate $[NH_3]$, $[NH_4^+]$ and $[OH^-]$ in a solution originally 0.200 M in NH_3.

37. Using the same reaction and value for K as in Problem 36, determine $[OH^-]$ if $[NH_3]$ = 0.500 M and $[NH_4^+]$ = 0.750 M.

38. Given the following reaction at 25°C:

$$2SO_2Cl(g) \rightleftharpoons 2SO_2(g) + Cl_2(g)$$

Calculate the equilibrium constant if the equilibrium concentrations are $[SO_2Cl] = 0.037\ M$, $[SO_2] = 0.591\ M$, and $[Cl_2] = 1.24\ M$.

39. Calculate the equilibrium concentration of Cl_2 for the following reaction at 25°C. The equilibrium constant is 2.3×10^2, and the equilibrium concentrations of PCl_3 and PCl_5 are 2.00 M and 0.04 M, respectively.

$$PCl_5(g) \rightleftharpoons PCl_3(g) + Cl_2(g)$$

40. The equilibrium constant for the reaction

$$SbCl_3(g) + Cl_2(g) \rightleftharpoons SbCl_5(g)$$

at 448°C is 40. What are the equilibrium concentrations $SbCl_3$, Cl_2, and $SbCl_5$ if $[Cl_2]_0 = 0.620\ M$ and $[SbCl_5]_0 = 0.180\ M$?

41. Using the same reaction and K as in Problem 40, calculate the equilibrium concentrations of all species if $[SbCl_5]_0 = 1.25\ M$ and $[Cl_2]_0 = [SbCl_3]_0 = 0$.

Section 6.8

42. Why must the temperature be specified when indicating the value of an equilibrium constant?

43. The following chemical process is at equilibrium:

$$2H_2(g) + O_2(g) \rightleftharpoons 2H_2O(g)$$

How would the process respond if the pressure were increased at a constant temperature?

44. The reaction of carbon disulfide with chloride is as follows:

$$CS_2(g) + 3Cl_2(g) \rightleftharpoons CCl_4(g) + S_2Cl_2(g) \qquad \Delta H° = -238\ \text{kJ}$$

Predict the effect of the following changes to the system on the direction of equilibrium:

a. The pressure on the system is doubled by halving the volume.
b. CCl_4 is removed as it is generated.
c. Heat is added to the system.

45. Given the following reaction at equilibrium,

$$Cl_2(g) + 3F_2(g) \rightleftharpoons 2ClF_3(g)$$

a. Predict the effect if the pressure were reduced at constant temperature.
b. Predict the effect if the volume were reduced by increasing the pressure at constant temperature.

46. The reaction of nitrogen gas with hydrogen chloride is as follows:

$$N_2(g) + 6HCl(g) \rightleftharpoons 2NH_3(g) + 3Cl_2(g) \qquad \Delta H = +461\ \text{kJ}$$

Predict the effect of each of the following changes to the system on the direction of equilibrium:

a. Triple the volume of the system.
b. The amount of nitrogen is doubled.
c. Heat is added to the system.

47. Using the following equation, what is the effect on the equilibrium when the partial pressure of ammonia is increased?

$$NH_4Cl(s) \rightleftharpoons NH_3(g) + HCl(g)$$

Multiple-Choice Self-Test

1. Which of the following changes will change the position of equilibrium?
   A. allow more time to pass
   B. remove some products
   C. add a catalyst
   D. all of these

2. Calculate K from the following information: $k_f = 1.00 \times 10^2 /(M \times t)$, $k_r = 1.60 \times 10^{-2}/(M \times t)$.
   A. 6.25×10^3
   B. 1.60
   C. 4.00×10^{-5}
   D. 1.00×10^2

3. Calculate the equilibrium constant for the following reaction

   $$PCl_5(g) \rightleftharpoons PCl_3(g) + Cl_2(g)$$

   knowing that $[PCl_5] = 0.00325\ M$, $[PCl_3] = 2.52\ M$, and $[Cl_2] = 0.02175\ M$ at equilibrium
   A. 16.9
   B. 0.0296
   C. 33.7
   D. 7.82

4. What is the equilibrium constant of the reverse reaction for the previous problem?
   A. 0.059
   B. 3.82
   C. 5.81
   D. 0.128

5. Calculate K_p at 160°C for the following reaction.

   $$4NO_2(g) + 6H_2O(g) \rightleftharpoons 4NH_3(g) + 5O_2(g) \qquad K = 0.455$$

   A. 0.0128
   B. 16.16
   C. 5.97
   D. 2.20

6. A 0.250-L closed vessel at 487°C contains 0.500 g of PCl_5, 19.55 g of PCl_3, and 10.1 g of Cl_2 at equilibrium. Calculate K_p based on the following equation.

   $$PCl_5(g) \rightleftharpoons PCl_3(g) + Cl_2(g)$$

   A. 781
   B. 0.0292
   C. 33.65
   D. 2.10×10^3

7. 0.125 moles of oxygen gas is added to carbon in a 0.250-L container. The mixture equilibrates at 500 K. Calculate the equilibrium concentration of carbon monoxide knowing that $K = 0.086$ at 500 K.

   $$C(s) + O_2(g) \rightleftharpoons 2CO(g)$$

   A. 0.19
   B. 0.100
   C. 1.00
   D. 0.041

8. What is the correct equilibrium expression for the following reaction?

   $$HCl(aq) + H_2O(l) \rightleftharpoons H_3O^+(aq) + Cl^-(aq)$$

   A. $\dfrac{[H_3O^+][Cl^-]}{[HCl]}$
   B. $\dfrac{[H_3O^+][Cl^-]}{[H_2O][HCl]}$
   C. $\dfrac{[H_2O][HCl]}{[H_3O^+]}$
   D. $\dfrac{[Cl^-]}{[HCl][H_2O]}$

9. For a certain reaction, $Q = 2.33$, while $K = 3.54$. What do you expect to happen?
   A. The reaction will proceed forward.
   B. The reaction will proceed backward.
   C. The reaction will proceed away from equilibrium.
   D. The direction cannot be determined.

10. 3.0 moles of each of the reactants and products for the following reaction are placed in a 2.00-L chamber. Predict the direction of the reaction knowing that $K = 33.3$.

$$PCl_5(g) \rightleftharpoons PCl_3(g) + Cl_2(g)$$

A. The reaction will proceed to the right.
B. The reaction will proceed to the left.
C. The reaction is at equilibrium.
D. The direction is unpredictable.

11. Calculate the amount of $COCl_2$ heated to 575 K in a 0.220-L vessel when the equilibrium concentration of CO and Cl_2 is 0.0367 *M*. $K = 2.50 \times 10^{-3}$.

$$COCl_2(g) \rightleftharpoons CO(g) + Cl_2(g)$$

A. 11.2 g B. 0.128 g C. 0.580 g D. 1.36 g

12. For the following reaction, calculate the concentration of C(*aq*) at equilibrium when 0.135 M B solution is allowed to react with A.

$$A(l) + B(aq) \rightleftharpoons C(aq) + D(aq) \qquad K = 1.5 \times 10^{-10}$$

A. 0.01353 *M* B. 4.5×10^{-6} *M* C. 2.03×10^{-11} *M* D. 2.03 *M*

13. Which of the following changes will not affect the equilibrium position of the following equation?

$$A(g) + 4B(s) \rightleftharpoons 2C(g) + D(g) + E(s)$$

A. removal of A
B. increase in pressure
C. addition of E
D. addition of C

14. What would you change to increase the yield of the following reaction?

$$A(g) + 4B(s) \rightleftharpoons 2C(g) + D(g) + E(s)$$

A. Increase in pressure
B. Decrease in pressure
C. Increase in temperature
D. Decrease in temperature

Answers to Exercises

1. a. $K = \dfrac{[H_2O]^2}{[H_2]^2[O_2]}$ b. $K = \dfrac{[Cl^-]^2[Fe^{3+}]^2}{[Cl_2][Fe^{2+}]^2}$ c. $K = \dfrac{[Cu(NH_3)_4{}^{2+}]}{[Cu^{2+}][NH_3]^4}$

2. a. $K = \dfrac{[NO_2]^2}{[NO]^2[O_2]}$ b. $K = \dfrac{1}{[Ag^+][I^-]}$ c. $K = \dfrac{1}{[Fe^{3+}][OH^-]^3}$

3. a. $K = [Zn^{2+}]^2[Fe(CN)_6{}^{4-}]$ b. $K = \dfrac{1}{[H^+][OH^-]}$

4. $K_p = 1.6 \times 10^{12}\ atm^{-1}$

5. $K = 0.059$

6. 1.4

7. a. $K = \dfrac{[NH_3]^2}{[N_2][H_2]^3}$ b. $K = \dfrac{[ICl]^2}{[Cl_2]}$ c. $K = \dfrac{[BF_3]^2}{[F_2]^3}$

8. a. $K_p = 1.1 \times 10^3\ atm^{1/4}$ b. $K_p = 6.3 \times 10^{-13}$ atm c. $K_p = 7.9 \times 10^{-7}\ atm^{1/2}$

9. a. $K = 17$ b. $K = 4.1$ c. $K = 4.9 \times 10^3$

10. a. $K = [CO_2]$ b. $K = \frac{[H_2][CN^-]^2[Zn^{2+}]}{[HCN]^2}$ c. $K = [H_2O]^2[CO_2]^2$

11. $[H^+] = 1.0 \times 10^{-3}\ M$

12. $K = 1.45 \times 10^1$

13. $H^+ + CH_3COO^- \rightleftharpoons CH_3COOH$

14. a. $K = \frac{[H_2]^2[S_2]}{[H_2S]^2}$ c. $K = \frac{[NH_3]^2[H_2O]^4}{[NO_2]^2[H_2]^7}$

 b. $K = \frac{[NO_2]^4[H_2O]^6}{[NH_3]^4[O_2]^7}$ c. $K = \frac{[PCl_5]}{[PCl_3][Cl_2]}$

15. $K' = K^{-1/5}$

16. $K' = 71.9$

17. $K = 10^{10}$ where $K = \frac{[\text{products}]}{[\text{reactants}]}$

18. $k' = 1.29$

19.

	Speed of reaction	**Position of equilibrium**
i. **Catalyst**	Increases	No effect
ii. **Pressure**	Depends on stoichiometry	Reaction proceeds to side with fewer moles of gas—no effect on *K*
iii. **Temperature**	Generally increases	Generally increases *K* as temperature increases
iv. **Concentration**	Generally increases	Affects position of equilibrium but not the value of *K*

20. $K = K_p(RT)^{-\Delta n} = K_p(RT) \Rightarrow K = 3.9 \times 10^{13}$

21. $K_p = 1.79 \times 10^{-4}$; $K_c = 5.89 \times 10^{-1}$

22. partial pressure $CO_2 = 0.201$ atm; partial pressure $CO = 0.549$ atm

23. $K_p = 1.5 \times 10^8\ \text{atm}^{-1}$

24. partial pressures of $PCl_5 = 0.023$ atm, $PCl_3 = 0.177$ atm, $Cl_2 = 0.277$ atm

25. partial pressures of $N_2O_4 = 0.109$ atm, $NO_2 = 0.111$ atm

26. a. to the left
 b. to the right
 c. no shift—the equilibrium is at equilibrium

27. a. $Q < K$; system shifts to the right
 b. $Q = K$; system is at equilibrium
 c. $Q > K$; system shifts to the left

28. $Q = 0.065$

29. a. $Q = 4 \times 10^{-3}$, system shifts to the right.
 b. $Q = 54$, system shifts to the left.
 c. $Q = 8.5$, system shifts to the left.

30. [NO] = 6.9×10^{-5} *M*; [N_2] = [O_2] = 3.4×10^{-3} *M*

31. partial pressures of PCl_3 = Cl_2 = 0.0125 atm, PCl_5 = 7.3×10^{-5} atm

32. [NO] = 1.4×10^{-2} *M*; [N_2] = 0.850 *M*; [O_2] = 0.560 *M*

33. partial pressures of PCl_5 = 0.485 atm, PCl_3 = 0.804 atm, Cl_2 = 1.30 atm

34. [H^+] = 5.07×10^{-5} *M*

35. [H_2] = [F_2] = 0.167 *M*; [HF] = 4.2 *M*

36. [NH_3] = 0.198 *M*; [OH^-] = [NH_4^+] = 1.9×10^{-3} *M*

37. [OH^-] = 1.2×10^{-5} *M*

38. 3.2×10^2

39. 4.6 *M*

40. [$SbCl_3$] = 7.3×10^{-3} *M*; [Cl_2] = 0.627 *M*; [$SbCl_5$] = 0.173 *M*

41. [$SbCl_3$] = [Cl_2] = 0.16 *M*; [$SbCl_5$] = 1.09 *M*

42. Temperature must be specified when indicating the value of an equilibrium constant because a change in temperature will change the equilibrium constant.

43. The reaction would move left to right (fewer moles on the right side).

44. a. The reaction shifts to the right.
    b. The reaction shifts to the right.
    c. The reaction shifts to the left.

45. a. The reaction moves to the left.
    b. The reaction moves to the right.

46. a. The reaction shifts to the left.
    b. The reaction shifts to the right.
    c. The reaction shifts to the right.

47. Adding $NH_3(g)$ drives the reaction to the left.

Answers to Multiple-Choice Self-Test

1. B	4. A	7. A	9. A	11. A	13. C
2. A	5. A	8. A	10. A	12. B	14. B
3. A	6. D				

CHAPTER 7

Acids and Bases

The Bottom Line: Chapter 7

Acid-base chemistry has profound effects on our bodies (i.e., blood acidity), environmental concerns (acid rain), and industrial processes (fertilizers, polymers, etc.). Equilibrium concepts provide the necessary background to understand acid-base chemistry.

7.1 The Nature of Acids and Bases

Your textbook introduces two concepts of acids and bases. The **Arrhenius concept** says that an **acid supplies H^+ to an aqueous solution** and a **base supplies OH^- to an aqueous solution**. This concept is limiting because there are many bases that do not contain OH^-. A more global description of acids and bases is the **Brønsted-Lowry definition**, which says an **acid is a proton (H^+) donor** and a **base is a proton acceptor**.

The remainder of this section and the next several sections are devoted to the reactions of acids in water. The general reaction for an acid, HA, in water is

$$\underset{\text{acid 1}}{HA} + \underset{\text{base 2}}{H_2O} \rightleftharpoons \underset{\text{conjugate base 1}}{A^-} + \underset{\text{conjugate acid 2}}{H_3O^+}$$

HA donates a proton to H_2O, giving A^- and H_3O^+. HA and A^- are **conjugate pairs**. H_2O and H_3O^+ are also conjugate pairs.

Example 7.1 - Conjugate Pairs

Write the dissociation reaction for each of the following acids in water, and identify the conjugate acid-base pairs.

a. formic acid (HCOOH)
b. perchloric acid ($HClO_4$)
c. the hydrated iron(III) ion $[Fe(H_2O)_6]^{3+}$

Solution

a. $$\underset{\text{acid 1}}{HCOOH(aq)} + \underset{\text{base 2}}{H_2O(l)} \rightleftharpoons \underset{\text{conjugate base 1}}{COOH^-(aq)} + \underset{\text{conjugate acid 2}}{H_3O^+(aq)}$$

Your textbook points out it is common practice to eliminate the solvent, water, from the equation (it has a constant concentration). This leads to the more common form,

$$HCOOH(aq) \rightleftharpoons COOH^-(aq) + H^+(aq)$$

b. long form: $$\underset{\text{acid 1}}{HClO_4(aq)} + \underset{\text{base 2}}{H_2O(l)} \rightleftharpoons \underset{\text{conjugate base 1}}{ClO_4^-(aq)} + \underset{\text{conjugate acid 2}}{H_3O^+(aq)}$$

common form: $\mathbf{HClO_4(aq) \rightleftharpoons ClO_4^-(aq) + H^+(aq)}$

c. long form: $[Fe(H_2O)_6]^{3+}(aq) + H_2O(l) \rightleftharpoons [Fe(H_2O)_5OH]^{2+}(aq) + H_3O^+(aq)$

$$\underset{\text{acid 1}}{\uparrow} \qquad \underset{\text{base 2}}{\uparrow} \qquad \underset{\text{conjugate base 1}}{\uparrow} \qquad \underset{\text{conjugate acid 2}}{\uparrow}$$

common form: $\mathbf{[Fe(H_2O)_6]^{3+}(aq) \rightleftharpoons [Fe(H_2O)_5OH]^{2+}(aq) + H^+(aq)}$

Equilibrium expressions for acid dissociations are written using the same concepts as for any other chemical equations. However, we use the **short form** of the acid dissociation when writing these expressions. Recall from your textbook that the equilibrium constant, K, is known as K_a for acid dissociations.

7.2 Acid Strength

The key idea of this section is that **the strength of an acid is indicated by the equilibrium position of the dissociation reaction**. If the equilibrium lies **far to the left** (as indicated by the value of K_a), the acid **does not dissociate** very much and is called **weak**.

Interactive Example 7.1 in your textbook illustrates that:

THE STRONGER THE ACID, THE WEAKER ITS CONJUGATE BASE.
THE STRONGER THE BASE, THE WEAKER ITS CONJUGATE ACID.

Example 7.2 A - Relative Acid And Base Strengths

Using Table 7.2 in your textbook along with your knowledge of strong acids, and knowing that K_a for $H_2O = 1 \times 10^{-14}$ at 25°C, arrange the following acids in order of their strength. Then arrange their conjugate bases in order.

$$HOC_6H_5,\ H_2O,\ HSO_4^-,\ NH_4^+,\ HNO_3$$

Solution

Acid strength is reflected by K_a (except for very strong acids, which have a K_a too large to measure accurately). Also, K_a for all but very strong acids is <<1, indicating that the equilibrium lies far to the left. **HNO_3** is a very strong acid. According to the values of K_a from Table 7.2, the list of acid strengths should read

$$HNO_3,\ HSO_4^-,\ NH_4^+,\ HOC_6H_5,\ H_2O$$

strongest ⟷ **weakest**

The list of conjugate base strength must be just the opposite:

$$OH^-,\ OC_6H_5^-,\ NH_3,\ SO_4^{-2},\ NO_3^-$$

strongest ⟷ **weakest**

Your textbook points out that water is an **amphoteric** substance (it can act as an acid OR a base). For the autoionization reaction

$$2H_2O \rightleftharpoons OH^- + H_3O^+$$

$$K_a = K_w = 1 \times 10^{-14} \text{ at } 25°C$$

Critical point: IN AQUEOUS SOLUTION, THE ION PRODUCT $[H^+][OH^-]$ IS ALWAYS EQUAL TO 1×10^{-14} at 25°C. Therefore, if you know $[H^+]$ in a solution, you always know $[OH^-]$. The reverse must also hold true.

Review the relationship between $[H^+]$ and $[OH^-]$ for acidic, basic, and neutral solutions given in your textbook; then try the next example.

Example 7.2 B - Conversion Between $[H^+]$ And $[OH^-]$

At 10°C, K_w for the autoionization of water equals 2.9×10^{-15}. Calculate $[H^+]$ or $[OH^-]$ as necessary under each of the following conditions.

a. Calculate $[H^+]$ if $[OH^-] = 9.3 \times 10^{-4}$ *M*. Is the solution acidic or basic?
b. Calculate $[H^+]$ and $[OH^-]$ for a neutral solution.
c. Calculate $[OH^-]$ if $[H^+] = 6.7 \times 10^{-11}$ *M*. Is the solution acidic or basic?

Solution

$$K_w = [H^+][OH^-]$$

(Remember, $[H_2O]$ is essentially constant so it is incorporated into the equilibrium constant.)

a. $[H^+] = K_w / [OH^-] = 2.9 \times 10^{-15} / 9.3 \times 10^{-4} = \mathbf{3.1 \times 10^{-12}}$ ***M***
$[OH^-] > [H^+]$, so this solution is **basic**.

b. For a neutral solution, $[H^+] = [OH^-]$. If we let $x = [H^+] = [OH^-]$, then

$$K_w = (x)(x) = x^2 = 2.9 \times 10^{-15}$$
$$x = [H^+] = [OH^-] = \mathbf{5.4 \times 10^{-8}\ M}$$

c. $[OH^-] = K_w / [H^+] = 2.9 \times 10^{-15} / 6.7 \times 10^{-11} = \mathbf{4.3 \times 10^{-5}}$ ***M***
$[OH^-] > [H^+]$, so this solution is **basic**.

7.3 The pH Scale

This is the first time in our chemistry course that we need to make use of **logarithms**. (See Appendix 1, part A1.2 in your textbook if you need a review.) As your textbook points out,

$$\mathbf{pH = -log[H^+]}$$
$$\mathbf{pOH = -log[OH^-]}$$

In fact, the pH is not **exactly** equal to $-\log[H^+]$, but is close enough for our purposes.

Note the discussion on the number of significant figures present at the beginning of the section in your textbook. When you are comfortable with that and you understand the equations below, please try the following problem.

$$pX = -\log[X]$$
$$\mathbf{[X] = 10^{-pX}}$$

Example 7.3 A - Converting Between p And Concentration

Calculate the "p" or "[ ]" as necessary for each of the following. (Remember to use the proper number of significant figures!)

a. Calculate $[Cl^-]$ if pCl = 7.32.
b. Calculate pAg if $[Ag^+] = 0.034$ *M*.
c. Calculate pNO_3 if $[NO_3^-] = 15$ *M*.
d. Calculate $[NH_4^+]$ if the $pNH_4 = 11.87$.

Solution

a. $[Cl^-] = 10^{-pCl} = 10^{-7.32}$ (Enter 7.32 into your calculator, press the +/– key, and press either **10^x** or **inv log.**)

$$[Cl^-] = 4.8 \times 10^{-8}\ M$$

b. pAg = –log(0.034) (Enter **0.034** into your calculator, press the **log** key and press the +/– key.)

$$pAg = 1.47$$

c. $pNO_3 = -\log(15) = \mathbf{-1.18}$

d. $[NH_4^+] = 10^{-11.87} = \mathbf{1.3 \times 10^{-12}}\ M$

For the next problem, recall that $K_w = [H^+][OH^-]$ or pK_w = pH + pOH. At 25°C, $\mathbf{K_w = 1.0 \times 10^{-14}}$, and $\mathbf{pK_w = 14}$. Therefore, at 25°C, **the sum of pH and pOH must always equal 14.0.**

Example 7.3 B - Converting Among pH, pOH, [H+], and [OH–]

Fill in the blanks in the following table.

	pH	pOH	$[H^+]$	$[OH^-]$	acid, base, or neutral?
solution a	6.88	______	______	______	______
solution b	______	______	______	8.4×10^{-14}	______
solution c	______	3.11	______	______	______
solution d	______	______	1.0×10^{-7}	______	______

Solution

a. $\mathbf{[H^+]} = 10^{-pH} = 10^{-6.88} = \mathbf{1.3 \times 10^{-7}}\ M$
$\mathbf{[OH^-]} = K_w / [H^+] = 1.0 \times 10^{-14} / 1.3 \times 10^{-7} = \mathbf{7.6 \times 10^{-8}}\ M$
$\mathbf{pOH} = -\log [OH^-] = -\log(7.6 \times 10^{-8}) = \mathbf{7.12}$

b. $\mathbf{pOH} = -\log (8.4 \times 10^{-14}) = \mathbf{13.08}$
$\mathbf{pH} = 14 - pOH = 14 - 13.08 = \mathbf{0.92}$
$\mathbf{[H^+]} = 10^{-pH} = 10^{-0.92} = \mathbf{0.12}\ M$

c. $\mathbf{pH + pOH = 14} \Rightarrow 14 - 3.11 = \mathbf{pH = 10.89}$
$\mathbf{[OH^-]} = 10^{-3.11} = \mathbf{7.8 \times 10^{-4}}\ M$
$\mathbf{[H^+]} = K_w / 7.8 \times 10^{-4} = \mathbf{1.3 \times 10^{-11}}\ M$

d. $\mathbf{[OH^-]} = 1.0 \times 10^{-14} / 1.0 \times 10^{-7} = \mathbf{1.0 \times 10^{-7}}\ M$
$\mathbf{pOH} = -\log (1.0 \times 10^{-7}) = \mathbf{7.00}$
$\mathbf{pH} = 14.00 - 7.00 = \mathbf{7.00}$

The completed table is:

	pH	pOH	$[H^+]$	$[OH^-]$	acid, base, or neutral?
solution a	6.88	7.12	1.3×10^{-7}	7.6×10^{-8}	acid
solution b	0.92	13.08	0.12	8.4×10^{-14}	acid
solution c	10.89	3.11	1.3×10^{-11}	7.8×10^{-4}	base
solution d	7.00	7.00	1.0×10^{-7}	1.0×10^{-7}	neutral

7.4 Calculating the pH of Strong Acid Solutions

Your textbook introduces you to the idea of multiple equilibria (considering the equilibrium position of several reactions in the system) in this section. In order to properly assess acid-base problems in aqueous solution, you must always

a. **recognize that autoionization of water is ALWAYS occurring in an aqueous solution, and**
b. **be able to determine whether autoionization will contribute significantly to the acid-base character of a solution.**

Calculating the pH of strong acid solutions is in general fairly straightforward because the dissociation equilibrium lies so far to the right—that is, **the acid completely dissociates**. The autoionization is negligible as a contributor of H^+ to the solution. (See the discussion regarding Le Châtelier's principle in this section of your textbook.) The rare exception to this is when your strong acid is exceptionally dilute ($< 10^{-6}$ *M*). In that case water can contribute a **relatively** large proportion of H^+ to the solution.

The bottom line is that $[H^+]$ at equilibrium is $\approx$ [strong acid]$_0$, except in very dilute solutions.

Example 7.4 - Practice With Strong Acids

A solution is prepared by adding 15.8 g of HCl to enough water to make a total volume of 400. mL. What is the pH of the solution? How much hydrogen ion is contributed by the autoionization of water?

Solution

Let's first find $[HCl]_0$.

$$\frac{\text{mol HCl}}{\text{L}} \,\Big\|\, \frac{15.8\text{ g}}{0.400\text{ L}} \times \frac{1\text{ mol}}{36.5\text{ g}} = \mathbf{1.08}\ \boldsymbol{M}$$

HCl is a strong acid, so it completely dissociates.

$$\mathbf{pH} = -\log[H^+] = -\log[HCl]_0 = -\log(1.08) = \mathbf{-0.033}$$

The next question had to do with $[H^+]$ due to the autoionization of water. We know that **the only source of OH^- is from the autoionization of water**. We can find $[OH^-]$ because

$$\mathbf{[OH^-]} = K_w / [H^+] = 1.0 \times 10^{-14} / 1.08 = \mathbf{9.3 \times 10^{-15}}\ \boldsymbol{M}$$

In addition, $[H^+]$ due to H_2O autoionization must equal $[OH^-]$ due to H_2O autoionization (1:1 stoichiometry). Therefore,

$$\mathbf{[H^+]_{H_2O} = 9.3 \times 10^{-15}}\ \boldsymbol{M}$$

We can see Le Châtelier's principle at work here. If the autoionization of water were not suppressed, $[H^+]_{H_2O}$ would equal 1.0×10^{-7} *M*.

7.5 Calculating the pH of Weak Acid Solutions

A succinct strategy for solving weak acid problems is proposed just before Example 7.3 in your textbook. The key points of the strategy are:

- Although there are often several reactions that can produce H^+, usually **only one predominates**. You can make the proper judgment based on the values of the equilibrium constants for the reactions.

- You must test any assumptions you make regarding the extent of dissociation of a weak acid (i.e.; [HA] = $[HA]_0$).

Let's work the following problem using the procedure given in your textbook.

Example 7.5 A - pH Of A Weak Acid

Calculate the pH of a 0.500 *M* aqueous solution of formic acid, HCOOH ($K_a = 1.77 \times 10^{-4}$).

Solution

Step 1: We recognize that HCOOH is a weak acid. The dissociation equilibrium lies far to the left. The same is true for the autoionization of H_2O ($K_w = 1.0 \times 10^{-14}$). The major species in solution are therefore **HCOOH and H_2O**.

Step 2: Both HCOOH and H_2O can produce H^+.

$$HCOOH(aq) \rightleftharpoons COOH^-(aq) + H^+(aq) \qquad K_a = 1.77\times 10^{-4}$$

$$H_2O(l) \rightleftharpoons H^+(aq) + OH^-(aq) \qquad K_w = 1.0 \times 10^{-14}$$

Step 3: The value of K_a for the dissociation of HCOOH is far greater than the value of K_w for the autoionization of water. Therefore HCOOH dissociation will predominate as a source of H^+. (Technically, we have **assumed** that H_2O contributes a negligible amount of H^+. We will not test this assumption.)

Step 4: $K_a = \dfrac{[COOH^-][H^+]}{[HCOOH]}$

Steps 5, 6, and 7:

	HCOOH	⇌	$COOH^-$	+	H^+
initial	0.500 *M*		0 *M*		0 *M*
change	$-x$		$+x$		$+x$
equilibrium	$0.500 - x$		x		x

Step 8: $K_a = 1.77 \times 10^{-4} = \dfrac{x^2}{0.500 - x}$

Step 9: The equilibrium lies far to the left. Assume that the extent of dissociation, x, **is negligible relative to 0.500 *M*.**

$$\mathbf{(0.500 - x = 0.500)}$$

$$1.77 \times 10^{-4} = \frac{x^2}{0.500}$$

$$x = [H^+] = [COOH^-] = 9.41 \times 10^{-3}\ M$$

Step 10: Comparing x to $[HCOOH]_o$,

$$\frac{x}{[HCOOH]_0} \times 100\% = \frac{9.41\times 10^{-3}}{0.500} \times 100\% = \mathbf{1.9\%}$$

This value is less than 5%, so **our assumption of negligible dissociation is valid**.

Step 11: $[H^+] = 9.41 \times 10^{-3}\ M \quad \Rightarrow \quad \mathbf{pH = 2.03}$

Example 7.5 B - Practice With Weak Acids

The value for $K_a = 7.45 \times 10^{-4}$ for citric acid ($C_6H_{10}O_8$). (We'll call it "HCA.") Calculate the pH of a 0.200 *M* HCA solution.

Solution

Our reactions of interest are:

$$HCA(aq) \rightleftharpoons CA^-(aq) + H^+(aq)$$
$$H_2O(l) \rightleftharpoons H^+(aq) + OH^-(aq)$$

K_a for HCA is much larger than K_w for water. Therefore the dissociation of HCA is the significant equilibrium. This equilibrium lies far to the left.

$$K_a = \frac{[CA^-][H^+]}{[HCA]}$$

	HCA	$\rightleftharpoons$	CA^-	+	H^+
initial	0.200 *M*		0 *M*		0 *M*
change	$-x$		$+x$		$+x$
equilibrium	$0.200 - x$ ($\approx$0.200)		x		x

$$7.45 \times 10^{-4} = \frac{x^2}{0.200}$$

$$x = [H^+] = [CA^-] = 1.22 \times 10^{-2}\ M$$

$$\frac{x}{[HCA]} \times 100\% = \frac{1.22 \times 10^{-2}}{0.200} \times 100\% = 6.1\%$$

The percent of dissociation is greater than 5%; therefore, we **CANNOT neglect *x* in comparison to 0.200.** (i.e., We are NOT making this assumption; therefore, we do not have to test it.) We must therefore solve the equation **using the quadratic equation**.

$$7.45 \times 10^{-4} = \frac{x^2}{0.200 - x}$$

Clearing the denominator,

$$1.49 \times 10^{-4} - 7.45 \times 10^{-4}(x) = x^2$$

Setting the equation equal to zero,

$$\underset{\overset{\uparrow}{a}}{1}x^2 + \underset{\overset{\uparrow}{b}}{7.45 \times 10^{-4}}(x) - \underset{\overset{\uparrow}{c}}{1.49 \times 10^{-4}} = 0$$

$$x = \frac{-b \pm \sqrt{b^2 - 4ac}}{2a}$$

$$x = \frac{-7.45 \times 10^{-4} \pm \sqrt{(7.45 \times 10^{-4})^2 - 4(1)(-1.49 \times 10^{-4})}}{2(1)}$$

$$x = \frac{-7.45 \times 10^{-4} + 0.02442}{2} = 1.18 \times 10^{-2}\ M$$

$$[H^+] = 1.18 \times 10^{-2}\ M \quad \Rightarrow \quad \mathbf{pH = 1.93}$$

We may wish to check our results.

$$K_a = \frac{(1.18\times10^{-2})^2}{0.200 - 1.18\times10^{-2}} = \frac{1.392\times10^{-4}}{0.1882} = \mathbf{7.4 \times 10^{-4}}$$

This is O.K. within round-off error.

With regard to calculating the pH of a mixture of weak acids, the basic question remains: **Which is the dominant equilibrium among the several that are followed?**

If you can resolve that, then the problem reduces to the pH of what is effectively one species in solution.

Example 7.5 C - The pH Of A Mixture Of Weak Acids

Calculate the pH of a mixture of 2.00 *M* formic acid (HCOOH, $K_a = 1.77 \times 10^{-4}$) and 1.50 *M* hypobromous acid (HOBr, $K_a = 2.06 \times 10^{-9}$). What are the concentrations of both the hypobromite ion (OBr^-) and hydroxide (OH^-) ion at equilibrium?

Solution

Like most other problems in equilibrium chemistry, this one is solved by **making proper assumptions**. Looking at the three sources of H^+,

$$HCOOH(aq) \rightleftharpoons H^+(aq) + COOH^-(aq) \qquad K_a = 1.77 \times 10^{-4}$$
$$HOBr(aq) \rightleftharpoons H^+(aq) + OBr^-(aq) \qquad K_a = 2.06 \times 10^{-9}$$
$$H_2O(l) \rightleftharpoons H^+(aq) + OH^-(aq) \qquad K_w = 1.0 \times 10^{-14}$$

We see that **the dissociation of formic acid is by far the most important as a supplier of H^+**. Le Châtelier's principle will dictate that the dissociation of hypobromous acid and the autoionization of water will be suppressed. We have therefore reduced the problem to finding the pH of 2.00 *M* formic acid.

$$K_a = 1.77 \times 10^{-4} = \frac{[H^+][COOH^-]}{[HCOOH]}$$

	HCOOH	⇌	H^+	+	$COOH^-$
initial	2.00 *M*		0 *M*		0 *M*
change	$-x$		$+x$		$+x$
equilibrium	$2.00 - x$ ($\approx$2.00)		x		x

$$1.77 \times 10^{-4} = \frac{x^2}{2.00}$$

$$x = 1.88 \times 10^{-2}\ M = [H^+] = [COOH^-]$$

To test the 5% rule,

$$\frac{0.0188}{2.00} \times 100\% = 0.94\%. \text{ Our assumption was O.K.}$$

$$[H^+] = 0.0188 \quad \Rightarrow \quad \mathbf{pH = 1.73}$$

To find $[OBr^-]$, use its equilibrium expression:

$$K_a = \frac{[OBr^-][H^+]}{[HOBr]}$$

We know the K_a, $[H^+]$, and [HOBr] (which ≈ $[HOBr]_0$ because of negligible dissociation).

$$2.06 \times 10^{-9} = \frac{x(0.0188)}{1.50}$$

$$\mathbf{x = [OBr^-] = 1.6 \times 10^{-7}\ \mathit{M}}$$

To find $[OH^-]$, if pH = 1.73, pOH = 14 − 1.73 = 12.27

$$[OH^-] = 10^{-12.27} = \mathbf{5.4 \times 10^{-13}\ \mathit{M}}$$

Example 7.5 D - Percent Dissociation

Determine the percent dissociation of the formic acid solution given in the previous problem.

Solution

$$\text{percent dissociated} = \frac{\text{amount dissociated } (M)}{\text{initial concentration } (M)} \times 100\%$$

The amount dissociated = x = $[COOH^-]$

$$\textbf{\% dissociated} = \frac{0.0188}{2.00} \times 100\% = \mathbf{0.94\%}$$

You will note that we determined this as part of the previous problem!

Finally, your textbook deals with determination of K_a from the percent dissociated. This type of problem is straightforward if you recognize that **you can calculate equilibrium concentrations from the percent of dissociation** as is shown in our next example.

Example 7.5 E - K_a From Percent Dissociation

In a 0.500 *M* solution, uric acid ($HC_5H_3N_4O_4$) is 1.6% dissociated. Calculate the value of K_a for uric acid.

Solution

$$\text{\% dissociated} = \frac{\text{amount dissociated}}{\text{initial concentration}} \times 100\%$$

The amount dissociated = $[C_5H_3N_4O_4^-]$ = $[H^+]$

$$1.6\% = \frac{[C_5H_3N_4O_4^-]}{0.500} \times 100\%$$

$$[C_5H_3N_4O_4^-] = [H^+] = 8.0 \times 10^{-3}\ M$$

We can now substitute into our equilibrium expression.

$$\mathbf{K_a} = \frac{[C_5H_3N_4O_4^-][H^+]}{[HC_5H_3N_4O_4]} = \frac{(8.0 \times 10^{-3})^2}{0.500 - 0.008} = \mathbf{1.3 \times 10^{-4}}$$

7.6 Bases

The key to determining the pH of basic solutions is to recognize that, **in an equilibrium sense**, bases work in the same way that acids do. Just as there are both strong and weak acids, there are both strong and weak bases.

Strong bases completely dissociate. Using LiOH in water as an example,

$$LiOH(s) \rightarrow Li^+(aq) + OH^-(aq)$$

Therefore one can consider that $[OH^-] \approx [LiOH]_0$. Once you know $[OH^-]$, you can use K_w to calculate $[H^+]$ and pH.

Your textbook points out that all alkali hydroxides are strongly basic. Alkaline earth hydroxides are also strongly basic, but they are somewhat less soluble than alkali hydroxides.

Example 7.6 A - pH Of A Strong Base

Calculate the pH of a solution made by putting 4.63 g of LiOH into water and diluting it to a total volume of 400. mL.

Solution

As we discussed earlier, LiOH is a strong base. The equilibrium concentration of OH^- will be equal to the initial concentration of LiOH.

$$\mathbf{[LiOH]_0} = \frac{4.63 \text{ g LiOH}}{0.400 \text{ L}} \times \frac{1 \text{ mol LiOH}}{23.95 \text{ g}} = \mathbf{0.482}\ \boldsymbol{M}$$

$$[LiOH]_0 = \mathbf{[OH^-] = 0.482}\ \boldsymbol{M}$$

$$pOH = 0.316$$

Recall that $pK_w = 14.00 = pH + pOH$. Therefore

$$\mathbf{pH} = 14 - pOH = 14.00 - 0.32 = \mathbf{13.68}$$

$[H^+] = 10^{-13.68} = 2.1 \times 10^{-14}\ M$. The only source of H^+ was the autoionization of water.

Weak bases react with water ("undergo hydrolysis") as described by the following equation.

$$\underset{\text{base 1}}{B(aq)} + \underset{\text{acid 2}}{H_2O(l)} \rightleftharpoons \underset{\text{acid 1}}{BH^+(aq)} + \underset{\text{base 2}}{OH^-(aq)}$$

For **weak bases**, as with weak acids, **the position of equilibrium lies far to the left**. The strategy for solving for the pH of weak bases (via pOH) is the same as for weak acids. Same steps. Same assumption. Same "5% test" as is shown in the following example.

Example 7.6 B - pH Of A Weak Base

Calculate the pH of a 0.350 M solution of methylamine, CH_3NH_2 ($K_b = 4.38 \times 10^{-4}$).

Solution

We must proceed with the same problem-solving strategy as with weak acids. Given the value of K_b, the extent of equilibrium for the base hydrolysis will be far to the left.

$$CH_3NH_2(aq) + H_2O(l) \rightleftharpoons CH_3NH_3^+(aq) + OH^-(aq)$$

Although the autoionization of water can supply OH^- to the solution, the value of K_w is small relative to the K_b of methylamine, so $[OH^-]$ due to water can be neglected.

$$K_b = \frac{[CH_3NH_3^+][OH^-]}{[CH_3NH_2]}$$

	CH_3NH_2	$\rightleftharpoons$	$CH_3NH_3^+$	+	OH^-
initial	0.350 *M*		0 *M*		0 *M*
change	$-x$		$+x$		$+x$
equilibrium	$0.350 - x$ (≈ 0.350)		x		x

(The assumption, which must be tested, is that $[CH_3NH_2] = [CH_3NH_2]_0$. We will use the "5% rule" to verify this later.)

$$K_b = 4.38 \times 10^{-4} = \frac{x^2}{0.350}$$

$$x = [OH^-] = [CH_3NH_3^+] = \mathbf{0.0124}\ \boldsymbol{M}$$

$$\frac{0.0124}{0.350} \times 100\% = 3.5\% \text{ hydrolysis, which passes the 5\% test.}$$

$$\textbf{pOH} = -\log(0.0124) = \mathbf{1.91}$$
$$\textbf{pH} = 14.00 - 1.91 = \mathbf{12.09}$$

Does the Answer Make Sense?

There are really two ways to answer the question. As a double check of our math,

$$K_b = \frac{(0.0124)^2}{0.350} = 4.39 \times 10^{-4}$$

So far so good. The more significant question relates to the value for pH. We needed to have a pH > 7 for a base. We have that in this case. Therefore, the answer makes sense.

7.7 Polyprotic Acids

A **polyprotic acid** can furnish **more than one proton** to a solution. Table 7.4 in your textbook lists a number of examples along with K_a values.

Please note that in every case, $K_{a_1} \gg K_{a_2}$ for these acids. This means that, for most of these polyprotic acids, the first proton comes off **relatively easily**. The second (and third, where applicable) does not. Another way of saying this is that the second and third dissociation reactions are generally so far "to the left" that we can neglect them. The beauty of this is that most polyprotic acid pH problems **reduce to finding the pH from a single, dominant equation**.

As you do the examples in your text and in this study guide, you will see that our problem-solving strategy is essentially the same as with our previous acid problems.

Example 7.7 A - pH Of Oxalic Acid

Using the information in Table 7.4 of your textbook, **calculate the pH** of a 1.40 *M* $H_2C_2O_4$ (oxalic acid) solution and the **equilibrium concentrations of $H_2C_2O_4$, $HC_2O_4^-$, $C_2O_4^{2-}$, and OH^-.**

Solution

pH, [$H_2C_2O_4$], and [$HC_2O_4^-$]

The major species in solution are **$H_2C_2O_4$ and H_2O**. There are a number of equilibria that will occur. Based on the values of K_{a_1}, K_{a_2}, and K_w, by far the **most significant equilibrium** will be the dissociation of $H_2C_2O_4$.

$$H_2C_2O_4(aq) \rightleftharpoons HC_2O_4^-(aq) + H^+(aq)$$

$$K_{a_1} = \frac{[H^+][HC_2O_4^-]}{[H_2C_2O_4]}$$

	$H_2C_2O_4$	$\rightleftharpoons$	$HC_2O_4^-$	+	H^+
initial	1.40 *M*		0 *M*		0 *M*
change	$-x$		$+x$		$+x$
equilibrium	$1.40 - x$		x		x

As always, we will test our **negligible dissociation** assumption ($1.40 - x \approx 1.40$) later on.

$$6.5 \times 10^{-2} = \frac{[H^+][HC_2O_4^-]}{[H_2C_2O_4]} = \frac{x^2}{1.40}$$

$$\boldsymbol{x = [H^+] = [HC_2O_4^-] = 0.302}$$

Testing the 5% rule,

$$\frac{0.302}{1.40} \times 100\% = 21.5\%!$$

Therefore, $[H_2C_2O_4] \neq [H_2C_2O_4]_0$ but rather equals $1.40 - x$.

$$6.5 \times 10^{-2} = \frac{x^2}{1.40 - x}$$

Clearing the fraction and setting the equation equal to zero so that we can use the quadratic formula,

$$\boldsymbol{x^2 + 0.065(x) - 0.091 = 0} \qquad a = 1, b = 0.065, c = -0.091$$

Solving,

$$x = \frac{-0.065 + \sqrt{(0.065)^2 - 4(1)(-0.091)}}{2(1)}$$

$$x = \frac{-0.065 + 0.6068}{2} = \boldsymbol{0.271\ M}$$

$$\boldsymbol{[H^+] = [HC_2O_4^-] = 0.27\ M}$$

$$\boldsymbol{[H_2C_2O_4] = 1.40 - 0.271 = 1.13\ M}$$

$$\textbf{pH} = -\log(0.27) = \boldsymbol{0.57}$$

Checking our math,

$$K_{a_1} = \frac{(0.27)^2}{1.1} = \mathbf{0.066}\text{. O.K., within round-off error.}$$

$[C_2O_4^{2-}]$ and $[OH^-]$

We can find $[C_2O_4^{2-}]$ by using the second dissociation equilibrium of oxalic acid,

$$\mathbf{HC_2O_4^-(\mathit{aq}) \rightleftharpoons H^+(\mathit{aq}) + C_2O_4^{2-}(\mathit{aq})} \qquad K_{a_2} = \mathbf{6.1 \times 10^{-5}}$$

We know $[HC_2O_4^-]$, $[H^+]$ and K_{a_2}. We can therefore substitute into our equilibrium expression,

$$K_{a_2} = \frac{[H^+][C_2O_4^{2-}]}{[HC_2O_4^-]} \quad \Rightarrow \quad 6.1 \times 10^{-5}\,M = \frac{(0.27)[C_2O_4^{2-}]}{0.27}$$

$$\mathbf{[C_2O_4^{2-}] = K_{a_2} = 6.1 \times 10^{-5}\,\mathit{M}}$$

Note that $[C_2O_4^{2-}]$ is smaller than $[HC_2O_4^-]$ by a factor of 10^4. We can therefore conclude that the dissociation of $HC_2O_4^-$ was not a significant contributor of H^+ to the reaction.

We can find $[OH^-]$ by using $pK_w = pH + pOH$.

$$\mathbf{pOH} = 14 - pH \quad \Rightarrow \quad \mathbf{pOH} = 14 - 0.57 = \mathbf{13.43}$$

$$\mathbf{[OH^-]} = 10^{-13.43} = \mathbf{3.7 \times 10^{-14}\,\mathit{M}}$$

The autoionization of water was not important here (as evidenced by the low $[OH^-]$).

Keep in mind that even though the problem was long, it was all *based on the assumption* that only the first acid dissociation equilibrium was important.

Example 7.7 B - Phosphoric Acid

Using data from Table 7.4 in your textbook, calculate the pH, $[PO_4^{3-}]$, and $[OH^-]$ in a 6.0 *M* phosphoric acid (H_3PO_4) solution. Also calculate the fraction of phosphate species that is PO_4^{3-} ("$f_{PO_4^{3-}}$")

Solution

In order to calculate $[PO_4^{3-}]$, we must use K_{a_3}. We must therefore know $[HPO_4^{2-}]$ and $[H^+]$. To know $[HPO_4^{2-}]$, we must use K_{a_2}, which means knowing $[H_2PO_4^-]$. This can be determined using K_{a_1}. **The bottom line** is that we must proceed as in the previous problem by determining pH and working our way down. We can find $[OH^-]$ by determining the pH and using pK_w as always.

We will make the usual simplifying assumption that leads to

$$K_{a_1} = \frac{[H^+][H_2PO_4^-]}{[H_3PO_4]} \quad \Rightarrow \quad 7.5 \times 10^{-3} = \frac{x^2}{6.0}$$

$$\mathbf{\mathit{x} = [H^+] = [H_2PO_4^-] = 0.212\,\mathit{M}}$$

We must now run the 5% test,

$$\frac{0.212}{6.0} \times 100\% = 3.5\%,$$

which passes the test of "negligible dissociation."

To find $[HPO_4^{2-}]$,

$$K_{a_2} = \frac{[H^+][HPO_4^{2-}]}{[H_2PO_4^-]} \Rightarrow 6.2 \times 10^{-8} = \frac{(0.212)[HPO_4^{2-}]}{0.212}$$

$$[HPO_4^{2-}] = 6.2 \times 10^{-8} = K_{a_2}$$

To find $[PO_4^{3-}]$,

$$K_{a_3} = \frac{[H^+][PO_4^{3-}]}{[HPO_4^{2-}]} \Rightarrow 4.8 \times 10^{-13} = \frac{(0.212)[PO_4^{3-}]}{6.2 \times 10^{-8}}$$

$$\mathbf{[PO_4^{3-}] = 1.4 \times 10^{-19}}\ \boldsymbol{M}$$

In Interactive Example 7.7 in your textbook some time is spent discussing the "fraction of species" ("*f*") for the components of carbonic acid. A slightly different form of the equations presented will give you the same results as in that example. (Try it!) It also works for Example 7.9 in your textbook. If we call the denominator of each of the equations below "*A*", so

$$\text{"}A\text{"} = [H^+]^2 + K_{a_1}[H^+] + K_{a_1}K_{a_2}$$

Then

$$f_{H_2CO_3} = \frac{[H^+]^2}{A} \qquad f_{HCO_3^-} = \frac{K_{a_1}[H^+]}{A} \qquad f_{CO_3^{2-}} = \frac{K_{a_1}K_{a_2}}{A}$$

Note how each species contributes to "A," and the total of the three fractions equals one. If we extend that to the phosphoric acid example, in which we have one more acid proton,

$$f_{PO_4^{3-}} = \frac{K_{a_1}K_{a_2}K_{a_3}}{[H^+]^3 + K_{a_1}[H^+]^2 + K_{a_1}K_{a_2}[H^+] + K_{a_1}K_{a_2}K_{a_3}}$$

$$= \frac{(7.5\times10^{-2})\times(6.2\times10^{-8})\times(4.8\times10^{-13})}{(0.212)^3 + (7.5\times10^{-3}\times[0.212]^2) + (7.5\times10^{-3}\times 6.2\times10^{-8}\times 0.212) + (7.5\times10^{-3}\times 6.2)}$$

$$= \frac{(2.23\times10^{-22})}{(9.53\times10^{-3}) + (3.37\times10^{-4}) + (9.86\times10^{-11}) + (2.23\times10^{-22})}$$

$$= \frac{2.23\times10^{-22}}{9.87\times10^{-3}}$$

$$\boldsymbol{f_{PO_4^{3-}} = 2 \times 10^{-20}}$$

We can find $[OH^-]$ in the usual way,

$$[OH^-] = K_w / [H^+] = 1.0 \times 10^{-14}/0.212$$

$$\mathbf{[OH^-] = 4.7 \times 10^{-14}}\ \boldsymbol{M}$$

Note the discussion regarding sulfuric acid along with Examples 7.10 and 7.11 in your textbook. Why does your textbook single out sulfuric acid for special consideration?

7.8 Acid-Base Properties of Salts

Salts are ionic compounds. They dissociate in water and may exhibit acid-base behavior. The **key question** in deciding whether a salt will act as an acidic, basic, or neutral species in solution is **"What are the acid-base properties, and strengths, of each component of the salt?"**

For example, sodium nitrite, $NaNO_2$, completely dissociates to give Na^+ and NO_2^- ions. These are the main species in solution in addition to H_2O. What are the acid-base properties of each of these species? Remember your conjugate acid-base relationships.

- Strong acids and bases have weak conjugates. Na^+ and other alkali and alkaline earth metals exhibit no acid-base properties.
- The nitrite ion, NO_2^-, is the conjugate base of the moderately weak acid HNO_2 ($K_a = 4.6 \times 10^{-4}$). Remember that "weak" is a relative term. HNO_2 is weak compared to HNO_3, which has a $K_a >> 1$. But HNO_2 is far stronger than HCN ($K_a = 6.2 \times 10^{-10}$). **The NO_2^- ion will act as a very weak base**. (Your textbook points out that bases compete for protons against OH^-!)
- Water will have relatively little acid-base effect. Therefore **an aqueous solution of $NaNO_2$ should be slightly basic**.

Your textbook proves the relationship that, for an aqueous solution,

$$K_w = K_a \times K_b$$

You will generally find the K_a for a weak acid or K_b for a weak base in chemical data tables (such as Tables 7.2, 3, and 4 in your textbook). You will have to **calculate K_a or K_b for the ion of a salt** from given values for the conjugate neutral acid or base.

Noting Table 7.6, you must decide how each part of the salt (cation and anion) will react with water. That table includes a sticky case, such as $NH_4C_2H_3O_2$, where **both cation and anion exhibit acid-base behavior**. When this occurs, the overall pH of the solution is determined by the relative K_a and K_b values for the ions.

Example 7.8 A - Predicting Acid-Base Behavior

Using data from Tables 7.2, 7.3, and 7.4 in your textbook, predict whether each of the following will create an acid, base, or neutral aqueous solution.

a. Na_3PO_4
b. KI
c. HC_5H_5NCl (pyridinium chloride)
d. NH_4F

Solution

a. Na^+ exhibits no acid-base behavior. PO_4^{3-} is the most basic form of H_3PO_4 ($K_{b_1} = K_w / K_{a_3}$). This solution will be a **fairly strong base**.

b. K^+ exhibits no acid-base behavior. I^- also exhibits no acid-base behavior. This solution will be **neutral**.

c. Cl^- exhibits no acid-base behavior. Pyridinium ion, $HC_5H_5N^+$, is the conjugate acid of pyridine, C_5H_5N ($K_a = K_w/K_b$). This solution will be **acidic**.

d. NH_4^+ is the conjugate acid of NH_3 ($K_a = K_w/K_b = 5.6 \times 10^{-10}$). F^- is the conjugate base of HF ($K_b = K_w/K_a = 1.4 \times 10^{-11}$). $K_a > K_b$; therefore the solution will be a **weak acid**.

Example 7.8 B - pH Of A Salt

Calculate the pH of a 0.500 *M* $NaNO_2$ solution. (K_a for $HNO_2 = 4.0 \times 10^{-4}$)

Solution

As discussed in the previous example, $NaNO_2$ will act as a base in water due to the hydrolysis of NO_2^-.

$$NO_2^-(aq) + H_2O(l) \rightleftharpoons HNO_2(aq) + OH^-(aq)$$

$$K_b = \frac{[HNO_2][OH^-]}{[NO_2^-]}$$

$$\mathbf{K_b} = K_w / K_a = 1.0 \times 10^{-14} / 4.0 \times 10^{-4} = \mathbf{2.5 \times 10^{-11}}$$

We may now proceed as with any other weak base problem.

	NO_2^-	$\rightleftharpoons$	HNO_2	+	OH^-
initial	0.500 *M*		0 *M*		0 *M*
change	$-x$		$+x$		$+x$
equilibrium	$0.500 - x$ ($\approx$0.500)		x		x

$$2.5 \times 10^{-11} = \frac{x^2}{0.500}$$

$$x = [OH^-] = [HNO_2] = 3.5 \times 10^{-6}$$

This clearly passes the 5% rule (K_b is so small)!

$$pOH = -\log(3.3 \times 10^{-6}) = 5.45$$

$$\mathbf{pH} = 14 - 5.45 = \mathbf{8.55}$$

7.9 Acid Solutions in Which Water Contributes to the H^+ Concentration

We have, up to this point, neglected the autoionization of water as a contributor to $[H^+]$ or $[OH^-]$. The two general cases in which autoionization **IS** important are the focuses of Sections 7.9 and 7.10. This section deals with the importance of autoionization when determining the pH of dilute solutions of very weak acids. The key to deriving the expression for $[H^+]$ is to understand the **material balance equation**,

$$[HA]_0 = [HA] + [A^-]$$

and the **charge balance equation**

$$[H^+] = [A^-] + [OH^-].$$

As you go through the derivation, note also the condition under which the expression reduces to the conventional weak acid expression for $[H^+]$.

Example 7.9 - pH Of A Dilute Weak Acid

Calculate the pH of a 3.0×10^{-5} *M* ammonium chloride solution (K_b for $NH_3 = 1.8 \times 10^{-5}$).

a. Neglecting the autoionization of water
b. Taking autoionization into account

Solution

a. Using the simple form,

$$K_a = \frac{[H^+]^2}{[NH_4^+]_0 - [H^+]}$$

Neglecting $[H^+]$ relative to $[NH_4^+]_0$,

$$K_a = \frac{[H^+]^2}{[NH_4^+]_0} \quad \text{or} \quad [H^+] = \sqrt{K_a[NH_4^+]_0}$$

$$\text{or} \quad [H^+] = (5.5 \times 10^{-10} \times 3.0 \times 10^{-5})^{1/2} = 1.3 \times 10^{-7}$$

pH = 6.89

b. The product of $K_a[NH_4^+]_o$ was **NOT** >> K_w; therefore we must include K_w in our expression.

$$[H^+] = \sqrt{K_a[NH_4^+]_0 + K_w} = [(5.5 \times 10^{-10}) \times (3.0 \times 10^{-5}) + (1.0 \times 10^{-14})]^{1/2}$$

$$[H^+] = 1.6 \times 10^{-7}$$

pH = 6.79

7.10 Strong Acid Solutions in Which Water Contributes to the H^+ Concentration

The best argument for including autoionization in these calculations is to calculate the pH of a 1×10^{-8} *M* HCl solution. Neglecting autoionization, we would conclude that the pH = 8!

Example 7.10 - pH Of A Very Dilute Strong Acid

Please calculate the pH of a 1.0×10^{-8} *M* HCl solution.

Solution

$$[Cl^-] = \frac{[H^+]^2 - K_w}{[H^+]}$$

Substituting,

$$1.0 \times 10^{-8} = \frac{[H^+]^2 - 1 \times 10^{-14}}{[H^+]}$$

Setting up the quadratic,

$$[H^+]^2 - 1.0 \times 10^{-8}\,[H^+] - 1.0 \times 10^{-14} = 0$$

$$[H^+] = \frac{-b \pm \sqrt{(b^2 - 4ac)}}{2a}$$

$$[H^+] = \frac{-(1.0 \times 10^{-8}) \pm \sqrt{1.0 \times 10^{-16} - 4(1)(-1.0 \times 10^{-14})}}{2}$$

$$[H^+] = \frac{-1.0\times10^{-8} \pm \sqrt{(4.01\times10^{-14})}}{2}$$

$$[H^+] = \frac{2.1\times10^{-7}}{2} = 1.05 \times 10^{-7}$$

$$\textbf{pH = 6.98}$$

7.11 Strategy for Solving Acid-Base Problems: A Summary

Your textbook makes the key point here that we cannot merely memorize which formula to use to solve a given acid-base problem. There are too many variations and possible conditions. When doing such problems, we do not merely solve, we **PROBLEM SOLVE**.

Exercises

Section 7.1

1. List four strong acids and four strong bases.

2. Write the dissociation reaction for each of the following acids in water. Identify the conjugate acid-base pair in each case.
   a. C_6H_5COOH (benzoic acid)
   b. H_3BO_3 (boric acid)
   c. $H_2PO_4^-$ (dihydrogen phosphate)
   d. HNO_3 (nitric acid)

3. Write the dissociation reaction for each of the following acids in water, and identify the conjugate acid-base pairs:
   a. $HC_2H_2ClO_2$
   b. HCN
   c. NH_4Cl

4. Write the equilibrium expressed for each of the reactions in Problem 2.

5. Write equilibrium expressions for each of the equations in Problem 3.

6. What is the conjugate base of the bicarbonate ion, HCO_3^- ? of formic acid, HCOOH? Which is the stronger base? Why?

7. Would the following salts act as acids or as bases or neither when dissolved in liquid H_2O?
   a. $NaNO_3$
   b. NaOH
   c. NaCN
   d. NaCl
   e. $NaNH_2$

Section 7.2

8. The values for K_a for the acids in Problem 2 are:

Substance	K_a
C_6H_5COOH	6.14×10^{-5}
H_3BO_3	5.83×10^{-10}
$H_2PO_4^-$	6.3×10^{-8}
HNO_3	$>>1$

Put the acids in order from strongest to weakest.

9. Predict which one of the bases in each pair is stronger.

   a. HCO_3^- or CO_3^{2-}
   b. NO_3^- or NO_2^-

10. Put the conjugate bases of the acids in Problem 8 in order from strongest to weakest.

11. a. Calculate the H^+ ion concentration in a solution that has an OH^- ion concentration of 1.0×10^{-5} moles/liter.
    b. Repeat (a) for $[OH^-] = 3.3 \times 10^{-10}$ moles/liter.
    c. Are the solutions described in (a) and (b) acidic, basic, or neutral?

Section 7.3

12. The pH of a solution is 11.93. What is $[H^+]$? $[OH^-]$? pOH?

13. For each of the following solutions at 25°C, calculate $[H^+]$ given $[OH^-]$ or $[OH^-]$ given $[H^+]$. Is the solution an acid or base?

    a. $[OH^-] = 1 \times 10^{-4}\ M$
    b. $[H^+] = 1 \times 10^{-6}\ M$
    c. $[H^+] = 1 \times 10^{-9}\ M$

14. A solution has $[OH^-] = 3.6 \times 10^{-1}\ M$. Is this solution strongly or weakly acidic or basic?

15. Calculate the pH for each of the following solutions at 25°C.

    a. $[H^+] = 3.8 \times 10^{-7}\ M$
    b. $[H^+] = 7.2 \times 10^{-4}\ M$
    c. $[H^+] = 4.1 \times 10^{-13}\ M$

16. Calculate the pOH for each of the following solutions at 25°C.

    a. $[OH^-] = 2.0 \times 10^{-12}\ M$
    b. $[OH^-] = 3.4 \times 10^{-11}\ M$
    c. $[OH^-] = 9.2 \times 10^{-3}\ M$

17. At 100°C, $K_w = 4.9 \times 10^{-13}$.

    a. What is the pH of a neutral solution at 100°C?
    b. Calculate the pH at 100°C if $[OH] = 6.3 \times 10^{-12}\ M$.

18. Calculate the pH and pOH for each of the following solutions at 25°C.

    a. $[H^+] = 2.0 \times 10^{-12}\ M$
    b. $[OH^-] = 3.9 \times 10^{-3}\ M$
    c. $[H^+] = 7.7 \times 10^{-13}\ M$
    d. $[OH^-] = 5.3 \times 10^{-9}\ M$

19. If $[Cl^-] = 0.9\ M$, what is pCl?

20. What is the pH of a solution that has an H^+ ion concentration of 1.0×10^{-5} mol/L? What if $[H^+] = 5.0 \times 10^{-5}\ M$?

Section 7.4

21. Calculate the pH of a $7.0 \times 10^{-2}\ M$ HCl solution.

22. Calculate the pH of a $2.8 \times 10^{-5}\ M$ HNO_3 solution.

23. Fill in the missing information in the following table:

	pH	pOH	$[H^+]$	$[OH^-]$	Acid, base, or neutral?
solution a	5.64				
solution b				3.9×10^{-6} *M*	
solution c			0.027 *M*		
solution d		1.7			

24. The pH of a solution of $HClO_4$ is 3.11. What is $[H^+]$?

25. The pOH of a 400-mL solution of HNO_3 is 12.44. How many grams of HNO_3 are in the solution?

26. If 0.10 mol of HCl is added to enough water to produce 1.0 L of solution, calculate the concentrations of H^+ and OH^- and the pH of the solution.

27. One liter of solution was prepared from water and 3.5×10^{-6} mol of HCl. Calculate the $[H^+]$ and $[OH^-]$ and the pH of the solution.

28. The gastric juice in our stomachs contains enough hydrochloric acid to make the hydrogen ion concentration about 0.01 mol/L. Calculate the approximate pH of gastric juice.

Section 7.5

29. A solution is made by dissolving 18.4 g of HNO_3 in enough water to make 662 mL of solution. Calculate the pH of the solution.

30. The K_a of chloroacetic acid ($ClCH_2COOH$) is 1.36×10^{-3}. Calculate the pH, the pOH, the $[H^+]$, and the $[OH^-]$ of a 1.00 *M* solution of chloroacetic acid.

31. What is the amount of hydrogen ion due to water in Problem 30? What is the percent dissociation of chloroacetic acid?

32. Calculate the pH of a 0.237 *M* solution of benzoic acid, C_6H_5COOH ($K_a = 6.14 \times 10^{-5}$).

33. A total of 0.0560 g of acetic acid is added to enough water to make 50 mL of solution. Calculate $[H^+]$, $[CH_3COO^-]$, $[CH_3COOH]$, and the pH at equilibrium. (K_a for acetic acid is 1.8×10^{-5}.)

34. Calculate K_a of a weak acid, HW, if a solution with an initial concentration of 0.200 *M* has a pH of 3.15.

35. Calculate the K_a of a 0.060 *M* weak monoprotic acid with a pH of 3.44.

36. Calculate the pH of a solution that contains 0.250 *M* H_2SO_4 and 1.00 *M* CH_3COOH ($K_a = 1.8 \times 10^{-5}$).

37. Calculate the original molarity of a solution of formic acid, HCOOH ($K_a = 1.9 \times 10^{-4}$), whose pH is 3.26 at equilibrium.

38. Calculate the pH of a solution that contains 1.23 *M* benzoic acid, C_6H_5COOH ($K_a = 6.4 \times 10^{-5}$), and 0.713 *M* arsenious acid, H_3AsO_3 ($K_a = 6.0 \times 10^{-10}$). What is the concentration of the benzoate ion (CH_3COO^-), hydroxide ion (OH^-), and the arsenite ion ($H_2AsO_3^-$)?

39. What is the percent dissociation of the benzoic acid in Problem 38?

40. Calculate the percent dissociation of 0.20 *M* benzoic acid whose $K_a = 6.4 \times 10^{-5}$.

41. Calculate the pH of a 0.20 *M* NH_4Cl solution ($K_a = 5.6 \times 10^{-10}$).

42. Calculate the H^+ and OH^- ion concentrations in a 0.010 *M* HCN solution.

43. What are the concentrations of species and the percent dissociation in 0.10 *M* $HC_2H_3O_2$?

Section 7.6

44. The pH of a 300-g solution of NaOH is 12.97. The density of the solution is 1.10 g/mL. How many grams of NaOH are in the solution?

45. Add 0.0150 mol of LiOH to sufficient water to make 1 L of solution. Calculate the $[LiOH]_0$, $[H^+]$, and $[OH^-]$, and the pH of the solution.

46. How many grams of KOH are necessary to prepare 800 mL of a solution of pH 11.56?

47. How many grams of $Ba(OH)_2$ are necessary to prepare 400 mL of a solution of pH 12.46?

48. Calculate the pH of a 2.8×10^{-4} *M* $Ba(OH)_2$ solution, assuming complete dissociation of the $Ba(OH)_2$.

49. How many grams of NaOH are needed to prepare a 546-mL solution with a pH of 10.00?

50. The pH of a 0.30 *M* solution of a weak base is 10.66. Calculate the K_b of the base.

51. A solution of ammonia, $K_b = 1.8 \times 10^{-5}$, has a pH of 11.22. Calculate the molarity of the solution.

52. Calculate the pH of a solution made by putting 11.17 g of KOH into water and diluting it to a volume of 600 mL.

53. Calculate the pH of a 0.500 *M* solution of dimethylamine, $(CH_3)_2NH$ ($K_b = 5.9 \times 10^{-4}$).

54. What is the percent of hydrolysis of the base in Problem 53? What is the concentration of hydrogen as a result of the autoionization of water?

55. Calculate the pH of a 0.76 *M* KOH solution.

56. Calculate how many grams of $HONH_2$ are needed to dissolve in enough water to make 250 mL of solution with a pH of 10.00. ($K_b = 1.1 \times 10^{-8}$)

57. Calculate the pH of a 0.15 *M* solution of CH_3COONa ($K_b = 5.6 \times 10^{-10}$).

58. Calculate the H^+ and OH^- ion concentrations in each of the following strong electrolyte solutions:

    a. 0.0050 *M* HNO_3
    b. 0.0050 *M* KOH

Section 7.7

59. What is the pH of a 0.100 *M* solution of arsenic acid?

$$H_3AsO_4 \rightleftharpoons H_2AsO_4^- + H^+ \qquad K_a = 6.0 \times 10^{-3}$$
$$H_2AsO_4^- \rightleftharpoons HAsO_4^{2-} + H^+ \qquad K_a = 1.05 \times 10^{-7}$$
$$HAsO_4^{2-} \rightleftharpoons AsO_4^{3-} + H^+ \qquad K_a = 3.0 \times 10^{-12}$$

    What is the concentration of AsO_4^{3-} ion? (Hint: Use the quadratic equation.)

60. Calculate the concentrations of CO_3^{2-}, HCO_3^-, and H^+ in a 0.025 *M* H_2CO_3 solution. (See Table 7.4 in your text for K_a values of carbonic acid.)

61. Calculate the equilibrium concentration of PO_4^{3-} in a solution prepared by adding 716.2 g of H_3PO_4 to water and diluting to a total volume of 750 mL. (See Table 7.4 in your textbook for K_a values of phosphoric acid.)

62. Calculate the concentrations of H_2CO_3, HCO_3^-, CO_3^{2-}, H^+, and OH^- and the pH at equilibrium in a solution that is initially 0.010 *M* H_2CO_3.

63. Calculate the concentrations of $C_2H_2O_4$, $C_2HO_4^-$, $C_2O_4^{2-}$, and H^+ in a 0.10 *M* of oxalic acid solution.

Section 7.8

64. Using data from Table 7.2 in your textbook, determine the values for K_b of

 a. CN^-
 b. NH_3
 c. $C_6H_5O^-$

65. Arrange the bases in Problem 64 in order from strongest to weakest.

66. Calculate the pH of a 0.450 *M* solution of sodium propionate, $NaOCH_2CH_3$ (K_a for propionic acid, $HOCH_2CH_3$, is 1.34×10^{-5}).

67. Calculate the pH of a 0.20 *M* K_2S solution (K_{a_2} for $H_2S = 1 \times 10^{-13}$).

68. Determine the equilibrium concentration of HCN in a 0.0500 *M* NaCN solution (K_a for HCN is 6.2×10^{-10}).

69. Predict whether an aqueous solution of ammonium acetate (CH_3COONH_4) will be acidic, basic, or neutral.

70. For 1 *M* solutions of each of the following, state whether the solution has a pH above 7, below 7, or approximately equal to 7. Also state whether the solution is acidic, basic, or neutral.

 a. $NaHCO_3$ c. HNO_3 e. Na_2S
 b. $CuCl_2{\bullet}4H_2O$ d. NH_4NO_3 f. KOH

Multiple-Choice Self-Test

1. The conjugate acid of NH_3 is:

 A. NH_2^- B. NHOH C. NH_4^+ D. NH_2^+

2. Which one of the following acids would produce the weakest conjugate base?

 A. sulfuric acid B. ammonium ion C. phenol D. acetic acid

3. What is K_b if K_a of an acid is 2.0×10^{-3}?

 A. 5.0×10^{11} B. 1×10^{-3} C. 2.0×10^{-11} D. 5.0×10^{-12}

4. The pOH of a 2.0 *M* HCl solution is:

 A. 12 B. 14 C. 13.7 D. 0.3

5. The pOH of a 0.300 *M* HI solution is:

 A. 13.5 B. 0.522 C. 0.150 D. 3.5

6. What is the final pH of a sulfuric acid solution prepared by mixing together 392.0 g of H_2SO_4 with enough water to reach 2.0 liters? After it is diluted by a factor of 15?

 A. −0.30, 0.88 B. 2.0, 3.13 C. 12, 12.3 D. 0.30, 4.5

7. 0.276 g of HCN is dissolved in 50.0 mL of solution. What is the pH of this acidic solution?

 A. 2.0 B. 4.7 C. 5.0 D. 6.5

8. A 0.50 *M* HX solution is 0.30% ionized. What is its pH?

 A. 0.30 B. 0.0016 C. 0.65 D. 2.8

9. What is the percent ionization of a 0.55 M HW solution with a $K_a = 1.5 \times 10^{-2}$?
   A. 15.0% B. 30.0% C. 0.30% D. 65.0%

10. What is the pOH of a 0.50 ethylamine ($C_2H_5NH_2$) solution? $pK_b = 5.3$.
   A. 11.2 B. 2.8 C. 0.28 D. 0.0016

11. Calculate the pH of a 0.10 M H_2S solution. $K_{a_1} = 1.3 \times 10^{-7}$, $K_{a_2} = 1.0 \times 10^{-13}$.
   A. 4.2 B. 6.1 C. 3.9 D. 1.1

12. What is the pH of a 0.1 M phosphoric acid solution? $K_{a_1} = 7.5 \times 10^{-3}$, $K_{a_2} = 6.2 \times 10^{-8}$, $K_{a_3} = 4.8 \times 10^{-13}$.
   A. 4.0 B. 1.5 C. 2.0 D. 1.0

13. Which of the following salts would produce an acidic solution?
   A. NaCl B. CH_3NH_3Cl C. CH_3NHNa D. $NaNO_3$

14. Which one of the following substances is most acidic?
   A. CH_3Cl B. CH_2ClO C. H_3CCOOH D. HI

15. Which one of the following oxides is acidic?
   A. CaO B. K_2O C. SO_2 D. MgO

16. A solution prepared by mixing water with an oxide of which one of the following groups of elements would yield an acid?
   A. Group IA B. Group IB C. Group VIA D. Group VIIA

Answers to Exercises

1. Acids: HCl, HNO_3, $HClO_4$, H_2SO_4. Bases: NaOH, KOH, $Ba(OH)_2$, LiOH.

2.

	acid 1		base 2		base 1		acid 2
a.	C_6H_5COOH	+	H_2O	$\rightleftharpoons$	$C_6H_5COO^-$	+	H_3O^+
b.	H_3BO_3	+	H_2O	$\rightleftharpoons$	$H_2BO_3^-$	+	H_3O^+
c.	$H_2PO_4^-$	+	H_2O	$\rightleftharpoons$	HPO_4^{2-}	+	H_3O^+
d.	HNO_3	+	H_2O	$\rightleftharpoons$	NO_3^-	+	H_3O^+

3.

	acid 1		base 2		base 1		acid 2
a.	$HC_2H_2ClO_2(aq)$	+	$H_2O(l)$	$\rightleftharpoons$	$C_2H_2ClO_2^-(aq)$	+	$H_3O^+(aq)$
b.	$HCN(aq)$	+	$H_2O(l)$	$\rightleftharpoons$	$CN^-(aq)$	+	$H_3O^+(aq)$
c.	$NH_4Cl(aq)$	+	$H_2O(l)$	$\rightleftharpoons$	$NH_3Cl^-(aq)$	+	$H_3O^+(aq)$

4. a. $K = \dfrac{[C_6H_5COO^-][H^+]}{[C_6H_5COOH]}$ c. $K = \dfrac{[HPO_4^{2-}][H^+]}{[H_2PO_4^-]}$

   b. $K = \dfrac{[H_2BO_3^-][H^+]}{[H_3BO_3]}$ d. $K = \dfrac{[NO_3^-][H^+]}{[HNO_3]}$ (K>>1)

5. a. $K_a = \dfrac{[C_2H_2ClO_2^-][H^+]}{[HC_2H_2ClO_2]}$ b. $K_a = \dfrac{[CN^-][H^+]}{[HCN]}$ c. $K_a = \dfrac{[NH_3Cl^-][H^+]}{[NH_4Cl]}$

6. Bicarbonate ion: HCO_3^- Conjugate base: CO_3^{2-}
Formic acid: $HCHO_2$ Conjugate base: HCO_2^-

By looking at the pK_a's of each, this can be determined. The higher the pK_a, the weaker the acid. The weaker the acid, the stronger the base. pK_a formic acid = 3.74, pK_a bicarbonate ion = 10.32; therefore bicarbonate ion is a weaker acid and a stronger base than formic acid.

7. a. neutral c. base e. base
b. strong base d. neutral

8. HNO_3, C_6H_5COOH, $H_2PO_4^-$, H_3BO_3
strongest ⟵⟶ weakest

9. a. CO_3^{2-} b. NO_2^-

10. $H_2BO_3^-$, HPO_4^{2-}, $C_6H_5COO^-$, NO_3^-
strongest ⟵⟶ weakest

11. a. $[H^+] = 1.0 \times 10^{-9}\ M$
b. $[H^+] = 3.03 \times 10^{-5}\ M$
c. To determine the acidity or basicity, use the conversion: pH = −log [H^+]. A pH greater than 7 indicates the solution is basic. A pH less than 7 indicates the solution is acidic. The solution in part (a) has a pH of 9.0 and is basic. The solution in part (b) has a pH of 4.5 and is basic.

12. $[H^+] = 1.2 \times 10^{-12}\ M$, $[OH^-] = 8.5 \times 10^{-3}\ M$, pOH = 2.07

13. a. $[H^+] = 1 \times 10^{-10}$, the solution is basic.
b. $[OH^-] = 1 \times 10^{-8}$, the solution is acidic.
c. $[OH^-] = 1 \times 10^{-5}$, the solution is basic.

14. strongly basic

15. a. pH = 6.42 b. pH = 3.14 c. pH = 12.39

16. a. pOH = 11.70 b. pOH = 10.47 c. pOH = 2.04

17. a. pH = 6.15 b. pH = 1.11

18. a. pH = 11.70, pOH = 2.30 c. pH = 12.11, pOH = 1.89
b. pH = 11.59, pOH = 2.41 d. pH = 5.72, pOH = 8.28

19. pCl = 0.05

20. a. pH = 5.00 b. pH = 4.30

21. pH = 1.15

22. pH = 4.55

23.

	pH	pOH	[H^+]	[OH^-]	Acid, base, or neutral?
solution a	5.64	8.36	$2.3 \times 10^{-6}\ M$	$4.4 \times 10^{-9}\ M$	acid
solution b	8.59	5.41	$2.6 \times 10^{-9}\ M$	$3.9 \times 10^{-6}\ M$	base
solution c	1.57	12.43	$0.027\ M$	$3.7 \times 10^{-13}\ M$	acid
solution d	12.3	1.7	$5. \times 10^{-13}\ M$	$2. \times 10^{-2}\ M$	base

24. $[HClO_4] = 7.8 \times 10^{-4}\ M$

25. 0.69 g HNO_3

26. $[H^+] = 0.10\ M$, $[OH^-] = 1.0 \times 10^{-13}\ M$, pH = 1.00
27. $[H^+] = 3.5 \times 10^{-6}\ M$, $[OH^-] = 2.9 \times 10^{-9}\ M$, pH = 5.45
28. pH = 2.0
29. pH = 0.355
30. pH = 1.43, pOH = 12.57, $[OH^-] = 2.7 \times 10^{-13}\ M$, $[H^+] = 3.7 \times 10^{-2}\ M$
31. $[H^+]_{H_2O} = [OH^-]_{total} = 2.7 \times 10^{-13}\ M$, 3.7%
32. pH = 2.42
33. $[H^+] = [CH_3COO^-] = 5.8 \times 10^{-4}\ M$, $[CH_3COOH] = 0.018\ M$, pH = 3.24
34. $K_a = 2.5 \times 10^{-6}$
35. $K_a = 2.2 \times 10^{-6}$
36. pH = 0.53
37. $1.6 \times 10^{-3}\ M$
38. pH = 2.05, $[C_6H_5COO^-] = [H^+] = 8.9 \times 10^{-3}\ M$, $[OH^-] = 1.1 \times 10^{-12}\ M$, $[H_2AsO_3^-] = 4.92 \times 10^{-8}\ M$
39. percent dissociation = 0.72%
40. 1.8%
41. pH = 4.98
42. $[OH^-] = 4.55 \times 10^{-9}\ M$, $[H^+] = 2.2 \times 10^{-6}\ M$
43. $[H^+] = [C_2H_3O_2^-] = 1.3 \times 10^{-3}\ M$; $[HC_2H_3O_2] = 0.10\ M$; % dissociation = 1.3%
44. 1.02 g NaOH
45. $[LiOH]_0 = [OH^-] = 0.015$, $[H^+] = 6.7 \times 10^{-13}\ M$, pH = 12.18
46. 0.16 g KOH
47. 1.0 g $Ba(OH)_2$
48. pH = 10.75
49. 2.2×10^{-3} g
50. 7.0×10^{-7}
51. 0.15 M
52. pH = 13.52
53. pH = 12.23
54. percent hydrolysis = 3.4%, $[H^+]_{H_2O} = [H^+]_{total} = 5.9 \times 10^{-13}\ M$
55. pH = 13.88
56. 7.5 g
57. pH = 8.96
58. a. $[OH^-] = 2.0 \times 10^{-12}\ M$, $[H^+] = 0.0050\ M$
    b. $[OH^-] = 0.0050\ M$, $[H^+] = 2.0 \times 10^{-12}$ M

59. pH = 1.66, $[AsO_4^{3-}] = 1.5 \times 10^{-17}\ M$

60. $[CO_3^{2-}] = 5.6 \times 10^{-11}\ M$, $[HCO_3^-] = [H^+] = 1.0 \times 10^{-4}\ M$

61. Calculate $[H^+]$, and work your way down: $[H^+] = 0.27$, (pH = 0.57); $[PO_4^{3-}] = 1.1 \times 10^{-19}\ M$

62. $[H_2CO_3] = 0.010\ M$, $[HCO_3^-] = [H^+] = 6.6 \times 10^{-5}\ M$, $[CO_3^{2-}] = 5.6 \times 10^{-11}\ M$,
$[OH^-] = 1.5 \times 10^{-10}\ M$, pH = 4.18

63. $[C_2H_2O_4] = 0.046\ M$, $[C_2HO_4^-] = [H^+] = 0.054\ M$, $[C_2O_4^{2-}] = 6.1 \times 10^{-5}\ M$

64. a. $K_b = 1.6 \times 10^{-5}$ b. $K_b = 1.8 \times 10^{-5}$ c. $K_b = 6.3 \times 10^{-5}$

65. $C_6H_5O^- > NH_3 > CN^-$

66. pH = 9.26

67. pH = 13.0

68. $[HCN] = 9.0 \times 10^{-4}\ M$

69. neutral (pH = 7.0)

70. a. basic, >7
b. neutral, =7
c. acidic, <7
d. acidic, <7
e. basic, >7
f. basic, >7

Answers to Multiple-Choice Self-Test

1. C	4. C	7. C	10. B	13. B	15. C
2. A	5. A	8. D	11. C	14. D	16. C
3. D	6. A	9. A	12. B		

CHAPTER 8

Applications of Aqueous Equilibria

The Bottom Line: Chapter 8

This chapter covers a great deal of material. The beauty is that **you already have the knowledge to handle the material.** Very little is "new." Most of the material represents **applications of equilibrium** theory.

8.1 Solutions of Acids or Bases Containing a Common Ion

The important concept in this section is how the addition of a **common ion** to a solution can depress the dissociation of an acid. This is based on Le Châtelier's principle. To illustrate the point, acetic acid ($HC_2H_3O_2$) dissociates in water as follows:

$$HC_2H_3O_2(aq) \xrightleftharpoons{K_a} H^+(aq) + C_2H_3O_2^-(aq)$$

Let us add sodium acetate ($NaC_2H_3O_2$), which completely dissociates to form $Na^+(aq)$ and $C_2H_3O_2^-(aq)$. The acetate ion, $C_2H_3O_2^-$, is *common* to the solution. (See the above equation.) It is called the *common ion. The common ion imposes a stress on the system.* The system responds by shifting **to the left**. *Therefore there is much less dissociation than before.* The general effect of a common ion is to depress the **amount of dissociation.**

Because the dissociation is depressed, we can assume that the equilibrium concentration of acetic acid equals the initial concentration. The same is true of the acetate ion.

Let's see how this applies to the pH of a weak acid.

Example 8.1 - The Common Ion Effect

Calculate the pH and the percent dissociation of the acid in each of the following solutions.

a. 0.200 M $HC_2H_3O_2$ ($K_a = 1.8 \times 10^{-5}$)
b. 0.200 M $HC_2H_3O_2$ in the presence of 0.500 M $NaC_2H_3O_2$

Solution

a. The major species are acetic acid, $\mathbf{HC_2H_3O_2}$, and $\mathbf{H_2O}$. As we saw in the previous chapter, K_w for the autoionization of water is insignificant compared to the K_a for acetic acid dissociation. We need to calculate based only on the acid dissociation.

$$HC_2H_3O_2(aq) \rightleftharpoons H^+(aq) + C_2H_3O_2^-(aq)$$

$$K_a = \frac{[H^+][C_2H_3O_2^-]}{[HC_2H_3O_2]}$$

Using techniques from Chapter 7,

$$1.8 \times 10^{-5} = \frac{x^2}{0.200}$$

$$x = [H^+] = [C_2H_3O_2^-] = \mathbf{1.9 \times 10^{-3}}\ M, \quad \mathbf{pH = 2.72}$$

$$\textbf{\% dissociation} = \frac{1.9\times10^{-3}}{0.200} \times 100\% = \mathbf{0.95\%}$$

b. **The difference here is that the presence of the acetate ion, $C_2H_3O_2^-$,** ($NaC_2H_3O_2$ is a strong electrolyte, as discussed previously) will *depress the dissociation of $HC_2H_3O_2$* as predicted by Le Châtelier's principle.

We can assume (and test) that because there is so little dissociation,

$$[HC_2H_3O_2]_0 \approx [HC_2H_3O_2] = 0.200\ M$$

and

$$[C_2H_3O_2^-] \approx [C_2H_3O_2^-] = 0.500\ M$$

This makes our problem remarkably easy to solve. Only $[H^+]$ is unknown,

$$K_a = \frac{[H^+][C_2H_3O_2^-]}{[HC_2H_3O_2]} \quad \Rightarrow \quad 1.8 \times 10^{-5} = \frac{[H^+](0.500)}{0.200}$$

$$\mathbf{[H^+] = 7.2 \times 10^{-6}\ M, \quad pH = 5.14}$$

Note that the **pH increased markedly** due to the presence of the **basic salt**. The extent of dissociation is now reflected by $[H^+]$ because the only significant source of H^+ ions is the dissociation of $HC_2H_3O_2$.

$$\textbf{\% dissociation} = \frac{7.2\times10^{-6}}{0.200} \times 100\% = \mathbf{3.6 \times 10^{-3}\ \%}$$

Therefore the dissociation was substantially depressed, and our assumption was very good.

8.2 Buffered Solutions

A buffered solution resists change in its pH upon the addition of a strong acid or strong base.

The last section set us up for doing buffer problems by introducing us to the common ion effect on equilibrium. We saw that the **extent of equilibrium is depressed** by the addition of a common ion.

A buffered solution generally contains a weak acid and its salt or a weak base and its salt.

Try the following buffer calculation, and think about

1. the common ion effect
2. the dominant equilibrium

Example 8.2 A - pH Of A Buffered Solution

Calculate the pH of a solution that contains 0.250 *M* formic acid, HCOOH ($K_a = 1.8 \times 10^{-4}$), and 0.100 *M* sodium formate, NaCOOH.

Solution

The major species in solution are **HCOOH, Na^+, $COOH^-$, and H_2O.**

Na^+ has no acid-base properties.

H_2O has $K_w = 1.0 \times 10^{-14}$ and is thus weak relative to HCOOH

$COOH^-$ has a $K_b = K_w/K_a = 5.6 \times 10^{-11}$. There will be very little hydrolysis of $COOH^-$. It will be depressed further by the common ion effect.

HCOOH has $K_a = 1.8 \times 10^{-4}$. Although this *acid dissociation is depressed* by the presence of the common ion, $COOH^-$, *it will be the dominant equilibrium.*

$$HCOOH(aq) \rightleftharpoons H^+(aq) + COOH^-(aq)$$

	HCOOH	⇌	H^+	+	$COOH^-$
initial	0.250 *M*		0 *M*		0.100 *M*
change	$-x$		$+x$		$+x$
equilibrium	$0.250 - x$ (≈0.250 *M*)		x		$0.100 + x$ (≈0.100 *M*)

As shown in our previous chapter, it is a virtual certainty that we can ignore x relative to the initial concentrations of HCOOH and $COOH^-$. We will test this with the 5% rule.

Keep in mind that buffer problems are among the easiest to solve because the **initial and final concentrations** of the acid and conjugate bases **do not change significantly**, and they are usually known.

$$K_a = \frac{[H^+][COOH^-]}{[HCOOH]} \quad \Rightarrow \quad 1.8 \times 10^{-4} = \frac{[H^+](0.100)}{0.250}$$

$$\mathbf{[H^+] = 4.5 \times 10^{-4}\ \mathit{M}, \qquad pH = 3.35}$$

The actual percent dissociation of HCOOH, which is equal to $[H^+]$, is small enough to easily pass the 5% rule.

The real utility of buffers comes when we add a strong acid or base. A buffered solution should *resist pH change* when a substantial amount of H^+ or OH^- ions are added. ("Substantial amount" will vary from solution to solution depending upon the **buffer capacity**. We will consider that in Section 8.3.)

If we add a strong base to a solution that contains a weak acid, the reaction is complete.

$$\mathbf{HA(\mathit{aq}) + OH^-(\mathit{aq}) \rightarrow A^-(\mathit{aq}) + H_2O}$$

Similarly, adding a strong acid to a weak base gives a complete reaction,

$$\mathbf{A^-(\mathit{aq}) + H^+(\mathit{aq}) \rightarrow HA(\mathit{aq})}$$

Notice that, in these equations, unless **excess** strong acid or base is added, **NO STRONG ACID OR BASE BUILDS UP.** Only weak species remain. This is why a buffer works so well.

Example 8.2 B - Practice With Buffers

A solution is prepared by adding 31.56 g NaCN and 22.30 g HCN to 600.0 mL of water (K_a for HCN = 6.2×10^{-10}).

a. What is the pH of this solution?
b. What is the pH after the addition of 50.0 mL of 3.00 *M* HCL?
c. What is the pH after a **further** addition of 80.0 mL of 4.00 *M* NaOH?

Solution

We have the following major species in solution: **Na^+, H_2O, HCN,** and **CN**:

HCN is a weak acid, $\mathbf{K_a = 6.2 \times 10^{-10}}$

CN^- its conjugate base, has $K_b = K_w/K_a = 1.0 \times 10^{-14}/6.2 \times 10^{-10} = \mathbf{1.6 \times 10^{-5}}$

H_2O with its low K_w, does not affect the acid-base characteristics of the system

Because CN^- is the **strongest acid-base substance in our solution**, its behavior will dominate. The base hydrolysis reaction is

$$\mathbf{CN^-(aq) + H_2O(l) \rightleftharpoons HCN(aq) + OH^-(aq)}$$

and

$$K_b = \frac{[HCN][OH^-]}{[CN^-]}$$

As always in problems involving buffered solutions, we can make our usual assumption regarding the lack of reaction of CN^- due to the presence of the common species, CN^- (in the form of HCN).

After calculating the pH of this solution (part a), *we need to assess the pH after addition of HCl*, a strong acid, for *part b*. In general terms, *the pH should go down, but only a small amount* because we have a buffer. The exception to this is if we *exceed* the buffer capacity. We must use our *two-part approach*, taking into account the initial, complete reaction of CN^- with H^+ (from HCl).

$$\mathbf{CN^-(aq) + H^+(aq) \rightarrow HCN(aq)}$$

As long as we have some CN^- ion left, we will maintain our buffered solution. We may then use our *equilibrium expression* to assess pH (via pOH).

In part c, we must consider the addition of a strong base, OH^- ion, to the system. It will react completely with our weak acid, HCN, to form CN^-.

$$\mathbf{HCN(aq) + OH^-(aq) \rightarrow CN^-(aq) + H_2O(l)}$$

As long as we have some HCN left, we will maintain a buffered solution. *The pH of the solution should rise, but not much,* if we maintain our buffer. We may use our *equilibrium expression* to determine the pOH, and then pH.

a. Let's calculate [HCN] and $[CN^-]$.

$$\mathbf{[HCN]} = \frac{\text{mol HCN}}{\text{L}} = \frac{1\text{ mol}}{27.0\text{ g}} \times \frac{22.30\text{ g}}{0.600\text{ L}} = \mathbf{1.38\ M}$$

$$\mathbf{[CN^-]} = \frac{\text{mol NaCN}}{\text{L}} = \frac{1\text{ mol}}{49.0\text{ g}} \times \frac{31.56\text{ g}}{0.600\text{ L}} = \mathbf{1.07\ M}$$

Using the equilibrium expression for the buffered solution, as discussed above,

$$K_b = \frac{[HCN][OH^-]}{[CN^-]} \quad \Rightarrow \quad 1.6 \times 10^{-5} = \frac{1.38[OH^-]}{1.07}$$

$$\mathbf{[OH^-] = 1.25 \times 10^{-5}\ M, \quad pOH = 4.90, \quad pH = 9.10}$$

b. **Stoichiometry Part**

mmoles $\mathbf{HCN_{initial}}$ = 1.38 mmol/mL × 600 mL = **828 mmol**
mmoles $\mathbf{CN^-_{initial}}$ = 1.07 mmol/mL × 600 mL = **642 mmol**
mmoles $\mathbf{H^+}$ added = 3.00 mmol/mL × 50.0 mL = **150 mmol**

	$CN^-(aq)$	+	$H^+(aq)$	→	$HCN(aq)$
initial (mmol)	642		150		828
final (mmol)	492		≈0		978

$$[CN^-] = 492 \text{ mmol}/(600 + 50) \text{ mL} = \mathbf{0.757}\ M$$

$$[HCN] = 978 \text{ mmol}/650 \text{ mL} = \mathbf{1.50}\ M$$

Equilibrium Part (Use the **equilibrium** expression.)

$$K_b = \frac{[HCN][OH^-]}{[CN^-]} \quad \Rightarrow \quad 1.6 \times 10^{-5} = \frac{1.50[OH^-]}{0.757}$$

$$\mathbf{[OH^-] = 8.1 \times 10^{-6}\ M, \quad pOH = 5.09, \quad pH = 8.91}$$

c. **Stoichiometry Part** (from part b)

$$\text{mmol } HCN_{initial} = 978 \text{ mmol}$$
$$\text{mmol } CN^-_{\ initial} = 492 \text{ mmol}$$

We are adding 80.0 mL × 4.00 mmol H^+/mL = **320 mmol OH^-.** This will react with HCN, as discussed in our strategy section.

	$HCN(aq)$	+	$OH^-(aq)$	→	$CN^-(aq)$	+	$H_2O(l)$
initial (mmol)	978		320		492		
final (mmol)	658		≈0		812		

$$[HCN] = 658 \text{ mmol}/(650 + 80) \text{ mL} = \mathbf{0.901}\ M$$
$$[CN^-] = 812 \text{ mmol}/730 \text{ mL} = \mathbf{1.11}\ M$$

Equilibrium Part (Use the **equilibrium** expression.)

$$K_b = \frac{[HCN][OH^-]}{[CN^-]} \quad \Rightarrow \quad 1.6 \times 10^{-5} = \frac{9.01[OH^-]}{1.11}$$

$$\mathbf{[OH^-] = 2.0 \times 10^{-5}\ M, \quad pOH = 4.71, \quad pH = 9.29}$$

Does the Answer Make Sense?

The initial pH was basic (9.10). This makes sense because CN^- is a stronger base than HCN is an acid.

The pH dropped (to 8.91) after addition of a strong acid. This makes sense.

The pH rose (to 9.30) after the subsequent addition of a strong base. This makes sense. In neither part b nor part c did we exceed our buffer capacity, so the pH changes were not drastic.

Example 8.2 C - pH And pKa

The K_a of propionic acid, $HC_3H_5O_2$, is 1.34 × 10^{-5} (K_a = 4.87). What is the pH when $[HC_3H_5O_2] = [C_3H_5O_2^-]$?

Solution

Let's look at the problem from a strictly mathematical point of view. If we use the equation for the acid dissociation of propionic acid,

$$\mathbf{HC_3H_5O_2(aq) \rightleftharpoons C_3H_5O_2^-(aq) + H^+(aq)}$$

$$K_a = \frac{[C_3H_5O_2^-][H^+]}{[HC_3H_5O_2]}$$

If $[HC_3H_5O_2] = [C_3H_5O_2^-]$, $\mathbf{K_a = [H^+] = 1.34 \times 10^{-5}}$ ***M*** and **pK_a = pH**; therefore **pH = 4.87.**

If we use the equation for the base hydrolysis of the propionate ion, $C_3H_5O_2^-$,

$$C_3H_5O_2^-(aq) + H_2O(l) \rightleftharpoons HC_3H_5O_2(aq) + OH^-(aq)$$

$$K_b = \frac{[HC_3H_5O_2][OH^-]}{[C_3H_5O_2^-]}$$

If $[C_3H_5O_2^-] = [HC_3H_5O_2]$,

$$[OH^-] = K_b = K_w/K_a = 1.0 \times 10^{-14}/1.34 \times 10^{-5} = \mathbf{7.46 \times 10^{-10}}$$

$$pOH = 9.13, \qquad \mathbf{pH = 4.87}$$

Although we normally would use the K_a expression to solve the pH in this situation, two important points need to be made from this problem:

1. **If [HA] = [A⁻], then pH = pK_a, and pOH = pK_b.**
2. In buffer problems **uniquely, either the K_a or K_b expressions can be used,** and the same pH will result. This is because **BOTH [HA] and [A⁻] are explicitly known.**

8.3 Exact Treatment of Buffered Solutions

Your textbook deals with the importance of water as a contributor of $[H^+]$ to the calculation of pH of buffered solutions. The overarching question of interest is, "How important is it to include *all* of the equilibria that occur in the expression for the pH of a buffered solution?"

1. What does the textbook conclude about the overarching question?
2. To what concentration is the conclusion valid?
3. Can buffers present at this concentration be effective?
4. Is it important to know how to derive the expressions introduced in this chapter? Why or why not?

8.4 Buffer Capacity

Your textbook emphasizes the two important measures involved in buffered solutions.

1. The pH is determined by the **ratio** of [HA]/[A⁻].
2. The **buffer capacity** is determined by the **magnitudes** of [HA] and [A⁻]. The more you have, the more strong acid or base can be neutralized.

Example 8.5 in your textbook compares the pH change brought about by adding a strong acid to solutions with different buffering capacities. Let's try a problem where we **exceed** the buffering capacity of a solution.

Example 8.4 A - Buffering Capacity

Calculate the pH of a 0.500-L solution that contains 0.15 *M* HCOOH (K_a = 1.8 × 10^{-4}) and 0.20 *M* NaCOOH. Then calculate the pH of the solution after the addition of 10.0 mL of 12.0 *M* NaOH.

Solution

We can solve the problems using the same strategy as in the previous sections. The major species are **HCOOH, Na^+, $COOH^-$,** and **H_2O.** The dominant equilibrium will involve the acid dissociation of HCOOH.

$$HCOOH(aq) \rightleftharpoons H^+(aq) + COOH^-(aq)$$

$$K_a = \frac{[H^+][COOH^-]}{[HCOOH]} \quad \Rightarrow \quad 1.8 \times 10^{-4} = \frac{[H^+](0.20)}{0.15}$$

$$[H^+] = 1.4 \times 10^{-4}\ M, \qquad pH = 3.87$$

When we add the strong base, we have the reaction between OH^- ions and HCOOH,

$$HCOOH(aq) + OH^-(aq) \rightarrow COOH^-(aq) + H_2O(l)$$

We must proceed with our two-step approach.

1. ***Stoichiometry Part***

mmoles $HCOOH_{initial}$ = 0.15 mmol/mL × 500 mL = **75 mmol**
mmoles $COOH^-_{initial}$ = 0.20 mmol/mL × 500 mL = **100 mmol**
mmoles OH^- added = 12.00 mmol/mL × 10.0 mL = **120 mmol**

We have more OH^- than is needed to neutralize the HCOOH. HCOOH is the **limiting reactant** in this solution.

	$HCOOH(aq)$	+	$OH^-(aq)$	→	$COOH^-(aq)$	+	$H_2O(l)$
initial (mmol)	75		120		100		
final (mmol)	≈0		45		175		

We have 45 mmoles of excess OH^-! There is no **"equilibrium part" in this problem!** OH^- ion is a far stronger base than $COOH^-$ and will thus determine the pH.

$$[OH^-] = 45\ \text{mmol}/(500 + 10)\ \text{mL} = \mathbf{0.088\ M}$$

$$pOH = 1.05 \qquad \mathbf{pH = 12.95}$$

When we *exceeded the buffer capacity* of the solution, *the pH changed drastically*. If we would have doubled the amounts of formic acid and sodium formate, we would not have exceeded the buffer capacity of the solution.

All other things being equal, the closer the $[HA]/[A^-]$ ratio is to 1, the better will be the buffer capacity of the solution. This will happen when $[H^+] = K_a$ or $pH = pK_a$. This means that, when selecting a buffer that will be used to maintain the pH of a solution in a particular small range,

the pK_a of the buffer should be as close as possible to the desired pH of the solution.

Example 8.4 B - Selecting A Buffer

We wish to buffer a solution at pH = 10.07. Which one of the following bases (and conjugate acid salts) would be most useful?

a. NH_3 ($K_b = 1.8 \times 10^{-5}$)
b. $C_6H_5NH_2$ ($K_b = 4.2 \times 10^{-10}$)
c. N_2H_4 ($K_b = 9.6 \times 10^{-7}$)

Solution

K_a for $NH_4^+ = K_w/K_b = 1.0 \times 10^{-14}/1.8 \times 10^{-5} = 5.6 \times 10^{-10}$

pK_a (NH_4) = 9.26

K_a for $C_6H_5NH_3^+ = K_w/K_b = 1.0 \times 10^{-14}/4.2 \times 10^{-10} = 2.4 \times 10^{-5}$

pK_a ($C_6H_5NH_3^+$) = 4.62

K_a for $N_2H_5^+ = K_w/K_b = 1.0 \times 10^{-14}/9.6 \times 10^{-7} = 1.0 \times 10^{-8}$

pK_a ($N_2H_5^+$) = 7.98

NH_3 would be the best choice for a buffer system. (Please keep in mind that a good deal of thought must go not only into the mathematics, but the **chemistry as well.** That is, how will NH_3 affect your experiment?)

8.5 Titrations and pH Curves

As a precursor to the calculations in this chapter, you should be able to define the following terms, first discussed in Section 4.9 in your textbook: titrant, buret, indicator, titration curve, and endpoint.

For a titration to be feasible, it must be **complete** and **fast**. To be complete, it should have a value of **$K \geq 10^7$ or so.**

Strong Acid - Strong Base

Let's consider the titration of an **HCl** solution with **NaOH**. We say the process is complete; however, we can calculate an equilibrium constant for the reaction.

$$H^+(aq) + OH^-(aq) \rightleftharpoons H_2O(l)$$

The reaction is the reverse of the autoionization of water; therefore

$$K = 1/K_w = \mathbf{1 \times 10^{14}}$$

This reaction is certainly complete! So much so, in fact, that this type of titration is less an equilibrium problem than a stoichiometry one.

Example 8.5 A - Strong Acid - Strong Base Titration

Calculate the pH after the following **total** volumes of **0.2500 *M* HCl** have been **added to 50.00 mL of 0.1500 *M*** **NaOH.**

a. 0.00 mL
b. 4.00 mL
c. 29.50 mL
d. 30.00 mL
e. 30.50 mL
f. 40.00 mL

Solution

a. **0.00 mL:** We have a 0.1500 *M* NaOH solution. Because it completely dissociates, an OH^- ion is a strong base.

$[OH^-] = 0.1500\ M$, pOH = 0.82, **pH = 13.18**

b. **4.00 mL total:** As discussed previously, the reaction between H^+ and OH^- is complete. Calculating moles of each reactant is always the preferred way to begin.

mmol OH^-_{initial} = 0.1500 mmol/mL × 50.00 mL = **7.500 mmol**
mmol H^+ added = 0.2500 mmol/mL × 4.00 mL = **1.000 mmol**

	$H^+(aq)$	+	$OH^-(aq)$	→	$H_2O(l)$
initial (mmol)	1.000		7.500		excess
final (mmol)	≈0		6.500		excess

The relatively small amount of H^+ ions added were neutralized by OH^- ions. We have 6.500 mmol of OH^- ion excess, which determines the pH.

$$\mathbf{[OH^-]} = 6.500\text{ mmol}/(50.00 + 4.00)\text{ mL} = \mathbf{0.1203}\ \boldsymbol{M}$$
$$\text{pOH} = 0.92, \quad \mathbf{pH = 13.08}$$

The pH has not declined much because we still have excess of OH^- ion.

c. **29.50 mL total:** Proceeding as above,

mmol $OH^-_{initial}$ = 7.500 mmol
mmol H^+ added = 0.2500 mmol/mL × 29.50 mL = **7.375 mmol**

	$H^+(aq)$	+	$OH^-(aq)$	→	$H_2O(l)$
initial (mmol)	7.375		7.500		excess
final (mmol)	≈0		0.125		excess

$$\mathbf{[OH^-]} = 0.125\text{ mmol}/(50.00 + 29.50)\text{ mL} = \mathbf{1.572 \times 10^{-3}}\ \boldsymbol{M}$$
$$\text{pOH} = 2.80, \quad \mathbf{pH = 11.20}$$

The pH has begun to come down somewhat, but the solution is still quite basic.

d. **30.00 mL total:** As above,

mmol $OH^-_{initial}$ = 7.500 mmol
mmol H^+ added = 0.2500 mmol/mL × 30.00 mL = **7.500 mmol**

	$H^+(aq)$	+	$OH^-(aq)$	→	$H_2O(l)$
initial (mmol)	7.500		7.500		excess
final (mmol)	≈0		≈0		excess

We have added **exactly** enough H^+ ions to neutralize all the OH^- ions. We have only water, Na^+, and Cl^- in the solution. Of these, only **water** has any acid-base properties. This neutral water solution, as always, has $K_w = 1.0 \times 10^{-14}$. The **pH = 7.00**.

CRITICAL POINT: THE ONLY CASE IN WHICH THE pH AT THE EQUIVALENCE POINT WILL BE EQUAL TO 7.00 WILL BE THAT OF A STRONG ACID-STRONG BASE TITRATION!

Also notice the rather sharp pH change (4.20 units) within 0.500 mL of the equivalence point. This is typical of a strong acid-strong base titration.

e. **30.50 mL total:** Proceeding as always,

	$H^+(aq)$	+	$OH^-(aq)$	→	$H_2O(l)$
initial (mmol)	7.625		7.500		excess
final (mmol)	0.125		≈0		excess

We have **passed** the equivalence point. (This is called the **post-equivalence point region**.) We have excess H^+ ion. The solution is now acidic.

$$\mathbf{[H^+]} = 0.125\text{ mmol}/80.50\text{ mL} = \mathbf{1.55 \times 10^{-3}}\ \boldsymbol{M}$$
$$\mathbf{pH = 2.81}$$

Being just 0.500 mL beyond the equivalence point caused a sharp drop in the pH. Generally, the stronger the species being titrated, the sharper the pH drop.

f. **40.00 mL total:** We are now well past the equivalence point.

	$H^+(aq)$	+	$OH^-(aq)$	→	$H_2O(l)$
initial (mmol)	10.00		7.500		excess
final (mmol)	2.50		≈0		excess

$$[H^+] = 2.50 \text{ mmol}/90.00 \text{ mL} = \mathbf{0.0278}\ M$$
$$\mathbf{pH = 1.56}$$

If you plot the mL of HCl added vs. pH for these data, your plot should look like that in Figure 8.2 in your textbook.

Weak Acid - Strong Base

When doing calculations for this type of titration, it is useful to relate our thinking back to our sections on **buffered solutions (8.2 and 8.4).**

$$HA(aq) + OH^-(aq) \rightarrow A^-(aq) + H_2O(l)$$

You will be generating a buffer (weak acid and conjugate base) as a result of the addition of OH^- ions to the solution. You will have a buffered solution until you exceed the buffer capacity of the solution, and then the pH will rise dramatically.

Let's use the titration of formic acid, HCOOH ($K_a = 1.8 \times 10^{-4}$), with NaOH as an illustrative example. **The equilibrium constant for the titration can be calculated.** The reaction

$$\mathbf{HCOOH(aq) + OH^-(aq) \rightleftharpoons COOH^-(aq) + H_2O(l)}$$

can be expressed as a **combination** of

1. $HCOOH(aq) \rightleftharpoons H^+(aq) + COOH^-(aq)$ $\qquad K_a = 1.8 \times 10^{-4}$

and

2. $H^+(aq) + OH^+(aq) \rightleftharpoons H_2O(l)$ $\qquad 1/K_w = 1.0 \times 10^{14}$

K for the overall reaction $= K_1 \times K_2 = K_a \times 1/K_w = K_a/K_w$

$$K = 1.8 \times 10^{-4}/1.0 \times 10^{-14} = \mathbf{1.8 \times 10^{10}}$$

This is why we say that the titration goes completely.

Example 8.5 B - Weak Acid–Strong Base Titration

Calculate the pH after the following **total** volumes of 0.4000 M NaOH are added to 50.00 mL of 0.200 M HCOOH.

a. 0.00 mL
b. 5.00 mL
c. 12.50 mL
d. 24.50 mL
e. 25.00 mL
f. 25.50 mL
g. 40.00 mL

Solution

a. **0.00 mL:** We have a weak acid solution, 0.2000 M HCOOH. We have done these kinds of problems many times and will present a **shortened version here.** The equilibrium of interest is

$$HCOOH(aq) \rightleftharpoons H^+(aq) + COOH^-(aq)$$

$$K_a = \frac{[H^+][COOH^-]}{[HCOOH]} \quad \Rightarrow \quad 1.8 \times 10^{-4} = \frac{x^2}{0.2000}$$

$$[H^+] = 6.0 \times 10^{-3}\ M, \qquad \mathbf{pH = 2.22}$$

In titration problems, if the titration is feasible (large K), our "5% test" will virtually always hold true.

b. **5.00 mL:** As pointed out previously, the titration reaction of interest is

$$\mathbf{HCOOH}(aq) + \mathbf{OH^-}(aq) \rightleftharpoons \mathbf{COOH^-}(aq) + \mathbf{H_2O}(l)$$

The reaction is essentially complete ($K = 1.8 \times 10^{10}$). After this reaction is completed, we will have a **buffered solution** containing HCOOH and $COOH^-$ ion. We must then calculate the pH of the buffered solution.

Stoichiometry Part:

mmol $HCOOH_{initial}$ = 0.2000 mmol/mL × 50.00 mL = **10.00 mmol**
mmol OH^- added = 0.4000 mmol/mL × 5.00 mL = **2.000 mmol**

	HCOOH(*aq*)	+	OH^-(*aq*)	→	$COOH^-$(*aq*)	+	$H_2O(l)$
initial (mmol)	10.00		2.000		≈0		excess
final (mmol)	8.00		≈0		2.00		excess

Equilibrium Part:

Formic acid is a stronger acid than formate ion is a base. Therefore we can use the acid dissociation equilibrium to solve the problem (but in this so-called **"buffer region"** the base hydrolysis would work as well). (See Section 8.2 if you need a review of buffered solutions.)

[HCOOH] = 8.00 mmol/(50.00 + 5.00) mL = **0.145 *M***
[$COOH^-$] = 2.00 mmol/55.00 mL = **0.0364 *M***

$$\mathbf{HCOOH}(aq) \rightleftharpoons \mathbf{H^+}(aq) + \mathbf{COOH^-}(aq)$$

$$K_a = \frac{[H^+][COOH^-]}{[HCOOH]} \quad \Rightarrow \quad 1.8 \times 10^{-4} = \frac{[H^+](0.0364)}{0.145}$$

$$[H^+] = 7.2 \times 10^{-4}\ M, \qquad \mathbf{pH = 3.14}$$

The pH does not rise much in the buffer region of a titration.

c. **12.50 mL:** We shall proceed as in part b.

Stoichiometry Part:

mmol $HCOOH_{initial}$ = 10.00 mmol
mmol OH^- added = 0.4000 mmol/mL × 12.50 mL = **5.000 mmol**

	HCOOH(*aq*)	+	OH^-(*aq*)	→	$COOH^-$(*aq*)	+	$H_2O(l)$
initial (mmol)	10.00		5.000		≈0		excess
final (mmol)	5.00		≈0		5.00		excess

Equilibrium Part:

[HCOOH] = 5.00 mmol/(50.00 + 12.50) mL = **0.0800 *M***
[$COOH^-$] = 5.00 mmol/62.50 mL = **0.800 *M***

$$K_a = \frac{[H^+][COOH^-]}{[HCOOH]} \quad \Rightarrow \quad 1.8 \times 10^{-4} = \frac{[H^+](0.0800)}{0.0800}$$

$$\mathbf{[H^+] = K_a} = 1.8 \times 10^{-4}\ M, \qquad \mathbf{pH = 3.74}$$

Recall from Section 8.4 that when [HA] = [A^-], pK_a = pH. This is the situation at this point in our titration. This is called the **titration midpoint**. We are **halfway** to the equivalence point.

The titration midpoint is an especially important experimental point because, if we know pH, we can find pK_a for the weak acid.

d. **24.50 mL:** Proceeding as always,

mmol $HCOOH_{initial}$ = 10.00 mmol
mmol OH^- added = 0.4000 mmol/mL × 24.50 mL = **9.80 mmol**

	$HCOOH(aq)$	+	$OH^-(aq)$	→	$COOH^-(aq)$	+	$H_2O(l)$
initial (mmol)	10.00		9.80		≈0		excess
final (mmol)	0.20		≈0		9.80		excess

We are near the limit of our buffer capacity.

[HCOOH] = 0.20 mmol/(50.00 + 24.50) mL = **2.67×10^{-3} *M***
[$COOH^-$] = 9.80 mmol/74.50 mL = **0.132 *M***

$$K_a = \frac{[H^+][COOH^-]}{[HCOOH]} \quad \Rightarrow \quad 1.8 \times 10^{-4} = \frac{[H^+](0.1.32)}{2.67 \times 10^{-3}}$$

$[H^+] = 3.6 \times 10^{-6}\ M$, **pH = 5.44**

e. **25.00 mL:**

mmol $HCOOH_{initial}$ = 10.00 mmol
mmol OH^- added = 10.00 mmol

	$HCOOH(aq)$	+	$OH^-(aq)$	→	$COOH^-(aq)$	+	$H_2O(l)$
initial (mmol)	10.00		10.00		≈0		excess
final (mmol)	≈0		≈0		10.00		excess

We have reached the equivalence point. We now have a solution that contains a **base** that can undergo hydrolysis,

$$COOH^-(aq) + H_2O(l) \rightleftharpoons HCOOH(aq) + OH^-(aq)$$

$$K_b = K_w/K_a = \frac{1.0 \times 10^{-14}}{1.8 \times 10^{-4}} = 5.6 \times 10^{-11} = \frac{[HCOOH][OH^-]}{[COOH^-]}$$

[$COOH^-$] = 10.00 mmol/(50.00 + 25.00) mL = **0.133 *M***

Solving this weak base problem in the usual way,

$$5.6 \times 10^{-11} = \frac{x^2}{0.133} \quad \Rightarrow \quad x = [OH^-] = 2.73 \times 10^{-6}\ M$$

pOH = 5.56, **pH = 8.44**

We have a mildly basic solution, and that is reflected in the pH value.

f. **25.50 mL:**

mmoles $HCOOH_{initial}$ = 10.00 mmol
mmoles OH^- added = 10.20 mmol

We have neutralized all the weak acid and have **0.20 mmol of strong base remaining.** This is the **post-equivalence point region.** The pH will be determined by the excess strong base.

$[OH^-]$ = 0.20 mmol/(50.00 + 25.50) mL = **2.65×10^{-3} *M***
pOH = 2.58, **pH = 11.42**

Notice that the pH has risen sharply now that we have **surpassed the buffer capacity** of our formic acid/formate-buffered solution.

g. **40.00 mL:** You should be able to show we have **6.00 mmol of excess OH^- ions** and

$$[OH^-] = 6.00 \text{ mmol}/90.00 \text{ mL} = 0.0667\ M$$
$$\text{pOH} = 1.18, \qquad \textbf{pH = 12.82}$$

Summing up, note that our "pH break" around the equivalence-point region was somewhat less sharp than with the strong acid-strong base titration. The smaller the titration equilibrium constant, the less sharp the break at the equivalence point. If you plot a titration curve, it should look similar to Figure 8.3 in your textbook.

Weak Base - Strong Acid

These kinds of titrations are handled using the same strategies as those involving weak acids and strong bases. We have the **buffer region, equivalence point, and post-equivalence point** as in the previous titrations. Think about the dominant reactions in each part of the problem, be careful regarding math, and **check to make sure that your answer makes sense.**

8.6 Acid-Base Indicators

The purpose of an **indicator** is to allow you to know when you have reached the equivalence point in a titration. Acid-base indicators are normally organic acids. Thus when you add an indicator to a solution and perform a titration, **you are titrating the indicator along with your substance of interest.** This requires some small extra amount of titrant.

The volume of titrant at which you visually detect the equivalence point is called the **endpoint. The volume at the equivalence point is never exactly equal to the endpoint, but it is usually quite close.**

Two criteria are used to choose an indicator for a given titration:

- The pK_a of the indicator should be within ±1 unit of the pH of the solution at the equivalence point.
- The color change at the endpoint should be clearly distinguishable.

Example 8.6 - Proper Indicator Choice

A 0.100 M NH_3 solution is being titrated with 0.200 M HCl. Using data from Figure 8.8 in your textbook, select a suitable indicator for the titration.

Solution

The reaction of interest is

$$NH_3(aq) + H^+(aq) \rightarrow NH_4^+(aq)$$

The important question is, "What is the pH at the equivalence point?" We can estimate it by noting that, since the HCl solution is **twice as concentrated** as the NH_3 solution, it will take **1/2 the volume** of HCl solution to neutralize the NH_3 solution. That is,

$$\textbf{Volume at the equivalence point} = V_0 + \tfrac{1}{2}V_0 = \tfrac{3}{2}V_0$$

where V_0 is the original volume of NH_3 solution.

Because the volume is 3/2 as much, the NH_4^+ solution will be 2/3 as concentrated as the original NH_3 solution.

$$[NH_4^+] = 0.100\ M \times 2/3 = \mathbf{0.067\ M}$$

$$K_{a_{NH_4^+}} = K_w / K_{b_{NH_3}} = 5.6 \times 10^{-10} = \frac{[H^+][NH_3]}{[NH_4^+]} = \frac{x^2}{0.0667}$$

$$x = [H^+] = 6.1 \times 10^{-6}\ M, \qquad \mathbf{pH = 5.21}$$

The indicator should have a pK_a close to 5.2. **Bromcresol green is a suitable choice.** The color change, blue to yellow (base to acid), is easy to distinguish.

8.7 Titration of Polyprotic Acids

This section in your textbook seems to be so very long, but it is surprisingly straightforward. Table 8.4 in your textbook summarizes the pH's at each point in the titration. When does the pH = pK_a? What are some of the important assumptions made when dealing with polyprotic acids?

8.8 Solubility Equilibria and Solubility Product

As with many of the previous sections, solubility equilibria use principles you have used before. This section deals with **solubility**, the amount of a salt that can be dissolved in water. Your textbook points out that the **solubility of a salt is variable. The solubility product (K_{sp}) is constant** (at a given temperature).

The solubility equilibrium expression is set up as any other. For example, the equilibrium expression for the dissolution of Ag_2S in water is

$$Ag_2S(s) \rightleftharpoons 2Ag^+(aq) + S^{2-}(aq) \qquad K_{sp} = 1.6 \times 10^{-49}$$

$$K_{sp} = [Ag^+]^2[S^{2-}]$$

Remember that the pure solid, Ag_2S, is not included in the equilibrium expression.

Example 8.8 A - K_{sp} From Solubility Data

Silver sulfide (Ag_2S) has a **solubility of 3.4×10^{-17} *M*** at 25°C. Calculate K_{sp} for Ag_2S.

Solution

The key here is *how we define solubility*. For the reaction

$$Ag_2S(s) \rightleftharpoons 2Ag^+(aq) + S^{2-}(aq),$$

one mole of S^{2-} is produced for *every mole of Ag_2S* that dissolves. Therefore the solubility equals the concentration of S^{2-} in solution.

$$\text{solubility} = s = [S^{2-}] = \mathbf{3.4 \times 10^{-17}\ M}$$

The stoichiometry of the reaction indicates that *2 moles of Ag^+ ion are produced for each mole of S^{2-} ion.* Therefore,

$$[Ag^+] = 2[S^{2-}] = \mathbf{6.8 \times 10^{-17}\ M}$$

$$\mathbf{K_{sp}} = [Ag^+]^2\,[S^{2-}] = (6.8 \times 10^{-17})^2(3.4 \times 10^{-17}) = \mathbf{1.6 \times 10^{-49}}$$

The keys to solving solubility problems are to

- properly define solubility, and
- properly use the reaction stoichiometry.

Example 8.8 B - Solubility From K_{sp} Data

Calculate the solubility of each of the following in moles per liter and grams per liter.

a. $NiCO_3$ $K_{sp} = 1.4 \times 10^{-7}$
b. $Ba_3(PO_4)_2$ $K_{sp} = 6 \times 10^{-39}$
c. $PbBr_2$ $K_{sp} = 4.6 \times 10^{-6}$

Solution

We need to properly write dissolution reactions and equilibrium expressions for each. We must then carefully define solubility for each salt.

a. $\mathbf{NiCO_3(s) \rightleftharpoons Ni^{2+}(aq) + CO_3^{2-}(aq)}$

$s = [Ni^{2+}] = [CO_3^{2-}]$
$\mathbf{K_{sp} = [Ni^{2+}]\,[CO_3^{2-}]}$
$1.4 \times 10^{-7} = (s)(s) = s^2 \quad \Rightarrow \quad \mathbf{s = 3.7 \times 10^{-4}\ M}$

This means that **3.7×10^{-4} mol/L of $NiCO_3$** will go into solution.

$g/L = 3.7 \times 10^{-4}\ mol/L \times 118.7\ g/mol = s = \mathbf{0.44\ g/L\ NiCO_3}$

b. $\mathbf{Ba_3(PO_4)_2(s) \rightleftharpoons 3Ba^{2+}(aq) + 2PO_4^{3-}(aq)}$

Solubility = s = **moles of $Ba_3(PO_4)_2$** that go into solution. The stoichiometry of the reaction dictates that

$[Ba^{2+}] = 3s \qquad [PO_4^{3-}] = 2s$
$\mathbf{K_{sp} = [Ba^{2+}]^3[PO_4^{3-}]^2}$
$6 \times 10^{-39} = (3s)^3(2s)^2 = 108s^5$
$s^5 = 5.55 \times 10^{-41} \quad \Rightarrow \quad \mathbf{s = 9 \times 10^{-9}\ M}$
$g/L = 9 \times 10^{-9}\ mol/L \times 601.8\ g/mol = s = \mathbf{5 \times 10^{-6}\ g/L}$

c. $\mathbf{PbBr_2(s) \rightleftharpoons Pb^{2+}(aq) + 2Br^-(aq)}$

$[Pb^{2+}] = s \qquad [Br^-] = 2s$
$\mathbf{K_{sp} = [Pb^{2+}][Br^-]^2}$
$4.6 \times 10^{-6} = (s)(2s)^2 \quad \Rightarrow \quad \mathbf{s = 1.0 \times 10^{-2}\ M}$
$g/L = 1.0 \times 10^{-2}\ mol/L \times 367\ g/mol = s = \mathbf{3.67\ g/L}$

Your textbook points out that **relative solubilities** among different compounds **cannot** be measured simply by comparing K_{sp} values. You must take the **composition of the salt** into account, as illustrated by the next example.

Example 8.8 C - Relative Solubilities

Which of the following compounds is the most soluble?

$AgCl$; $K_{sp} = 1.5 \times 10^{-10}$
Ag_2CrO_4; $K_{sp} = 9.0 \times 10^{-12}$
Ag_3PO_4; $K_{sp} = 1.8 \times 10^{-18}$

Solution

$$\mathbf{AgCl(s) \rightleftharpoons Ag^+(aq) + Cl^-(aq)}$$

$s = [Ag^+] = [Cl^-]$
$K_{sp} = [Ag^+][Cl^-] = s^2 = 1.5 \times 10^{-10}$
$\mathbf{s = 1.2 \times 10^{-5}\ M}$

$$Ag_2CrO_4(s) \rightleftharpoons 2Ag^+(aq) + CrO_4^{2-}(aq)$$

$s = [CrO_4^{2-}]$ $\quad 2s = [Ag^+]$
$K_{sp} = [Ag^+]^2[CrO_4^{2-}] = (2s)^2(s) = 4s^3 = 9.0 \times 10^{-12}$
$\mathbf{s = 1.3 \times 10^{-4}\ M}$

$$Ag_3PO_4(s) \rightleftharpoons 3Ag^+(aq) + PO_4^{3-}(aq)$$

$s = [PO_4^{3-}]$ $\quad 3s = [Ag^+]$
$K_{sp} = [Ag^+]^3[PO_4^{3-}] = (3s)^3(s) = 27s^4 = 1.8 \times 10^{-18}$
$\mathbf{s = 1.6 \times 10^{-5}\ M}$

Ag_2CrO_4 is the most soluble.

Common Ion Effect

We have encountered this before. Recall that **Le Châtelier's principle** predicts that **adding a common ion to the solution shifts the equilibrium to the left** in solubility equations. In other words, **adding a common ion** (that doesn't react with other species in the solution) **reduces the solubility.**

Example 8.8 D - Common Ion Effect

Calculate the solubility of SrF_2 ($K_{sp} = 7.9 \times 10^{-10}$) in

a. pure water
b. 0.100 M $Sr(NO_3)_2$
c. 0.400 M NaF

Solution

The reaction of interest is

$$SrF_2(s) \rightleftharpoons Sr^{2+}(aq) + 2F^-(aq)$$

$$K_{sp} = [Sr^{2+}][F^-]^2$$

a. In pure water, solubility of SrF_2 equals

$\mathbf{s = [Sr^{2+}]}$ $\quad \mathbf{2s = [F^-]}$
$7.9 \times 10^{-10} = (s)(2s)^2 = 4s^3$
$\mathbf{s = 5.8 \times 10^{-4}\ M}$

b. $Sr(NO_3)_2$ is a strong electrolyte. This means that a 0.100 M solution will supply **0.100 M Sr^{2+} ion** to the solution

$$\mathbf{[Sr^{2+}] = 0.100 + s \approx 0.100\ M}$$

We are assuming that s is negligible relative to 0.100 M. We can test this with the 5% test.

$[F^-] = 2s$
$K_{sp} = 7.9 \times 10^{-10} = (0.100)(2s)^2 = 0.4s^2$
$\mathbf{s = 4.4 \times 10^{-5}\ M}$

A solubility of 4.4×10^{-5} M is < 5% of 0.100. The solubility was thus *decreased by a factor of 13* due to the presence of the common ion.

c. NaF is a strong electrolyte. The dissociation of 0.400 M NaF will give

$[F^-] = 0.400\ M + 2s \approx \mathbf{0.400}$
$[Sr^{2+}] = s$

$K_{sp} = 7.9 \times 10^{-10} = s(0.400)^2 = 0.16s$
$\mathbf{s = 4.9 \times 10^{-9}\ M}$

The solubility was markedly *decreased* by the presence of the common F^- ion.

Although the rigorous solutions are beyond the scope of your textbook, in general, substances (such as acids or bases) that can combine with one of the ions of a salt will increase the solubility of the salt by removing product, thus forcing the reaction to the right.

8.9 Precipitation and Qualitative Analysis

This section asks the musical question, "If two solutions are mixed, will a precipitate form?" Then, "If it forms, what will be the concentration of each ion in solution?"

Your textbook introduces Q, **the ion product** of the initial ion concentrations. **If $Q \geq K_{sp}$, precipitation will occur.**

Example 8.9 A - The Ion Product

A 200.0-mL solution of $1.3 \times 10^{-3}\ M$ $AgNO_3$ is mixed with 100.0 mL of a $4.5 \times 10^{-5}\ M$ Na_2S solution. Will precipitation occur?

Solution

Na^+ and NO_3^- ions are completely soluble, so we would presume the precipitate would be **Ag_2S.** The reaction of interest is

$$2Ag^+(aq) + S^{2-}(aq) \rightleftharpoons Ag_2S(s)$$

The ion product is

$$Q = [Ag^+]_0^2[S^{2-}]_0$$

$$[Ag^+]_0 = 1.3 \times 10^{-3}\ M \times \frac{200.0 \text{ mL (original volume)}}{(200.0 + 100.0) \text{ mL (total soln volume)}} = \mathbf{8.7 \times 10^{-4}\ M}$$

$$[S^{2-}]_0 = 4.5 \times 10^{-5}\ M \times \frac{100.0 \text{ mL}}{300.0 \text{ mL}} = \mathbf{1.5 \times 10^{-5}\ M}$$

$$Q = (8.7 \times 10^{-4})^2\,(1.5 \times 10^{-5}) = \mathbf{1.1 \times 10^{-11}}$$

$K_{sp} = 1.6 \times 10^{-49}$. $Q > K_{sp}$, so **precipitation will occur.**

Once we determine precipitation will occur, we are faced with the problem of determining the **equilibrium concentrations** of each of our ions of interest.

The general strategy involves **assuming** that, because K_{sp} is so low, **if a precipitate forms, it will do so quantitatively.** We then can use the equilibrium expression involving the solubility of the salt (with a common ion) to solve for the equilibrium concentration of each ion. Let's illustrate this with the next example.

Example 8.9 B - Solubility From Mixing Solutions

Calculate the equilibrium concentration of each ion in a solution obtained by mixing 50.0 mL of $6.0 \times 10^{-3}\ M$ $CaCl_2$ with 30 mL of 0.040 M NaF. (K_{sp} for $CaF_2 = 4.0 \times 10^{-11}$)

Solution

First, let's verify that precipitation in fact occurs,

$$[\mathbf{Ca^{2+}}]_0 = 6.0 \times 10^{-3}\ M \times \frac{50.0\text{ mL}}{(50.0 + 30.0)\text{ mL}} = \mathbf{3.7_5 \times 10^{-3}}\ \boldsymbol{M}$$

$$[\mathbf{F^-}]_0 = 0.040\ M \times \frac{30.0\text{ mL}}{80.0\text{ mL}} = \mathbf{0.015}\ \boldsymbol{M}$$

$$Ca^{2+}(aq) + 2F^-(aq) \rightleftharpoons CaF_2(s)$$

$$Q = [Ca^{2+}][F^-]^2 = (3.7_5 \times 10^{-3})(0.015)^2 = \mathbf{8.4 \times 10^{-7}}$$

$$\boldsymbol{Q > K_{sp}}\text{, so precipitation occurs.}$$

We can assume a **quantitative (stoichiometric) reaction between Ca^{2+} and F^-** to form CaF_2 solid. We then can see how much of which ion remains in excess in the solution.

mmol $Ca^{2+}_{initial}$ = 6.0 × 10^{-3} mmol/mL × 50.0 mL = **0.30 mmol**
mmol $F^-_{initial}$ = 0.040 mmol/mL × 30.0 mL = **1.2 mmol**

	$Ca^{2+}(aq)$	+	$2F^-(aq)$	$\rightleftharpoons$	$CaF_2(s)$
initial (mmol)	0.30		1.2		0
final (mmol)	≈0		0.60		0.30

$[Ca^{2+}] = s$
$[F^-]$ = 0.60 mmol/80.0 mL = **$7.5 \times 10^{-3}\ M + 2s \approx 7.5 \times 10^{-3}\ M$**

$$K_{sp} = [Ca^{2+}][F^-]^2$$

$$4.0 \times 10^{-11} = s(7.5 \times 10^{-3})^2$$

$$\mathbf{s = 7.1 \times 10^{-7}\ M = [Ca^{2+}]}$$

$$\mathbf{7.5 \times 10^{-3}\ M = [F^-]}$$

Remember again that, although the solubility of different ions can change depending upon what is in solution, **the solubility product (K_{sp}) remains the same** at a given temperature.

8.10 Complex Ion Equilibria

Your textbook introduces several new terms here. You should be able to define **complex ion, ligand,** and **formation constant.** The ultimate goal of this section is to demonstrate that **introducing a Lewis base** into a solution **enhances the solubility** of an otherwise insoluble salt.

The key idea here is that, to avoid being drowned in a sea of equations, **we must make (and test) simplifying assumptions where possible!**

Let's do the following example to demonstrate the idea of **complex ion formation.**

Example 8.10 A - Complex Ion Formation

Calculate the concentrations of Ag^+ and $Ag(CN)_2^-$ in a solution prepared by mixing **100.0 mL of 5.0 × 10^{-3} *M* $AgNO_3$** with **100.0 mL of 2.00 *M* KCN.**

$$Ag^+(aq) + 2CN^-(aq) \rightleftharpoons Ag(CN)_2^-(aq) \qquad \beta = 1.3 \times 10^{21}$$

Solution

Note that β equals the equilibrium constant for the reaction, and it is overwhelmingly large. This means that **virtually all the Ag^+ ion will end up as $Ag(CN)_2^-$.** The other important point to note is that **our ligand, CN^-, is present in huge excess.** Therefore,

$$[CN^-] \approx [CN^-]_0$$

The equilibrium expression is

$$\beta = \frac{[Ag(CN)_2^-]}{[Ag^+][CN^-]^2}$$

To get equilibrium concentrations,

$$[CN^-] \approx [CN^-]_0 = 1.00\ M$$

(after mixing, the volume doubles, so the concentration is halved). Virtually all the silver is present as $Ag(CN)_2^-$. Therefore,

$$[Ag(CN)_2^-] \approx [Ag^+]_0 = 2.5 \times 10^{-3}\ M$$

(Remember the dilution!) Rearranging our overall equilibrium expression,

$$[Ag^+] = \frac{[Ag(CN)_2^-]}{\beta[CN^-]^2} = \frac{2.5\times10^{-3}}{1.3\times10^{21}(1.00)^2} = 1.9 \times 10^{-24}\ M$$

The important question in all of this is, "How does complex ion formation affect solubility?"

Using our first example (8.10 A), we note that most of the silver was complexed as $Ag(CN)_2^-$. That means that, as free Ag^+ ion is being produced by an insoluble salt (such as AgI, $K_{sp} = 1.5 \times 10^{-16}$), it is being **removed by complexing with CN^- ions.** This is pulling the dissolution of AgCl(*s*) **to the right.**

FORMATION OF A COMPLEX ION
INCREASES THE SOLUBILITY OF AN "INSOLUBLE" SALT.

Example 8.10 B - Solubility And Complex Ion Formation

Calculate the solubility of AgI(*s*) in 1.00 *M* CN^- ion (K_{sp} for AgI = 1.5×10^{-16}).

Solution

We have, until now, said that the solubility of AgI equals

$$s = [Ag^+] = [I^-]$$

However, Ag^+ ion forms complexes that enhance the solubility. A more precise statement is that

$$s = [I^-] = [\text{all Ag species}] = [Ag^+] + [AgCN] + [Ag(CN)_2^-]$$

We know, based on the overall formation constant, β, that virtually all the Ag^+ ion is present as $Ag(CN)_2^-$. Therefore,

$$s = [I^-] \approx [Ag(CN)_2^-]$$

The overall equilibrium of AgI(*s*) and CN^-(*aq*) can be represented as

$$AgI(s) + 2CN^-(aq) \rightleftharpoons Ag(CN)_2^-(aq) + I^-(aq)$$

$$K = K_{sp}\beta = \frac{[Ag(CN)_2^-][I^-]}{[CN^-]^2} = 1.5 \times 10^{-16}\,(1.3 \times 10^{21})$$

$$\mathbf{K = 1.95 \times 10^5}$$

$\mathbf{[CN^-] = 2.00 - 2x}$	(original concentration − amount complexed)
$\mathbf{[Ag(CN)_2^-] = x}$	(amount formed, x − solubility, s)
$\mathbf{[I^-] = x}$	(x − solubility, s)

$$K = \frac{x^2}{(2.00 - 2x)^2}$$

$$\sqrt{K} = \frac{x}{2.00 - 2x}$$

Getting rid of the denominator,

$$883 - 883x = x$$

$$\mathbf{x = solubility = 0.999}\ M$$

Note how our solubility has increased from 1.2×10^{-8} M to 0.999 M due to the presence of the ligand.

Exercises

Section 8.1

1. Qualitatively, what is the effect on pH of adding sodium benzoate to an aqueous solution of benzoic acid? Why does this change occur?

2. Calculate the pH and the percent dissociation of the acid in each of the following solutions.
   a. 0.150 M C_6H_5COOH (benzoic acid, $K_a = 6.14 \times 10^{-5}$)
   b. 0.150 M C_6H_5COOH in the presence of 0.350 M C_6H_5COONa

3. Calculate the pH and the percent hydrolysis of the base in each of the following solutions.
   a. 0.438 M NH_3 ($K_b = 1.8 \times 10^{-5}$)
   b. 0.438 M NH_3 to which has been added 0.300 M NH_4Cl

4. A solution is prepared from 0.150 mol of CH_3COOH, 0.0100 mol of CH_3COO^- (from CH_3COONa), and enough water to make a total volume of 1.00 L. Calculate the value of $[H^+]$ at equilibrium (acetic acid, $K_a = 1.8 \times 10^{-5}$).

5. A solution contains 0.350 mol of $(CH_3)_3N$, 0.050 mol of $(CH_3)_3NH^+$, and enough water to make a total volume of 1.00 L. Calculate the value of $[OH^-]$ (trimethylamine, $K_b = 6.25 \times 10^{-5}$).

Section 8.2

6. Determine the pH in a solution containing 0.085 M NH_3 and 0.247 M NH_4Cl (K_b for $NH_3 = 1.8 \times 10^{-5}$).

7. Calculate the pH of a buffer system with 0.10 M Na_2HPO_4/0.15 M KH_2PO_4.

8. Is a 6 M HCl solution a good buffer? Is it biologically useful?

9. What is the pH of a 200-mL solution containing 21.46 g of benzoic acid and 37.68 g of sodium benzoate,

   a. initially?
   b. after 30.0 mL of 5.00 *M* HCl has been added?
   c. Has this buffer exceeded its buffer capacity after addition of the acid?

10. A buffer is prepared by adding 20.5 g of CH_3COOH and 17.8 g of CH_3COONa to enough water to make 5.00×10^2 mL of solution. Calculate the pH of this buffer solution.

11. Calculate the pH of 1.00 L of a buffer solution containing 0.10 *M* HCN and 0.12 *M* CN^-.

12. What must be the ratio of acetic acid to sodium acetate to prepare a buffer whose pH = 4.81 (K_a of $CH_3COOH = 1.8 \times 10^{-5}$).

13. What is the pH of the buffer containing 0.35 *M* NH_4Cl and 0.15 *M* NH_3?

14. Calculate the pH if 0.01 mol of HCl is added to the buffer in Problem 11.

15. A solution contains 0.300 moles of acetic acid and 0.200 moles of sodium acetate in a total volume of 500. mL. How much 6.00 *M* NaOH must be added so the pH of the solution equals the pK_a of acetic acid?

16. Calculate the pH if 0.0200 mol of NaOH is added to the original buffer in Problem 11.

17. A solution contains 0.216 moles of a base and 0.614 moles of the conjugate acid in a total volume of 800 mL. The pH of the solution is 9.65. What is the value for K_b of the base?

18. How many grams of sodium acetate must be dissolved in a 0.200 *M* acetic acid solution to make a 400.-mL buffer solution with a pH of 4.56? (Assume the volume of the solution remains constant.)

19. The pH of a bicarbonate-carbonic acid buffer is known to be 8.00. Calculate the ratio of the concentration of carbonic acid to that of the bicarbonate ion (K_a for $H_2CO_3 = 4.3 \times 10^{-7}$).

Section 8.4

20. What would the pH of the solution in Problem 18 be if we added 40.0 mL of 0.30 *M* HCl?

21. We wish to buffer a solution at pH = 4.58. Which of the following acids (and conjugate base salts) would be most useful?

   a. CH_3COOH ($K_a = 1.8 \times 10^{-5}$)
   b. C_6H_5COOH ($K_a = 6.14 \times 10^{-5}$)
   c. $ClCH_2COOH$ ($K_a = 1.36 \times 10^{-3}$)
   d. C_6H_5OH ($K_a = 1.6 \times 10^{-10}$)

Section 8.5

22. Calculate *K* for the titration of HCl with NaOH.

23. A volume of 17.8 mL of a 0.344 *M* H_2SO_4 solution is required to completely neutralize 20.0 mL of a KOH solution. Calculate the concentration (in molarity) of the KOH solution.

24. If 37.42 mL of 0.1078 *M* NaOH solution is required to completely neutralize 25.00 mL of an HCl solution, what is the concentration (in molarity) of the acid solution?

25. Calculate the pH after the following **total** volumes of 0.3000 *M* NaOH have been added to 40.00 mL of 0.6000 *M* HCl.

    a. 0.00 mL
    b. 5.00 mL
    c. 40.00 mL
    d. 79.40 mL
    e. 80.00 mL
    f. 80.50 mL
    g. 90.00 mL

26. A sample of 50.0 mL of a commercial vinegar solution (which contains acetic acid) is titrated with a 1.00 *M* NaOH solution. What is the concentration of acetic acid in vinegar if 5.75 mL of the base was required for the titration?

27. Calculate the pH after the following **total** volumes of 0.3200 *M* HCl have been added to 25.00 mL of 0.1600 *M* NaOH.

    a. 0.00 mL
    b. 1.00 mL
    c. 12.40 mL
    d. 12.50 mL
    e. 12.60 mL
    f. 15.00 mL

28. If a 25.00-mL sample of base "X" requires 18.34 mL of 0.100 *M* HCl to reach the equivalence point, what is the concentration of base "X"? (Assume that the acid-base stoichiometry is 1:1.)

29. If 12.5 mL of 0.500 *M* H_2SO_4 exactly neutralizes 50.0 mL of NaOH, what is the concentration of the NaOH solution?

30. Calculate the pH at the equivalence point for the titration of 0.10 *M* NH_3 with 0.10 *M* HCl.

31. Calculate the pH at the equivalence point for the titration of 0.01 *M* CH_3COOOH with 0.10 *M* NaOH.

32. A solution containing 100.0 mL of 0.1350 *M* CH_3COOH ($K_a = 1.8 \times 10^{-5}$) is being titrated with 0.5400 *M* NaOH. Calculate the pH:

    a. initially
    b. halfway to the equivalence point
    c. at the equivalence point
    d. 5 mL past the equivalence point

33. Calculate *K* for the titration in Problem 32.

34. Calculate the number of milliliters of a 0.153 *M* sodium hydroxide solution that must be added to a 20.0-mL sample of a solution of hydrochloric acid, whose pH is 0.747, to reach the equivalence point.

35. The equivalence point for a solution of acid "A" occurs after 46.7 mL of base has been added. After 34.2 mL of base had been added (earlier in the titration), the pH was 4.64. What is the K_a for acid "A"?

36. A solution containing 50.00 mL of 0.1800 *M* NH_3 ($K_b = 1.8 \times 10^{-5}$) is being titrated with 0.3600 *M* HCl. Calculate the pH:

    a. initially
    b. after the addition of 5.00 mL of HCl
    c. after the addition of a total volume of 12.50 mL of HCl
    d. after the addition of a total volume of 25.00 mL of HCl
    e. after the addition of 26.00 mL of HCl

37. Calculate the equilibrium constant for the titration of benzoic acid (C_6H_5COOH) with NH_3. Is this titration feasible? Why or why not?

38. A solution containing 2.049 g of a weak acid required 43.88 mL of 0.1207 *M* NaOH to reach the equivalence point. What is the molar mass of the acid?

Section 8.6

39. Using information found in Figure 8.8 in your textbook, select a suitable indicator for the titrations in Problems 25 and 27 **other than phenolphthalein.**

40. Pick the best indicator for the titration in Problem 32.

41. Pick a suitable indicator for the titration in Problem 36.

42. Pick a suitable indicator for the titration in Problem 35.

43. What indicator would you choose for the titration of an aqueous solution of HCN (concentration unknown) with a 0.100 *M* solution of NaOH (K_a for HCN = 6.2×10^{-10})?

Section 8.8

44. Write products and equilibrium expressions for the following dissolution reactions.

    a. $PbI_2(s) \rightleftharpoons$
    b. $Sr_3(PO_4)_2(s) \rightleftharpoons$
    c. $MnS(s) \rightleftharpoons$

45. Calculate [I^-] in an AgI solution with [Ag^+] = 1.2×10^{-8} *M*. (AgI has $K_{sp} = 1.5 \times 10^{-16}$.)

46. Manganese sulfide (MnS) has a K_{sp} of 2.3×10^{-13}. Calculate the solubility of MnS.

47. Calculate the solubility product (K_{sp}) of silver sulfate if its molar solubility is 1.5×10^{-2} mol/L.

48. The K_{sp} for manganese(II) carbonate is 1.8×10^{-11}. Calculate the solubility of $MnCO_3$.

49. Strontium phosphate, $Sr_3(PO_4)_2$, has a K_{sp} of 1×10^{-31}. Calculate the solubility of $Sr_3(PO_4)_2$.

50. Calculate the K_{sp} for calcium sulfate if its solubility is 0.67 g/L.

51. The solubility of AgBr is 7.1×10^{-7} *M*. Calculate K_{sp} for AgBr.

52. Calculate the solubility (in g/L) of $Fe(OH)_3$. The K_{sp} for iron(III) hydroxide [$Fe(OH)_3$] is 1.8×10^{-15}.

53. The solubility of SrF_2 is 5.8×10^{-4} *M*. Calculate K_{sp} for SrF_2.

54. The solubility of the ionic compound M_2X_3, which has a molar mass of 288 g, is 3.6×10^{-17} g/L. Calculate the K_{sp} of the compound.

55. Calculate the concentration of Ag^+ in a saturated solution of Ag_2CrO_4 ($K_{sp} = 9.0 \times 10^{-12}$).

56. What is the solubility of silver chloride (in g/L) ($K_{sp} = 1.6 \times 10^{-10}$) in a 6.5×10^{-3} *M* silver nitrate solution?

57. Calculate the number of grams of ZnS ($K_{sp} = 2.5 \times 10^{-22}$) that will dissolve in 3.0×10^2 mL of 0.050 *M* $Zn(NO_3)_2$.

58. Calculate the solubility of $Co(OH)_2$ ($K_{sp} = 2.5 \times 10^{-16}$) at pH 11.50.

59. Calculate the solubility of $PbCO_3$ ($K_{sp} = 1.5 \times 10^{-15}$) in

    a. pure water
    b. 0.0400 *M* $Pb(NO_3)_2$

60. If 50.0 mL of a solution of 2.0×10^{-3} *M* $CaCl_2$ is mixed with 100.0 mL of 5.0×10^{-2} *M* NaF, will precipitation occur?

61. Determine if a precipitate will form when exactly 200 mL of 0.0040 *M* $BaCl_2$ is added to 600 mL of 0.0080 *M* K_2SO_4.

Section 8.9

62. Calculate the number of moles of Ag_2CrO_4 ($K_{sp} = 9.0 \times 10^{-12}$) that will dissolve in 1.00 L of 0.010 *M* K_2CrO_4 solution. What will be the ion concentrations at equilibrium?

63. Calculate the equilibrium concentration of each ion in a solution obtained by mixing 75.0 mL of 2.5×10^{-2} *M* $AgNO_3$ with 25.0 mL of 3.2×10^{-4} *M* KI.

64. Will $BaCO_3$ ($K_{sp} = 1.6 \times 10^{-9}$) precipitate if a sample of 20.0 mL of 0.10 *M* $Ba(NO_3)_2$ is added to 50.0 mL of 0.10 *M* Na_2CO_3?

65. Write equations for the stepwise and overall formation of $Cd(NH_3)_4^{2+}$.

66. What is the molar solubility of silver chloride in a 1.0 *M* NH_3 solution?

Section 8.10

67. Calculate the concentrations of Ag^+ and $Ag(NH_3)_2^+$ in a solution with an initial Ag^+ concentration of 4.0×10^{-3} *M* and an initial NH_3 concentration of 0.500 *M* ($K_1 = 2.1 \times 10^3$, $K_2 = 8.2 \times 10^3$).

Multiple-Choice Self-Test

1. Calculate the pH of a solution prepared by mixing 40.0 mL of a 0.02 *M* HCl solution with 200.0 mL of 0.20 *M* HCN solution. Assume volumes to be additive. K_a for HCN = 1.0×10^{-10}.
   A. 2.4 B. 10 C. 2.0 D. 5.6

2. Calculate the pH of a solution prepared by mixing 60.0 mL of a 0.200 *M* NaOH solution with 60.0 mL of a 0.200 *M* CH_3NH_2 solution. Assume the volumes to be additive. K_b for CH_3NH_2 = 4.3×10^{-5}.
   A. 11.8 B. 13.0 C. 1.0 D. 0.7

3. 40.0 g of NaF and 40.0 g of HF are mixed in 900 mL of solution. What is the pH of this buffer pair? $K_a = 7.2 \times 10^{-4}$.
   A. 2.82 B. 4.90 C. 1.60 D. 3.14

4. A student is asked to prepare a buffer solution composed of KH_2PO_4 and K_2HPO_4 with a pH of 6.00. What ratio of **base to acid** is required to prepare this buffer solution? $K_a = 6.2 \times 10^{-8}$.
   A. 10.0: 1 B. 0.545: 1 C. 1.12: 1 D. 0.062: 1

5. How many grams of NaF must be added to 120 mL of 1.00 *M* HF solution, in order to adjust the pH to 3.40? Assume no change in volume upon addition of solid NaF. $K_a = 7.2 \times 10^{-4}$.
   A. 0.69 B. 4.34 C. 1.64 D. 6.89

6. 4.00 mL of a 0.80 *M* NaOH solution is added to 130.0 mL of 0.20 *M* HCN/0.20 *M* NaCN buffer solution. Calculate the change in pH of this buffer solution.
   A. 0.11 B. 1.2 C. −1.2 D. 0.46

7. 40.0 mL of 0.320 *M* benzoic acid is titrated with 60.0 mL of 0.20 *M* NaOH. Calculate the pH of the resulting solution at the equivalence point. $K_a = 6.3 \times 10^{-5}$.
   A. 5.5 B. 8.5 C. 7.7 D. 3.5

8. 50.0 mL of 0.02 *M* HCN is titrated with 100.0 mL of 0.010 *M* LiOH. Calculate the pH of the resulting solution at the equivalence point. $K_a = 4.9 \times 10^{-9}$.

A. 9.9 B. 4.1 C. 7.7 D. 5.2

9. 1.024 g of an unknown monoprotic strong acid is dissolved in 70.0 mL of solution. The solution is then titrated to the equivalence point with 80.0 mL of 0.200 *M* KOH. Calculate the molar mass of this acid.

A. 128 g/mol B. 81.0 g/mol C. 64 g/mol D. 211 g/mol

10. How many mL of a 1.06 *M* calcium hydroxide solution are required to titrate 70.0 mL of 0.88 *M* phosphoric acid solution to the third equivalence point?

A. 87.2 B. 70.0 C. 22.6 D. 43.9

11. Calculate the molar solubility of $BaCO_3$ in water. $K_{sp} = 8.1 \times 10^{-9}$.

A. 9.0×10^{-5} B. 4.9×10^{-9} C. 8.5×10^{-9} D. 4.5×10^{-5}

12. Calculate the molar solubility of AgCl in 0.11 *M* NaCl solution.

A. 6.6×10^{-11} B. 1.5×10^{-9} C. 7.2×10^{-10} D. 2.0×10^{-11}

13. Calculate the molar solubility of Ag_2CrO_4 in a 0.08 *M* Na_2CrO_4 solution.

A. 4.8×10^{-7} B. 7.00×10^{-12} C. 1.1×10^{-5} D. 2.4×10^{-7}

14. How many milligrams of FeS per liter will dissolve in a 0.20 *M* Na_2S solution? $K_{sp} = 4.9 \times 10^{-18}$.

A. 1.6×10^{-13} B. 4.6×10^{-9} C. 8.6×10^{-12} D. 2.4×10^{-18}

15. Calculate the formation constant, K_f, for the following complex ion:

$$Ag^+ + 2Br^- \rightleftharpoons [AgBr_2]^-$$

in which the concentrations of free silver, free bromide, and complex ion are 1.56×10^{-6}, 0.20, and 0.20 *M*, respectively.

A. 3.2×10^6 B. 7.8×10^6 C. 1.3×10^7 D. 7.8×10^{-7}

16. Calculate the concentration of free zinc ions in a solution prepared by mixing 2.00 g of $ZnCl_2$ with 16.00 g of NaOH in 1000 mL of solution.

$$Zn^{2+} + 4OH^- \rightleftharpoons [Zn(OH)_4]^{2-}$$

A. 2.38×10^{-16} B. 1.2×10^{-14} C. 7.9×10^{-18} D. 3.81×10^{-16}

Answers to Exercises

1. The pH is raised due to the common-ion effect.

2. a. pH = 2.52, 2% dissociated.
   b. pH = 4.58, less than 2.0×10^{-2} % dissociated.

3. a. pH = 11.45, 0.6% hydrolyzed.
   b. pH = 9.42, less than 6×10^{-3} % hydrolyzed.

4. $[H^+] = 2.7 \times 10^{-4}$ *M*

5. $[OH^-] = 4.4 \times 10^{-4}$ *M*

6. pH = 8.79

7. pH = 7.03

8. Yes, a 6 *M* HCl solution is a superior buffer because it resists changes in pH when an acid or base is added. However, it does not meet the traditional definition of a buffer, which contains a conjugate acid/base pair or pairs. It is not biologically useful as it is far too acidic to be biologically safe.

9. a. pH = 4.37
   b. pH = 3.73
   c. No, the buffer has not exceeded its capacity.

10. pH = 4.55

11. pH = 9.29

12. The ratio must be 0.86 to 1.00.

13. pH = 8.88

14. pH = 9.21

15. A total of 8.33 mL of 6.0 *M* NaOH must be added.

16. pH = 9.45

17. $K_b = 1.27 \times 10^{-4}$

18. 4.3 g

19. 0.023

20. pH = 4.39

21. CH_3COOH would be the most useful ($pK_a = 4.75$).

22. $K = 1/K_w = 1 \times 10^{14}$

23. 0.612 *M* KOH

24. 0.1614 *M* HCl

25. a. pH = 0.22
    b. pH = 0.30
    c. pH = 0.82
    d. pH = 2.82
    e. pH = 7.00
    f. pH = 11.10
    g. pH = 12.36

26. 0.115 *M*

27. a. pH = 13.20
    b. pH = 13.15
    c. pH = 10.93
    d. pH = 7.00
    e. pH = 3.07
    f. pH = 1.70

28. 7.34×10^{-2} *M*

29. 0.250 *M*

30. 5.28

31. 8.35

32. a. pH = 2.81
    b. pH = 4.74
    c. pH = 8.89
    d. pH = 12.32

33. $K = 1.8 \times 10^{9}$

34. 23.4 mL

35. 6.3×10^{-5}

36. a. pH = 11.26 c. pH = 9.26 e. pH = 2.32
    b. pH = 9.86 d. pH = 5.09

37. $K = K_aK_b/K_w = 1.1 \times 10^5$. This titration is not feasible. The value for K is too low (below 10^7 or so).

38. molar mass = 386.9 g/mol

39. Bromthymol blue or thymol blue are suitable choices.

40. Thymol blue or phenolphthalein are suitable choices.

41. Bromcresol green is a suitable choice.

42. methyl orange

43. alizarin yellow R

44. a. $PbI_2(s) \rightleftharpoons Pb^{2+}(aq) + 2I^-(aq)$ $K_{sp} = [Pb^{2+}][I^-]^2$
    b. $Sr_3(PO_4)_2(s) \rightleftharpoons 3Sr^{2+}(aq) + 2PO_4^{3-}(aq)$ $K_{sp} = [Sr^{2+}]^3[PO_4^{3-}]^2$
    c. $MnS(s) \rightleftharpoons Mn^{2+}(aq) + S^{2-}(aq)$ $K_{sp} = [Mn^{2+}][S^{2-}]$

45. 1.3×10^{-8} M

46. solubility = 4.8×10^{-7} M

47. $K_{sp} = 1.4 \times 10^{-5}$

48. 4.2×10^{-6} M

49. solubility = 2×10^{-7} M

50. $K_{sp} = 2.4 \times 10^{-5}$

51. $K_{sp} = 5.0 \times 10^{-13}$

52. 9.7×10^{-3}

53. $K_{sp} = 7.8 \times 10^{-10}$ (actual = 7.9×10^{-10})

54. 3.3×10^{-93}

55. $[Ag^+] = 2.6 \times 10^{-4}$ M

56. solubility = 3.6×10^{-6} g/L

57. 1.5×10^{-19} g ZnS

58. solubility = 2.5×10^{-11} M

59. a. solubility = 3.9×10^{-8} M
    b. 3.8×10^{-14} M

60. Yes, $Q = 7.4 \times 10^{-7}$ (>4.0×10^{-11}, the K_{sp} of CaF_2)

61. Yes, $Q = 6.0 \times 10^{-6}$. $BaSO_4$ will precipitate out of solution until $[Ba^{2+}][SO_4^{2-}] > 1.5 \times 10^{-9}$, the K_{sp} of $BaSO_4$.

62. 1.5×10^{-5} moles; solubility = 1.5×10^{-5} mol/L; $[Ag^+] = 3.0 \times 10^{-5}$ M; $[CrO_4^{2-}] = 0.010$ M

63. $[Ag^+] = 1.9 \times 10^{-2}$ M; $[I^-] = 8.0 \times 10^{-15}$ M.

64. Yes, $BaCO_3$ will precipitate out of solution.

65. stepwise: $Cd^{2+} + NH_3 \rightleftharpoons Cd(NH_3)^{2+}$ K_1

$Cd(NH_3)^{2+} + NH_3 \rightleftharpoons Cd(NH_3)_2^{2+}$ K_2

$Cd(NH_3)_2^{2+} + NH_3 \rightleftharpoons Cd(NH_3)_3^{2+}$ K_3

$Cd(NH_3)_3^{2+} + NH_3 \rightleftharpoons Cd(NH_3)_4^{2+}$ K_4

overall: $Cd^{2+} + 4NH_3 \rightleftharpoons Cd(NH_3)_4^{2+}$ $\beta_4 = K_1K_2K_3K_4$

66. solubility = 0.50 M

67. $[Ag^+] = 9.6 \times 10^{-10}$; $[Ag(NH_3)_2^+] = 4.0 \times 10^{-3}$

Answers to Multiple-Choice Self-Test

1. A	4. D	7. B	10. A	13. C	15. A
2. B	5. B	8. A	11. A	14. A	16. A
3. A	6. A	9. C	12. B		

CHAPTER 9

Energy, Enthalpy, and Thermochemistry

The Bottom Line: Chapter 9

This chapter serves as an introduction to the chemistry of energy production and exchange. Additional, more advanced material will be presented in Chapter 10.

9.1 The Nature of Energy

Your textbook describes a number of new terms for you. The terms, along with brief definitions, are given below:

- **Thermodynamics:** The study of energy and its interconversions.
- **Energy:** The capacity to produce work or heat.
- **Kinetic Energy**: The energy of motion. Kinetic Energy = 1/2 mass × (velocity)^2.
- **Potential Energy**: Energy that can be converted to useful work.
- **Heat**: Involves transfer of energy between two objects.
- **Work**: Force × distance
- **State Function**: A property that is independent of pathway. That is, it does not matter how you get there, the difference in the value is the same. For example, you can drive from New York to Los Angeles via many different routes. No matter which one you take, you are still going from New York to Los Angeles. The actual distance between the two cities is the same. **Energy** is a state function; **work** and **heat** are not.

Four more definitions will set the stage for thermodynamics. The **universe** is composed of the **system** and the **surroundings.**

- **System:** That on which we are focusing.
- **Surroundings:** Everything else in the universe.
- **Exothermic:** Energy (as heat) flows **out of the system.**
- **Endothermic:** Energy (as heat) flows **into the system.**

Work

We will take a different approach to this topic than your textbook does. Let's look at work from the point of view of **units**. We learned in the review chapter on gas laws that

$$\text{Force} = \text{mass} \times \text{acceleration} = \text{kg} \times \text{m s}^{-2}.$$
$$\text{Work} = \text{force} \times \text{distance} = \text{kg m s}^{-2} \times \text{m} = \text{kg m}^2\ \text{s}^{-2}.$$
$$\mathbf{1\ Joule = 1\ kg\ m^2\ s^{-2}}.$$

Example 9.1 A - The Units Of Work

If pressure = force/area, what are the units of pressure × volume?

Solution

$$P = F/A = \text{kg m s}^{-2} / \text{m}^2 = \text{kg m}^{-1}\ \text{s}^{-2}\ (= 1\ \text{Pascal})$$
$$P \times V = \text{kg m}^{-1} \cdot \text{s}^{-2} \times \text{m}^3 = \text{kg m}^2\ \text{s}^{-2}\ (= 1\ \text{Joule})$$

Therefore, $P \times V$ has the same units as force × distance (work), and both are measures of **energy**.

Conclusion

For an ideal gas,

$$\textbf{WORK} = P\Delta V$$

This equation holds at constant pressure.

The **sign conventions** for work are as follows.

- When the **system expands,** it is doing positive work on the surroundings; therefore it is doing **negative work on the system.**
- When the **system contracts,** the surroundings have done work on the system; therefore there is **positive work done on the system.**

From the point of view of the system, then,

$$w = -P\Delta V$$

Example 9.1 B - Work

Calculate the work (with the proper sign) associated with the contraction of a gas from 75 L to 30 L (work is done "on the system") at a constant **external** pressure of 6.0 atm in

a. L atm, and
b. Joules (1 L atm = 101.3 J).

Solution

Keep in mind that system compression is positive work, and system expansion is negative work.

a. The work is positive (compression).

$$\Delta V = \text{change in volume} = V_{\text{final}} - V_{\text{initial}} = 30\text{ L} - 75\text{ L} = \mathbf{-45\ L}$$

$$w = -P\Delta V = -6.0\text{ atm }(-45\text{ L}) = \mathbf{+270\ L\ atm}$$

b. $w\text{ (in Joules)} = +270\text{ L atm} \times \dfrac{101.3\text{ J}}{1\text{ L atm}} = \mathbf{+2.7 \times 10^4\ J}$

The First Law

The law of conservation of energy, also called the **first law of thermodynamics,** is described in your textbook. It states that **energy can be converted from one form to another, but can neither be created nor destroyed.** Another way of stating the first law is that

THE ENERGY OF THE UNIVERSE IS CONSTANT.

We know that energy can be changed through **work**. As chemical bonds are made and broken, energy is converted between potential energy (stored in chemical bonds) and thermal energy (kinetic energy) as heat.

The change in the internal energy of the system, which is equal in size but opposite in sign to that of the surroundings, is equal to the **sum of the heat and work.**

$$\Delta E = q + w$$

Your textbook points out that the **SIGN** of the energy change must be viewed from the point of view of the **SYSTEM.**

ΔE = − means the system loses energy.
ΔE = + means the system gains energy.

Example 9.1 C - The First Law

Calculate the change in energy of the system if 38.9 J of work is done *by* the system with an associated heat loss of 16.2 J.

Solution

The most important part of problem solving in thermodynamics is getting the signs correct.

q = "−," because heat is **lost**.

w = "−," because work is done **by** the system.

$$\Delta E = q + w = -16.2\text{ J} + (-38.9\text{ J}) = \mathbf{-55.1\text{ J}}$$

The system has **lost** 55.1 J of energy.

Example 9.1 D - Practice With Heat And Work

A piston is compressed from a volume of 8.3 L to 2.8 L against a constant pressure of 1.9 atm. In the process, there is a heat gain by the system of 350 J. Calculate the change in energy of the system.

Solution

$$w = -P\Delta V = -1.9\text{ atm }(-5.5\text{ L}) = +\ 10.45\text{ L atm}$$

$$10.45\text{ L atm} \times \frac{101.3\text{ J}}{\text{L atm}} = \mathbf{+1059\text{ J}}$$

$$q = +350\text{ J}$$

$$\Delta E = q + w = +1059 + 350 = 1409\text{ J} = \mathbf{1400\text{ J}}\text{ (to 2 significant figures)}$$

9.2 Enthalpy

Your textbook derives and defines a term called **enthalpy** (H). It is a **state function,** so the change in H is independent of pathway. That is,

$$\Delta H = H_{\text{products}} - H_{\text{reactants}}$$

The change in enthalpy (ΔH) of the system is equal to the energy flow as heat **at constant pressure**.

$$\Delta H = q_{\text{p}}$$

If $\Delta H > 0$, the reaction is **endothermic.** (Heat is absorbed by the system.)
If $\Delta H < 0$, the reaction is **exothermic.** (Heat is given off by the system.)

Example 9.2 - Enthalpy

Upon adding solid potassium hydroxide pellets to water, the following reaction takes place:

$$\text{NaOH}(s) \rightarrow \text{NaOH}(aq)$$

For this reaction at constant pressure, ΔH = −43 kJ/mol. Answer the following questions regarding the addition of 14 g of NaOH to water:

a. Does the beaker get warmer or colder?
b. Is the reaction exo- or endothermic?
c. What is the enthalpy change for the dissolution?

Solution

a. If $\Delta H < 0$, then heat is given off by the system. The beaker therefore gets **warmer**.

b. If heat is given off by the system, the reaction is **exothermic**.

c. $\text{kJ} \parallel \dfrac{-43\text{ kJ}}{\text{mol}} \times \dfrac{1\text{ mol NaOH}}{40.0\text{ g NaOH}} \times 14\text{ g NaOH} = \mathbf{-15\text{ kJ} = \Delta H}$

9.3 Thermodynamics of Ideal Gases

Your textbook sets up this section by recommending that we simplify our conditions as much as possible by studying the thermodynamics of an ideal gas. Test your understanding of the material in this section by answering the following questions:

1. What conditions are most favorable if a gas is to behave ideally?
2. Why is C_v equal to $^3/_2R$, the result anticipated for C_p?
3. Probe that $C_p = {^5/_2}R$.
4. Why are C_v values for polyatomic gases considerably greater than $^3/_2R$?
5. Suggest what is meant by nontranslational motions. What conclusion does this support?
6. Prove that $\Delta H = nC_p\Delta T$.
7. In Example 9.2 in your textbook, why are the w and q different for the two different pathways, but the ΔE is the same?

9.4 Calorimetry

Calorimetry is the experimental technique used to determine the heat exchange (***q***) associated with a reaction.

At constant pressure, $q = \Delta H$.
At constant volume, $q = \Delta E$.

In both cases, however, **heat gain or loss** is being determined. The amount of heat exchanged in a reaction depends upon:

1. **The net temperature change** during the reaction.
2. **The amount of substance.** The more you have, the more heat can be exchanged.
3. The **heat capacity** (C) of a substance.

$$C = \frac{\text{heat absorbed}}{\text{increase in temperature}} = \text{J/°C}$$

Some substances can absorb more heat than others for a given temperature change.

There are three ways of expressing heat capacity:

1. **Heat capacity** (as above) = J/°C.
2. **Specific heat capacity** = heat capacity per gram of substance (J/°C g or J/K g).
3. **Molar heat capacity** = heat capacity per mole of substance (J/°C mol or J/K mol)

You can solve calorimetry problems very well using dimensional analysis. Before we solve numerical problems, let's do a problem involving interpretation of specific heat capacities.

Example 9.4 A - Specific Heat Capacity

Look at Table 9.3 in your textbook. Based on the values for specific heat capacity, which requires less heat to raise its temperature, water or aluminum? Why is this important in cooking?

Solution

The specific heat capacity of water is **4.18 J/°C g**. This means it takes 4.18 J of energy to raise the temperature of one gram of water 1°C.

The specific heat capacity of aluminum is **0.89 J/°C g**. This means it takes 0.89 J of energy to raise the temperature of one gram of aluminum 1°C.

In other words, it takes almost 5 times as much energy (4.18/0.89) to raise the temperature of an equivalent amount of water by 1°C. Therefore, **aluminum requires less heat** to cause an equal rise in temperature. This is important in cooking because pots made of aluminum get hot fairly quickly and transfer this heat very well to the food in the pot. Note that iron pots have an even lower specific heat capacity than aluminum (but they are more difficult to take care of)!

Your textbook discusses doing **constant-pressure** calorimetry using a "coffee cup calorimeter." In this case, $\Delta H = q_p$ in units of joules. Remember that you may use dimensional analysis to solve calorimetry problems.

Example 9.4 B - Constant Pressure Calorimetry

Recall from Example 9.2 that the enthalpy change (ΔH) per mole of NaOH is −43 kJ/mol when

$$\text{NaOH}(s) \rightarrow \text{NaOH}(aq)$$

If 10.0 g of solid NaOH is added to 1.00 L of water (specific heat capacity = 4.18 J/°C g) at 25.0°C in a constant pressure calorimeter, what will be the final temperature of the solution? (Assume the density of the final solution is 1.05 g/mL.)

Solution

Where are we going? In this problem, we are asked to solve for the temperature of the solution. To solve this, we need to know three things:

1. **Mass of the solution** = 1.00 L × 1050 g/L = 1050 g.
2. **Specific heat capacity of the solution** = 4.18 J/°C g.
3. **The enthalpy of the reaction** = $-43 \text{ kJ/mol} \times 10.0 \text{ g NaOH} \times \frac{1 \text{ mol NaOH}}{40.0 \text{ g NaOH}} \times \frac{1000 \text{ J}}{1 \text{ kJ}}$

   $= -10{,}750$ J.

The **change in temperature** is ΔT. We can solve using dimensional analysis. (Keep in mind the temperature will **rise** because heat is evolved.)

$$°\text{C} \parallel \frac{1°\text{C g}}{4.18 \text{ J}} \times \frac{1}{1050 \text{ g}} \times 10{,}750 \text{ J} = 2.4°\text{C}$$

$$\underset{\frac{1}{\text{specific heat capacity}}}{\uparrow} \qquad \underset{\frac{1}{\text{soln mass}}}{\uparrow} \qquad \underset{\Delta H}{\uparrow}$$

The **final temperature** will equal 25.0°C + 2.4°C = **27.4°C**.

Constant-volume calorimetry is discussed in your textbook. The **bomb calorimeter** is used for this application. In this case, because $\Delta V = 0$, no work is done, and $\Delta E = q_v$ in units of joules. Here, as well, dimensional analysis works well.

Each bomb calorimeter is different. The **heat capacity (J/°C)** of the bomb and its parts must be determined using a known substance before the energy (or heat) of combustion can be determined.

Example 9.4 C - Constant Volume Calorimetry

The heat of combustion of glucose, $C_6H_{12}O_6$, is 2800 kJ/mol. A sample of glucose weighing 5.00 g was burned with excess oxygen in a bomb calorimeter. The temperature of the bomb rose 2.4°C. **What is the heat capacity of the calorimeter?**

A 4.40-g sample of propane (C_3H_8) was then burned with excess oxygen in the same bomb calorimeter. The temperature of the bomb increased 6.85°C. Calculate $\Delta E_{\text{combustion}}$ of propane.

Solution

Where are we going? There are two parts to this problem. First we are asked to calculate the heat capacity of the bomb calorimeter using the data for glucose. Second, we are asked to use this heat capacity to determine the energy (heat) of combustion of propane, C_3H_8. Remember that the energy of combustion is expressed in **kJ/mol.**

How do we determine the heat capacity of the calorimeter and, ultimately, the energy of the combustion of propane? A useful beginning is to convert grams of substance to moles of substance.

a. ***Heat Capacity of the Calorimeter***

$$\text{moles } C_6H_{12}O_6 = 5.00\text{ g} \times \frac{1\text{ mol}}{180.0\text{ g}} = 2.78 \times 10^{-2}\text{ moles}$$

$$\textbf{heat capacity} = \frac{\text{kJ}}{°\text{C}} \parallel \frac{2800\text{ kJ}}{\text{mol}} \times 2.78 \times 10^{-2}\text{ mol} \times \frac{1}{2.4°\text{C}} = \mathbf{32.4\ kJ/°C}$$

b. ***Energy of Combustion of Propane***

$$\text{moles } C_3H_8 = 4.40\text{ g} \times \frac{1\text{ mol}}{44.0\text{ g}} = 0.100\text{ moles}$$

$$\Delta E_{\text{combustion}} = \frac{\text{kJ}}{\text{mol}} \parallel \frac{32.4\text{ kJ}}{°\text{C}} \times 6.85°\text{C} \times \frac{1}{0.100\text{ mol}} = \mathbf{-2200\ kJ/mol}$$

(Note: We add the **negative** sign because heat is **evolved.**)

9.5 Hess's Law

The critical point that is made in this section is that **enthalpy changes are state functions.** The implication is that **it does not matter if ΔH for a reaction is calculated in one step or a series of steps. This idea is called Hess's law.**

By using ΔH values of known reactions, we can use Hess's law to solve for enthalpies of reactions whose values we do not know.

Example 9.5 A - Hess's Law

Given the following reactions and ΔH values,

a. $2N_2O(g) \rightarrow O_2(g) + 2N_2(g)$ $\Delta H_a = -164$ kJ
b. $2NH_3(g) + 3N_2O(g) \rightarrow 4N_2(g) + 3H_2O(l)$ $\Delta H_b = -1012$ kJ

Calculate ΔH for

$$4NH_3(g) + 3O_2(g) \rightarrow 2N_2(g) + 6H_2O(l).$$

Solution

The idea is to manipulate equations a and b so that they add up to the desired equation. There are three ways we can manipulate equations:

1. **We can reverse the entire equation**. By doing this, the products become reactants and vice versa.
2. **We can multiply the entire equation by a factor** such as 3, 2, 1/2, or 1/3.
3. **We can do both No. 1 and No. 2.**

The most important thing to keep in mind is that **WHEN YOU MANIPULATE AN EQUATION, YOU MUST MANIPULATE THE ΔH VALUE IN EXACTLY THE SAME WAY!**

If you multiply an equation by 2, you must multiply ΔH by 2. If you reverse the equation, you must multiply ΔH by -1. (An exothermic reaction becomes endothermic and vice versa.)

An Effective Route To Solving Hess's Law Problems

In my experience, the best way to solve Hess's law problems is to **find a substance that appears only once in the reactants.** Modify that reaction so the substance appears where it should be, and in the correct amount, as in the final reaction. The entire substance equation must therefore be correct. In our example, NH_3 appears only once in the reactants. (N_2O appears in **both** equations...STAY AWAY FROM N_2O!)

We have $2NH_3$ on the left-hand side. We want $4NH_3$ on that side. Therefore, we must multiply equation b and ΔH_b by +2, which gives

$$\mathbf{4NH_3(g) + 6N_2O(g) \rightarrow 8N_2(g) + 6H_2O(l) \qquad \Delta H_b = -2024\ kJ}$$

Oxygen appears only once in the reactants. Therefore, if we modify equation a to get the correct amount of O_2 in the proper place, we should be done. (We have modified both equations.) We need to **reverse** equation a and **multiply** it by 3 to get $3O_2$ on the left side. This will agree with the desired reaction. Remember to multiply ΔH_a by -3 as well! This gives,

$$\mathbf{3O_2(g) + 6N_2(g) \rightarrow 6N_2O(g) \qquad \Delta H_a = +492\ kJ}$$

Let's get the final ΔH by adding our new a and b,

$4NH_3(g) + 6N_2O(g) \rightarrow 8N_2(g) + 6H_2O(l)$	$\Delta H_b = -2024$ kJ
$3O_2(g) + 6N_2(g) \rightarrow 6N_2O(g)$	$\Delta H_a = +492$ kJ
$3O_2(g) + 6N_2(g) + 4NH_3(g) + 6N_2O(g) \rightarrow 6N_2O(g) + 8N_2(g) + 6H_2O(l)$	$\Delta H = -1532$ kJ

and, canceling common terms,

$$\mathbf{3O_2(g) + 4NH_3(g) \rightarrow 2N_2(g) + 6H_2O(l) \qquad \Delta H = -1532\ kJ}$$

Getting the correct final reaction serves as your check of correctness.

Example 9.5 B - Practice With Hess's Law

Given the following reactions and ΔH values,

a. $B_2O_3(s) + 3H_2O(g) \rightarrow B_2H_6(g) + 3O_2(g)$ $\Delta H = +2035$ kJ
b. $2H_2O(l) \rightarrow 2H_2O(g)$ $\Delta H = +88$ kJ
c. $H_2(g) + \frac{1}{2}O_2(g) \rightarrow H_2O(l)$ $\Delta H = -286$ kJ
d. $2B(s) + 3H_2(g) \rightarrow B_2H_6(g)$ $\Delta H = +36$ kJ

Calculate ΔH for

$$2B(s) + {}^3/_2O_2(g) \rightarrow B_2O_3(s)$$

Solution

Keep the strategy in mind! Boron and B_2O_3 appear only once. Work with them first.

1. Working with $B(s)$, we see that **equation d is correct as is.**
2. Working with $B_2O_3(s)$, we must **multiply equation a by −1** (thus reversing the equation).

Now we must work with things that are present more than once.

3. Working with $O_2(g)$, we see that we **must multiply equation c by −3** to guarantee a total of ${}^3/_2O_2(g)$ on the left side when all equations are added together. (Remember, we already have $3O_2(g)$ on the left side from equation a.)

Three of our equations are now set. Equation b involves $H_2O(l)$ and $H_2O(g)$. Because we multiplied equation c by −3, we have $3H_2O(l)$ on the left-hand side to cancel. Therefore, we must **multiply equation b by −3/2.**

Checking the Equations and Determining ΔH

a. $B_2H_6(g) + 3O_2(g) \rightarrow B_2O_3(s) + 3H_2O(g)$ $\Delta H = -2035$ kJ
b. $3H_2O(g) \rightarrow 3H_2O(l)$ $\Delta H = -132$ kJ
c. $3H_2O(l) \rightarrow 3H_2(g) + {}^3/_2O_2(g)$ $\Delta H = +858$ kJ
d. $2B(s) + 3H_2(g) \rightarrow B_2H_6(g)$ $\Delta H = +36$ kJ

$$B_2H_6(g) + 3O_2(g) + 3H_2O(g) + 3H_2O(l) + 2B(s) + 3H_2(g) \rightarrow B_2O_3(s) + 3H_2O(g) + 3H_2O(l) + 3H_2(g) + {}^3/_2O_2(g) + B_2H_6(g)$$

Canceling, we get

$2B(s) + {}^3/_2O_2(g) \rightarrow B_2O_3(s)$ **$\Delta H = -1273$ kJ**

Look at Interactive Example 9.6 in your textbook. Some of the equations here were turned around and multiplied by an integer value, but notice the ΔH values ultimately agree with those in your textbook. Hess's law illustrates that enthalpy is a state function.

9.6 Standard Enthalpies of Formation

Look at your textbook where the **standard enthalpy of formation** (ΔH_f°) of a compound is defined. Here are some important points worth going over:

1. ΔH_f° is always given **per mole** of compound formed.
2. ΔH_f° involves formation of a **compound** from its **elements** with the substances in their **standard states**.

3. Your textbook lists the following standard-state conditions:

For an element:

- It is the form in which the element exists at 25°C and 1 atmosphere. Your textbook notes that "standard pressure" is 100,000 Pa = 1 bar. This is slightly different than one atmosphere, but the thermodynamic values are essentially the same at both pressures.

For a compound:

- For a gas, it is a pressure of exactly 1 atmosphere.
- For a substance in solution, it is a concentration of exactly 1 *M*.
- For a pure solid or liquid, it is the pure solid or liquid.

4. ΔH_f° for an element in its standard state (such as Ba(*s*) or $N_2(g)$) equals 0.

Example 9.6 A - Standard Enthalpies Of Formation

By consulting Appendix 4 of your textbook, and from your knowledge of standard states, list the standard enthalpy of formation for each of the following substances.

a. $Al_2O_3(s)$
b. Ti(*s*)
c. $P_4(g)$
d. $SO_4^{2-}(aq)$
e. $F_2(g)$

Solution

a. −1676 kJ/mol
b. 0 kJ/mol (The solid is the standard state of titanium.)
c. 59 kJ/mol
d. −909 kJ/mol
e. 0 kJ/mol (The gaseous diatom is the standard state of fluorine.)

The key to calculating standard enthalpy changes in reactions is to remember "products minus reactants." More correctly,

$$\Delta H^\circ_{\text{reaction}} = \sum n_p \Delta H_f^\circ(\text{products}) - \sum n_r \Delta H_f^\circ(\text{reactants}).$$

This reads "the sum of the ΔH_f° for *n* moles of each of the products minus the sum of the ΔH_f° for *n* moles of each of the reactants."

Note: Keep in mind that, just as in Hess's law problems, when you multiply the substance by an integer coefficient in a balanced equation, you must multiply the ΔH_f° value by that integer as well!

Let's work the following example together.

Example 9.6 B - Calculating Standard Enthalpies Of Formation

Using the data in Appendix 4 of your textbook, calculate ΔH° for the following reaction:

$$2C_3H_6(g) + 9O_2(g) \rightarrow 6CO_2(g) + 6H_2O(l)$$

Solution

The ΔH° for the reaction $= \sum n_p \Delta H_f^\circ(\text{products}) - \sum n_r \Delta H_f^\circ(\text{reactants})$

$\sum n_p \Delta H_f^\circ$ (products) $= 6 \times \Delta H_f^\circ[\text{for } H_2O(l)] + 6 \times \Delta H_f^\circ[\text{for } CO_2(g)]$

$= 6 \text{ mol} \times -286 \text{ kJ/mol} + 6 \text{ mol} \times -393.5 \text{ kJ/mol}$

$= \mathbf{-4077\ kJ}$

$$\sum n_r \Delta H_f^\circ(\text{reactants}) = 2 \times \Delta H_f^\circ[\text{for } C_3H_6(g)] + 9 \times \Delta H_f^\circ[\text{for } O_2(g)]$$
$$= 2 \text{ mol} \times 20.9 \text{ kJ/mol} + 9 \text{ mol} \times 0 \text{ kJ/mol}$$
$$= \mathbf{+41.8 \text{ kJ}}$$

Finally,

$$\Delta H^\circ = (-4077 \text{ kJ}) - (+41.8 \text{ kJ}) = \mathbf{-4119 \text{ kJ}}$$

9.7 Present Sources of Energy

Carefully read Section 9.7 in your textbook. This material deals with the application of thermodynamics to the "real world" and as such is perhaps the most interesting material in the chapter! When you are done, use the following review questions to test your understanding of the material.

1. What is **petroleum**?
2. From what was petroleum formed?
3. What gases make up **natural gas**?
4. How are hydrocarbons separated from one another?
5. What are the major uses of the various petroleum fractions?
6. What is **pyrolitic cracking**?
7. There are two reasons that lead additives are a problem in gasoline. Discuss them.
8. What is **hydraulic fracturing**, and how does it work? Why is this technique useful at this time?
9. What are the four stages of **coal**? How do they differ?
10. What are some of the problems with the use of coal as an energy source?
11. What is the greenhouse effect? Why will increased amounts of CO_2 exacerbate this effect?

9.8 New Energy Sources

One of the impressive things about your textbook is the extent to which it deals with the application of chemistry to important issues of the day as the following section-related questions demonstrate.

1. What is meant by **coal gasification**?
2. Outline the process of coal gasification.
3. Define **synthetic gas**.
4. Why is hydrogen a theoretically good choice for use as a fuel?
5. Why is hydrogen a poor **practical** choice for use as a fuel? (List reactions and thermodynamic values.)
6. What is the most common method of producing ethanol?
7. What are some of the advantages and disadvantages of alcohols in fuels?
8. Why might biodiesel be useful?

Exercises

Section 9.1

1. A system does 3 J of work on the surroundings, and 12 J of work are added to the system.
   a. What is the energy change of the system?
   b. Of the surroundings?

2. One hundred joules of work are required to compress a gas. At the same time, the gas gives off 23 J of heat to the surroundings. What is the energy change of the system?

3. A gas expands from 10 L to 20 L against a constant pressure of 5 atm. During this time it absorbs 2 kJ of heat. Calculate the work done in kJ.

4. A piston expands from 11.2 L to 29.1 L against 1.00 atm of pressure. This is done without any transfer of heat.

   a. Calculate the change in energy of the system.
   b. Calculate the change in energy for the above change if, in addition, the system absorbs 1,037 J of heat from the surroundings.

5. If the internal energy of a thermodynamic system decreases by 300 J when 75 J of work is done *on* the system, how much heat was transferred, and in which direction–to or from the system?

6. How much work is done by a system in which pressure is kept constant, but the volume changes from 20 L to 0.5 L against a constant pressure of 1.00 atm?

Section 9.2

7. A gas is compressed from 27.9 L to 16.3 L against a constant pressure of 3.4 atm. During this process, 122 J of heat is gained by the system. Calculate the change in energy of the system.

8. If 596 J of heat are added to 29.6 g of water at 22.9°C in a coffee-cup calorimeter, what will be the final temperature of the water?

9. A 5.037-g piece of iron heated to 100°C is placed in a coffee-cup calorimeter that initially contains 27.3 g of water at 21.2°C. If the final temperature is 22.7°C, what is the specific heat capacity of the iron (J/g°C)?

10. One liter of an ideal gas at 0°C and 10 atm was allowed to expand to 1.89 L against a constant external pressure of 1 atm at a constant temperature. The enthalpy change (ΔH) for this process is −901 J. Calculate q, w, and ΔE.

11. The heat capacity of a bomb calorimeter was determined by burning 6.79 g of methane (heat of combustion = −802 kJ/mol) in the bomb. The temperature changed by 10.8°C.

    a. What is the heat capacity of the bomb?
    b. A 12.6-g sample of acetylene, C_2H_2, produced a temperature increase of 16.9°C in the same calorimeter. What is the heat of combustion of acetylene (kJ/mol)?

12. A sample of C_6H_5COOH (benzoic acid) weighing 1.221 g was placed in a bomb calorimeter and ignited in a pure O_2 atmosphere. A temperature rise from 25.24°C to 31.67°C was noted. The heat capacity of the calorimeter was 5.020 kJ/°C, and the combustion products were CO_2 and H_2O. Calculate the ΔH in kJ/mol for the reaction.

13. When 1.50 L of 1 *M* Na_2SO_4 solution at 30.0°C is added to 1.50 L of 1 *M* $Ba(NO_3)_2$ solution at 30.0°C in a calorimeter, a white solid ($BaSO_4$) forms. The temperature of the mixture increases to 42.0°C. Assuming that the specific heat capacity of the solution is 6.37 J/°C g and that the density of the final solution is 2.00 g/mL, calculate the enthalpy change per mole of $BaSO_4$ formed.

14. Calculate the heat energy necessary to convert 10.0 g of water (from ice just melted) at 0.0°C to water at 20.0°C, assuming the specific heat capacity remains constant at 4.184 J/g°C.

Section 9.5

15. Calculate ΔH for

$$N_2(g) + O_2(g) \rightarrow 2NO(g)$$

given:

a. $N_2(g) + 2O_2(g) \rightarrow 2NO_2(g)$ $\quad \Delta H_{298} = 66.4$ kJ/mol
b. $2NO(g) + O_2(g) \rightarrow 2NO_2(g)$ $\quad \Delta H_{298} = -114.1$ kJ/mol

16. Given

a. $2H_2(g) + C(s) \rightarrow CH_4(g)$ $\quad \Delta H_{298} = -74.81$ kJ/mol
b. $2H_2(g) + O_2(g) \rightarrow 2H_2O(l)$ $\quad \Delta H_{298} = -571.66$ kJ/mol
c. $C(s) + O_2(g) \rightarrow CO_2(g)$ $\quad \Delta H_{298} = -393.52$ kJ/mol

Calculate ΔH for

$$CH_4(g) + 2O_2(g) \rightarrow CO_2(g) + 2H_2O(l)$$

17. Given the following thermochemical data, calculate the $\Delta H°$ for:

$$Ca(s) + 2H_2O(l) \rightarrow Ca(OH)_2(s) + H_2(g)$$

a. $H_2(g) + {}^1/_2O_2(g) \rightarrow H_2O(l)$ $\quad \Delta H° = -285$ kJ
b. $CaO(s) + H_2O(l) \rightarrow Ca(OH)_2(s)$ $\quad \Delta H° = -64$ kJ
c. $Ca(s) + {}^1/_2O_2(g) \rightarrow CaO(s)$ $\quad \Delta H° = -635$kJ

Section 9.6

18. Using standard heats of formation (Appendix 4 in your textbook), calculate ΔH for the following reactions.

a. $2H_2O_2(l) \rightarrow 2H_2O(l) + O_2(g)$ $\quad \Delta H_f^\circ$ [for $H_2O_2(l)$] = −187.8 kJ/mol
b. $HCl(g) \rightarrow H^+(aq) + Cl^-(aq)$
c. $2NO_2(g) \rightarrow N_2O_4(g)$
d. $C_2H_2(g) + H_2(g) \rightarrow C_2H_4(g)$
e. $2NaOH(s) + CO_2(g) \rightarrow Na_2CO_3(s) + H_2O(g)$

19. Calculate the standard change in enthalpy for the following thermite reaction by using enthalpies of formation:

$$2Al(s) + Fe_2O_3(s) \rightarrow Al_2O_3(s) + 2Fe(s)$$

NOTE: This reaction occurs when a mixture of powdered aluminum and iron(III) oxide are ignited with a magnesium fuse.

20. Calculate the value for $\Delta H°$ for the reaction:

$$CaCO_3(s) \rightarrow CaO(s) + CO_2(g)$$

21. Calculate $\Delta H°$ (in kJ/mol) for the combustion of methane, CH_4.

22. Calculate the $\Delta H°$ for these reactions:

a. $C_2H_4(g) + H_2(g) \rightarrow C_2H_6(g)$
b. $C_2H_4(g) + 3O_2(g) \rightarrow 2CO_2(g) + 2H_2O(g)$
c. $C_2H_4(g) + 3O_2(g) \rightarrow 2CO_2(g) + 2H_2O(l)$

Multiple-Choice Self-Test

1. When zinc reacts with hydrochloric acid, hydrogen gas is released. In this system the release of the hydrogen gas is counteracted by an outside force, which results in a smaller volume by the end of the reaction. The work done by the outside force:

   A. is negative on the system
   B. is positive on the surroundings
   C. is positive on the system
   D. is zero

2. A piston performs work of 210 L atm on the surroundings while the cylinder in which it is placed expands from 10 to 25 L. At the same time, 45 J of heat is transferred from the surroundings to the system. Against what pressure was the piston working?

   A. 14 atm B. 11 atm C. 17 atm D. 254 atm

3. As a system increases in volume, it releases 52.5 J of energy in the form of heat to the surroundings. The piston is working against a pressure of 10.25 atm. The final volume of the system is 58.0 L. What was the initial volume of the system if the energy of the system decreased by 102.5 J?

   A. 62.9 L B. 53.1 L C. 48 L D. 68 L

4. A 500.0-g sample of an element at 195°C is dropped into an ice-water mixture. 109.5 g of ice melts, and an ice-water mixture remains. Calculate the specific heat of the element, and determine which element it is.

   A. Zn B. Ba C. Pb D. Ag

5. Benzoic acid, $C_7H_6O_2$, is a standard used in determining the heat capacity of a calorimeter. $\Delta H°$ of combustion of the benzoic acid is 3.22×10^3 kJ/mol. 0.5 g of benzoic acid is burned in a calorimeter containing 1000.0 g of water. The change in temperature of the calorimeter is 3°C. Calculate the heat capacity of the calorimeter in J/K.

   A. 450 J/K B. 210 J/K C. 4025 J/K D. 2307 J/K

6. Silane, SiH_4, is highly combustible and creates a fire hazard.

   $$SiH_4(g) + 2O_2(g) \rightarrow SiO_2(s) + 2H_2O(g) \qquad \Delta H° = ?$$

   Calculate $\Delta H°$ for this reaction based on the following information:

   $$Si(s) + 2H_2(g) \rightarrow SiH_4(g) \qquad \Delta H° = 34 \text{ kJ/mol}$$
   $$Si(s) + O_2(g) \rightarrow SiO_2(s) \qquad \Delta H° = -911 \text{ kJ/mol}$$
   $$H_2(g) + {}^1\!/_2O_2(g) \rightarrow H_2O(g) \qquad \Delta H° = -242 \text{ kJ/mol}$$

   A. −1429 kJ B. −733 kJ C. 733 kJ D. −1143 kJ

7. Ethane, C_2H_6, may be produced by using the following method

   $$C(s) + CH_4(g) + H_2(g) \rightarrow C_2H_6 \qquad \Delta H° = -10 \text{ kJ/mol}$$

   Calculate the $\Delta H°$ for the following reaction given the information below:

   $$C(s) + 2H_2(g) \rightarrow CH_4(g) \qquad \Delta H° = ?$$
   $$2CH_4(g) + 2CH_4(g) \rightarrow 2C_2H_6(g) + 2H_2(g) \qquad \Delta H° = 130 \text{ kJ}$$

   A. −150 kJ B. −75 kJ C. 70 kJ D. 10 kJ

8. How much heat will be evolved if 56.08 g of calcium oxide reacts with sulfuric acid according to the following reaction?

$$CaO(s) + H_2SO_4(g) \rightarrow CaSO_4(s) + H_2O(l) \qquad \Delta H° = ?$$

$Ca(s) + ½O_2(g) \rightarrow CaO(s)$ $\quad \Delta H° = -152$ kJ/mol
$Ca(s) + S(s) + 2O_2(g) \rightarrow CaSO_4(s)$ $\quad \Delta H° = -1434$ kJ/mol
$H_2(g) + ½O_2(g) \rightarrow H_2O(l)$ $\quad \Delta H° = -286$ kJ/mol
$H_2(g) + S(s) + 2O_2(g) \rightarrow H_2SO_4(g)$ $\quad \Delta H° = -814$ kJ/mol

A. −2960 kJ B. −754 kJ C. −10744 kJ D. 754 kJ

9. Calculate the heat of reaction for the following reaction given the information below:

$$2KIO_3(s) + 12HCl(g) \rightarrow 2ICl(l) + 2KCl(s) + 6H_2O(l) + 4Cl_2(g) \qquad \Delta H° = ?$$

The heats of formation for the reactants and products, respectively, are:

−501.0, −92.0, −24.0, −435.0, and −286.0 kJ/mol

A. −2634 kJ B. −528 kJ C. −3227 kJ D. −152 kJ

10. Carbon dioxide and water in the atmosphere have all the following effects except:

A. Increasing the temperature of the earth
B. Allowing visible light to escape the earth
C. Absorbing infrared radiation
D. Resulting in longer daylight in summer

Answers to Exercises

1. a. 9 J
   b. −9 J
2. 77 J
3. $w = -5$ kJ
4. a. −1.81 kJ
   b. −776 J
5. −375 J was transferred *from* the system
6. 1.98 kJ
7. +4100 J (rounding of 4117 to 2 sig figs)
8. 27.7°C
9. 0.44 J/g°C
10. $q = -901$ J; $w = -90.1$ J; $\Delta E = -991$ J
11. a. 31.5 kJ/°C
    b. −1100 kJ/mol
12. −3230 kJ/mol
13. $\Delta H = -306$ kJ/mol
14. 837 J
15. 180.5 kJ/mol
16. −890.37 kJ
17. −414 kJ

18. a. −196 kJ
    b. −75 kJ
    c. −57 kJ
    d. −175 kJ
    e. −125 kJ

19. −850 kJ

20. 178.5 kJ

21. $\Delta H^\circ = -890.37$ kJ

22. a. −137 kJ
    b. −1320 kJ
    c. −1410 kJ

Answers to Multiple-Choice Self-Test

1. C
2. A
3. B
4. A
5. B
6. A
7. B
8. B
9. B
10. D

CHAPTER 10

Spontaneity, Entropy, and Free Energy

The Bottom Line: Chapter 10

Recall the first law of thermodynamics from Chapter 9. It says that energy can neither be created nor destroyed - just changed from one form to another. The law describes energy changes in chemical reactions but does not answer the more fundamental question, "Can we predict **IF** reactions will occur using thermodynamics?" This chapter shows how such predictions can be made.

10.1 Spontaneous Processes and Entropy

Your textbook makes several points regarding **spontaneous** reactions.

- **Spontaneous** means occurring **without outside intervention**.
- Rate of reaction is **irrelevant** to spontaneity. Spontaneous means it will happen, not necessarily quickly, or even in Earth's lifetime.
- Spontaneous processes **increase the entropy of the universe**. (The entropy of a **system can decrease** if that of the **surroundings increases**.)

⇒ Entropy is a complex mathematical function that describes the number of possible arrangements (**positional probability**) of the states of a substance.
⇒ Each arrangement available to a substance is called a **microstate**.
⇒ Gases (in general) have a much higher positional entropy than liquids or solids.
Let's have some practice with microstates and entropy.

Example 10.1 A - Microstates And Entropy

You have 3 identical atoms, "A," "B," and "C." They can go back and forth from one "dash" to the other via the "wall" as shown.

AB	‖	C	(for example)
side 1	$\rightleftharpoons$	side 2	

a. List all the possible microstates that these three atoms can have on our two dashes. In other words, list all the ways they can arrange themselves.

b. How much more probable is arrangement II than arrangement I?

arrangement I	•••	‖	___
arrangement II	••	‖	•

c. Does this make sense in terms of our understanding of entropy?

Solution

a. Microstates

ABC	‖	___(I)	A	‖	BC
AB	‖	C (II)	B	‖	AC

AC	B (II)	C	AB
BC	A (II)		ABC

b. Arrangement I can happen **1 way**. Arrangement II can happen 3 ways. Arrangement II is thus 3 times as likely as arrangement I.

c. The "atoms" are **more likely** to spread than be confined to one side of the wall. That agrees with our concept of entropy as positional probability.

Your textbook considers the statistical probability that more or less even mixing among atoms will occur. A formula is given to calculate the number of microstates,

$$\frac{N!}{L!R!}$$

Where N = the total number of molecules,
L = the number of molecules in the left-hand bulb of a two-bulb system, and
R = the number of molecules in the right-hand bulb.

Note how the likelihood of even, or near even, mixing increases as N increases.

Example 10.1 B - Positional Probability

Which of the following pairs is likely to have the higher positional probability per mole at a given temperature?

a. Solid or gaseous phosphorus
b. $CH_4(g)$ or $C_3H_8(g)$
c. $KOH(s)$ or $KOH(aq)$

Solution

a. Gaseous phosphorus will have the higher positional probability. It is not as constrained by intramolecular bonds as solid phosphorus is.

b. All other things being equal, larger molecules containing many single bonds have more positional possibilities than smaller ones. $C_3H_8(g)$ has the higher positional probability.

c. In general (all other things being equal), liquids have **slightly** higher entropies than solids do. **However**, in this case, the positional probabilities of $KOH(aq)$ are constrained due to hydrogen bonding interactions. $KOH(s)$ has the higher positional probability.

10.2 The Isothermal Expansion and Compression of an Ideal Gas

Many issues are raised in this section. The most fundamental one deals with the isothermal expansion and contraction of an ideal gas. Let's test your understanding.

1. Define **isothermal.**
2. Why is no work done in the **free expansion** example?
3. Prove that work is *not* a state function.
4. Justify the use of the integral for the case of an infinite-step expansion.
5. Define **reversible process** from a thermodynamic standpoint.
6. What does the reversible process tell us about heat exchange?
7. What are the implications of *saying* that "all real processes are reversible?"

10.3 The Definition of Entropy

Your textbook derives an expression relating entropy, which considers the motions of particles, to volume, and then temperature. These are macroscopic properties. The next section considers how actual entropy values change for systems as their temperatures change. Let's test our understanding.

1. Zumdahl notes that each of the two vessels has an average entropy. Why?
2. How does Zumdahl define the three possible microstates?
3. Could there be alternative ways of defining microstates? Propose one alternative way.
4. Reproduce the mathematical example that demonstrates the idea that entropy is additive and the number of microstates is multiplicative. Extend this to a situation in which you have three microstates in each of **three** vessels.

10.4 Entropy and Physical Changes

A reversible process can only occur at equilibrium. Melting and boiling are two such points, therefore the entropy can easily be calculated, as shown in Interactive Example 10.3 in your textbook. Note not just the equation, but the **sign** of the entropy value. Does this make sense?

10.5 Entropy and the Second Law of Thermodynamics

The following questions will test your understanding of the "second law."

1. State the second law of thermodynamics.
2. State the second law in terms of the system and surroundings.
3. Under what circumstance can the **entropy of the system decrease** for a spontaneous process?
4. How do the first and second laws fundamentally differ?

10.6 The Effect of Temperature on Spontaneity

Exothermic reactions **give off energy to the surroundings**. Therefore, **random motions** of particles in the surroundings **increase**. When random motions increase, **positional probabilities increase**.

The key point from all this is that **exothermic reactions increase the entropy of the surroundings (ΔS_{surr})**.

The **magnitude** of the increase in ΔS_{surr} **depends on the temperature**. (See the money-related discussion in your textbook.)

Recall from Chapter 9 that we think of **heat flow** in terms of **the system.**

Exothermic reaction (at constant pressure), $\Delta H = -$
Endothermic reaction (at constant pressure), $\Delta H = +$

$$\Delta S_{surr} = \frac{-\Delta H}{T} \text{ (in Kelvins)}$$

Example 10.6 A - ΔS_{surr} And The Heat Of Reaction

Calculate ΔS_{surr} for each of the following reactions at 25°C and 1 atm.

a. $C_3H_8(g) + 5O_2(g) \rightarrow 3CO_2(g) + 4H_2O(g)$ $\Delta H = -2045$ kJ
b. $(NH_4)_2Cr_2O_7(s) \rightarrow N_2(g) + Cr_2O_3(s) + 4H_2O(g)$ $\Delta H = -315$ kJ
c. $H_2O(l) \rightarrow H_2O(g)$ $\Delta H = +44$ kJ

Solution

a. $\Delta S_{surr} = \dfrac{-\Delta H}{T}$

$T = (25 + 273) = 298 \text{ K}$

$\Delta S_{surr} = \dfrac{-(-2045 \text{ kJ})}{298 \text{ K}} = 6.86 \text{ kJ/K} = \mathbf{6860 \text{ J/K}}$

There is an increase, because the reaction is exothermic.

b. $\Delta S_{surr} = \dfrac{-(-315 \text{ kJ})}{298 \text{ K}} = 1.06 \text{ kJ/K} = \mathbf{1060 \text{ J/K}}$

There is an increase here as well because the reaction is exothermic.

c. $\Delta S_{surr} = \dfrac{-(44 \text{ kJ})}{298 \text{ K}} = -0.15 \text{ kJ/K} = \mathbf{-150 \text{ J/K}}$

Energy must be taken from the surrounding to the system in this reaction. Therefore, ΔS_{surr} will be negative.

The relationship between **entropy change** and reaction spontaneity is summarized in Table 10.4 in your textbook. Consider the information in that table and try the following example.

Example 10.6 B - Reaction Spontaneity

Determine if the values for entropy in each of the following will produce a spontaneous process. Also, which of the following processes is endothermic (from the perspective of the system)?

a. $\Delta S_{sys} = 30 \text{ J/K}$ $\Delta S_{surr} = 50 \text{ J/K}$
b. $\Delta S_{sys} = -27 \text{ J/K}$ $\Delta S_{surr} = 40 \text{ J/K}$
c. $\Delta S_{sys} = 140 \text{ J/K}$ $\Delta S_{surr} = -85 \text{ J/K}$
d. $\Delta S_{sys} = 60 \text{ J/K}$ $\Delta S_{surr} = -85 \text{ J/K}$

Solution

$$\mathbf{\Delta S_{univ} = \Delta S_{sys} + \Delta S_{surr}}$$

- A reaction will be spontaneous if $\mathbf{\Delta S_{univ} > 0}$.
- A reaction will be endothermic if $\mathbf{\Delta S_{univ} < 0}$.

a. $\Delta S_{univ} = 30 + 50 = 80$ J/K, **spontaneous, exothermic**.
b. $\Delta S_{univ} = -27 + 40 = 13$ J/K, **spontaneous, exothermic**.
c. $\Delta S_{univ} = 140 + (-85) = 55$ J/K, **spontaneous, endothermic**.
d. $\Delta S_{univ} = 60 + (-85) = -25$ J/K, **not spontaneous, endothermic**.

Notice also from Table 10.4 that because temperature is related to the entropy of the surroundings, there are circumstances under which **temperature plays the deciding factor in whether a reaction will be spontaneous**.

10.7 Free Energy

Free energy is a mathematical term that describes **unequivocally** whether a reaction will be spontaneous. It is experimentally useful because **it reflects ΔS_{univ}.** Your textbook states two important relationships.

$$\Delta G = \Delta H - T\Delta S$$

(When no subscript appears, it is assumed that we are referring to the system.)

$$\Delta S_{univ} = \frac{-\Delta G}{T}$$

The first equation gives us an explicit way of calculating free energy. It further says that there are circumstances under which **temperature will determine whether a reaction is spontaneous**. (See Table 10.6 in your textbook.) Note also that since temperature in Kelvins will always be greater than zero, that $-\Delta G = +\Delta S_{univ}$, **always**.

The case of the temperature dependency of ice melting, as described in your textbook, is a perfect example of this. You have two opposing entropy factors. On the one hand, the reaction is **endothermic**, which **opposes the process** ($\Delta S_{surr} = -$). On the other hand, melting **increases the positional probability of the system** ($\Delta S_{sys} = +$). The temperature will determine which process will dominate (whether ice melts).

The second equation says that the ΔG must be **negative** (< 0) in order for a reaction to proceed.

Example 10.7 A - Free Energy And Spontaneity

Given the values for ΔH, ΔS, and T, determine whether each of the following sets of data represent spontaneous or nonspontaneous processes.

	ΔH (kJ)	ΔS (J/K)	T (K)
a.	40	300	130
b.	40	300	150
c.	40	−300	150
d.	−40	−300	130
e.	−40	300	150

Solution

$$\Delta G = \Delta H - T\Delta S$$

For a reaction to be spontaneous, ΔG must be < 0. When you do the calculations, make sure that you either **change** ΔH to joules **or** ΔS to kilojoules!

a. $\Delta G = 40 \text{ kJ} - 0.300 \text{ kJ/K } (130 \text{ K})$ = +1 kJ, **nonspontaneous**
b. $\Delta G = 40 \text{ kJ} - 0.300 \text{ kJ/K } (150 \text{ K})$ = −5 kJ, **spontaneous**
c. $\Delta G = 40 \text{ kJ} - (-0.300 \text{ kJ/K})(150 \text{ K})$ = +85 kJ, **nonspontaneous**
d. $\Delta G = -40 \text{ kJ} - (-0.300 \text{ kJ/K})(130 \text{ K})$ = −1 kJ, **spontaneous**
e. $\Delta G = -40 \text{ kJ} - 0.300 \text{ kJ/K } (150 \text{ K})$ = −85 kJ, **spontaneous**

Note that temperature is NOT important to having a spontaneous reaction **only** when the reaction is exothermic and there is an increase in entropy.

Example 10.7 B - Free Energy And Temperature

You know that the boiling point of water is 373 K. See how this compares to the minimum temperature for a reaction that you determine thermodynamically for the phase change:

$$H_2O(l) \rightarrow H_2O(g)$$

where ΔH = 44 kJ and ΔS = 119 J/K.

Solution

The criterion for spontaneity is $\Delta G < 0$. This means that $\Delta H - T\Delta S < 0$.

Adding $T\Delta S$ to both sides, $\Delta H < T\Delta S$.

Dividing both sides by ΔS, $\frac{\Delta H}{\Delta S} < T$.

Using the data from our problem, $T > \frac{\mathbf{4.4 \times 10^4\ J}}{\mathbf{119\ J/K}}$, $T >$ **370 K**, which is close to the actual value.

10.8 Entropy Changes in Chemical Reactions

This section in your textbook begins by reminding us that $\mathbf{\Delta S_{surr}}$ is related to **heat flow** from the system. BUT, $\mathbf{\Delta S_{sys}}$ is related to the positional **probabilities** for each of the reactants. For example,

$$2H_2(g) + O_2(g) \rightarrow 2H_2O(g)$$

$\Delta S = -89$ kJ (at 25°C). We lose entropy because **3 total moles** of the gases on the left have **more** positional possibilities than **2 moles** of vapor on the right.

- **For a chemical reaction involving only the gas phase, entropy is related to the total number of moles on either side of the equation. A decrease means lower entropy, an increase means higher entropy.**
- **For a chemical reaction involving different phases, the production of a gas will (in general) increase the entropy much more than an increase in the number of moles of a liquid or solid.**

For example,

$$2HNO_3(aq) + Na_2CO_3(s) \rightarrow 2NaNO_3(aq) + H_2O(l) + \mathbf{CO_2(g)}$$
$$\Delta S = +88 \text{ kJ (at 25°C)}.$$

Example 10.8 A - The Sign Of Entropy Changes

Predict the sign of $\Delta S°$ for each of the following reactions:

a. $(NH_4)_2Cr_2O_7(s) \rightarrow Cr_2O_3(s) + 4H_2O(l) + N_2(g)$
b. $Mg(OH)_2(s) \rightarrow MgO(s) + H_2O(g)$
c. $PCl_5(g) \rightarrow PCl_3(g) + Cl_2(g)$

Solution

a. There are many more moles on the right-hand side. In addition, a gas is formed on the right-hand side. **The change in entropy will be positive.**

b. A gas is formed on the right-hand side. **The change in entropy will be positive.**

c. More moles of gas are present on the right-hand side. **The change in entropy will be positive.**

The **third law of thermodynamics** says that **the entropy of a perfect crystal at 0 K is zero.** This means that the absolute entropy of substances can be explicitly measured. (See Appendix 4 in your textbook.)

As was true with ΔH° (a state function), ΔS° (also a state function) can be determined as the difference between the sum of the entropy of products minus the sum of the entropy of the reactants.

$$\Delta S^\circ_{\text{reaction}} = \sum n_p S^\circ_{\text{products}} - \sum n_r S^\circ_{\text{reactants}}$$

Example 10.8 B - Entropy Of Reaction

Calculate ΔS° for each of the following reactions, using data from Appendix 4 in your textbook.

a. $N_2O_4(g) \rightarrow 2NO_2(g)$
b. $Fe_2O_3(s) + 2Al(s) \rightarrow 2Fe(s) + Al_2O_3(s)$
c. $4NH_3(g) + 5O_2(g) \rightarrow 4NO(g) + 6H_2O(g)$

Solution

$$\Delta S^\circ_{\text{reaction}} = \sum n_p S^\circ_{\text{products}} - \sum n_r S^\circ_{\text{reactants}}$$

a. $\Delta S^\circ_{\text{rxn}} = 2S^\circ_{NO_2(g)} - S^\circ_{N_2O_4(g)}$
$= 2$ mol (240 J/K mol) − 1 mol (304 J/K mol)
$=$ **176 J/K** (The entropy should increase due to an increase in the number of moles of gas.)

b. $\Delta S^\circ_{\text{rxn}} = [2S^\circ_{Fe(s)} + S^\circ_{Al_2O_3(s)}] - [S^\circ_{Fe_2O_3(s)} + 2S^\circ_{Al(s)}]$
$=$ [2 mol (27 J/K mol) + 1 mol (51 J/K mol)] − [1 mol (90 J/K mol) + 2 mol (28 J/K mol)]
$=$ **−41 J/K** (This is fairly small because there were no net phase changes.)

c. $\Delta S^\circ_{\text{rxn}} = [4S^\circ_{NO(g)} + 6S^\circ_{H_2O(g)}] - [4S^\circ_{NH_3(g)} + 5S^\circ_{O_2(g)}]$
$=$ [4 mol (211 J/K mol) + 6 mol (189 J/K mol)] − [4 mol (193 J/K mol) + 5 mol (205 J/K mol)]
$=$ **181 J/K** (We increased the number of moles of gas, which is reflected by the increase in entropy.)

10.9 Free Energy and Chemical Reactions

The **standard free energy change (ΔG°)** is the free energy change that occurs if reactants **in their standard states (1 atm, 25°C)** are converted to products in their standard states.

Your textbook points out that ΔG° *cannot be measured directly*, but is an important value because it represents a *standard set of conditions* at which to compare properties of reactions (as we will see later).

Three methods of calculating ΔG° are introduced in this section.

1. $\Delta G^\circ = \Delta H^\circ - T\Delta S^\circ$
2. **By manipulating known equations, as in Hess's law problems for ΔH°.**
3. $\Delta G^\circ = \sum n_p \Delta G^\circ(\text{products}) - \sum n_r \Delta G^\circ(\text{reactants})$

The following examples illustrate each of these methods.

Example 10.9 A - Standard Free Energy From Entropy And Enthalpy

Using data for **$\Delta H°$ and $\Delta S°$**, calculate **$\Delta G°$** for the following reactions at 25°C and 1 atm.

a. $Cr_2O_3(s) + 2Al(s) \rightarrow Al_2O_3(s) + 2Cr(s)$
b. $C_3H_8(g) + 5O_2(g) \rightarrow 3CO_2(g) + 4H_2O(g)$

Solution

$$\Delta G° = \Delta H° - T\Delta S°$$

We need to calculate $\Delta H°$ and $\Delta S°$ for each reaction.

$$\Delta H° = \sum n_p \Delta H_f^{\circ}\text{(products)} - \sum n_r \Delta H_f^{\circ}\text{(reactants)}$$

$$\Delta S° = \sum n_p S°_{\text{products}} - \sum n_r S°_{\text{reactants}}$$

a. **$Cr_2O_3(s) + 2Al(s) \rightarrow Al_2O_3(s) + 2Cr(s)$**

$\Delta H°$ = [1 mol(−1676 kJ/mol) + 2 mol(0 kJ/mol)] − [1 mol(−1128 kJ/mol) + 2 mol(0 kJ/mol)]

↑ Al_2O_3 ↑ Cr ↑ Cr_2O_3 ↑ Al

= **−548 kJ**

$\Delta S°$ = [1 mol(51 J/K mol) + 2 mol (24 J/K mol)] − [1 mol(81 J/K mol) + 2 mol(28 J/K mol)]

↑ Al_2O_3 ↑ Cr ↑ Cr_2O_3 ↑ Al

= −38 J/K

$\Delta G°$ = $\Delta H° - T\Delta S°$ = −548 kJ − 298 K (−0.038 kJ/K)
= **−537 kJ; the reaction is spontaneous.**

c. **$C_3H_8(g) + 5O_2(g) \rightarrow 3CO_2(g) + 4H_2O(g)$**

$\Delta H°$ = [3 mol(−393.5 kJ/mol) + 4 mol(−242 J/mol)] − [5 mol(0 kJ/mol) + 1mol(−104 kJ/mol)]

↑ CO_2 ↑ H_2O ↑ O_2 ↑ C_3H_8

= −2044 kJ

$\Delta S°$ = [3 mol(214 J/K mol) + 4 mol(189 J/K mol)] − [5 mol(205 J/K mol) + 1 mol(270 J/K mol)]

↑ CO_2 ↑ H_2O ↑ O_2 ↑ C_3H_8

= 103 J/K

$\Delta G°$ = $\Delta H° - T\Delta S°$ = −2044 kJ − 298 K (0.103 kJ/K)
= **−2075 kJ, this combustion is highly spontaneous.**

Example 10.9 B - Standard Free Energy By Combining Equations

Given the following data:

(Equations 1)	$S(s) + {}^{3}/{}_{2}O_2(g) \rightarrow SO_3(g)$	$\Delta G°$ = −371 kJ
(Equations 2)	$2SO_2(g) + O_2(g) \rightarrow 2SO_3(g)$	$\Delta G°$ = −142 kJ

Calculate $\Delta G°$

(Goal Equation) $S(s) + O_2(g) \rightarrow SO_2(g)$

Is the reaction spontaneous?

Solution

We solve this problem just as we would a Hess's law problem (see Chapter 9). We must manipulate equations 1 and 2 so that we can combine them to get the reaction of interest. (Remember what you do to the equations must also be done to the $\Delta G°$ values!)

One mole of S(*s*) appears only once on the left-hand side of the **goal equation**, and **one mole** on the left-hand side of equation 1.

Equation 1 must remain as it is.

One mole of $SO_2(g)$ appears on the **right-hand side** of the goal equation. **Two moles** of $SO_2(g)$ are present on the **left-hand side** of equation 2. We must therefore

multiply equation 2 by −1/2

to get one mole of $SO_2(g)$ on the right-hand side. (Remember to multiply $\Delta G°$ by −1/2 as well!)

$$S(s) + {}^{3}/_{2}O_2(g) \rightarrow SO_3(g) \qquad \Delta G° = -371\text{ kJ}$$
$$SO_3(g) + SO_2(g) \rightarrow {}^{1}/_{2}O_2(g) \qquad \Delta G° = +71\text{ kJ}$$
$$S(s) + O_2(g) \rightarrow SO_3(g) \qquad \mathbf{\Delta G° = -300\text{ kJ}}$$

Example 10.9 C - Standard Free Energy From "Products – Reactants"

Calculate $\Delta G°$ for the reaction

$$C_3H_8(g) + 5O_2(g) \rightleftharpoons 3CO_2(g) + 4H_2O(g)$$

using $\Delta G°$ data from Appendix 4. Compare the answer with that from our Example 10.9 A, part b.

Solution

$$\begin{aligned}\Delta G° &= \sum n_p \Delta G°(\text{products}) - \sum n_r \Delta G°(\text{reactants}) \\ &= [3\Delta G°_{f[CO_2(g)]} + 4\Delta G°_{f[H_2O(g)]}] - [\Delta G°_{f[C_3H_8(g)]} + 5\Delta G°_{f[O_2(g)]}] \\ &= [3\text{ mol}(-394\text{ kJ/mol}) + 4\text{ mol}(-229\text{ kJ/mol})] - [1\text{ mol}(-24\text{ kJ/mol}) + 5\text{ mol}(0\text{ kJ/mol})] \\ &= \mathbf{-2074\text{ kJ}}\end{aligned}$$

The $\Delta G°$ values agree within about 1%. Also note that as with $\Delta H°$, $\Delta G°$ for elements in their standard states equals 0.

10.10 The Dependence of Free Energy on Pressure

We have until now assumed standard conditions. This section deals with **free energy at nonstandard pressures.**

For an ideal gas, enthalpy is not pressure dependent. **Entropy, however is affected by pressure.** More positions are possible at lower pressure than higher pressure, therefore

$$S_{\text{low pressure}} > S_{\text{high pressure}}$$

Your textbook derives the relationship

$$\mathbf{\Delta G = \Delta G° + RT\ln(Q)}$$

Where R = 8.3145 J/K mol
T = temperature in Kelvins
Q = reaction quotient (the mass action expression relating to **initial** quantities)

It is important to learn to use **and interpret** the results from this equation. Through a similar equation (to be introduced in the next section) we can relate equilibrium constants to ΔG.

Example 10.10 - Relating Free Energy And Pressure

Calculate ΔG at 700 K for the following reaction:

$$C(s, \text{graphite}) + H_2O(g) \rightleftharpoons CO(g) + H_2(g)$$

Initial pressures are $P_{H_2O} = 0.85$ atm, $P_{CO} = 1.0 \times 10^{-4}$ atm, $P_{H_2} = 2.0 \times 10^{-4}$ atm.

Solution

Where are we going? We are asked to solve for the free energy of the system, ΔG, *not* ΔG°, because the reaction is done at pressures other than 1 atmosphere. However, we can use the standard free energy, ΔG°, to find ΔG.

Our first task is to determine $\Delta G°$.

$$\begin{aligned}\Delta G° &= [\Delta G^\circ_{f[CO(g)]} + \Delta G^\circ_{f[H_2(g)]}] - [\Delta G^\circ_{f[C(s,\,\text{graphite})]} + \Delta G^\circ_{f[H_2O(g)]}] \\ &= [1\text{ mol}(-137\text{ kJ/mol}) + 1\text{ mol}(0\text{ kJ/mol})] - [1\text{ mol}(0\text{ kJ/mol}) + 1\text{ mol}(-229\text{ kJ/mol})] \\ &= \mathbf{+92\ kJ}\end{aligned}$$

This tells us that the reaction is not spontaneous **under standard conditions**. To evaluate at 700 K and the given pressures.

$$Q = \frac{P_{H_2}P_{CO}}{P_{H_2O}} = \frac{2.0\times10^{-4}\text{ atm }(1.0\times10^{-4}\text{ atm})}{0.85\text{ atm}} = \mathbf{2.35 \times 10^{-8}\ atm}$$

$$\begin{aligned}\Delta G &= \Delta G° + RT\ln(Q) = 9.2 \times 10^4\text{ J} + 8.3145\text{ J/K mol }(700\text{ K})\ln(2.35 \times 10^{-8}) \\ &= 9.2 \times 10^4\text{ J} + (-1.022 \times 10^5\text{J}) \\ &= -1.02 \times 10^4\text{ J} \\ &= \mathbf{-10\ kJ}\end{aligned}$$

Under conditions of high temperature and low product pressure, the reaction becomes spontaneous.

The final subheading in Section 10.10 of your textbook addresses "The Meaning of ΔG for a Chemical Reaction." It stresses that a spontaneous reaction will not necessarily go to completion. Rather, there may be some intermediate point that reflects the lowest possible ΔG value for the reaction.

10.11 Free Energy and Equilibrium

Equilibrium occurs at the lowest value of free energy available to the system. In Chapter 6, we defined equilibrium as occurring when the forward rate of reaction is equal to the reverse rate of reaction. That is still valid. Thermodynamically, your textbook defines equilibrium as occurring when

$$G_{\text{forward reaction}} = G_{\text{reverse reaction}}$$

In other words, $\mathbf{\Delta G = 0}$ **at equilibrium.**

If $\Delta G < 0$, it means that $G_{\text{reactants}} > G_{\text{products}}$ and the reaction will go **to the right until** $G_{\text{reactants}} = G_{\text{products}}$.
If $\Delta G > 0$, it means that $G_{\text{reactants}} < G_{\text{products}}$ and the reaction will go **to the left until** $G_{\text{reactants}} = G_{\text{products}}$.

Table 10.7 in your textbook mathematically relates $\Delta G°$ to K. Remember that if $K = 1$ (the equilibrium condition), $\ln K = 0$ and $\Delta G = \Delta G°$. At equilibrium, therefore,

$$\Delta G° = -RT\ln(K)$$

Example 10.11 A - Free Energy And The Direction Of Reaction

Using the same reaction as in our Example 10.10,

$$C(s, \text{graphite}) + H_2O(g) \rightleftharpoons CO(g) + H_2(g)$$

where T = **700 K** and $\Delta G°$ = **92 kJ**, determine the direction of reaction if the following initial pressure of each gas is

$$P_{H_2O} = \mathbf{0.67\ atm},\ P_{CO} = \mathbf{0.23\ atm},\ P_{H_2} = \mathbf{0.51\ atm}$$

Solution

The goal is to calculate ΔG.

$$\Delta G = \Delta G° + RT\ln(K)$$

$$Q = \frac{P_{H_2}P_{CO}}{P_{H_2O}} = \frac{0.51\text{ atm }(0.23\text{ atm})}{0.67\text{ atm}} = \mathbf{0.175\ atm}$$

$$\Delta G = 9.2 \times 10^4\text{ J} + 8.3145\text{ J/K mol }(700\text{ K})\ln(0.175)$$

$$\Delta G = 82000\text{ J} = \mathbf{82\ kJ}$$

This reaction will go very far to the left.

Example 10.11 B - Free Energy And Equilibrium Constant

Given the values for $\Delta G°$ that you calculated in Example 10.9A, calculate K for the following reaction (at 25°C):

$$Cr_2O_3(s) + 2Al(s) \rightleftharpoons Al_2O_3(s) + 2Cr(s)$$

Solution

$\Delta G° = -537$ kJ

$$\Delta G° = -RT\ln(K)$$

$$\ln(K) = \frac{\Delta G°}{-RT},\text{ or } K = e^{-\Delta G°/RT}$$

$$\ln(K) = \frac{-5.37\times10^5\text{ J}}{-8.3145\text{ J/K mol }(298\text{ K})} = 216.73$$

$K = e^{216.73} = \mathbf{1.3 \times 10^{94}}$, this reaction is quite spontaneous.

Example 10.11 C - Summing It All Up

We have previously (Chapter 8) studied the weak base-strong acid titration

$$\mathbf{NH_3(aq) + H^+(aq) \rightleftharpoons NH_4^+(aq)}$$

We said that you can look at it as the sum of two reactions (for purposes of calculating K).

1. $NH_3(aq) + H_2O(l) \rightleftharpoons NH_4^+(aq) + OH^-(aq)$ $(K_b = 1.8 \times 10^{-5})$
2. $H^+(aq) + OH^-(aq) \rightleftharpoons H_2O(l)$ $(1/K_w = 1.0 \times 10^{+14})$

Please do the following:

a. Calculate K for the titration reaction using the two equilibrium expressions.
b. Calculate $\Delta H°$, $\Delta S°$, $\Delta G°$, and K for the titration reaction using data from Appendix 4 of your textbook.
c. How do the calculated K values compare?
d. Is the titration spontaneous?

Solution

a. Summing the reactions requires multiplying the equilibrium constants

$$\mathbf{K_{titration} = K_b \times 1/K_w = 1.8 \times 10^9}$$

b. $\Delta H°$ and $\Delta S° = S(\text{products}) - S(\text{reactants})$

$$\mathbf{\Delta H°} = [1\text{ mol}(-132\text{ kJ/mol})] - [1\text{ mol}(-80\text{kJ/mol}) + 1\text{ mol}(0\text{ kJ/mol})]$$

(↑ NH_4^+, ↑ NH_3, ↑ H^+)

$$= \mathbf{-52\ kJ}$$

$$\mathbf{\Delta S°} = [1\text{ mol}(113\text{ J/K mol})] - [1\text{ mol}(111\text{ J/K mol}) + 1\text{ mol}(0\text{ J/K mol})]$$
$$= \mathbf{2\ J/K}$$

$$\mathbf{\Delta G°} = \Delta H° - T\Delta S° = -52000\text{ J} - 298\text{ K }(2\text{ J/K}) = -52596\text{ J}$$
$$= \mathbf{-53\ kJ}$$

$$\ln(K) = -\frac{\Delta G}{RT} = +\frac{52596\text{ J}}{8.3145\text{ J/K mol}(298\text{ K})} = 21.23$$

$$\mathbf{K} = e^{21.23} = \mathbf{1.7 \times 10^9}$$

c. The K values are comparable (within about 5%).

d. The titration is spontaneous (as we know from experience).

Your textbook discusses the temperature dependence of K by introducing the **van't Hoff** equation

$$\ln\left(\frac{K_2}{K_1}\right) = \frac{-\Delta H°}{R}\left[\frac{1}{T_2} - \frac{1}{T_1}\right]$$

You may assume that $\Delta H°$ is constant with temperature, so if we know K_1, T_1, and T_2, we can find K_2.

Example 10.11 D - Van't Hoff Equation

At 700°C, the value of K_p is 1.50 and $\Delta H°$ is −818.3 kJ/mol for

$$C(s) + CO_2(g) \rightleftharpoons 2CO(g)$$

What is the value of K_p at 1000°C?

Solution

We know that

$\Delta H = -818.3$ kJ/mol

$T_1 = 973$ K $\qquad T_2 = 1273$ K

$K_1 = 1.50 \qquad K_2 = ?$

$$\ln\left(\frac{K_2}{K_1}\right) = \frac{-\Delta H°}{R}\left[\frac{1}{T_2} - \frac{1}{T_1}\right]$$

$$\ln(K_2) - \ln(1.50) = \frac{+818{,}300\ \text{J mol}^{-1}}{8.3148\ \text{J K}^{-1}\ \text{mol}^{-1}}\left[\frac{1}{1273\ \text{K}} - \frac{1}{973\ \text{K}}\right]$$

$$\ln(K_2) - 0.405 = +98{,}415\ (-2.422 \times 10^{-4})$$

$$\ln(K_2) - 0.405 = -23.84$$

$$\ln(K_2) = -23.44$$

$$K_2 = e^{-23.44} = \mathbf{6.6 \times 10^{-11}}$$

Note that K decreased as temperature increased for the exothermic reaction so the answer makes sense.

10.12 Free Energy and Work

Your textbook makes the critical point that a thermodynamically favorable reaction might be made faster via a catalyst, but a catalyst would not be effective in one in which $\Delta G = +$. What are the two solutions to q_p and why are they different?

10.13 Reversible and Irreversible Processes: A Summary

Your textbook uses the example of a battery to illustrate the tradeoff between work and heat. The goal is to run the battery at low current to minimize frictional heating loss. Why is it necessary to use more work to charge a battery than is released when discharging the battery? Finally, note the final thought in the chapter, relating to the energy crisis, "As we use energy, we degrade its usefulness." Why?

10.14 Adiabatic Processes

Your textbook defines an adiabatic process as that in which no energy as heat flows into or out of the system.

- What is the change in energy, ΔE, in terms of work and heat in an adiabatic process?

Let's continue by reviewing the key questions posed by your textbook:

- Where does the energy for system expansion come from if not heat?
- What happens to the temperature of the gas as a result of an expansion in which $q = 0$?

The answers to these questions lead to the conclusion, proved in your textbook, that the energy comes from the gas itself. As a result of losing kinetic energy to enable the expansion, the gas attains a lower temperature.

Example 10.16 in your textbook demonstrates that an adiabatic expansion (less work available) does not increase the volume as much as an isothermal expansion, in which the work term is much larger.

Example 10.14 - Adiabatic Expansion

We have 2.38 mol of a monatomic, ideal gas at 32°C and an initial pressure of 8.00 atm. Determine the final pressure and volume of the gas if the external pressure is lowered to 2.00 atm in a reversible manner. Also, what is the work for the process?

Solution

We are considering an adiabatic expansion in which we need pressure and volume. As derived in your textbook, we can use

$$P_1V_1^{\gamma} = P_2V_2^{\gamma}$$

$$\gamma = 5/3$$

$$P_1V_1^{5/3} = P_2V_2^{5/3}$$

We calculate V_1 from the ideal gas law.

$$V_1 = \frac{nRT_1}{P_1} = \frac{2.38 \text{ mol}\left(0.08206 \dfrac{\text{L atm}}{\text{K mol}}\right)(305 \text{ K})}{8.00 \text{ atm}} = 7.45 \text{ L}$$

Now we can solve for V_2.

$$V_2^{5/3} = \frac{P_1V_1^{5/3}}{P_2} = \frac{(8.00 \text{ atm})(7.45 \text{ L})^{5/3}}{2.00 \text{ atm}}$$

$$\mathbf{V_2 = 17.1 \text{ L}}$$

The work in an adiabatic process is equal to the change in energy. In order to calculate ΔE, we need the temperature lowering of the gas as a result on the expansion.

$$T_2 = \frac{P_2V_2}{nR} = \frac{2.00 \text{ atm}(17.1 \text{ L})}{(2.38 \text{ mol})\left(0.08206 \dfrac{\text{L atm}}{\text{K mol}}\right)} = 175 \text{ K}$$

$$\boldsymbol{\Delta E = w} = nC_v\Delta T = (2.38 \text{ mol})(3/2)(8.3148 \text{ J/K mol})(175\text{K} - 305 \text{ K}) = \mathbf{-3{,}560 \text{ J}}$$

Energy flows out of the system with the expansion work, so the sign for ΔE is negative.

Exercises

Section 10.1

1. Given 8 molecules in the two-bulb set up described in Table 10.1 in your textbook, calculate the relative probability of finding all 8 molecules in the left-hand bulb. What does this tell you regarding entropy and probability?

2. Which of the following pairs of substances is likely to have the higher positional entropy?
   a. $HCl(aq)$ or $HCl(g)$
   b. $P_4(s)$ or $P_4O_{10}(g)$
   c. $O_2(g)$ or $P_4O_{10}(g)$
   d. $H_2O(s)$ or $H_2O(l)$
   e. $NO_2(g)$ or $N_2O_4(g)$
   f. $Ar(g)$ at 5 atm or $Ar(g)$ at 0.30 atm.

3. Predict the sign of the entropy change for each of the following processes.
   a. Potassium hydroxide pellets are dissolved in water.
   b. Solid ammonium dichromate is burned to give solid chromium oxide, water vapor, and nitrogen gas.
   c. Saturated calcium acetate is mixed with ethanol to form a gel.

4. Predict the sign of the entropy change for each of the following reactions:

   a. $Ag^+(aq) + Cl^-(aq) \rightarrow AgCl(s)$
   b. $NH_4Cl(s) \rightarrow NH_3(g) + HCl(g)$
   c. $H_2(g) + Br_2(g) \rightarrow 2HBr(g)$

Section 10.4

5. When water freezes, is there an increase in entropy? Explain.

Section 10.6

6. Calculate ΔS_{surr} for each of the following reactions at 25°C and 1 atm:

   a. $Br(l) \rightarrow Br(g)$ $\Delta H = +31$ kJ
   b. $2C_2H_6(g) + 7O_2(g) \rightarrow 4CO_2(g) + 6H_2O(g)$ $\Delta H = -2857$ kJ

7. A chemical reaction gives a change in entropy of the universe of −48 J/K. Is the process spontaneous? Why or why not?

8. Which of the following values represent spontaneous processes? Which ones are exothermic (from the point of view of the system)?

   a. $\Delta S_{sys} = +358$ J/K, $\Delta S_{surr} = -358$ J/K
   b. $\Delta S_{sys} = -358$ J/K, $\Delta S_{surr} = -52$ J/K
   c. $\Delta S_{sys} = -358$ J/K, $\Delta S_{surr} = +463$ J/K
   d. $\Delta S_{sys} = +358$ J/K, $\Delta S_{surr} = -463$ J/K

Section 10.7

9. Given the following values for ΔH, ΔS, and T, determine whether each of the following sets of data represent spontaneous or nonspontaneous processes.

	ΔH(kJ)	ΔS(J/K)	T(K)
a.	−16	50	300
b.	12	40	300
c.	−5	−20	200
d.	−5	20	200
e.	−5	−20	500

10. Given the following ΔH and ΔS values, determine the temperature at which the reactions would be spontaneous:

    a. $\Delta H = 10.5$ kJ; $\Delta S = 30$ J/K
    b. $\Delta H = 1.8$ kJ; $\Delta S = 113$ J/K
    c. $\Delta H = -11.7$ kJ; $\Delta S = -105$ J/K

11. Predict the sign of the entropy change for each of the following processes:

    a. evaporating a beaker of ethanol at room temperature
    b. cooling nitrogen gas from 80°C to 20°C
    c. freezing liquid bromine below its melting point (−7.2°C)

12. The heat of fusion for actinium is 10.50 kJ/mol. The entropy of fusion is 9.6 J/K mol. Calculate the melting point of actinium.

13. At a constant temperature of 298 K, calculate ΔS_{sys} and ΔS_{univ} for the free expansion of 3.0 L of an ideal gas at 1.0 atm to 11.0 L.

14. The heat of vaporization for protactinium is 481 kJ/mol. The entropy of vaporization is 109 J/K mol. Calculate the boiling point of protactinium. Compare with the actual value of approximately 4500 K.

15. If the molar heat of vaporization of ethanol is 39.3 kJ/mol and its boiling point is 78.3°C, calculate ΔS for the vaporization of 0.50 mol ethanol.

16. The normal boiling point of diethyl ether is 308 K. The enthalpy of vaporization is 27.2 kJ/mol. Calculate ΔS for the vaporization of 1.0 mol of diethyl ether under these conditions.

17. The melting point of silicon is 1683 K. The heat of fusion is 46.4 kJ/mol. Calculate the entropy of fusion of silicon.

18. Determine whether the following chemical change is spontaneous.

$$SO_2(g) + NO_2(g) \rightarrow SO_3(g) + NO(g)$$

Section 10.8

19. Predict the sign of $\Delta S°$ for each of the following reactions:

    a. $Sr(g) + {}^1/_2O_2(g) \rightarrow SrO(c)$ (c = crystalline)
    b. $2Al(s) + 3F_2(g) \rightarrow 2AlF_3(s)$

20. Using Appendix 4, calculate the standard enthalpy changes for the following reactions at 25°C:

    a. $CaCO_3(s) \rightarrow CaO(s) + CO_2(g)$
    b. $N_2(g) + 3H_2(g) \rightarrow 2NH_3(g)$
    c. $H_2(g) + Cl_2(g) \rightarrow 2HCl(g)$

21. Using data from Appendix 4 in your textbook, calculate $\Delta S°$ for each of the following reactions:

    a. $CH_4(g) + N_2(g) \rightarrow HCN(g) + NH_3(g)$
    b. $2Ag_2O(s) \rightarrow 4Ag(s) + O_2(g)$
    c. $Cd(s) + {}^1/_2O_2(g) \rightarrow CdO(s)$

22. Using data for $\Delta H°$ and $\Delta S°$ in Appendix 4 in your textbook, calculate $\Delta G°$ (at 25°C) for each of the reactions in problem 21.

23. Calculate $\Delta G°$ for each of the reactions in problem 21 using $\Delta G°$ data from Appendix 4 in your textbook. How do these compare with your answers from problem 22?

Section 10.9

24. Using Appendix 4, calculate $\Delta G°$ for the combustion of ethane (C_2H_6):

$$2C_2H_6(g) + 7O_2(g) \rightarrow 4CO_2(g) + 6H_2O(l)$$

25. Calculate $\Delta G°$ for the following reactions at 25°C:

    a. $2MgO(s) \rightarrow 2Mg(s) + O_2(g)$
    b. $CH_4(g) + 2O_2(g) \rightarrow CO_2(g) + 2H_2O(l)$

26. Calculate $\Delta G°$ for the following reactions:

    a. $2Mg(s) + O_2(g) \rightarrow 2MgO(s)$
    b. $2C_2H_2(g) + 5O_2(g) \rightarrow 4CO_2(g) + 2H_2O(l)$
    c. $N_2(g) + O_2(g) \rightarrow 2NO(g)$

27. Given the following data:

a. $N_2(g) + 2O_2(g) \rightarrow 2NO_2(g)$ $\Delta G° = 104$ kJ
b. $2NO(g) + O_2(g) \rightarrow 2NO_2(g)$ $\Delta G° = -70$ kJ

Calculate $\Delta G°$ for $N_2(g) + O_2(g) \rightarrow 2NO(g)$.

28. Using Appendix 4, calculate $\Delta G°$ for each of the following reactions at 298 K:

a. $2Cu_2O(s) + O_2(g) \rightarrow 4CuO(s)$
b. $C_2H_5OH(l) \rightarrow C_2H_4(g) + H_2O(g)$

29. Calculate $\Delta G°$ for the following reaction at 25°C (use Appendix 4):

$$2H_2(g) + O_2(g) \rightleftharpoons 2H_2O(l)$$

30. Given the following data:

a. $2H_2(g) + C(s) \rightarrow CH_4(g)$ $\Delta G° = -51$ kJ
b. $2H_2(g) + O_2(g) \rightarrow 2H_2O(l)$ $\Delta G° = -474$ kJ
c. $C(s) + O_2(g) \rightarrow CO_2(g)$ $\Delta G° = -394$ kJ

Calculate $\Delta G°$ for $CH_4(g) + 2O_2(g) \rightarrow CO_2(g) + 2H_2O(l)$.

31. Calculate ΔG at 600 K for the following reaction:

$$P_4(g) + 5O_2(g) \rightleftharpoons P_4O_{10}(s)$$

where the initial pressures are $P_{P_4} = 0.52$ atm and $P_{O_2} = 2.1 \times 10^{-3}$ atm.

32. Determine the free energy of formation for the following change:

$$H_2O(l) + SO_3(g) \rightarrow H_2SO_4(l)$$

33. Calculate $\Delta G°$ for the reaction:

$$C_2H_4(g) + 3O_2(g) \rightarrow 2CO_2(g) + 2H_2O(l)$$

34. Calculate the equilibrium constant, K, at 25°C for each of the reactions in problem 21.

Section 10.10

35. The value of the equilibrium constant for a given reaction is $K = 6 \times 10^{-23}$. What does that indicate about the spontaneity of the reaction?

Section 10.11

36. The value of the equilibrium constant for a given reaction is $K = 8 \times 10^{58}$. What does this tell us regarding the speed of the reaction?

37. We said in Chapter 7 that at 25°C, $K_w = 1.0 \times 10^{-14}$. Calculate K_w thermodynamically and compare it to our Chapter 7 value for the reaction:

$$H_2O(l) \rightleftharpoons H^+(aq) + OH^-(aq)$$

38. Calculate K_p for the following reaction at 25°C.

$$2H_2O(g) \rightleftharpoons 2H_2(g) + O_2(g)$$

Multiple-Choice Self-Test

1. A state of higher entropy means:

   A. A lower number of possible arrangements
   B. A higher number of possible arrangements
   C. Lower probabilities to reach a possible state
   D. An exothermic process

2. Which of the following processes has the lowest probability of being achieved?

   A. A feather flying away from the ground
   B. A rock rolling down the hill
   C. Water freezing into ice at 273 K
   D. A piece of paper flying away from the ground

3. Which of the following processes must be spontaneous?

   A. $\Delta S_{surr} > 0, \Delta S_{sys} > 0$
   B. $\Delta S_{surr} > 0, \Delta S_{sys} < 0$
   C. $\Delta S_{univ} < 0, \Delta S_{sys} > 0$
   D. $\Delta S_{surr} < 0$

4. Heat is released during a particular process. This means that:

   A. The process is spontaneous under all conditions
   B. $\Delta S_{surr} > 0$
   C. The process tends to be spontaneous
   D. $\Delta S_{sys} > 0$

5. Which of the following processes would you expect to be spontaneous?

   A. $\Delta S_{surr} = 25$ J/K, $\Delta S_{sys} = -27$ J/K
   B. $\Delta S_{surr} = 25$ J/K, $\Delta S_{sys} = 27$ J/K
   C. $\Delta S_{univ} = -20$ J/K, $\Delta S_{sys} = -20$ J/K
   D. $\Delta S_{surr} = -80$ J/K, $\Delta S_{sys} = 20$ J/K

6. Which of the following conditions must be met for a process to be spontaneous?

   A. $\Delta G < 0$ B. $\Delta H < 0$ C. $\Delta S_{surr} > 0$ D. $\Delta S_{sys} > 0$

7. Which of the following system conditions would allow a process to be spontaneous at all temperatures?

   A. $\Delta S > 0, \Delta H < 0$ B. $\Delta S > 0, \Delta H > 0$ C. $\Delta S < 0, \Delta H > 0$ D. $\Delta S < 0, \Delta H < 0$

8. Calculate one temperature that would allow the following process to be spontaneous:

$$\Delta S = 30 \text{ J/K}, \Delta H = 120 \text{ kJ}$$

   A. 400°C B. 4002 K C. 360°C D. 3600 K

9. A particular process involved the expansion of a gas at 20 atm. The gas released energy in terms of heat to the surrounding equal to 148300 J. If the free energy of this process was −300 J at 298 K, and the entropy change for the surrounding was 500 J/K, by how many times did the volume of the gas change?

   A. 50 times B. 1500 units C. 15.0 times D. 25 times

10. Calculate the standard absolute entropy, in J/mol K, of Mg:

$$2NO_2(g) + 2MgO(s) \rightarrow Mg(NO_3)_2(s) + Mg(s) \qquad \Delta S° = -462.1 \text{ J/K}$$

$S°_{NO_2(g)} = 239.9$ J/mol K $\quad S°_{MgO(s)} = 27.0$ J/mol K $\quad S°_{Mg(NO_3)_2(s)} = 39.2$ J/mol K

A. 32.5 B. 234.3 C. 27.8 D. 0.00

11. Calculate $\Delta S°$ for the following reaction:

$$H_3AsO_4(aq) \rightarrow 3H^+(aq) + AsO_4^{3-}(aq)$$

$S°_{H^+(aq)} = 0.00$ J/mol K $\quad S°_{H_3AsO_4(aq)} = 44.0$ J/mol K $\quad S°_{AsO_4^{3-}(aq)} = -38.9$ J/mol K

A. 5.10 J/K B. −82.9 J/K C. −5.1 J/K D. −72.7 J/K

12. Under standard conditions (all gases at $P = 1$ atm), will the following reaction take place under sunlight?

$$3Cl_2(g) + 2CH_4(g) \rightarrow CH_3Cl(g) + CH_2Cl_2(g) + 3HCl(g)$$

$\Delta G°(CH_4) = -50.72$ kJ/mol $\quad \Delta G°(CH_3Cl) = -57.37$ kJ/mol
$\Delta G°(CH_2Cl_2) = -68.85$ kJ/mol $\quad \Delta G°(HCl) = -95.30$ kJ/mol

A. Yes C. Under sunlight only
B. No D. Not in any conditions

13. Calculate the free energy change ΔG, in kJ, for the following reaction at 298 K:

$$\underset{0.2 \text{ atm}}{N_2(g)} + \underset{0.2 \text{ atm}}{2O_2(g)} \rightarrow \underset{0.39 \text{ atm}}{2NO_2(g)}$$

A. 109 B. 37.32 C. 13.98 D. −37.32

14. The value of K_{eq} goes down by a factor of 100.0. Compute the value, in kJ, of the change in $\Delta G°$ for the reaction.

A. 1.37 B. 11.41 C. 0.382 D. −11.41

15. If the free energy is equal to 10.0 kJ, and it is four times as big as the standard free energy, what must the value of Q be under standard conditions?

A. 20.6 B. 0.003 C. 3.25 D. 1.00

16. Calculate the standard free energy of a specific reaction if the formation constant is 1.47×10^9.

$$Zn^{2+}(aq) + 4NH_3(aq) \rightleftharpoons [Zn(NH_3)_4]^{2+}(aq)$$

A. −54.0 kJ B. 23.5 kJ C. 54.0 kJ D. 2.51 kJ

Answers to Exercises

1. Probability = $1/2^8 = 1/256$. This says that the probability dictates that molecules will be distributed more evenly throughout the bulbs—in other words, a higher degree of entropy.

2. a. $HCl(g)$
   b. $P_4O_{10}(g)$
   c. $P_4O_{10}(g)$ (though the difference is quite small)
   d. $H_2O(l)$
   e. $N_2O_4(g)$
   f. $Ar(g)$ at 0.30 atm

3. a. positive b. sharply positive c. negative

4. a. ΔS = negative
   b. ΔS = positive
   c. We cannot accurately predict the sign of ΔS but we know that the change is very small.

5. No, entropy decreases as molecules are less randomly distributed, as in freezing.

6. a. $\Delta S_{surr} = -100$ J/K (2 sig figs) b. $\Delta S_{surr} = 9590$ J/K

7. The entropy of the universe decreased. The process is therefore **nonspontaneous.**

8. Only process "c" is spontaneous. Process "c" is the only exothermic process.

9. a. spontaneous
   b. nonspontaneous
   c. spontaneous
   d. spontaneous
   e. nonspontaneous

10. a. spontaneous at 350 K or above
    b. spontaneous at 16 K or above
    c. spontaneous at 111 K or below

11. a. $\Delta S > 0$ (positive) b. $\Delta S < 0$ (negative) c. $\Delta S < 0$ (negative)

12. melting point = 1100 K

13. $\Delta S_{sys} = \Delta S_{univ} = 0.027$ L atm/K = 2.7 J/K

14. boiling point = 4410 K

15. 56 J/K

16. $\Delta S = 88$ J/K

17. entropy of fusion = 27.6 J/K mol

18. yes, spontaneous ($\Delta G° = -36$ kJ)

19. a. negative
    b. negative

20. a. 178 kJ b. −92 kJ c. −184 kJ

21. a. +17 J/K b. +133 J/K c. −100 J/K

22. a. +159 kJ b. +22 kJ c. −228 kJ

23. a. +159 kJ b. +22 kJ c. −228 kJ

24. $\Delta G° = -2932$ kJ

25. a. $\Delta G° = 1138$ kJ b. $\Delta G° = -817$ kJ

26. a. −1138 kJ b. −2468 kJ c. 174 kJ

27. $\Delta G° = 174$ kJ

28. a. $\Delta G° = -216$ kJ b. $\Delta G° = 14$ kJ

29. $\Delta G° = -474$ kJ

30. $\Delta G° = -817$ kJ

31. $\Delta G° = -2541$ kJ

32. $\Delta G° = -82$ kJ

33. $\Delta G° = -1466$ kJ

34. a. $K = 1.4 \times 10^{-28}$ b. $K = 1.4 \times 10^{-4}$ c. $K = 9.2 \times 10^{39}$

35. The reaction is not spontaneous.

36. The equilibrium constant says nothing about the rate of reaction.

37. $K = 9.5 \times 10^{-15}$. This is very close to our value of K_w of 1.0×10^{-14}.

38. $K_p = 5.3 \times 10^{-81}$

Answers to Multiple-Choice Self-Test

1. B	4. B	7. A	10. A	13. A	15. A
2. D	5. B	8. B	11. B	14. B	16. A
3. A	6. A	9. A	12. A		

CHAPTER 11

Electrochemistry

The Bottom Line: Chapter 11

Your textbook defines electrochemistry as the study of the interchange of chemical and electrical energy. In this chapter we are concerned with the use of chemical reactions to generate an electric current and the use of electric current to produce chemical reactions.

11.1 Galvanic Cells

We discussed reduction-oxidation (redox) reactions in Section 4.10. The following terms should be reviewed: **reduction, oxidation, reducing agent, oxidizing agent, half-reaction. Review the half-reaction method of balancing redox equations** (Section 4.11), and go over some of the problems at the end of that chapter in this study guide and in your textbook. The rest of this discussion assumes you are able to balance redox equations.

A **galvanic cell** is a device in which **chemical energy is converted to electrical energy**. There is a need to physically separate the oxidizing and reducing agents in galvanic cells so that energy of reaction can be used. Figures 11.2 and 11.3 in your textbook show the important features of a galvanic cell. You should know the function of the following:

- **cathode** - Reduction occurs here. Species undergoing reduction ("oxidizing agent") receive electrons from the cathode.
- **anode** - Oxidation occurs here. Species undergoing oxidation ("reducing agent") lose electrons here.
- **salt bridge** (or porous disk) - It allows exchange of ions to keep electric neutrality while electroactive solutions remain separated.

Also know the following terms:

- **electromotive force (emf)** - This is the force with which electrons are pulled through a wire.
- **volt(V)** - This is the unit of electrical potential. It equals 1 joule/coulomb.

In constructing galvanic cells, keep in mind the **direction of electron flow**. Species undergoing *reduction receive electrons from the cathode*. Species undergoing *oxidation donate electrons to the anode*. The direction of electron flow is therefore **from the anode to the cathode**.

Example 11.1 - A Bit Of Review

In each of the following equations, identify the species that would receive electrons from the cathode, and the species that would lose electrons at the anode in the galvanic cells.

a. $Mg(s) + 2H^+(aq) \rightarrow Mg^{2+}(aq) + H_2(g)$
b. $MnO_4^-(aq) + 5Fe^{2+}(aq) + 8H^+(aq) \rightarrow Mn^{2+}(aq) + 5Fe^{3+}(aq) + H_2O(l)$

Solution

It is quite useful to separate balanced reactions into half-reactions. The reduction half-reaction receives electrons at the **cathode**. The **oxidation** half-reaction loses electrons at the **anode**.

a. oxidation (anode): $Mg(s) \rightarrow Mg^{2+}(aq) + 2e^-$
reduction (cathode): $2H^+(aq) + 2e^- \rightarrow H_2(g)$

b. oxidation (anode): $Fe^{2+}(aq) \rightarrow Fe^{3+}(aq) + e^-$
reduction (cathode): $MnO_4^-(aq) + 8H^+(aq) + 5e^- \rightarrow Mn^{2+}(aq) + 4H_2O(l)$

As in all galvanic cells, the direction of electron flow is from the anode to the cathode.

11.2 Standard Reduction Potentials

The electromotive force (emf) of a galvanic cell is a **combination** of the potentials of two half-reactions. Because the cathodic potentials are described relative to anodic reactions, we need one absolute standard against which all other half-reactions can be compared. The standard is the **standard hydrogen** electrode,

$$\mathbf{2H^+ + 2e^- \rightarrow H_2 \qquad E° = 0.00\ V\ (exactly,\ by\ definition)}$$

This is an arbitrary but necessary assignment. Table 11.1 in your textbook has a list of standard reduction potentials for many half-reactions. **Standard means 298 K and 1 atmosphere.**

Here are guidelines you can keep in mind for calculating E°_{cell}:

- Galvanic cells require $E^\circ_{cell} > 0$ V
- One of the tabulated reduction potentials will have to be reserved (to form an oxidation half-reaction) in every $E°$ calculation.
- To determine which reaction is to be reversed, the **sum** of the **oxidation and reduction half-reactions must be > 0 V in a galvanic cell.**
- When you **reverse** a reaction, **$E°$ gets the opposite sign.**
- When you **multiply** a reaction by a coefficient (for purposes of balancing), **the $E°$ is NOT changed!**

Let's have some practice with calculating E°_{cell} and determining the reduction and oxidation reactions in a galvanic cell.

Example 11.2 A - Cell emf

Using data from Table 11. 1 in your textbook, calculate the emf values (E°_{cell}) for each of the following reactions. State which are galvanic.

a. $Mg(s) + 2H^+(aq) \rightarrow Mg^{2+}(aq) + H_2(g)$
b. $Cu^{2+}(aq) + 2Ag(s) \rightarrow Cu(s) + 2Ag^+(aq)$
c. $2Zn^{2+}(aq) + 4OH^-(aq) \rightarrow 2Zn(s) + O_2(g) + 2H_2O(l)$

Solution

We need to split each reaction into two half-reactions and look up their reduction potentials. **(Remember that when a half-reaction is an oxidation, $E°$ must change sign!)**

a. reduction: $2H^+ + 2e^- \rightarrow H_2$ $\quad E° = 0.00$ V
oxidation: $Mg \rightarrow Mg^{2+} + 2e^-$ $\quad E° = +2.37$ V
E°_{cell} = **+2.37 V, galvanic**

b. reduction: $Cu^{2+} + 2e^- \rightarrow Cu$ $\quad E° = +0.34$ V
oxidation: $2Ag \rightarrow 2Ag^+ + 2e^-$ $\quad E° = -0.80$ V
$E°_{cell}$ = **−0.46 V, not galvanic - will not run in this direction**

c. reduction: $2Zn^{2+} + 4e^- \rightarrow 2Zn$ $\quad E° = -0.76$ V
oxidation: $4OH^- \rightarrow O_2 + 2H_2O + 4e^-$ $\quad E° = -0.40$ V
$E°_{cell}$ = **−1.16 V, not galvanic**

Another key point regarding the magnitude and sign of *E*° for half-reactions is that **THE MORE POSITIVE THE *E*° VALUE THE MORE LIKELY THE SPECIES IS TO BE REDUCED.** (It is a stronger oxidizing agent.) (With regard to oxidation of a species, the opposite is also true.) For example, Br_2 is more likely to be reduced than I_2 because Br_2 has the more positive *E*°.

Example 11.2 B - Reduction Strength

Place the following in order of increasing strength as oxidizing agents.

Fe^{2+}, ClO_2, F_2, AgCl

Solution

From looking at the *E*° values in Table 11.1 of your textbook, the order is **Fe^{2+} < AgCl < ClO_2 < F_2.**

Example 11.2 C - Composing Galvanic Cells

Given the following half-cells, decide which is the anode and which the cathode, balance the equations, write the overall cell reaction, and calculate $E°_{cell}$ if the cells are galvanic.

a. $Ni^{2+} + 2e^- \rightarrow Ni$ $\quad E° = -0.23$ V
$O_2 + 4H^+ + 4e^- \rightarrow 2H_2O$ $\quad E° = +1.23$ V

b. $Ce^{4+} + e^- \rightarrow Ce^{3+}$ $\quad E° = +1.70$ V
$Sn^{2+} + 2e^- \rightarrow Sn$ $\quad E° = -0.14$ V

Solution

The goal is to see which arrangement of half-reactions will make $E°_{cell} > 0$.

a. Reversing the Ni^{2+} reduction will create a galvanic cell. Note that to balance the overall reaction, it must also be multiplied by 2.

cathode:	$O_2 + 4H^+ + 4e^- \rightarrow 2H_2O$	*E*°	= 1.23 V
anode:	$2Ni \rightarrow 2Ni^{2+} + 4e^-$	*E*°	= 0.23 V
overall:	$O_2(g) + 4H^+(aq) + 2Ni(s) \rightarrow 2H_2O(l) + 2Ni^{2+}(aq)$	$E°_{cell}$	**= 1.46 V**

b. Reversing the Sn^{2+} half-reaction will create a galvanic cell. The cerium half-reaction must be multiplied by 2 to balance the overall reaction.

cathode:	$2Ce^{4+} + 2e^- \rightarrow 2Ce^{3+}$	*E*°	= 1.70 V
anode:	$Sn \rightarrow Sn^{2+} + 2e^-$	*E*°	= 0.14 V
overall:	$2Ce^{4+}(aq) + Sn(s) \rightarrow Sn^{2+}(aq) + 2Ce^{3+}(aq)$	$E°_{cell}$	**= 1.84 V**

We are now most of the way toward completely characterizing a galvanic cell. We have:

1. calculated cell potentials
2. described the direction of electron flow
3. designated the anode and cathode

We have not yet "4. *described the nature of each electrode and the ions present in each compartment.*" Normally the solid metal in a half-reaction will serve as the electrode. If a half-reaction contains only ions, a nonreacting conductor (usually platinum) will be the electrode.

Example 11.2 D - Describing Galvanic Cells

Describe a galvanic cell based on the two half-reactions below. Use the four-step analysis listed above.

$$Cu^{2+} + 2e^- \rightarrow Cu \qquad E° = 0.34\text{ V}$$
$$Cr_2O_7^{2-} + 14H^+ + 6e^- \rightarrow 2Cr^{3+} + 7H_2O \qquad E° = 1.33\text{ V}$$

Solution

Steps 1 and 3:

In order to have a galvanic cell, $E°$ must be > 0. We must reverse the copper reduction. We also need to balance the entire cell reaction electrically by multiplying the copper equation by 3.

anode: $3Cu \rightarrow 3Cu^{2+} + 6e^-$ $\qquad E° = -0.34\text{ V}$

cathode: $Cr_2O_7^{2-} + 14H^+ + 6e^- \rightarrow 2Cr^{3+} + 7H_2O$ $\qquad E° = +1.33\text{ V}$

overall: $\mathbf{3Cu(\mathit{s}) + Cr_2O_7^{2-}(\mathit{aq}) + 14H^+(\mathit{aq}) \rightarrow 3Cu^{2+}(\mathit{aq}) + 2Cr^{3+}(\mathit{aq}) + 7H_2O(\mathit{l})}$ $\qquad \mathbf{E°_{cell} = 0.99\text{ V}}$

Step 4:

A **copper electrode** will be in the **anode** side. A **platinum electrode** will be in the **cathode** side.

Step 2:

The electron flow (from anode to cathode) will be from **copper to platinum**.

11.3 Cell Potential, Electrical Work, and Free Energy

The first part of this section emphasizes that the **actual work** that can be achieved is **always less than** the **theoretical work** available. Nonetheless, your textbook presents the relationship between **free energy** (maximum is assumed, though not attainable) and **cell potential**.

$$\Delta G = -nFE$$

at standard conditions,

$$\Delta G° = -nFE°$$

where $\mathbf{\Delta G}$ = free energy (in Joules)
$\mathbf{n}$ = moles of electrons exchanged in the redox reaction
$\mathbf{F}$ = the Faraday, a constant (96,486 Coulombs per mole of electrons)
$\mathbf{E}$ = cell voltage (Joules/Coulomb)

Note that ΔG and E have **opposite signs. For a spontaneous process, ΔG is "−" and E is "+".**

Example 11.3 A - Cell Voltage And Free Energy

Calculate the $\Delta G°$ for the reaction

$$\mathbf{Zn^{2+}(}\boldsymbol{aq}\mathbf{) + Cu(}\boldsymbol{s}\mathbf{) \rightleftharpoons Zn(}\boldsymbol{s}\mathbf{) + Cu^{2+}(}\boldsymbol{aq}\mathbf{)}$$

Will zinc ions plate out on a copper strip?

Solution

Where are we going? There are two related questions, with the second being determined by the answer to the first. We are asked for the $\Delta G°$ of the reaction. We are then asked if (based on that answer) zinc ions will plate out on a copper strip. That is, will the reaction proceed; is it spontaneous. Which is true for a spontaneous reaction; a positive or negative ΔG?

The two pertinent half-reactions are

$$\begin{array}{ll} Zn^{2+} + 2e^- \rightarrow Zn & E° = -0.76\text{ V} \\ Cu \rightarrow Cu^{2+} + 2e^- & E° = -0.34\text{ V} \\ \hline & E°_{cell} = -1.10\text{ V} \end{array}$$

$$\Delta G° = -nFE° = -2\text{ mol e}^- (96{,}486\text{ C/mol e}^-)(-1.10\text{ J/C})$$

$$\Delta G° = 2.12 \times 10^5\text{ J} = \mathbf{212\ kJ}$$

Zinc ions will not plate out on a copper strip at standard conditions. (Note that this result is also indicated by the fact that $E°_{cell} < 0$.)

Example 11.3 B - Practice With Cell Voltage And Free Energy

Vanadium(V) can be reduced to vanadium(IV) by reaction with a **Jones reductor**, a Zn-Hg amalgam. The reactions of interest are:

$$\begin{array}{ll} VO_2^+ + 2H^+ + e^- \rightarrow VO^{2+} + H_2O & E° = +1.00\text{ V} \\ Zn^{2+} + 2e^- \rightarrow Zn & E° = -0.76\text{ V} \end{array}$$

Calculate $E°$ and $\Delta G°$ for the reaction.

Solution

For a spontaneous reduction of vanadium, Zn is oxidized. The reactions are (Watch the electron balance!):

$$\begin{array}{ll} 2VO_2^+ + 4H^+ + 2e^- \rightarrow 2VO^{2+} + 2H_2O & E° = +1.00\text{ V} \\ Zn \rightarrow Zn^{2+} + 2\,e^- & E° = +0.76\text{ V} \\ \hline \mathbf{2VO_2^+(aq) + 4H^+(aq) + Zn(s) \rightarrow 2VO^{2+}(aq) + 2H_2O(l) + Zn^{2+}(aq)} & E°_{cell} = \mathbf{1.76\ V} \end{array}$$

$$\Delta G° = -2\text{ mol e}^- (96{,}486\text{ C/mol e}^-)(1.76\text{ J/C})$$

$$\Delta G° = -3.40 \times 10^5\text{ J} = \mathbf{-340\ kJ}$$

11.4 Dependence of the Cell Potential on Concentration

This section considers the calculation of cell voltages at nonstandard concentrations (i.e., not 1 *M*). Your textbook introduces the **concentration cell**, a cell in which current flows due only to a **difference in concentration** of an ion in two different compartments of a cell. **Le Châtelier's principle** is applicable here. In a cell in which there is an equal concentration of metal ion on both sides, $E°_{cell} = 0$.

However, if the concentrations are different, a "stress" is put on the system that will be equalized by electron flow to allow reduction and oxidation to occur. (See Figures 11.11 and 11.12 in your textbook.) When the concentrations in the half-cells become equal, $E^\circ_{cell} = 0$, and the system is at equilibrium.

Example 11.4 A - Concentration Cells

A cell has on its left side a 0.20 *M* Cu^{2+} solution. The right side has a 0.050 *M* Cu^{2+} solution. The compartments are connected by Cu electrodes and a salt bridge. Designate the cathode, anode, and direction of current.

Solution

Current will flow in this cell until the concentration of Cu^{2+} is equal in both compartments. This means that the concentration of Cu^{2+} in the **left-hand side (0.20 *M*) must be reduced** by

$$Cu^{2+} + 2e^- \rightarrow Cu$$

The left hand-side will be the **cathode**. The right-hand side (0.050 *M* Cu^{2+}) will be the **anode**. Current will flow from right to left (anode to cathode).

The point of introducing concentration cells (which, as your textbook points out, produce a very small voltage) is really to illustrate the fact that **nonstandard concentrations produce a cell voltage that is different from that at standard concentrations**.

$$E_{cell}\text{ (nonstandard)} \neq E^\circ_{cell}$$

The Nernst equation is derived in your textbook. We use the equation to **calculate the cell voltage at nonstandard concentrations**.

$$E_{cell} = E^\circ_{cell} - \frac{0.0591}{n}\log(Q) \qquad \text{(at 25°C)}$$

where Q has its usual significance as $[\text{products}]_0 / [\text{reactants}]_0$

The mathematics of the Nernst equation are that **if $[\text{reactants}]_0$ are higher than $[\text{products}]_0$**, $\log(Q)$ will be < 0 and **E_{cell} will be > E°_{cell}**. This is consistent with Le Châtelier's principle and concentration cells. Also, when a battery is fully discharged it is at equilibrium, and $E_{cell} = 0$.

Example 11.4 B - The Nernst Equation

Calculate E_{cell} for a galvanic cell based on the following half-reactions at 25°C.

$$Cd^{2+} + 2e^- \rightarrow Cd \qquad E^\circ = -0.40\text{ V}$$
$$Pb^{2+} + 2e^- \rightarrow Pb \qquad E^\circ = -0.13\text{ V}$$

where $[Cd^{2+}] = 0.010$ *M* and $[Pb^{2+}] = 0.100$ *M*.

Solution

For the cell to be galvanic, E° must be greater than 0 V. This means that the cadmium must undergo oxidation, giving a net cell reaction

$$Cd(s) + Pb^{2+}(aq) \rightarrow Cd^{2+}(aq) + Pb(s) \qquad E^\circ_{cell} = +0.27\text{ V}$$

The Nernst equation for this cell is

$$E = E^\circ - \frac{0.0591}{2}\log\left[\frac{[Cd^{2+}]}{[Pb^{2+}]}\right] = 0.27 - 0.0296\log\left[\frac{0.010}{0.100}\right]$$

$$E = 0.27 - 0.0296\log(0.10) = 0.27 + 0.0296$$

$$\mathbf{E_{cell} = 0.30\text{ V}}$$

Example 11.4 C - Nernst Equation

Calculate E_{cell} for a galvanic cell based on the following half-reactions at 25°C.

(Equation 1) $FeO_4^{2-} + 8H^+ + 3e^- \rightarrow Fe^{3+} + 4H_2O$ $E° = +2.20$ V
(Equation 2) $O_2 + 4H^+ + 4e^- \rightarrow 2H_2O$ $E° = +1.23$ V

where $[FeO_4^{2-}] = 2.0 \times 10^{-3}$ *M* $[O_2] = 1.0 \times 10^{-5}$ atm
$[Fe^{3+}] = 1.0 \times 10^{-3}$ *M* pH = 5.2

Solution

The ferrate ion (FeO_4^{2-}) has a higher reduction potential, so it will be reduced while water is oxidized to oxygen. To balance electrically, equation 1 must be multiplied by 4 and equation 2 must be reversed and multiplied by 3. This gives

$4FeO_4^{2-} + 32H^+ + 12e^- \rightarrow 4Fe^{3+} + 16H_2O$ $E° = +2.20$ V
$6H_2O \rightarrow 3O_2 + 12H^+ + 12e^-$ $E° = -1.23$ V

$\mathbf{4FeO_4^{2-}(aq) + 20H^+(aq) \rightarrow 4Fe^{3+}(aq) + 3O_2(g) + 10H_2O(l)}$ $E^\circ_{cell} = \mathbf{+0.97\ V}$

$$E = E^\circ_{cell} - \frac{0.0591}{n}\log(Q)$$
$$= 0.97 - \frac{0.0591}{12}\log\left[\frac{[Fe^{2+}]^4[O_2]^3}{[FeO_4^{2-}]^4[H^+]^{20}}\right]$$
$$= 0.97 - \frac{0.0591}{12}\log\left[\frac{(1.0\times10^{-3})^4(1.0\times10^{-5})^3}{(2.0\times10^{-3})^4(6.31\times10^{-6})^{20}}\right]$$

↑
(pH=5.2)

Notice the extreme H^+ ion dependence! If you try to take 6.31×10^{-6} to the 20th power on a programmable calculator, it will handle it well. Simpler calculators will require you to split it up as follows.

$$(6.31 \times 10^{-6})^{20} = (6.31)^{20} \times (10^{-6})^{20}$$
$$= 1.0 \times 10^{16} \times 10^{-120}$$
$$= \mathbf{1.0 \times 10^{-104}}$$

$$E = 0.97 - 4.9 \times 10^{-3}\log\left[\frac{1.0\times10^{-27}}{1.6\times10^{-115}}\right]$$
$$= 0.97 - 4.9 \times 10^{-3}\log(6.3 \times 10^{87})$$
$$= 0.97 - 0.43$$
$$= \mathbf{0.54\ V}$$

Recall that **at equilibrium both *E* and Δ*G* = 0**. Your textbook uses this information, along with the relationship between Δ*G* and *K*, to derive a formula that **relates *E* to *K***.

$$\log(K) = \frac{nE°}{0.0591} \quad \text{(at 25°C)}$$

Remember, we are dealing with **equilibrium conditions** in this case.

Example 11.4 D - Equilibrium Constants And Cell Potential

Calculate the equilibrium constant for the reaction in Example 11.2 D,

$3Cu(s) + Cr_2O_7^{2-}(aq) + 14H^+(aq) \rightarrow 3Cu^{2+}(aq) + 2Cr^{3+}(aq) + 7H_2O(l)$ $\quad$ $\mathbf{E^\circ_{cell} = 0.99\ V}$

Solution

There are 6 moles of electrons transferred in this redox reaction.

$$\log(K) = \frac{nE^\circ}{0.0591} = \frac{6(0.99)}{0.0591} = 100.51$$

$$\mathbf{K} = 10^{100.51} = 10^{0.51} \times 10^{100} = \mathbf{3 \times 10^{100}}$$

Example 11.4 E - Summing It All Up

Consider the reaction

$$Ni^{2+}(aq) + Sn(s) \rightarrow Ni(s) + Sn^{2+}(aq)$$

Calculate the following: E°_{cell}, ΔG°, and K at 25°C. In addition, determine the minimum **ratio of $[Sn^{2+}]/[Ni^{2+}]$** necessary in order to make the reaction spontaneous as written.

Solution

a. The half-reactions are:

$Ni^2 + 2e^- \rightarrow Ni$	E°	$= -0.23$ V
$Sn \rightarrow Sn^{2+} + 2e^-$	E°	$= +0.14$ V
	$\mathbf{E^\circ_{cell}}$	**= −0.09 V**

The reaction is not spontaneous under standard conditions.

b. $\Delta G^\circ = -nFE^\circ = (-2 \text{ mol e}^-)(96{,}486 \text{ C/mol e}^-)(-0.09 \text{ J/C}) = 1.74 \times 10^4$ J

ΔG° = 20 kJ (rounded to 1 sig fig)

c. We can calculate K using either **log $K = nE^\circ/0.059$** or **ln $K = -\Delta G^\circ/RT$**. Let's do it both ways.

$$\log K = \frac{2(-0.09)}{0.0591} = -3.05 \quad \Rightarrow \quad \mathbf{K} = 9 \times 10^{-4}$$

$$\ln K = \frac{-17400 \text{ J}}{8.3148 \text{ J/K mol}(298 \text{ K})} = -7.022 \quad \Rightarrow \quad \mathbf{K} = 9 \times 10^{-4}$$

d. In order for the reaction to be spontaneous, $E > 0$. Using the Nernst equation,

$$E > 0 > E^\circ_{cell} - \frac{0.0591}{2}\log\left[\frac{[Sn^{2+}]}{[Ni^{2+}]}\right]$$

reducing,

$$0 > -0.09 - 0.0296\log\left[\frac{[Sn^{2+}]}{[Ni^{2+}]}\right]$$

$$\frac{0.09}{-0.0296}\ (= -3.05) > \log\left[\frac{[Sn^{2+}]}{[Ni^{2+}]}\right]$$

$$\frac{[Sn^{2+}]}{[Ni^{2+}]} < 9 \times 10^{-4}$$

The ratio of Sn^{2+} to Ni^{2+} must be less than 9×10^{-4} for this reaction to be spontaneous.

11.5 Batteries

The following review questions will help you test your knowledge of the material in this section.

1. Define battery.
2. Why is a lead-storage battery especially useful in automobiles?
3. Give the anode, cathode, and overall cell reactions in the lead-storage battery.
4. Calculate $E°$ for the lead-storage battery.
5. Describe the idea behind how the extent of battery discharge can be measured.
6. Why can "jump starting" a car be dangerous?
7. Describe the parts of a typical dry-cell battery.
8. Why does an alkaline dry cell last longer than an acid dry cell?
9. Give the anode, cathode, and overall cell reactions for the acid and alkaline dry cells.
10. Why is the nickel-cadmium battery rechargeable?
11. Define **fuel cell**.
12. Give the anode, cathode, and overall reactions for the hydrogen-oxygen fuel cell.

11.6 Corrosion

The following review questions will help you test your knowledge of the material in this section.

1. Define corrosion.
2. Why is corrosion so ubiquitous? (Consider the reduction potentials in Table 11.1.)
3. If most native metals **theoretically** oxidize in air, why don't these metals **actually** oxidize?
4. List the reactions and calculate the $E°$ for the corrosion of iron.
5. What is the role of water and salt in the rusting of iron? Why do cars rust more in wet, cold areas of the United States than in dry, warm areas?
6. What is the role of chromium in corrosion protection?
7. How does galvanizing iron help prevent rust?
8. How does magnesium act as **cathodic protection** for iron pipes?

11.7 Electrolysis

This section begins with the definition of **electrolysis** as **the process of forcing a current through a cell to produce a chemical change for which the cell potential is negative**. In order for electrolysis to occur, you must apply an external voltage that is **greater than** the potential of the galvanic cell if you want to force the reaction in the opposite (electrolytic) direction.

Example 11.7 A - Electrolytic Cells

What voltage is necessary to force the following electrolysis reaction to occur?

$$2I^{-}(aq) + Cu^{2+}(aq) \rightarrow I_2(s) + Cu(s)$$

Which process would occur at the anode? cathode? Assuming the iodine oxidation takes place at a platinum electrode, what is the direction of electron flow in this cell?

Solution

$$\begin{array}{lll} \text{anode (oxidation):} & 2I^- \rightarrow I_2 + 2e^- & E° = -0.54\text{ V} \\ \text{cathode (reduction):} & Cu^{2+} + 2e^- \rightarrow Cu & E° = +0.34\text{ V} \\ & & E°_{cell} = -0.20\text{ V} \end{array}$$

More than 0.20 V must be applied externally to make this reaction proceed. The direction of electron flow is always From Anode To CAThode ("FAT CAT"), so it is from the platinum electrode to the copper electrode.

If you put a potential on a system that contains one metal ion, and the **potential is above that at which the metal ion will reduce**, you will plate out that metal. The amount of metal reduced is directly related to the current, in **amps (Coulombs/sec)**, that flows in the system. In practice, this is not the case because (as pointed out several times in your textbook) **electrochemistry is not a perfectly efficient process**. However, for purposes of problem solving we will assume it is.

Dimensional analysis works wonders with electrolysis problems. Your goal is often to find **moles of electrons** of the metal.

$$\text{moles of } e^- = \underset{\text{Faraday}}{\underset{\uparrow}{\frac{1 \text{ mol } e^-}{96{,}486 \text{ C}}}} \times \underset{\text{amps}}{\underset{\uparrow}{\frac{\text{C}}{\text{s}}}} \times \underset{\text{time}}{\underset{\uparrow}{\text{s}}}$$

Example 11.7 B - Electrolysis

How many grams of copper can be reduced by applying a 3.00-A current for 16.2 min to a solution containing Cu^{2+} ions?

Solution

Our goal is to find grams. As always, we want to work through moles and, in this case, moles of electrons.

$$\text{current} \xrightarrow{\text{time}} \text{Coulombs} \Rightarrow \text{moles of } e^- \Rightarrow \text{moles of Cu} \Rightarrow \text{g of Cu}$$

$$\text{g Cu} \,\Big\|\, \frac{3.00 \text{ C}}{\text{s}} \times \frac{60 \text{ s}}{\text{min}} \times 16.2 \text{ min} \times \frac{1 \text{ mol } e^-}{96{,}486 \text{ C}} \times \frac{1 \text{ mol Cu}}{2 \text{ mol } e^-} \times \frac{63.54 \text{ g Cu}}{1 \text{ mol Cu}} = \mathbf{0.96 \text{ g Cu}}$$

Electrolysis can be used to separate a mixture of ions if the reduction potentials are fairly far apart. Remember that **the metal ion with the highest reduction potential is the easiest to reduce**.

Example 11.7 C - Order Of Reduction

Using Table 11.1 in your textbook, predict the order of reduction and which of the following ions will reduce first at the cathode of an electrolytic cell:

$$Ag^+, Zn^{2+}, IO_3^-$$

Solution

$$\begin{array}{ll} Ag^+ + e^- \rightarrow Ag & E° = +0.80\text{ V} \\ Zn^{2+} + 2e^- \rightarrow Zn & E° = -0.76\text{ V} \\ IO_3^- + 6H^+ + 5e^- \rightarrow 1/2 I_2 + 3H_2O & E° = +2.10\text{ V} \end{array}$$

The IO_3^- will reduce first, followed by Ag^+ and Zn^{2+} as the voltage is increased.

Example 11.7 D - Electrolysis Of Water

What volume of $H_2(g)$ and $O_2(g)$ is produced by electrolyzing water at a current of 4.00 A for 12.0 minutes (assuming ideal conditions)?

Solution

The overall reaction is

$$\mathbf{2H_2O(\mathit{l}) \rightarrow 2H_2(\mathit{g}) + O_2(\mathit{g})}$$

In actual practice, the actual ratio is not exactly 2:1 for a variety of reasons including oxygen solubility. We can, however, run through the calculation for practice.

We are asked to find the **volume** of hydrogen and oxygen. The **ideal gas law** relates moles of a gas to volume. Our strategy must therefore be to **find moles of each gas**, then use the ideal gas relationship that **1 mole of a gas occupies 22.4 L** (under ideal conditions) to find volume. Also, for every mole of water, **2 moles of electrons** are exchanged.

$$\text{mol } H_2 \;\|\; \frac{4.00\text{ C}}{\text{s}} \times \frac{60\text{ s}}{\text{min}} \times 12.0\text{ min} \times \frac{1\text{ mol e}^-}{96{,}486\text{ C}} \times \frac{1\text{ mol H}_2}{2\text{ mol e}^-} = 0.0149\text{ mol } H_2$$

$$\text{mol } O_2 \;\|\; 0.0149\text{ mol } H_2 \times \frac{1\text{ mol O}_2}{2\text{ mol H}_2} = 0.00746\text{ mol } O_2$$

$$\textbf{L H}_2 = \frac{22.4\text{ L}}{\text{mol}} \times 0.0149\text{ mol} = \mathbf{0.334\ L\ H_2 \Rightarrow 0.167\ L\ O_2}$$

11.8 Common Electrolytic Processes

Your textbook begins this section by noting that metals are "very good reducing agents. . . " What is the consequence of the ease with which they are oxidized? Listed below are other questions to guide your study of this section:

1. Why did the Hall-Heroult process allow the price of aluminum to drop precipitously?
2. In the Hall-Heroult process, the graphite rods must on occasion be replaced. Why?
3. Why is it necessary to go to such great lengths to electrolytically produce aluminum metal? That is, why can't it be electrolyzed from aqueous solution?
4. Speculate: What are some reasons that recycling of aluminum cans is an attractive alternative to producing aluminum from bauxite?
5. In the electrolysis of aqueous sodium chloride, why isn't sodium produced at the cathode?

Exercises

Section 11.1

1. In each of the following half-reactions, give the species being reduced and the number of electrons needed to balance the half-reaction:

   a. $AgBrO_3 + ?e^- \rightarrow Ag + BrO_3^-$
   b. $HCrO_4^- + 7H^+ + ?e^- \rightarrow Cr^{3+} + 4H_2O$
   c. $WO_3 + 6H^+ + ?e^- \rightarrow W + 3H_2O$

2. Identify the species in each of the following galvanic cell reactions that would receive electrons from the cathode and that would lose electrons at the anode:

   a. $Au^{3+}(aq) + Zn(s) \rightleftharpoons Au^+(aq) + Zn^{2+}(aq)$
   b. $3Pu^{6+}(aq) + 2Cr^{3+}(aq) \rightleftharpoons 2Cr^{6+}(aq) + 3Pu^{4+}(aq)$

Section 11.2

Use the following reactions and potentials taken from the Chemical Rubber Company Handbook of Physics and Chemistry, 53rd Edition, for Problems 3-8.

Reaction	Potential (V)
$2SO_4^{2-} + 4H^+ + 2e^- \rightleftharpoons S_2O_6^{2-} + 2H_2O$	−0.2
$AuBr_4^- + 3e^- \rightleftharpoons Au + 4Br^-$	+0.858
$O_3 + 2H^+ + 2e^- \rightleftharpoons O_2 + H_2O$	+2.07
$Ba^{2+} + 2e^- \rightleftharpoons Ba$	−2.09
$In^{3+} + e^- \rightleftharpoons In^{2+}$	−0.49

3. Select the strongest oxidizing agent from the listed reactions.

4. Select the strongest reducing agent from the listed reactions.

5. Using Table 11.1 in your textbook, arrange the following species in order of increasing strength as oxidizing agents. (Assume all species are in their standard states.)

$$Sn^{2+}, Au^{3+}, Fe^{3+}, Fe^{2+}, Ca^{2+}$$

6. Write a balanced equation and calculate the value of $E°$ for the reaction of $AuBr_4^-$ with In^{2+}.

7. Write a balanced equation and calculate the value of $E°$ for the reaction of Ba^{2+} with O_2 to form O_3. Would this reaction be spontaneous? Why or why not?

8. Write the balanced reaction for the galvanic cell that would yield the highest voltage given the reactions listed above.

9. A galvanic cell consists of an Ag electrode in a 1.0 *M* $AgNO_3$ solution and a Ni electrode in a 1.0 *M* $Ni(NO_3)_2$ solution. Calculate the standard emf of this electrochemical cell at 25°C. (Use Table 11.1 in your textbook.)

10. Regarding the following reaction:

$$F_2(g) + 2I^-(aq) \rightarrow 2F^-(aq) + I_2(s)$$

   a. List the species being oxidized.
   b. List the species being reduced.
   c. Calculate $E°$ for this cell. (See Table 11.1 in your textbook.)
   d. Which species receives electrons from the cathode?
   e. Which species donates electrons to the anode?

11. Answer the same questions that were posed in Problem 10 for the following reaction:

$$Hg_2^{2+}(aq) + Zn(s) \rightleftharpoons 2Hg(s) + Zn^{2+}(aq)$$

12. Answer the same questions that were posed in Problem 10 for the reaction:

$$Fe^{2+}(aq) + Ce^{4+}(aq) \rightleftharpoons Fe^{3+}(aq) + Ce^{3+}(aq)$$

13. Given the following half-reactions:

$$Co^{2+} + 2e^- \rightleftharpoons Co \qquad E° = -0.277\ V$$
$$Ce^{4+} + e^- \rightleftharpoons Ce^{3+} \qquad E° = 1.61\ V$$

   a. Write the overall equation for the galvanic cell.
   b. Calculate $E°$ for the cell.

14. Regarding the galvanic cell you composed for the reaction in Problem 13:

   a. Designate the anode and cathode.
   b. Describe the direction of electron flow.

15. Given the following half-reactions:

$$[PtCl_4]^{2-} + 2e^- \rightleftharpoons Pt + 4Cl^- \qquad E° = 0.755\ V$$
$$Fe^{3+} + 3e^- \rightleftharpoons Fe \qquad E° = -0.037\ V$$

a. Write the overall equation for the galvanic cell.
b. Calculate $E°$ for the cell.

16. Regarding the galvanic cell you composed for the equation in Problem 15, designate the anode and cathode.

17. Given the following half-reactions:

$$Ag + 2CN^- \rightleftharpoons Ag(CN)_2^- + e^- \qquad E° = +0.31\ V$$
$$Ag \rightleftharpoons Ag^+ + e^- \qquad E° = -0.80\ V$$

Calculate $E°$ for the reaction:

$$Ag^+(aq) + 2CN^-(aq) \rightleftharpoons Ag(CN)_2^-(aq)$$

18. Given the following half-reactions:

$$La^{3+} + 3e^- \rightleftharpoons La \qquad E° = -2.52\ V$$
$$Fe^{3+} + e^- \rightleftharpoons Fe^{2+} \qquad E° = 0.77\ V$$

a. Write the overall equation for the galvanic cell.
b. Calculate $E°$ for the cell.

19. Regarding the galvanic cell you composed in Problem 18, designate the anode and the cathode.

20. Given the following half-reactions:

$$PuO_2^+ + 4H^+ + e^- \rightleftharpoons Pu^{4+} + 2H_2O \qquad E° = +1.15\ V$$
$$Cu^{2+} + 2e^- \rightleftharpoons Cu \qquad E° = +0.34\ V$$

a. Write the overall equation for the galvanic cell.
b. Calculate $E°$ for the cell.

21. Regarding the galvanic cell you composed for the reaction in Problem 20:

a. Designate the anode and cathode.
b. Describe the direction of electron flow.
c. Describe the electrodes in each compartment.

Section 11.3

22. Calculate the standard free-energy change for the following reaction (Use Table 11.1 in your textbook.):

$$2Hg^{2+} + Mn \rightleftharpoons Hg_2^{2+} + Mn^{2+}$$

23. Calculate the standard free-energy change for the following reaction:

$$Na^+ + Cu \rightleftharpoons Na + Cu^+$$

24. Calculate the standard free-energy change for the following reaction:

$$Cr^{3+} + ClO_2^- \rightleftharpoons Cr^{2+} + ClO_2$$

25. Using data from Table 11.1 in your textbook, calculate $\Delta G°$ for the reaction:

$$2Al(s) + 6H^+(aq) \rightleftharpoons 2Al^{3+}(aq) + 3H_2(g)$$

Will the reaction be spontaneous, i.e., will hydrogen gas bubble from solution?

26. Calculate $\Delta G°$ for the following reaction (Use Table 11.1 in your textbook.):

$$2K^+ + Cu \rightleftharpoons 2K + Cu^{2+}$$

27. Calculate $\Delta G°$ for the following reaction:

$$3Hg_2^{2+} + 2Cr \rightleftharpoons 6Hg + 2Cr^{3+}$$

28. Calculate $\Delta G°$ for the following reaction:

$$Br_2 + Sn \rightleftharpoons 2Br^- + Sn^{2+}$$

29. Calculate $E°$ and $\Delta G°$ for the reaction:

$$UO_2^{2+}(aq) + 4H^+(aq) + 2Ag(s) + 2Cl^-(aq) \rightleftharpoons U^{4+}(aq) + 2AgCl(s) + 2H_2O(l)$$

Given the following half-reactions:

$$UO_2^{2+} + 4H^+ + 2e^- \rightleftharpoons U^{4+} + 2H_2O \qquad E° = +0.334\text{ V}$$
$$AgCl + e^- \rightleftharpoons Ag + Cl^- \qquad E° = +0.222\text{ V}$$

Section 11.4

30. A cell has on its left side a 1.0×10^{-3} *M* Zn^{2+} solution. The right side has a 0.030 *M* Zn^{2+} solution. The compartments are connected by Zn electrodes and a salt bridge. Designate the cathode, anode, and direction of current.

31. Calculate the emf for each of the following half-reactions:
   a. $Cr_2O_7^{2-}(0.350\ M) + 14H^+(0.0100\ M) + 6e^- \rightleftharpoons 2Cr^{3+}\ (1 \times 10^{-3}\ M) + 7H_2O \qquad E° = +1.33\text{ V}$
   b. $AuBr_2^-(0.084\ M) + e^- \rightleftharpoons Au + 2Br^-(0.1443\ M) \qquad E° = +0.959\text{V}$

32. For a galvanic cell based on the following half-reactions at 25°C:

$$Cl_2 + 2e^- \rightarrow 2Cl^- \qquad E° = 1.36\text{ V}$$
$$Ni^{2+} + 2e^- \rightarrow Ni \qquad E° = -0.23\text{ V}$$

Calculate E_{cell} where $[Cl_2] = 0.5$ atm, $[Cl^-] = 1.0\ M$, and $[Ni^{2+}] = 1.0\ M$.

33. For a galvanic cell based on the following half-reactions at 25°C:

$$Pb^{2+} + 2e^- \rightarrow Pb \qquad E° = -0.13$$
$$Zn^{2+} + 2e^- \rightarrow Zn \qquad E° = -0.76$$

Calculate E_{cell} where $[Pb^{2+}] = 1 \times 10^{-2}\ M$ and $[Zn^{2+}] = 3 \times 10^{-2}\ M$.

34. Calculate the cell voltage at 25°C for the equation in Problem 25 given the following equilibrium concentrations:

$$[Al^{3+}] = 0.025\ M,\ P_{H_2} = 1.0\text{ atm},\ \text{pH} = 3.50.$$

35. Calculate the cell voltage for the reaction:

$$Hg_2^{2+}(aq) + Cd(s) \rightleftharpoons 2Hg(s) + Cd^{2+}(aq)$$

where $[Hg_2^{2+}] = 1.0 \times 10^{-3}\ M$ and $[Cd^{2+}] = 5.0 \times 10^{-3}\ M$.

$$Cd^{2+} + 2e^- \rightleftharpoons Cd \qquad E° = -0.40\text{ V}$$
$$Hg_2^{2+} + 2e^- \rightleftharpoons 2Hg \qquad E° = +0.80\text{ V}$$

36. Calculate the equilibrium constant, *K*, for the following reaction at 25°C:

$$PbO_2 + 4H^+ + 2Hg + 2Cl^- \rightleftharpoons Pb^{2+} + 2H_2O + Hg_2Cl_2 \qquad E°_{cell} = 1.12\text{ V}$$

37. Given the data in Problem 35, calculate the equilibrium constant, *K*, for the reaction:

$$2Hg(s) + Cd^{2+}(aq) \rightleftharpoons Hg_2^{2+} + Cd(s)$$

How does the value of *K* in this problem relate to the value for the reverse reaction (as presented in Problem 35)?

38. Calculate *K* for the reaction of aluminum with hydrogen given in Problem 25.

39. Calculate the electromotive force at 25°C for a zinc-copper galvanic cell in which the zinc sulfate concentration is 0.010 *M*, and the copper sulfate concentration is 0.10 *M*.

40. Calculate the value of the equilibrium constant at 25°C for the reaction:

$$Zn(s) + 2H^+(aq) \rightarrow Zn^{2+}(aq) + H_2(g)$$

Section 11.7

41. Bismuth can be electrolytically reduced according to the following reaction:

$$BiO^+ + 2H^+ + 3e^- \rightleftharpoons Bi + H_2O$$

How many grams of bismuth can be reduced by applying a 5.60-A current for 28.3 minutes to a solution containing BiO^+ ions (assuming 100% efficiency)?

42. A sample containing nitric acid was titrated by electrolytically reducing water to form OH^- ion. How many moles of H^+ ion were originally in the sample if the hydroxide required 356.1 sec of generation at 9.07×10^{-3} A?

Multiple-Choice Self-Test

1. In the following reaction, which element or substance is the oxidizing agent?

$$PbSO_4(s) + H_2O(l) \rightarrow Pb(s) + PbO_2(s) + SO_4^{2-}(aq) + H^+(aq)$$

A. PbO_2 B. $PbSO_4$ C. Pb(s) D. H_2O

2. Calculate $E°$ for the following reaction:

$$Sn(s) + Sn^{4+}(aq) \rightarrow 2Sn^{2+}(aq)$$

$Sn^{2+}(aq) + 2e^- \rightarrow Sn(s)$ $E° = -0.14$ V
$Sn^{4+}(aq) + 2e^- \rightarrow Sn^{2+}(s)$ $E° = 0.15$ V

A. 0.01V B. 0.29 V C. 0.16 V D. −0.13 V

3. Calculate $E°$ for the following reaction:

$$Fe(s) + 2Fe^{3+}(aq) \rightarrow 3Fe^{2+}(aq)$$

$Fe^{3+}(aq) + e^- \rightarrow Fe^{2+}(aq)$ $E° = 0.77$ V
$Fe^{2+}(aq) + 2e^- \rightarrow Fe(s)$ $E° = -0.41$ V

A. 1.18 V B. −0.36 V C. 1.59 V D. −0.05 V

4. Calculate the potential for the reduction of iron(III) to iron(II) based on the following information:

$Fe^{3+}(aq) + 3e^- \rightarrow Fe$ $E° = -0.04$ V
$Fe^{2+}(aq) + 2e^- \rightarrow Fe$ $E° = -0.44$ V

A. −0.48 V B. −0.40 V C. 0.76 V D. 0.40 V

5. Calculate $\Delta G°$ for the following reaction:

$$Sn(s) + Sn^{4+}(aq) \rightarrow 2Sn^{2+}(aq)$$

$Sn^{2+}(aq) + 2e^- \rightarrow Sn(s)$ $E° = -0.14$ V
$Sn^{4+}(aq) + 2e^- \rightarrow Sn^{2+}(s)$ $E° = 0.15$ V

A. 55.0 B. −56.0 C. −227 D. −25.2

6. Calculate $\Delta G°$ for the following reaction:

$$Fe(s) + 2Fe^{3+}(aq) \rightarrow 3Fe^{2+}(aq)$$

$$Fe^{3+}(aq) + e^- \rightarrow Fe^{2+}(aq) \quad E° = 0.77\text{ V}$$
$$Fe^{2+}(aq) + 2e^- \rightarrow Fe(s) \quad E° = -0.41\text{ V}$$

A. −227 B. 112 C. −112 D. 30.9

7. Calculate $E°$ for a reaction that has a $\Delta G° = -770$ kJ and transfers 4 electrons.

A. 22.4 V B. 2.99 V C. 1.99 V D. 0.33 V

8. For the following reaction, calculate the half-reaction potential knowing that $\Delta G° = -53.0$ kJ, and

$$Fe^{2+}(aq) + AgCl(s) \rightarrow Ag(s) + Fe^{3+}(aq) + Cl^-(aq) \quad E° = -0.55\text{ V}$$
$$AgCl(s) + e^- \rightarrow Ag(s) + Cl^-(aq) \quad E° = ?$$

A. 0.77 V B. −0.77 V C. 0.22 V D. −0.53 V

9. Calculate the emf of a cell that utilizes the following reaction:

$$2Co^{3+}(aq) + 2H_2O(l) \rightarrow H_2O_2(aq) + 2H^+(aq) + 2Co^{2+}(aq)$$

at 298 K, when $[Co^{2+}] = 1.0\ M$, $[H^+] = 0.20\ M$, $[H_2O_2] = 1.3\ M$, and $[Co^{3+}] = 0.50\ M$.

A. −0.087 V B. 0.064 V C. 0.06 V D. −0.007 V

10. Calculate the K_{sp} for iron(III) sulfide given the following data:

$$FeS(s) + 2e^- \rightarrow Fe(s) + S^{2-}(aq) \quad E° = -1.01\text{ V}$$
$$Fe^{2+}(aq) + 2e^- \rightarrow Fe(s) \quad E° = -0.44\text{ V}$$

A. 5.54×10^{-20} B. 7.64×10^{-10} C. 8.90×10^{-26} D. 2.82×10^{-17}

11. Calculate the pH of the cathode compartment for the following reaction if emf = 3.01 V when $[Cr^{3+}] = 0.15\ M$, $[Al^{3+}] = 0.30\ M$, and $[(Cr_2O_7)^{2-}] = 0.55\ M$.

$$2Al(s) + (Cr_2O_7)^{2-}(aq) + 14H^+ \rightarrow 2Al^{3+}(aq) + 2Cr^{3+}(aq) + 7H_2O(l)$$

A. 0 B. 1 C. 1.5 D. 3.6

12. Which of the following elements would you expect to corrode most easily?

A. Ag B. Au C. Al D. Fe

13. You are asked to protect an iron surface from corrosion using cathodic protection. Which one of the following elements would you consider to be the best protector?

A. Al B. Mn C. Na D. Zn

14. An aqueous copper(II) chloride solution is electrolyzed for a period of 156 minutes using a current of 9.00 A. If inert electrodes are used in the process, how many grams of copper is removed from the solution?

A. 27.8 g B. 55.6 g C. 31.8 g D. 15.4 g

15. How long, in hours, does it take to remove all the polonium from an aqueous $PoCl_4$ solution that contains 1958 g of $PoCl_4$ if you use a current of 6.80 A?

A. 32.4 B. 16.2 C. 34.0 D. 88

16. An aqueous lead(II) chloride solution contains 927.0 g of lead chloride. What current, in A, will be necessary to remove all the lead from the solution in 48.0 hours?

A. 3.73 B. 5.00 C. 10.0 D. 7.47

Answers to Exercises

1. a. $Ag^+ + e^- \rightarrow Ag$
   b. $HCrO_4^- + 7H^+ + 3e^- \rightarrow Cr^{3+} + 4H_2O$ ($Cr^{6+} \rightarrow Cr^{3+}$)
   c. $WO_3 + 6H^+ + 6e^- \rightarrow W + 3H_2O$ ($W^{6+} \rightarrow W$)

2.

	Receives electrons	Loses electrons
a.	Au^{3+}	Zn
b.	Pu^{6+}	Cr^{3+}

3. O_3 is the strongest oxidizing agent.

4. Ba is the strongest reducing agent.

5. $Ca^{2+} < Fe^{2+} < Sn^{2+} < Fe^{3+} < Au^{3+}$

6. $AuBr_4^- + 3In^{2+} \rightleftharpoons Au + 4Br^- + 3In^{3+}$ $E° = +1.35$ V

7. $Ba^{2+} + O_2 + H_2O \rightleftharpoons Ba + O_3 + 2H^+$ $E° = -4.16$ V
   $E° <$ O; this reaction is nonspontaneous.

8. $Ba + O_3 + 2H^+ \rightleftharpoons Ba^{2+} + O_2 + H_2O$

9. 1.03 V

10. a. I^- is being oxidized
    b. F_2 is being reduced
    c. $E° = +2.33$ V
    d. F_2 receives electrons
    e. I^- donates electrons to the anode

11. a. Zn is being oxidized
    b. Hg_2^{2+} is being reduced
    c. $E° = +1.56$ V
    d. Hg_2^{2+} receives electrons
    e. Zn donates electrons

12. a. Fe^{2+} is being oxidized
    b. Ce^{4+} is being reduced
    c. $E° = +0.93$ V
    d. Ce^{4+} receives electrons
    e. Fe^{2+} donates electrons

13. a. $2Ce^{4+} + Co \rightleftharpoons Co^{2+} + 2Ce^{3+}$ b. $E° = 1.89$ V

14. a. Oxidation of Co takes place at the anode. Reduction of Ce^{4+} takes place at the cathode.
    b. Electrons flow from the anode to the cathode.

15. a. $3[PtCl_4]^{2-} + 2Fe \rightleftharpoons 3Pt + 12Cl^- + 2Fe^{3+}$ b. $E° = 0.792$ V

16. Oxidation of Fe takes place at the anode. Reduction of $[PtCl_4]^{2-}$ takes place at the cathode.

17. $E° = 1.11$ V

18. a. $3Fe^{3+} + La \rightleftharpoons 3Fe^{2+} + La^{3+}$ b. $E° = 3.29$ V

19. Oxidation of La takes place at the anode. Reduction of Fe^{3+} takes place at the cathode.

20. a. $2PuO_2^+ + Cu + 8H^+ \rightleftharpoons 2Pu^{4+} + Cu^{2+} + 4H_2O$ b. $E° = +0.81$ V

21. a. Oxidation of copper takes place at the anode. Reduction of PuO_2^+ takes place at the cathode.
b. Electrons flow from the anode to the cathode.
c. Copper can act as the anode. Platinum can act as the cathode.

22. $\Delta G° = -403$ kJ

23. $\Delta G° = 312$ kJ

24. $\Delta G° = 140$ kJ

25. $\Delta G° = -961$ kJ; Yes, the reaction will be spontaneous.

26. $\Delta G° = 629$ kJ

27. $\Delta G° = -886$ kJ

28. $\Delta G° = -237$ kJ

29. $E° = +0.112$ V, $\Delta G° = -21.6$ kJ

30. The right side will be the cathode. The left side will be the anode. The current will flow from left to right (anode to cathode).

31. a. $E = +1.11$ V b. +1.00 V

32. $E_{cell} = 1.58$ V

33. $E_{cell} = 0.62$ V

34. $E = +1.48$ V

35. $E = +1.18$ V

36. $K = 8 \times 10^{37}$ (using 0.0591)

37. $K = 3 \times 10^{-41}$. K for this problem = $1/K$ for Problem 35.

38. $K = 1 \times 10^{168}$

39. 1.13 V

40. 5.0×10^{25}

41. 6.87 g of Bi

42. 3.35×10^{-5} moles of H^+

Answers to Multiple-Choice Self-Test

1. B
2. B
3. A
4. C
5. B
6. A
7. B
8. C
9. C
10. A
11. A
12. C
13. D
14. A
15. D
16. A

CHAPTER 12

Quantum Mechanics and Atomic Theory

The Bottom Line: Chapter 12

I find the material in this chapter to be among the most interesting and useful in all of chemistry. The coverage of the electromagnetic spectrum leads into the electronic structure of atoms. From this information, we can rationalize and predict such properties of atoms as size, ionization energy, and the way in which they will form bonds (a discussion that continues in Chapters 13 and 14).

12.1 Electromagnetic Radiation

This topic is important for many reasons, as is pointed out in your textbook. From my point of view, the most fascinating aspect has to do with outer space. Electromagnetic radiation is the only source of information from celestial objects other than the Moon, Mars, and Venus. We can't directly touch the stars, but we can receive their radiation and learn from this. Also related to this is our ability to use electromagnetic radiation to determine metal ion content in water samples.

The definitions for **electromagnetic radiation, wavelength,** and **frequency** are given in Section 12.1 of your textbook. Be able to define these terms. You should also know the different regions of the electromagnetic spectrum and the wavelengths covered by each region. The relationship between frequency and wavelength is

$$\underset{\text{nu}}{\underset{\uparrow}{\nu}} = c/\,\underset{\text{lambda}}{\underset{\uparrow}{\lambda}}$$

wavelength (λ) is in meters.
frequency (ν) is in sec^{-1} or Hz.
speed of light (c) is in meters/sec.

Example 12.1 - Wavelength - Frequency Conversion

Photosynthesis (use of the sun's light in plants to convert CO_2 and H_2O into glucose and oxygen) uses light with a frequency of 4.54×10^{14} s^{-1}. To what wavelength does this correspond?

Solution

$$\nu = c/\lambda, \text{ so } \lambda = c/\nu$$

$$\lambda = \frac{2.9979\times10^{8}\ \text{m/s}}{4.54\times10^{14}\ \text{s}^{-1}} = 6.60 \times 10^{-7}\ \text{m} = 660\ \text{nm}$$

Does the Answer Make Sense?

As discussed in your textbook, the visible region goes from about 400 to 700 nm. Photosynthesis occurs at the far end (660 nm) of this, so the answer would seem to be reasonable.

12.2 The Nature of Matter

Your textbook begins this section by discussing the importance of the **blackbody radiation** profile and how that differs from the **ultraviolet catastrophe** effect. From this, two critically important equations are introduced. The first one is:

$$\Delta E = h\nu$$

Where $\boldsymbol{\Delta E}$ is the change in energy for a system (in Joules **PER PHOTON**),
$\boldsymbol{h}$ is Planck's constant (6.626×10^{-34} J s), and
$\boldsymbol{\nu}$ is the frequency of the wave (in s^{-1}).

Only certain specific amounts of energy can be gained or lost in a substance. These **quanta** have magnitudes that depend on the substance. Given the observed frequency change, the change in energy can be calculated for the absorption or emission of "**photons**" (the term used if the quanta of energy are viewed as **particles**).

From Section 12.1, you know that ν and λ are related by $\nu = c/\lambda$. Therefore,

$$\Delta E = h\nu = \frac{hc}{\lambda}$$

This is the second equation you should know. These two equations are among the most important in all of chemistry, as you shall see in succeeding chapters.

Let's have some practice at interconversion between energy, frequency, and wavelength.

Example 12.2 A - Interconverting Between Energy, Frequency, And Wavelength

Sodium atoms have a characteristic yellow color when excited in a flame. The color comes from the emission of light of 589.0 nm.

a. What is the frequency of this radiation?
b. What is the energy of this radiation per photon? Per mole of photons?

Solution

a. $\nu = \dfrac{c}{\lambda} = \dfrac{2.9979 \times 10^{8}\ \text{m/s}}{5.890 \times 10^{-7}\ \text{m}} = \mathbf{5.090 \times 10^{14}\ s^{-1}}$

b. $\Delta E = h\nu = 6.626 \times 10^{-34}\ \text{J s} \times 5.090 \times 10^{14}\ \text{s}^{-1} = \mathbf{3.373 \times 10^{-19}\ J}$

This value is *per photon.* There are $\dfrac{6.022 \times 10^{23}\ \text{photons}}{\text{mole photons}}$ (Avogadro's number). To convert ΔE per photon to ΔE per mole,

$$\frac{\text{J}}{\text{mole}} = \frac{3.373 \times 10^{-19}\ \text{J}}{\text{photon}} \times \frac{6.022 \times 10^{23}\ \text{photons}}{\text{mole}} = 2.031 \times 10^{5}\ \text{J/mole}$$

$$\mathbf{\Delta E = 203.1\ kJ/mole}$$

Example 12.2 B - Practice With Interconversions

It takes 382 kJ of energy to remove one mole of electrons from gaseous cesium. What is the wavelength associated with this energy?

Solution

The conversion between energy and wavelength requires the value of energy to be **per photon.** This is our first conversion. We may then convert directly to wavelength.

$$\frac{\text{kJ}}{\text{photon}} = \frac{3.82\times10^{5}\ \text{J}}{\text{mol}} \times \frac{1\ \text{mol}}{6.022\times10^{23}\ \text{photons}} = \mathbf{6.343\times10^{-19}\ J\ (per\ photon)}$$

$$\Delta E = \frac{hc}{\lambda},\ \text{so}\ \lambda = \frac{hc}{\Delta E}$$

$$\lambda = \frac{6.626\times10^{-34}\ \text{J s}(2.9979\times10^{8}\ \text{m/s})}{6.343\times10^{-19}\ \text{J}} = 3.13\times10^{-7}\ \text{m} = \mathbf{313\ nm}$$

This lies in the ultraviolet region.

De Broglie's Equation

Let's review your textbook's derivation of de Broglie's equation. The goal is to relate wavelength (λ) to mass (m) of a particle.

$$E = \text{mc}^2,\ \text{and}\ E = \frac{hc}{\lambda}$$

This gives us $m = \dfrac{h}{\lambda c}$ or, for a particle that is not moving at the speed of light (c) but rather at some velocity (v), **de Broglie's equation** is

$$m = \frac{h}{\lambda v}\ \text{or}\ \lambda = \frac{h}{mv}$$

The important point is that, **at least on an atomic scale,** electromagnetic radiation has mass, and particles with mass exhibit a characteristic wavelength. This is the **dual nature of light.**

Example 12.2 C - De Broglie's Equation

What is the wavelength of an electron (mass = 9.11×10^{-31} kg) traveling at 5.31×10^{6} m/s?

Solution

Before we "plug and chug," recall that 1 J = 1 kg m^2 s^{-2}.

$$\lambda = \frac{h}{mv} = \frac{6.626\times10^{-34}\ \text{J s}}{9.11\times10^{-31}\ \text{kg}\,(5.31\times10^{6}\ \text{m/s})} = \frac{6.626\times10^{-34}\ \text{kg m}^2\ \text{s}^{-2}\ \text{s}}{4.83_7\times10^{-24}\ \text{kg m s}^{-1}} = 1.37\times10^{-10}\ \text{m}$$

$$= \mathbf{0.137\ nm}$$

Keep in mind that your answer is probably reasonable if the dimensions work out!

The importance of the **dual nature of light** is outlined in the paragraphs on **diffraction** at the end of <u>Section 12.2 in your textbook</u>.

12.3 The Atomic Spectrum of Hydrogen

The key idea for Section 12.3 is the difference between a **continuous spectrum and a discrete or line spectrum.**

- **A continuous spectrum** contains all the wavelengths over which the spectrum is continuous.

- **A line spectrum** contains certain specific wavelengths that are characteristic of the substance emitting those wavelengths.

It is concluded in your textbook that the fact that hydrogen has a line spectrum shows that only certain energy transfers are allowed in hydrogen. **There are specific energy levels among which the hydrogen electron can shift.** These energy levels are said to be **quantized.**

12.4 The Bohr Model

At the beginning of Section 12.4, your textbook discusses Neils Bohr's reasoning for relating the energy levels to the observed wavelengths emitted by the hydrogen atom. Although Bohr's ultimate conclusions have since been enhanced, his work represented a great leap forward in 1913. The equation that is descriptive of Bohr's model is

$$E = -2.178 \times 10^{-18}\ \text{J}\left[\frac{Z^2}{n^2}\right]$$

where E is the energy (in Joules)
Z is the nuclear charge (1 for hydrogen's one proton)
n is an integer related to orbital position. The farther out from the nucleus, the higher the value of n. If an electron is given enough energy, it goes away from the nucleus. We say it is **ionized,** and $\boldsymbol{n} = \infty$.

The lowest energy state is $n = 1$, the ground state.
The highest energy state is $n = \infty$, where an electron is ionized.

Example 12.4 A - Changes In Energy In The Bohr Atom

Calculate the energy **change** corresponding to the excitation of an electron from the $n = 1$ to $n = 3$ electronic state in the hydrogen atom.

Solution

What are we trying to solve? We are asked to solve for the change in energy.

$$\Delta E = E_{\text{final}} - E_{\text{initial}} = E_3 - E_1$$

There are two possible approaches to the calculation. The first is to calculate each energy separately and take the difference. The second is to subtract the equations, then factor and calculate the change using one equation.

Approach 1:

$$\Delta E = -2.178 \times 10^{-18}\ \text{J}\left[\frac{1^2}{3^2}\right] - -2.178 \times 10^{-18}\ \text{J}\left[\frac{1^2}{1^2}\right]$$

Approach 2:

$$\Delta E = -2.178 \times 10^{-18}\ \text{J}\left[\frac{1^2}{3^2} - \frac{1^2}{1^2}\right] = -2.178 \times 10^{-18}\ \text{J}\left[-\frac{8}{9}\right]$$

Both approaches will yield

$$\boldsymbol{\Delta E = +1.936 \times 10^{-18}\ \text{J}}$$

The "+" sign is very important! It means energy was **absorbed** to excite the electron. In other words, the system has **gained energy.**

Example 12.4 B - Wavelength From Energy In The Bohr Atom

What wavelength of electromagnetic radiation is associated with the energy change in promoting an electron from the $n = 1$ to $n = 3$ level in the hydrogen atom? (Use the value of ΔE from the previous example.)

Solution

$$\Delta E = \frac{hc}{\lambda} \text{ so } \lambda = \frac{hc}{\Delta E}$$

$$\lambda = \frac{6.626\times10^{-34}\ \text{J s}(2.9979\times10^{8}\ \text{m/s})}{1.936\times10^{-18}\ \text{J}} = 1.026\times10^{-7}\ \text{m} = \mathbf{102.6\ nm}$$

This energy corresponds to radiation in the ultraviolet region of the spectrum.

(Note: Even if ΔE = "−" (as it will be when energy is released), you **must** assign a "+" sign to λ because you can't have a negative wavelength!)

12.5 The Quantum Mechanical Description of the Atom

As is pointed out in Section 12.5 of your textbook, the quantum mechanical model's key assumption regarding electron motion around the hydrogen atom is that **the electron is assumed to behave as a standing wave.** Only certain orbits are shaped such that the "wave" (electron) can fit. The **wave function** of an electron represents the allowed coordinates in which the electron *may* reside in the atom. Each wave function is called an **orbital.** Your textbook points out that Schrödinger was not certain that treating the electron as a wave would make any sense—the key would be if the model would fit experimental atomic data.

Read the discussion on the **Heisenberg uncertainty principle** in your textbook. The principle says that "there is a limit to just how precisely we can know both the position and momentum of a particle at a given time." It turns out that when the radiation used to locate a particle hits that particle, it changes its momentum. Therefore the position and momentum cannot both be measured exactly. As one is measured more precisely, the other is known less precisely. The relation is given by

$$\Delta x \cdot \Delta p \geq \frac{h}{4\pi}$$

where Δx is the uncertainty in the particle's position
Δp is the uncertainty in the particle's momentum
h is Planck's constant

The smallest possible uncertainty is $h/4\pi$.

The Heisenberg Uncertainty Principle tells us we cannot know exactly where an electron is around an atom at any given time. However, we can know the **probability** of it being within a certain region.

12.6 The Particle in a Box

This section is among the most mathematically sophisticated in the entire text. It deals with a model known as "the particle in a box," that demonstrates the existence of quantized electronic energy levels. The discussion gets fairly involved. Please use the following questions to help guide your study of the particle in a box model.

1. What is the difference between $\hat{H}$ and E?
2. Why is the particle in a box model used even though it is not an accurate physical model for the hydrogen atom?

3. What energy boundary conditions are imposed that allow us to say the particle is trapped in a one-dimensional box?
4. Why is the sine function introduced in this discussion?
5. Your textbook says that one property of the solutions to the Schrödinger equation lends support to using wave mechanics to describe the properties of matter. What is that property?
6. What is the importance of the boundary conditions on the box?
7. What is the importance of squaring the wave function?
8. What are **nodes**?
9. Why can't $n = 0$ in the particle in a box model?

12.7 The Wave Equation for the Hydrogen Atom

Let's consider the differences between the actual atom and the particle in a model. The keys are that the hydrogen atom (and hence the electron spin) exists in three dimensions. Further, due to the attraction between the electron and proton, the electron has potential energy, unlike the particle in a box.

Three quantum numbers arise from the three-dimensional model, and the principal energies are the same as with the Bohr model. Table 12.1 in your textbook gives actual mathematical solutions and their related quantum numbers.

12.8 The Physical Meaning of a Wave Function

The goal of this section is to try to make some physical sense of the mathematical wave functions. It is pointed out that any physical model resulting from a purely mathematical framework will not be ideal. Nonetheless, some good visual approximations can be made.

The square of the wave function is called the **probability distribution**, and it represents the probability (statistical likelihood) of finding the electron in a particular area around the nucleus.

This section in your textbook ends by noting that there is a probability (exceedingly small, but it does exist) of finding the electron anywhere. Therefore the (arbitrary) limit of the size of an orbital is "the radius of the sphere that encloses 90% of the total electron probability." In other words, at any time, t, there is a 90% chance of finding the electron in that orbital. It is also helpful to think of an orbital as a "three-dimensional electron density map" in which there is a finite probability of finding the electron.

12.9 The Characteristics of Hydrogen Orbitals

This section in your textbook begins by describing the meaning and values of quantum numbers.

Quantum Numbers

Name	Designation	Property of the Orbital	Possible Range of Values
principal quantum number	n	related to **size and energy** of the orbital	integers > 0 (1,2,3,…)
angular momentum quantum number	ℓ	related to the **shape** of the orbital	integers from 0 to $n-1$*

*$\ell = 0$ is called s, $\ell = 1$ is p, $\ell = 2$ is d, $\ell = 3$ is f, $\ell = 4$ is g $\ell = 5$ is h and so on through the alphabet. As your textbook outlines, this labeling system arose historically from early spectral line studies: s = "sharp," p = "principle," d = "diffuse," and f = "fundamental."

Name	Designation	Property of the Orbital	Possible Range of Values
magnetic quantum number	m_ℓ	related to the **position of the orbital in space relative to other orbitals**	integers from $-\ell$ to $+\ell$ and 0

Table 12.3 in your textbook gives the allowed values for the quantum number for the first four levels of the hydrogen atom. In examining the table, **the most important** thing you can do is **derive** the values of ℓ and m_ℓ that go with each value of n. For example, if $n = 2$, and if ℓ can go from 0 to $n-1$, then $\ell = 0$ or $\ell = 1$.

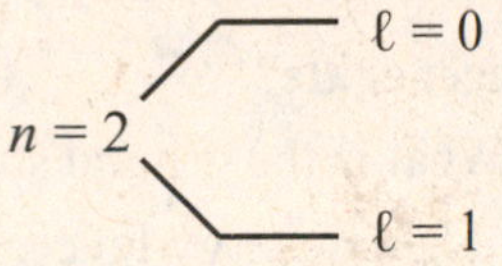

For $\ell = 0$ (s), m_ℓ can equal only 0 ($-\ell$ to $+\ell$).
For $\ell = 1$ (p), m_ℓ can equal -1, 0, or $+1$.

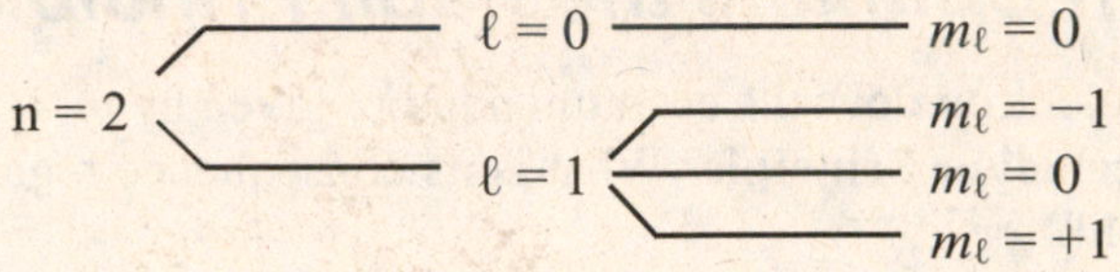

Therefore, there are 4 energy levels associated with $n = 2$, **one for 2*s* and 3 for 2*p*.**

Example 12.9 A - Quantum Numbers

If for a given atom $\ell = 7$, how many energy levels are possible? What is the subshell designation for this orbital?

Solution

If $\ell = 7$, m_ℓ can range from -7 to $+7$ and can include 0. Therefore, there are **15 possible energy levels.** If $\ell = 3$ is called f and $\ell = 4$ is a g, **$\ell = 7$ would be a *k* orbital.** (Remember that this example is hypothetical and that j is skipped because it is reserved as a symbol for angular momentum.)

Example 12.9 B - Practice With Quantum Numbers

Which of the following sets of quantum numbers are not allowed in the hydrogen atom? For each incorrect set, state why it is incorrect.

a. $n = 1, \ell = 0, m_\ell = 1$
b. $n = 2, \ell = 2, m_\ell = 1$
c. $n = 5, \ell = 3, m_\ell = 2$
d. $n = 6, \ell = -2, m_\ell = 2$
e. $n = 6, \ell = 2, m_\ell = -2$

Solution

a. Not allowed. (m_ℓ can't be greater than ℓ.)
b. Not allowed. (ℓ can't be greater than or equal to n.)
c. Allowed.
d. Not allowed. (ℓ can't be negative.)
e. Allowed.

The section ends with a discussion of orbital shapes and energies. Please use the following questions to test your understanding of the material.

1. What are nodes?
2. How many nodes are in a 5*s* orbital?
3. Draw a 2*p* orbital.
4. How do the three 2*p* orbitals differ from one another?
5. In what ways do *d* orbitals differ from *p* orbitals?
6. Why doesn't a 2*d* orbital exist?
7. What is the chemical meaning of a **degenerate**?
8. What is meant by the **ground state**? What is the ground state orbital for hydrogen?
9. How can a 1*s* electron be excited to a higher energy level?

12.10 Electron Spin and the Pauli Principle

1. What property of an electron does the quantum number given by m_s represent?
2. What is the **Pauli Exclusion Principle**? What is its consequence regarding the number of electrons that an orbital can hold?

Example 12.10 A - Practice With The Four Quantum Numbers

Which of the following sets of quantum numbers are not allowed? For each incorrect set, state why it is incorrect.

a. $n = 3, \ell = 3, m_\ell = 0, m_s = -½$
b. $n = 4, \ell = 3, m_\ell = 2, m_s = -½$
c. $n = 4, \ell = 1, m_\ell = 1, m_s = +½$
d. $n = 2, \ell = 1, m_\ell = -1, m_s = -1$
e. $n = 5, \ell = -4, m_\ell = 2, m_s = +½$
f. $n = 3, \ell = 1, m_\ell = 2, m_s = -½$
g. $n = 3, \ell = 2, m_\ell = -1, m_s = 1$

Solution

a. Not allowed. (ℓ cannot be equal to n.)
b. Allowed.
c. Allowed.
d. Not allowed. (m_s must be either +½ or −½.)
e. Not allowed. (ℓ must be a **positive** integer.)
f. Not allowed. (m_ℓ must be between −1 and +1.)
g. Not allowed. (m_s must be either +½ or −½.)

Remember the key message from Section 12.10—**an orbital can hold a maximum of two electrons,** and they must have opposite spins.

Example 12.10 B - Electrons In Orbital

If each orbital can hold a maximum of two electrons (of opposite spin), how many electrons can each of the following hold?

a. 2*s* b. 5*p* c. 4*f* d. 3*d* e. 4*d*

Solution

The key here is to figure out **how many orbitals** each contains. This is determined by the azimuthal quantum number, NOT by the principle quantum number. For example, a *p* orbital ($\ell = 1$) can have m_ℓ **= +1, 0, or −1.** *Each* m_ℓ *can have 2 electrons.* Therefore, *p* can have a total of 3 × 2, or 6 electrons. It doesn't matter if it is a 5*p*, 3*p*, or 2*p*. Each *p* can have up to 6 electrons.

a. $2s \Rightarrow \ell = 0, m_\ell = 0$ (1 value) × 2 electrons = **2 electrons**
b. $5p \Rightarrow \ell = 1, m_\ell = +1, 0, -1$ (3 values) × 2 electrons = **6 electrons**
c. $4f \Rightarrow \ell = 3, m_\ell = +3, +2, +1, 0, -1, -2, -3$ (7 values) × 2 electrons = **14 electrons**
d. $3d \Rightarrow \ell = 2, m_\ell = +2, +1, 0, -1, -2$ (5 values) × 2 electrons = **10 electrons**
e. $4d \Rightarrow$ same as 3*d*

12.11 Polyelectronic Atoms

1. What are the three energy contributions that must be considered when describing the helium atom?
2. What does your textbook mean by the **electron correlation problem?** How do we deal with the problem?
3. Why does it take more energy to remove an electron from He^+ than from He?
4. What is meant by **effective nuclear charge?**
5. Why is Z_{eff} less than Z_{actual}?
6. What is the **SCF Method**?
7. What is the key SCF assumption, and what are the consequences of that assumption?

12.12 The History of the Periodic Table

1. What was the original basis of the construction of the periodic table?
2. What are triads?
3. List the properties that Mendeleev used to predict ekasilicon's position in the periodic table.
4. Several atoms of element 116 have been observed. Based on the periodic table, Table 12.5 and Figure 12.25 in your textbook, what properties would you predict this element would have?
5. What is the important difference between Mendeleev's periodic table and the modern one?

12.13 The Aufbau Principle and the Periodic Table

Read the statement of the Aufbau principle in your textbook. Your textbook presents **orbital diagrams** for the first ten elements, **hydrogen through neon.** When constructing orbital diagrams and electron configurations, please keep the following in mind:

1. Electrons fill in order from lowest to highest energy.
2. The Pauli exclusion principle holds. An orbital can hold only two electrons.
3. Two electrons in the same orbital must have opposite spins.
4. You must know how many electrons can be held by each azimuthal quantum number (i.e., *s* can hold 2, 6 for *p*, 10 for *d*, 14 for *f*).
5. **Hund's rule** applies. The lowest energy configuration for an atom is the one having the maximum number of **unpaired** electrons for a set of **degenerate** orbitals. By convention, all unpaired electrons are represented as having parallel spins with the spin "up."

One excellent approach to address the order in which electrons fill as the atomic number increases is shown in Figure 12.28 in your textbook. For a slightly different approach, practice writing the following triangle. (Use the **penetration effect** discussed in Section 12.14 to justify the order of filling presented in the triangle)

$1s^2$
$2s^2\ 2p^6$
$3s^2\ 3p^6\ 3d^{10}$
$4s^2\ 4p^6\ 4d^{10}\ 4f^{14}$
$5s^2\ 5p^6\ 5d^{10}\ 5f^{14}$
$6s^2\ 6p^6\ 6d^{10}\ 6f^{14}$
$7s^2\ 7p^6\ 7d^{10}\ 7f^{14}$

To determine the order of filling, draw arrows from the upper right to the lower left.

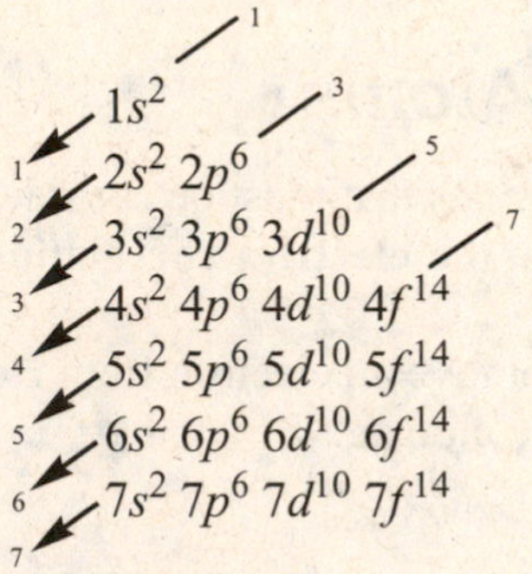

Get the final filling pattern by following the arrows in order:

$1s^2\ 2s^2\ 2p^6\ 3s^2\ 3p^6\ 4s^2\ 3d^{10}\ 4p^6\ 5s^2\ 4d^{10}\ 5p^6\ 6s^2\ 4f^{14}\ 5d^{10}\ 6p^6\ 7s^2 \ldots$

Let's apply this strategy toward determining the electron configuration for oxygen (8 electrons).

- The 1*s* and 2*s* levels hold a total of 4 electrons ($\mathbf{1s^2 2s^2}$).
- The 2*p* level can hold up to 6 electrons. However, we have only 4 remaining, which means we will have a $\mathbf{2p^4}$.

O: $\mathbf{1s^2 2s^2 2p^4}$

Let's translate this to an orbital diagram.

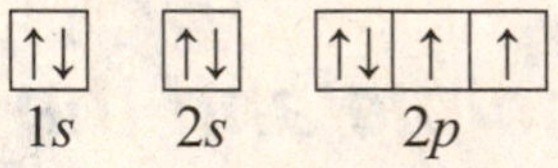

The first three 2*p* electrons occupy their own **degenerate** 2*p* orbitals. The fourth electron shares a degenerate orbital, but does so with **opposite spin**.

Example 12.13 A - Electron Configurations

Write electron configurations for each of the following neutral atoms (**use a configuration triangle**):

a. boron b. sulfur c. vanadium d. iodine

Solution

a. B (5 electrons): $1s^2 2s^2 2p^1$
b. S (16 electrons): $1s^2 2s^2 2p^6 3s^2 3p^4$
c. V (23 electrons): $1s^2 2s^2 2p^6 3s^2 3p^6 4s^2 3d^3$
d. I (53 electrons): $1s^2 2s^2 2p^6 3s^2 3p^6 4s^2 3d^{10} 4p^6 5s^2 4d^{10} 5p^5$

Example 12.13 B - Orbital Diagrams

Draw orbital diagrams for the following:

a. sodium b. phosphorus c. chlorine

Solution

Write the electron configuration for the atom. The only sticking points will be how unpaired electrons fill degenerate orbitals (singly, if possible). All the inner electrons will be paired. The outer electrons may be unpaired. You must deal with each atom separately.

a. Na (11 electrons): $1s^2 2s^2 2p^6 3s^1$

1s	2s	2p	3s
[↑↓]	[↑↓]	[↑↓][↑↓][↑↓]	[↑↓]

b. P (15 electrons): $1s^2 2s^2 2p^6 3s^2 3p^3$

1s	2s	2p	3s	3p
[↑↓]	[↑↓]	[↑↓][↑↓][↑↓]	[↑↓]	[↑][↑][↑]

c. Cl (17 electrons): $1s^2 2s^2 2p^6 3s^2 3p^5$

1s	2s	2p	3s	3p
[↑↓]	[↑↓]	[↑↓][↑↓][↑↓]	[↑↓]	[↑↓][↑↓][↑]

Look at the term "**core**" and "**valence**" electrons in your textbook. Know how to define these terms.

Notice in the problem we just completed that phosphorus and chlorine have the same core electronic structure. This **core** has the **same electron configuration as neon.** We can therefore write a shorthand version of electron configurations:

$$P = \underset{\text{core}}{[\text{Ne}]}\ \underset{\text{valence}}{3s^2 3p^3} \qquad Cl = [\text{Ne}]3s^2 3p^5$$

Using the same strategy,

$$\text{yttrium (39 electrons)} = \underset{\text{(longhand)}}{1s^2 2s^2 2p^6 3s^2 3p^6 4s^2 3d^{10} 4p^6 5s^2 4d^1}$$

$$= \underset{\text{(shorthand)}}{\mathbf{[Kr]5s^2 4d^1}}$$

Neon and krypton are both **noble gases (Group 8A)** and are atoms that have complete inner energy levels and an outer energy level with complete *s* and *p* orbitals.

Example 12.13 C - Shorthand Configurations

Write the shorthand configuration for the atoms in Example 12.13 A, and state for parts a, b, and d how many **valence electrons** the element has.

Solution

a. B: $[\text{He}]2s^2 2p^1$ (3 valence electrons)
b. S: $[\text{Ne}]3s^2 3p^4$ (6 valence electrons)
c. V: $[\text{Ar}]4s^2 3d^3$ (transition metal, 2 valence electrons)
d. I: $[\text{Kr}]5s^2 4d^{10} 5p^5$ (7 valence electrons … $n = 4$ level is complete)

Notice that: boron is in Group 3A
sulfur is in Group 6A

vanadium is a transition metal
iodine is in Group 7A

The group number indicates the number of valence electrons for nontransition metals. This should help you in determining electron configurations.

Also, sulfur is in Period 3. It is filling $n = 3$ electronic orbitals. Iodine is filling $n = 5$ electronic orbitals. It is in Period 5. Use the group and period locations to help do electron configurations quickly and correctly.

Two final ideas:

- With transition metals, if it is possible to have a $3d^5$ (half-filled) or $3d^{10}$ (completely filled) electronic configuration at the expense of a filled 4*s*, that will happen. Thus, Cu is **[Ar]$4s^1 3d^{10}$**, NOT [Ar]$4s^2 3d^9$. The same holds true for 4*s*/5*d* filling of lanthanides and actinides.
- Know what is meant by representative, *d*-transition, *f*-transition, and noble-gas elements.

Example 12.13 D - Practice With Electron Configurations

Write shorthand electron configurations for the following real and hypothetical atoms:

a. Sr   b. Mo   c. Ge   d. Q (hypothetical, 111 electrons

Solution

a. **Sr** is in Group 2A, Period 5. **[Kr]$5s^2$**
b. **Mo** is the fourth transition metal in Period 5, [Kr]$5s^2 4d^4$, BUT it can have a half-filled 4*d* if the configuration is $5s^1 4d^5$. Therefore Mo = **[Kr]$5s^1 4d^5$.**
c. **Ge** is in Group 4A, Period 4. **[Ar]$4s^2 3d^{10} 4p^2$.**
d. **Q** would be the ninth transition metal in Period 7, [Rn]$7s^2 6d^9$, BUT it can have a completed 6*d* if the configuration is $7s^1 6d^{10}$. Therefore the likely configuration is **[Rn]$7s^1 5f^{14} 6d^{10}$.**

12.14 Further Development of the Polyelectronic Model

Your textbook points out that we make an important simplification in dealing with **polyelectronic atoms**. Rather than take electron-proton and electron-electron interaction into account, we put all interactions under the umbrella of **one electron at a time interacting with the nucleus.** This is what leads to the idea of an **effective nuclear charge (Z_{eff})**. Also discussed is the idea that the orbitals determined via the SCF method have the **same types of boundary surfaces** but different **radial parts** than hydrogen orbitals.

The **penetration effect** is used to explain the order in which electrons fill energy levels. For example, 4*s* fills before 3*d* because an electron in a 4*s* orbital has greater likelihood of being near the nucleus ("penetrating") than a 3*d*.

12.15 Periodic Trends in Atomic Properties

Ionization Energy increases as successive electrons are removed from an atom because:

1. The value for Z_{eff} increases because there are fewer electron-electron repulsions and a higher positive to negative charge ratio than before.

2. Upon going from *f* to *d* to *p* to *s* electrons, there is a higher penetration effect. For example, the removal of a 3*p* will require more energy than the removal of a 3*d* from the same atom.

3. When you remove all the electrons from an energy level, you begin removing **core electrons**, which are more tightly bound to the nucleus than valence electrons.

Example 12.15 A - Ionization Energy

Examine Table 12.6 in your textbook. Justify the large increases in ionization energy at I_5 and I_7 for **sulfur.**

Solution

I_4 represents removal of the last 3*p* electron. The remaining electrons are 3*s* and core electrons. The 3*s* electron has a much greater penetration effect than the 3*p*. Therefore, I_5 is much larger than I_4.

I_7 represents ionization of a core (2*p*) electron. This electron is much closer to the nucleus, thus requiring more energy to ionize it.

Note that:

1. First ionization energy **increases** as we go **across a period**.
2. First ionization energy **decreases** as we go **down a group**.
3. Anomalies exist, such as the decrease from P to S. (Can you explain why?)

Electron Affinity is the **change in energy** associated with the **addition of an electron** to a **gaseous atom.** In keeping with thermodynamic convention, if the addition is **exothermic,** the energy change will be **negative.** Although electron affinity *generally* increases from left to right across a period, there are several exceptions. For example, the electron affinity of phosphorus is lower than that of sulfur. That is because P is $3p^3$ (half-filled) while S is $3p^4$. If you put an extra electron on phosphorus, it must share an orbital, thus forcing electron-electron repulsion. These repulsions already exist in the 3*p* orbitals of sulfur. The trend of electron affinities is less predictable than that of ionization energies.

The **atomic radius** of an atom **decreases** from left to right **across a period**. This is because the Z_{eff} increases. Atomic radius **increases** going **down a group.** This is because of increased orbital size. (See Figure 12.38 in your textbook.) You must consider **ionic radius** in terms of the following questions: What will adding an electron do to electron-electron repulsions? What will subtracting an electron do to the effective nuclear charge?

Example 12.15 B - Trends In Atomic Radius

Order the atoms or ions in the following groups from smallest to largest radius.

a. Cs, Si, F, Ca, Ga

b. Ca^{2+}, I^-, I, Li

Solution

a. Cs is in Period 6, Group 1
F is in Period 2, Group 7

F, Si, Ga, Ca, Cs
↑ smallest ... ↑ largest

b. I is large. Adding an electron forces extreme electron-electron repulsion, making it larger. Calcium is large, but taking its 4*s* electrons away markedly increases the Z_{eff}.

Ca^{2+}, Li, I, I^-
↑ smallest ... ↑ largest

12.16 The Properties of a Group: The Alkali Metals

This section begins with a review of the information in the periodic table.

1. Elements in a group exhibit similar properties. It is primarily the number of valence electrons that determine an atom's chemistry.
2. Electron configurations can be gleaned from the periodic table.
3. Learn the names of the different groups (halogens, alkali metals, lanthanides, etc.). See Figure 12.39 in your textbook.
4. Metals tend to lose electrons (have low ionization energies). Nonmetals tend to gain electrons. Metalloids (semimetals) have properties of both.

The focus of the chapter is on **Group I metals.**

Properties as We Go Down the Group

a. The first ionization energy decreases.
b. The atomic radius increases.
c. The density increases.
d. The reactivity increases (they lose electrons readily).
e. The melting and boiling points decrease.

Example 12.16 - The Alkali Metals

Explain the trends the alkali metals follow in properties "A" through "D" above based on your knowledge of electronic configurations and atomic structure.

Solution

a. **First ionization energy:** As we go down the group, the valence electron falls in a higher energy level. The nuclear attraction is less, making it easier to ionize the electron.
b. **Atomic radius**: Again, the valence electron occupies a higher energy level, thereby being farther from the nucleus.
c. **Density:** Atomic mass increases faster than atomic size.
d. **Reactivity**: The electrons are easier to ionize; therefore reactions that require less energy are possible.

Exercises

Section 12.1

1. The visible region of the spectrum goes from 400 nm to 700 nm. What is the frequency range of the visible spectrum?
2. List the regions of the electromagnetic spectrum and the wavelengths of radiation associated with each region.
3. Calculate the frequency of blue light of wavelength 4.5×10^2 nm.
4. Calculate the wavelength of green light of frequency 5.7×10^{14} Hz.

Section 12.2

5. Red light with a wavelength of 670.8 nm is emitted when lithium is heated in a flame.
   a. What is the frequency of this radiation?
   b. What is the energy of this radiation per photon? Per mole of photons?

6. It takes 6.72×10^{-18} J of energy to remove an electron from an unknown atom. What is the maximum wavelength of light that can do this?

7. A carbon-oxygen double bond in a certain organic molecule absorbs radiation that has a frequency of 6.0×10^{13} s^{-1}.

   a. To what region of the spectrum does this radiation belong?
   b. What is the wavelength of this radiation?
   c. What is the energy of this radiation per photon? Per mole of photons?
   d. A carbon-oxygen bond in a different molecule absorbs radiation with frequency equal to 5.4×10^{13} s^{-1}. Does this radiation give more or less energy?

8. Calculate the energy of a photon that is emitted at a wavelength of 5.69×10^{3} nm.

9. Many spectroscopists prefer using frequencies to wavelengths when describing electromagnetic radiation. Can you think of an advantage to the use of frequencies? (A main concern of spectroscopists is the energy of radiation that is either emitted or absorbed.)

10. Calculate the wavelength of a thoroughbred racehorse that weighs 600 pounds and is moving at a speed of 40 mi/hr.

11. What are the wavelengths associated with the following?

    a. an alpha particle (mass = 6.64×10^{-27} kg) traveling at 3.0×10^{6} m/s
    b. a 1000-kg automobile traveling at 100 km/hr

12. Calculate the energy associated with an electronic transition involving each of the following types of photons.

    a. red photons of wavelength 670. nm
    b. yellow photons of wavelength 580. nm
    c. violet photons of 450. nm
    d. X-ray photons of wavelength 0.154 nm

Section 12.4

13. Make a plot of energy vs. n for the Bohr hydrogen atom for $n = 1$ to $n = 50$.

    a. What is the energy of the Bohr hydrogen atom when $n = \infty$?
    b. What is the ionization energy for the Bohr hydrogen atom (i.e., the energy required to move an electron from $n = 1$ to $n = \infty$)?

14. Calculate the wavelength of light that must be absorbed by a hydrogen atom in its ground state to reach the excited state of $\Delta E = +2.914 \times 10^{-18}$ J.

15. How much energy is required to ionize a mole of hydrogen atoms?

16. Calculate the wavelength of light emitted in the spectral transition of $n = 4$ to $n = 2$ in the hydrogen atom.

17. What region of the spectrum would you look in to find the radiation associated with the $n = 4$ to $n = 1$ transition of the Bohr hydrogen atom?

18. What region of the spectrum would you look in to find the radiation associated with the spectral transition of $n = 3$ to $n = 2$ in the hydrogen atom?

Section 12.8

19. Use the wave mechanical model to explain the quantized nature of the orbits of a hydrogen atom.

20. A chemistry book lists the radius of the hydrogen orbital as 1 Å. Will the electron ever be farther than 1 Å from the nucleus?

21. Even on the planet Mars, the probability of finding an electron of an atom on the nose of the Mona Lisa is not zero. Explain.

Section 12.9

22. Which of the following sets of quantum numbers are allowed?

    a. $n = 7, \ell = 7, m_\ell = 0$
    b. $n = 7, \ell = 0, m_\ell = 1$
    c. $n = 7, \ell = 5, m_\ell = -3$
    d. $n = 3, \ell = -1, m_\ell = 0$
    e. $n = 0, \ell = 0, m_\ell = 0$

23. What is the maximum number of electrons allowed in the *n* states corresponding to the M, N, and O shells?

24. What is the maximum number of electrons that can be accommodated in the following?

    a. all orbitals with $n = 4$
    b. all the 4*f* orbitals
    c. all the 5*g* orbitals

Section 12.10

25. Account for the fact that a *p* subshell containing three electrons has one in each orbital rather than two in one orbital and the third in another.

Section 12.11

26. Account for the fact that 2*s* electrons are more strongly bound to the nucleus than are 2*p* electrons in the same atom.

Section 12.13

27. Write n, ℓ, m_ℓ, and m_s quantum numbers for the 5 electrons of a boron atom.

28. What is the electron configuration for calcium?

29. How many half-filled orbitals do each of the following have in the ground state?

    a. O
    b. B
    c. Ar
    d. Mn
    e. K
    f. Cf
    g. Zn

30. Indicate the higher of the two energy states in each of the following pairs:

    a. 3*d* or 4*s*
    b. 4*p* or 5*s*
    c. 4*s* or 4*p*

Section 12.15

31. Arrange the following atoms in order of increasing Z_{eff} for the highest-energy electron: Te, In, Mg, Ga, Xe, Ca.

32. The first and second ionization energies for argon are 1525 and 2665 kJ per mole. Calculate Z_{eff} for the first and second electrons removed.

33. Calculate the energy required to form one mole of sodium ions from a gas of sodium atoms if $Z_{eff} = 3.40$.

34. In which orbital would an electron have a greater likelihood of being near the nucleus: 4*f* or 6*s*?

35. Order the following groups from smallest to largest radius.

   a. Ar, Cl^-, K^+, S^{2-}
   b. C, Al, F, Si
   c. Na, Mg, Ar, P
   d. I^-, Ba^{2+}, Cs^+, Xe

36. Which of the following will have the most exothermic electron affinity? the least?

   a. Ge, Si, C
   b. Cl, Cl^-, Cl^+

37. Which group in the periodic table contains elements with the highest ionization energies? Which periods in the periodic table contain elements with the highest ionization energies?

38. When an electron is removed from a neutral nitrogen atom, it comes from a half-filled 2*p* orbital. When an electron is removed from a neutral oxygen atom, it comes from a filled 2*p* orbital and therefore leaves behind another electron in that orbital. What effect would you expect this difference to have on the relative values of the ionization energies of these two elements? Explain.

Section 12.16

39. Properties of the alkali metals are discussed in Section 12.16. List some properties you would expect for the alkaline earth metals, Be, Mg, Ca, Sr, and Ba.

40. Which elements are metalloids, and why are they called metalloids?

Multiple-Choice Self-Test

1. The frequency of an electromagnetic wave is 1.5×10^{14} hertz. Calculate its wavelength in meters.

   A. 2.0×10^{-6} m B. 6.6×10^{-9} m C. 5.0×10^{5} m D. 5.0×10^{-5} m

2. Calculate the wavelength of an electromagnetic wave with a frequency of 1.7×10^{14} Hz.

   A. 5.9×10^{6} m B. 0.67×10^{-15} m C. 0.33×10^{8} m D. 1.8×10^{-6} m

3. Carbon absorbs energy at a wavelength of 150 nm. The total amount of energy emitted by a carbon sample is 1.98×10^{5} J. Calculate the number of carbon atoms present in the sample, assuming that each atom emits one photon.

   A. 1.50×10^{23} B. 2.50×10^{19} C. 1.48×10^{20} D. 1.65×10^{5}

4. A particle has a velocity equal to 0.25 *c* and a wavelength of 1.3×10^{-16} m. Calculate the mass of the particle in kilograms. $c = 3.0 \times 10^{8}$ m/s

   A. 1.7×10^{-20} kg B. 6.8×10^{-26} kg C. 8.5×10^{-19} kg D. 3.3×10^{-28} kg

5. How many distinct magnetic quantum numbers are possible if the angular momentum quantum number is 6?

   A. 13 B. 7 C. 12 D. 3

6. Which of the following quantum number sets is unacceptable?

   A. 1, 0, 0 B. 6, 2, 0 C. 4, 3, 3 D. 4, 2, 3

7. The Pauli exclusion principle is violated by which one of the following electron systems?

   A. 1, 0, 0, ½ and 1, 0, 0, ½
   B. 5, 4, −2, ½ and 5, 4, −2, −½
   C. 4, 3, −3, ½ and 4, 3, −3, −½
   D. 3, 2, −2, ½ and 3, 2, −2, −½

8. Which one of the following elements does the configuration $1s^22s^22p^63s^23p^5$ describe?

A. Cl B. Ar C. K D. S

9. Which one of the following elements has a $6s^24f^{14}5d^3$ valence shell configuration?

A. Re B. Ta C. Mo D. Hf

10. What elements in the periodic table have the following electron configuration: [Noble gas]ns^2nd^5?

A. Fe, Ru, Os, Uno B. Mn, Tc, Re C. F, Cl, Br, I, At D. Co, Rh, Ir, Une

11. Place the following atoms, P, Kr, Mg, Li, in order of increasing first ionization energies.

A. P < Kr < Mg < Li
B. Mg < Li < P < Kr
C. Kr < P < Mg < Li
D. Li < Mg < P < Kr

12. Place the following atoms, Cl, F, Na, C, in order of decreasing electron affinity values.

A. C > Cl > F > Na
B. Cl > F > C > Na
C. F > Na > Cl > C
D. F > Cl > C > Na

13. At what step of ionization does arsenic exhibit a sudden, marked increase in its ionization energy?

A. sixth B. fifth C. fourth D. third

14. Place the following elements, Br, Kr, C, Se, Te, in order of increasing atomic size.

A. Br < Te < Kr < Se < C
B. C < Kr < Br < Se < Te
C. Te < Se < Br < Kr < C
D. Br < Kr < C < Se < Te

15. Which of the following elements has the lowest reducing ability?

A. Li B. Cs C. Na D. K

Answers to Exercises

1. 7.5×10^{14} s^{-1} to 4.3×10^{14} s^{-1}
2. See Figure 12.3 in your textbook.
3. 6.7×10^{14} Hz
4. 5.3×10^{-7} m
5. a. 4.469×10^{14} s^{-1} b. 2.961×10^{-19} J, 178.3 kJ
6. 29.6 nm
7. a. infrared
   b. 5×10^{-6} m
   c. 4.0×10^{-20} J, 24 kJ
   d. less
8. 3.49×10^{-20} J
9. direct proportionality between frequency and energy
10. 1.4×10^{-37} m
11. a. 3.3×10^{-14} m b. 2.4×10^{-38} m
12. a. 2.97×10^{-19} J
    b. 3.43×10^{-19} J
    c. 4.42×10^{-19} J
    d. 1.29×10^{-15} J
13. a. 0 b. 2.18×10^{-18} J
14. 6.817×10^{-8} m

15. 1.31×10^6 J

16. 486 nm

17. ultraviolet

18. visible

19. The key point is that the Schrödinger equation shows that electrons can reside in certain mathematically allowed locations.

20. yes

21. The probability of finding the electron on Mars mathematically approaches zero with increasing distance from the nucleus, but it never actually reaches zero.

22. c

23. 18, 32, and 50

24. a. 32 electrons b. 14 electrons c. 18 electrons

25. Mutually repelling electrons will occupy separate *p* orbitals. This behavior is summarized by Hund's rule, which states that the lowest energy configuration for an atom is the one having the maximum number of unpaired electrons allowed by the Pauli principle in a particular set of degenerate orbitals.

26. The average distance of 2*s* electrons is closer to the nucleus than that of 2*p* electrons.

27. 1, 0, 0, −½; 1, 0, 0, +½; 2, 0, 0, −½; 2, 0, 0, +½; 2, 1, 0, −½

28. Ca: $1s^22s^22p^63s^23p^64s^2$

29. a. 2 c. 0 e. 1 g. 0
    b. 1 d. 5 f. 4

30. a. 3*d* b. 5*s* c. 4*p*

31. Mg < Ca < Ga < In < Te < Xe

32. 3.23 and 4.28

33. 496 kJ

34. 6*s*

35. a. K^+, Ar, Cl^-, S^{2-} c. Ar, P, Mg, Na
    b. F, C, Si, Al d. Ba^{2+}, Cs^+, Xe, I^-

36. a. C, Ge b. Cl^+, Cl^-

37. 8A (noble gases), Periods I and II

38. Since electrons repel each other, an electron is more readily removed from an orbital containing two electrons; therefore, the ionization energy for the oxygen electron should be less than that of the nitrogen electron.

39. Examples are reaction with water to give bases and metallic character, among others.

40. Si, Ge, As, Sb, Te, Po, and At are called metalloids because they exhibit both metallic and nonmetallic properties under certain circumstances.

Answers to Multiple-Choice Self-Test

1. A	4. B	7. A	10. B	12. B	14. B
2. D	5. A	8. A	11. D	13. A	15. A
3. A	6. D	9. B			

CHAPTER 13

Bonding: General Concepts

The Bottom Line: Chapter 13

In this chapter you will use many of the concepts you learned in Chapter 12, especially electronic configurations. You will learn why different types of bonds form, the nature of those bonds, and a model to predict the three-dimensional structure of molecules formed from covalent bonds.

13.1 Types of Chemical Bonds

Your textbook says bonds are "*forces that hold groups of atoms together and make the atoms function as a unit.*" Bonds form because the energy of the system is lower than if bonds did not form.

Ionic bonding is due to electrostatic attraction. It results from the loss of an electron from an alkaline or alkaline earth metal and its gain by a nonmetal.

Your textbook gives a formula for the energy of an ion pair called the **energy of interaction.** Let's use that formula to calculate the stability gain when Mg^{2+} and O^{2-} interact.

Example 13.1 - Coulomb's Law

Calculate the energy of interaction (in kJ/mol) between Mg^{2+} and O^{2-} if the distance between the centers of Mg^{2+} and O^{2-} is 0.205 nm (2.05 Å).

Solution

$$E = 2.31 \times 10^{-19}\ \text{J nm}\left[\frac{Q_1 Q_2}{r}\right] = 2.31 \times 10^{-19}\ \text{J nm}\left[\frac{(+2)(-2)}{0.205\ \text{nm}}\right]$$

$$= -4.51 \times 10^{-18}\ \text{J/ion pair} = \mathbf{-4.51 \times 10^{-21}\ kJ/ion\ pair}$$

For a mole of ion pairs,

$$E = \frac{-4.51\times10^{-21}\ \text{kJ}}{\text{ion pair}} \times \frac{6.022\times10^{23}\ \text{ion pairs}}{\text{mole}} = -2710\ \text{kJ/mol}$$

Commentary

Note how much higher this value is than the value in the NaCl example in your textbook! Part of the reason is because the **radius is smaller** (owing to the higher charge on Mg and O). The more important contributor is the **higher charge** on each ion. That allows for more powerful electrostatic interaction, which gives a more stable bond.

Covalent Bonding occurs when bonds form between similar kinds of atoms for the same reason as between dissimilar atoms—the energy of the system is lowered as a result of the bond formation.

Look at Figure 13.1 in your textbook. There is a particular distance apart where the combination of repulsive and attractive forces allows the system to have a minimum energy. In this case, we have a **covalent bond** in which **electrons are shared by both nuclei approximately equally**. Examples are S_8, graphite and diamond.

In the case where there is **unequal sharing of electrons, a polar covalent** bond exists. Charges are not distributed equally in such a molecule. Positive and negative poles exist. Examples of polar covalent bonds are C—Cl, H—Cl, and O—H.

In summary, the nature of the bond will depend upon the ability of each atom in the bond to attract electrons to itself. This is called **electronegativity**.

13.2 Electronegativity

Recall that unequal sharing of electrons in a bond results in a polar covalent bond. Ionic bonds result from the transfer of electrons between atoms.

Your textbook says we can get a measure of the degree of ionic character of a bond by **measuring bond energies.** The more the electrostatic interaction that occurs between two atoms, the greater will be the difference in bond energies when compared to the average of the perfectly covalent bonds involving the atoms. This difference in bond energy is called Δ and is the relative **electronegativity** difference between the bonding atoms.

For the representative elements, **electronegativity decreases going down a group and increases going across a period**. Thus francium has the lowest electronegativity, and fluorine has the highest.

The greater the Δ, the more ionic character the bond has. If Δ = 0, the bond is perfectly covalent. There are no precise cut-offs.

Example 13.2 A - Electronegativities Of Atoms

Look at the periodic table at the front of your textbook. Based only on their positions in the table, place the following atoms in order of increasing electronegativity.

Sr, Cs, Se, O, Ba

Solution

Remember the trend: Electronegativity increases from the lower left to the upper right of the periodic table. Based on their positions, the order is:

Cs < Ba < Sr < Se < O

least electronegative — most electronegative

Try the next example, keeping in mind that the determinant of bond polarity is the value of Δ.

Example 13.2 B - Bond Polarity

Using Figure 13.3 in your textbook, calculate Δ for each of the following bonds, and order the set from most covalent to most ionic character.

a. Na—Cl b. Li—H c. H—C d. H—F e. Rb—O

Solution

a. Na = 0.9, Cl = 3.0, Δ = 3.0 − 0.9 = **2.1**
b. Li = 1.0, H = 2.1, Δ = 2.1 − 1.0 = **1.1**
c. H = 2.1, C = 2.5, Δ = 2.5 − 2.1 = **0.4**
d. H = 2.1, F = 4.0, Δ = 4.0 − 2.1 = **1.9**
e. Rb = 0.8, O = 3.5, Δ = 3,5 − 0.8 = **2.7**

In order of increasing ionic character:

H—C, Li—H, H—F, Na—Cl, Rb—O
most covalent ... most ionic

The information gained from electronegativities will be used in the next section to figure out the polarities of molecules.

13.3 Bond Polarity and Dipole Moments

Recall from Section 13.2 that it is possible to determine the **polarity of a bond** by the **size of Δ.** If a molecule is diatomic (two atoms), there is often only one bond, and that will determine whether the molecule is polar.

For instance, in Example 13.2, we determined that H—F was polar, with fluorine being the more electronegative atom. A **partial negative charge** (δ−) resides on the fluorine atom, and a **partial positive charge** (δ+) resides on the hydrogen atom.

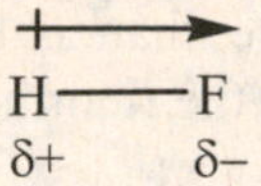

The **arrow** points to the **center of negative charge** while the **tail** is at the **center of positive charge**. A **dipole moment** means that the molecule has **two poles**.

The situation is clear-cut with HF. It becomes more difficult with three or more atoms in a molecule because the **individual dipoles can cancel each other out**. Look at Table 13.4 in your textbook. This shows how individual bond polarities can cancel each other out to yield a molecule with no dipole moment. Although you will be able to derive the geometries in Table 13.4 later on, you should memorize them for now.

Example 13.3 A - Dipole Moment

Does $CHCl_3$ (a tetrahedral molecule with carbon at the center) have a dipole moment? If so, show the orientation of the dipole moment.

Solution

Perform the following steps:

1. Look up the electronegativity of each atom.
2. Draw the molecule in three-dimensional space.
3. Determine the polarity of each bond and the net polarity on each atom.
4. Draw the dipoles, and determine the direction (if any) of the molecule dipole moment.

C = 2.5, H = 2.1, Cl = 3.0

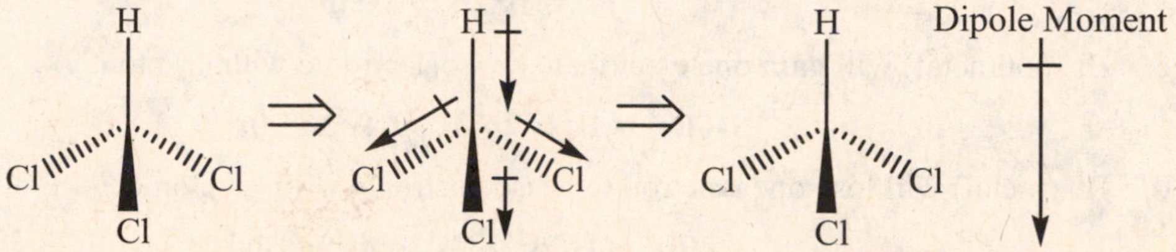

Example 13.3 B - Practice With Dipole Moments

For each of the following, determine the orientation of the dipole moment (if any).

a. HI b. N_2 c. CCl_2F_2 (carbon is the central atom)

Solution

a. H——I (2.1, 2.5) ⇒ H——I ($\delta+$, $\delta-$) ⇒ H——I (dipole arrow $+\!\!\longrightarrow$ pointing toward I)

b. N——N (3.0, 3.0) ⇒ No dipole moment ($\Delta = 0$)

c. C = 2.5, Cl = 3.0, F = 4.0 ⇒

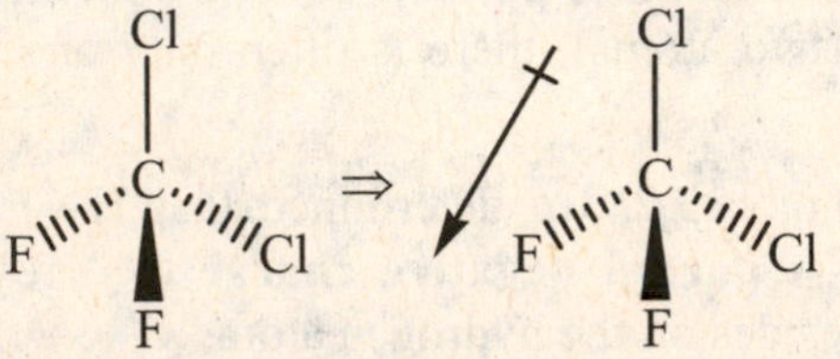

(Fluorine is more electronegative than chlorine; therefore the dipole moment on this molecule is tipped slightly toward the fluorine atoms.)

13.4 Ions: Electron Configurations and Sizes

Your textbook deals with only **nonmetals, representative metals,** and ionic bonds in this discussion. It has been observed that atoms that form bonds in stable compounds have a noble **gas electronic configuration.** (Each is isoelectronic with a noble gas.)

Example 13.4 A - Ionic Electron Configuration

Write the electronic configuration, and determine the charge on each of the following atoms when it forms its most stable ion (noble gas electronic configuration).

a. Mg b. P c. Br d. Rb

Solution

Determine how many electrons the atom must gain or lose (metals will lose, nonmetals will gain) to obtain the electronic structure of the **nearest noble gas**.

a. Magnesium is $1s^2 2s^2 2p^6 3s^2$. It will lose the two $3s$ electrons to have the electronic configuration of neon. Magnesium will therefore ionize to $\mathbf{Mg^{2+}}$.

$$\mathbf{Mg^{2+} = 1s^2 2s^2 2p^6} \text{ (isoelectronic with neon)}$$

b. P (nonmetal) will gain 3 electrons to be isoelectronic with argon.

$$\mathbf{P^{3-} = 1s^2 2s^2 2p^6 3s^2 3p^6}$$

c. Br (nonmetal) will gain one electron to be isoelectronic with krypton.

$$\mathbf{Br^- = 1s^2 2s^2 2p^6 3s^2 3p^6 4s^2 3d^{10} 4p^6}$$

d. Rb (metal) will lose one electron to be isoelectronic with krypton.

$$\mathbf{Rb^+ = 1s^2 2s^2 2p^6 3s^2 3p^6 4s^2 3d^{10} 4p^6}$$

Note that Br^-, Rb^+, and Kr are all **isoelectronic**.

To determine the formula of binary ionic compounds, remember that **chemical compounds are electrically neutral**. (The sum of the cation charges must equal the anion charges.)

For example, the formula of the ionic compound formed from combining magnesium and chlorine can be determined by:

1. assessing the charge on the ions, and
2. determining how many of each ion is required to combine to make the compound electrically neutral.

The ions are $\mathbf{Mg^{2+}}$ and $\mathbf{Cl^-}$. In order to maintain electronic neutrality,

$$Mg^{2+} + 2Cl^- \rightarrow MgCl_2.$$

Example 13.4 B - Formulas Of Binary Ionic Compounds

Determine the formula for each of the following sets of atoms when they combine to form a binary ionic compound.

a. K and Br b. Sr and F c. Al and Se d. Ba and O

Solution

a. $K^+ + Br^- \rightarrow$ **KBr**
b. $Sr^{2+} + 2F^- \rightarrow$ $\mathbf{SrF_2}$
c. $2Al^{3+} + 3Se^{2-} \rightarrow$ $\mathbf{Al_2Se_3}$
d. $Ba^{2+} + O^{2-} \rightarrow$ **BaO**

We have maintained electronic neutrality throughout.

We discussed ion size in our review of Chapter 12. As we go down a group, ion size increases (higher energy levels have a larger average distance). Because the ratio of protons to electrons becomes greater, **cations are smaller than their neutral atoms**. Because of electron-electron repulsion and less effective shielding, **anions are always larger than their neutral atoms**. *The larger the charge, the more pronounced the effect.*

Thus S^{2-} is larger than S.
Ca^{2+} is smaller than Ca.
S^{2-} is **isoelectronic** with Ca^{2+}. Because Ca^{2+} has 2 more protons than electrons while S^{2-} has more electrons than protons, S^{2-} is much larger than Ca^{2+}. Therefore, we can conclude that for an isoelectronic series **the more positive the nuclear charge (*Z*) the smaller the ion**.

Example 13.4 C - Ion Size

Order the following ions from smallest to largest.

a. O^{2-}, Na^+, Mg^{2+}, F^-
b. Se^{2-}, Te^{2-}, Rb^+, Mg^{2+}

Solution

a. These ions are all **isoelectronic with neon**. Therefore the smallest will have the highest positive charge. The ions will get larger until we have the highest negative charge.

$$\underset{\text{smallest}}{Mg^{2+}} < Na^+ < F^- < \underset{\text{largest}}{O^{2-}}$$

b. Because Se^{2-} and Te^{2-} are anions, they will be larger than Rb^{+} and Mg^{2+}. Because Te^{2-} is farther down the periodic table than Se^{2-}, it will be larger. Magnesium is higher on the periodic table and has a greater charge than rubidium; therefore it will have the smallest ion.

$$Mg^{2+} < Rb^{+} < Se^{2-} < Te^{2-}$$

smallest largest

13.5 Formation of Binary Ionic Compounds

The theme of this section is that there are **many separate processes** that go into **forming an ionic solid**. As we have said before, *the ionic compound forms because its energy is lower than if its elements remained separated.* However, not every part of the process is energetically favorable. Your textbook points out that it is the **lattice energy** (the energy released when an ionic solid is formed from its gaseous ions) that is the most favorable and more than makes up for some parts of the process that are energetically unfavorable. Let's examine the formation of the **KCl ionic solid** ("salt") from **K(*s*)** and **Cl_2(*g*)**.

Processes That Must Occur

1. K(*s*) must form K(*g*); **(Energy of Sublimation = +64 kJ).**
2. K(*g*) must form K^+(*g*); **(First Ionization Energy = +419 kJ).**
3. ½Cl_2(*g*) must form Cl(*g*); **(Bond Energy × ½ = 120 kJ).**
4. Cl(*g*) must form Cl^-(*g*); **(Electron Affinity = −349 kJ).**
5. K^+(*g*) must combine with Cl^-(*g*) to form KCl(*s*); **(Lattice Energy = −690 kJ).**

The net energy of formation (ΔH_f°) equals the sum of the energy changes, −436 kJ. So you see that the value for any of the processes that make up salt formation can be obtained if you understand the processes involved, and are given suitable data.

Example 13.5 - Calculation Of Lattice Energy

Given the following data, determine ΔH_f° for LiBr.

$$Li(s) + \tfrac{1}{2}Br_2(g) \rightarrow LiBr(s)$$

ionization energy for Li =	+520 kJ/mol
electron affinity for Br =	−324 kJ/mol
energy of sublimation for Li =	+161 kJ/mol
lattice energy =	−787 kJ/mol
bond energy of Br_2 =	+193 kJ/mol

Solution

While ΔH_f° will be the sum of all energy changes, be careful to **multiply the bond energy by ½**, because we have ½ of a mole of Br_2(*g*).

$$520 + (-324) + (+161) + (-787) + (+96) = -334 \text{ kJ} = \Delta H_f^\circ$$

Note that here, as in the previous example, the lattice energy is the most significant contributor to salt formation.

The remainder of the section makes the point that the **higher the charge** on each ion, the greater the lattice energy will be. This value counteracts the higher endothermic ionization energies, thus resulting in a **more energetically stable crystal**.

13.6 Partial Ionic Character of Covalent Bonds

If you can answer the questions posed in this review section, then you understand the material your textbook is trying to get across.

1. Why do we say that there are **no totally ionic** bonds?
2. How do we determine the **percent ionic character** of a bond?
3. Why is it ambiguous to say that NH_4Cl or Na_2SO_4 are ionic compounds?
4. How do we define (as an operating definition) an ionic compound (salt)?

13.7 The Covalent Chemical Bond: A Model

If you can answer the questions posed in this review section, then you understand the material your textbook is trying to get across.

1. (Review) What is a chemical bond?
2. (Review) Why do chemical bonds occur?
3. Why is it useful to think of each bond in a molecule as being **environment independent**?
4. If you were given the energy of stabilization of CCl_4, discuss how you would determine the bond energy of a C—H bond in $CHCl_3$.
5. What is **a model**?
6. Why do we develop models?
7. List the fundamental properties of a model.
8. What are the **limitations** of models?
9. Is a wrong model useless? Why or why not?

13.8 Covalent Bond Energies and Chemical Reactions

Recall from the last section that your textbook calculated the average bond energy for a C—H bond for methane (CH_4). The assumptions made were:

1. that the energy needed to break each C—H bond was the same (413 kJ), and
2. that each of the bonds was not sensitive to its environment.

Assumption No. 2 is not really correct, as is shown by the table (beginning of section 13.8 in your textbook) listing the energy required to break the C—H bond in molecules similar to methane in which the hydrogens have been replaced by bromine, chlorine and fluorine. In spite of this, average bond energies can help give us fair estimates of heats of reaction, and as such they are useful tools.

Look at Table 13.6 in your textbook. This table gives **average bond energies** for many covalent bonds. Notice that **multiple bonds** (bonds that involve **sharing more than two electrons** between atoms) require more energy to break than single bonds.

Your textbook uses the example of the combination of hydrogen and fluorine gas to make hydrogen fluoride. It states that

$$\Delta H = \sum D \text{ (bonds broken)} - \sum D \text{ (bonds formed)}$$

where ΔH = heat of reaction and D = bond energy per mole of bonds.

The key to calculating the heat of reaction from the average bond energy is to **carefully list all bonds broken and all bonds formed**. Also recognize that *it takes energy to break bonds* (endothermic, $\Delta H = +$) while *energy is released when bonds are formed* (exothermic, $\Delta H = -$).

Example 13.8 - Heat Of Reaction From Bond Energy

Using data from Table 13.6 in your textbook, calculate ΔH for the following reaction. Compare this to ΔH_f° calculated from the thermodynamic values in the appendix. (See Chapter 9 if you need a review of this.) (ΔH_f° [for CF_4] = −680 kJ/mol)

$$CH_4(g) + 4F_2(g) \rightarrow CF_4(g) + 4HF(g)$$

Solution

Let's make a list of bonds broken and bonds formed.

Bonds Broken	Energy per Bond (kJ)		Total Energy (kJ)
4 C—H	413 × 4 bonds	=	1652
4 F—F	154	=	616 (154 × 4 moles)
	Total Energy to Break Bonds	=	**2268 kJ**

Bonds Formed	Energy per Bond (kJ)		Total Energy (kJ)
4 C—F	485 × 4 bonds	=	1940
4 H—F	565	=	2260 (565 × 4 moles)
	Total Energy to Form Bonds	=	**4200 kJ**

ΔH = (Energy to break bonds) − (Energy to form bonds)
= 2268 kJ − 4200 kJ
= **−1932 kJ**

This reaction is exothermic.

The value for ΔH_f° = (ΔH_f° [for CF_4] + 4 ΔH_f° [for HF]) − (ΔH_f° [for CH_4] + 4 ΔH_f° [for F_2])
= (−680 + 4(−271)) − (−75 + 4(0))
= **−1689 kJ**

13.9 The Localized Electron Bonding Models

If you can answer the following review questions, you understand the important ideas in this brief section.

1. What are **lone pairs** and **bonding pairs**?
2. What are the parts of the LE model?

13.10 Lewis Structures

Lewis structures often are used to depict **bonding pairs and lone pairs** in molecules. We are concerned only with **valence** electrons because these are the ones (for period 1 and 2 atoms) that are involved in bond making and breaking.

Individual atoms are represented with Lewis structures by putting valence electrons (as dots or circles) around the atomic symbol. For example, magnesium ([Ne]$3s^2$) would be represented as

• Mg •

Mg^{2+} ([Ne]) would be Mg. It no longer has its valence electrons. Chlorine ([Ne]$3s^2 3p^5$) would be

••
: Cl •
••

and Cl^- ($[Ne]3s^23p^6$) would be

$$:\ddot{\underset{\cdot\cdot}{Cl}}:$$

Example 13.10 A - Lewis Dot Structures

Draw Lewis dot structures for the following atoms or ions.

a. N b. N^{3-} c. I d. Ba e. Ba^{2+}

Solution

a. $N = [He]2s^22p^3$ (5 valence electrons) $\cdot\ddot{\underset{\cdot}{N}}\cdot$

(I drew the $2s^2$ on the left and the three $2p$ electrons singly. You can draw a maximum of 2 electrons on any side, unless there is a triple bond involved.)

b. $N^{3-} = [He]\ 2s^2\ 2p^6$ (8 valence electrons) $:\ddot{\underset{\cdot\cdot}{N}}:$

c. $I = [Kr]\ 5s^2\ 4d^{10}\ 5p^5$ (7 valence electrons) $:\ddot{\underset{\cdot\cdot}{I}}\cdot$

d. $Ba = [Xe]\ 6s^2$ (2 valence electrons) $\cdot$**Ba**$\cdot$

e. $Ba^{2+} = [Xe]$ (0 valence electrons) **Ba**

EVERY PERIOD 1 AND 2 ELEMENT (with the exception of H, He, B, and Be) CAN FORM COMPOUNDS OF LOWEST ENERGY IF THEIR HIGHEST ENERGY LEVELS ARE FILLED (s^2p^6). THIS IS CALLED THE OCTET RULE. If an ion or atom observes the octet rule, we will say that it is "happy" (using terminology first coined by Nobel Prize winner in chemistry Roald Hoffman). Hydrogen is "happy" if its $1s$ orbital is filled. We will discuss boron and beryllium later on.

Your textbook discusses a strategy for drawing Lewis structures. I will propose a different strategy. Use the one you feel more comfortable with.

Kelter Strategy for Writing Lewis Structures

1. **Total number of valence electrons in the system:** Sum the number of valence electrons on all the atoms. Add the total negative charge if you have an anion. Subtract the charge if you have a cation.

e.g., CO_3^{2-}; C has 4 valence electrons = 4
O has 6 valence electrons × 3 atoms = 18
charge on the ion is −2 so add = 2
electrons in the system = 24

2. **Number of electrons if each atom is to be happy:** Atoms in our examples will need 8 electrons (octet rule) or 2 electrons (hydrogen). Using CO_3^{2-} as an example:

C needs 8 electrons = 8
O needs 8 electrons × 3 atoms = 24
electrons for happiness = 32

The −2 charge comes as a result of the electrons in the system. The charge is **never** counted toward happiness.

3. **The number of bonds in the system:** Covalent bonds are made by sharing electrons. You need 32 electrons, and you have 24. You are 8 electrons deficient. If you make 4 bonds (with 2 electrons per bond), you will make up the deficiency. Therefore,

$$\text{\# bonds} = \frac{\text{"\#2"} - \text{"\#1"}}{2} = \frac{32 - 24}{2} = \textbf{4 bonds}$$

4. **Draw the structure.** The central atom is carbon. The oxygen atoms surround it. Because there are 4 bonds, there will be two single bonds and one double bond. Each bond accounts for two electrons. Then complete the octets by putting electrons around each atom. Double-check your results by counting total electrons in the system.

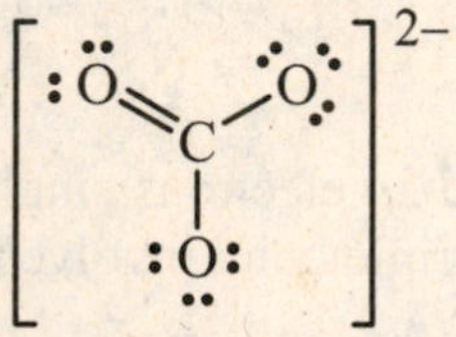

$$\left[\ddot{\text{O}} = \overset{}{\underset{|}{\text{C}}} - \ddot{\text{O}} \atop \ddot{\text{O}} \right]^{2-} = 24 \text{ electrons}$$

We will discuss resonance in this system later on.

Example 13.10 B - More Lewis Structures

Using the steps outlined above (or in your textbook), write Lewis structures that obey the octet rule for the following:

 a. Cl_2 b. CH_2Cl_2 c. H_2CO d. NH_3

Solution

a. 1. Total of valence electrons = 7 per Cl × 2 = **14**
 2. Total if happy = 8 per Cl × 2 = **16**
 3. # bonds = (16 − 14)/2 = **1 bond**

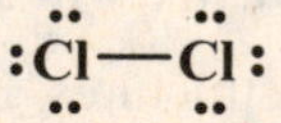

$$:\ddot{\text{Cl}} - \ddot{\text{Cl}}: \quad = 14 \text{ electrons}$$

b. 1. Total of valence electrons = 4 for the C = 4
 7 per Cl × 2 = 14
 1 per H × 2 = 2
 valence electrons = 20

 2. Total if happy = 8 for the C = 8
 8 per Cl × 2 = 16
 2 per H × 2 = 4
 electrons if happy = 28

 3. # bonds = (28 − 20)/2 = **4 bonds**

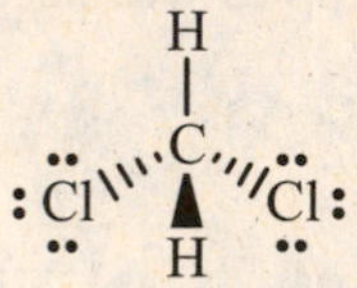

$$\underset{\text{H}}{\overset{\displaystyle \text{H} \atop |}{:\ddot{\text{Cl}} \cdots \overset{\blacktriangle}{\text{C}} \cdots \ddot{\text{Cl}}:}} \quad = 20 \text{ electrons (Hydrogen does NOT get a complete octet!)}$$

c. 1. Total of valence electrons = 4 for the C = 4
 6 for the O = 6
 1 per H × 2 = 2
 valence electrons = 12

2. Total if happy =

8 for the C	=	8
8 for the O	=	8
2 per H × 2	=	4
electrons if happy	=	**20**

3. # bonds = (20 − 12)/2 = **4 bonds**

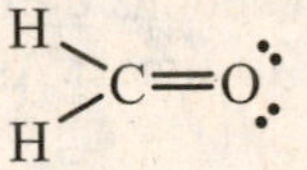

= 12 electrons

(*Only C, O, N, S, and P commonly have double bonds. Only C and N commonly have <u>stable</u> triple bonds.)

d. 1. Total of valence electrons =

5 for the N	=	5
1 per H × 3	=	3
valence electrons	=	**8**

2. Total if happy =

8 for the N	=	8
2 per H × 3	=	6
electrons if happy	=	**14**

3. # bonds = (14 − 8)/2 = **3 bonds**

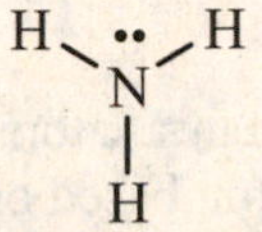

= 8 electrons

13.11 Resonance

Recall the example of CO_3^{2-} that we did at the beginning of Section 13.10. We determined the Lewis structure to be

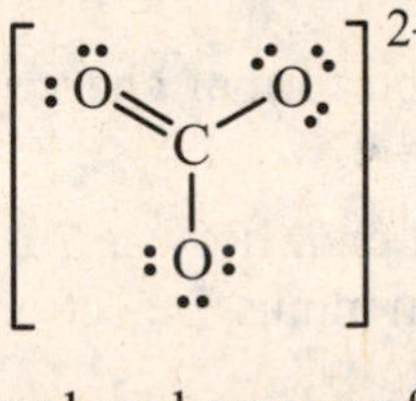

However, the double bond could have been placed on any of the three oxygens:

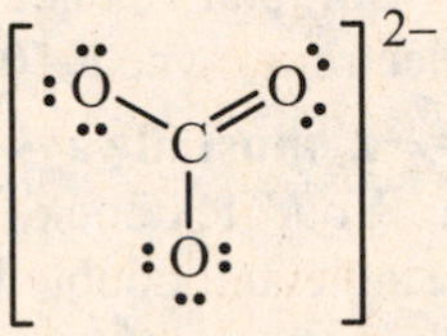

OR

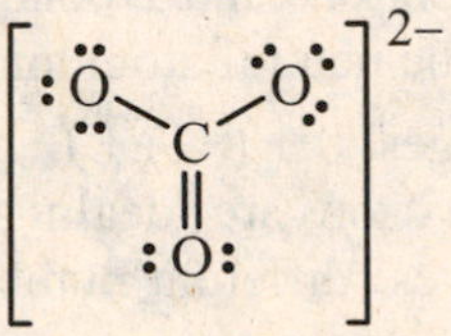

Measurements in bond lengths suggest that **all three C—O bond lengths are equivalent**. Electrons move around the entire molecule. Therefore the actual structure is a **time-average** of all these structures. These structures are called **resonance** structures. The Lewis structure of the molecule can be drawn any of three ways. The double bond seems to resonate between the carbon and oxygen atoms.

13.12 Exceptions to the Octet Rule

Although your textbook deals with boron in this section, we will focus on the more general case of **central atoms that can exceed the octet rule.** Your textbook presents a fascinating discussion on **hyperconjugation** in which central atoms such as P and S can maintain the octet by a combination of ionic and covalent bonding. There is some evidence to support this model. However, we, along with your textbook, will determine structures assuming exceptions do occur. (Your textbook points out that in vibrant science, scientists will disagree from time to time!) As a first step in this discussion, your textbook discusses formal charge.

Formal Charge

Let's review your textbook's discussion on formal charge and how it is used to decide on likely resonance structures. Formal charge is the difference between the number of valence electrons on the free atom and the number of valence electrons assigned to the atom in a molecule. Formal charge is a **computational device** based on a **localized electron (LE) model** and as such is not perfectly correct. To determine formal charge (a somewhat more realistic estimate of charge distribution in a molecule), we need to know:

1. How many electrons an atom "owns."

   Electrons Owned = **# valence electrons** around the atom
   + **# bonds** (which equals ½ # shared electrons)

2. The **formal charge** on an atom.

   Formal Charge = **# valence electrons** on the neutral atom
   − **# electrons owned** by the atom based on the resonance structure you drew

Let's look at CO_3^{2-} again.

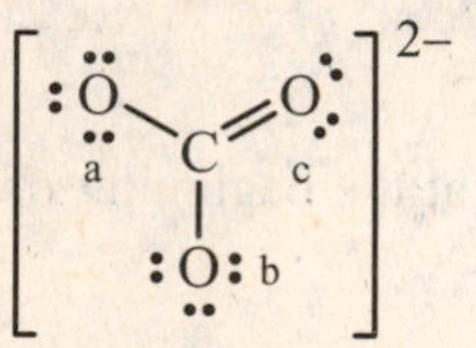

- Carbon owns 4 electrons (4 bonds). The formal charge on carbon = **4 valence electrons** on the neutral atom minus **4 electrons owned = 0**.
- Oxygen$_a$ has 6 valence electrons and 1 bond. It owns **7 total valence electrons**. The formal charge = 6 valence electrons on the neutral atom minus 7 electrons owned = **−1**.
- Oxygen$_b$ has the same formal charge as oxygen$_a$ = **−1**.
- Oxygen$_c$ has 4 valence electrons and 2 bonds. It owns 6 total valence electrons. The formal charge = 6 valence electrons on the neutral atom minus 6 electrons owned = **0**.

The sum of the formal charges, 0 + (−1) + (−1) + 0 = **−2, must always equal the charge on the ion** (or molecule, if that's what you are dealing with). Your textbook says that if you can write **nonequivalent** Lewis structures (different numbers of single and double bonds) for a molecule or ion, those with formal charges closest to zero and with any negative formal charges on the most electronegative atoms are considered to best describe the bonding.

Example 13.12 A - Formal Charges

Assign formal charges to each atom in the following resonance structures of CO_2.

1. 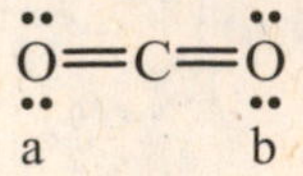

2. $: O \equiv C - \ddot{O}:$
 a b

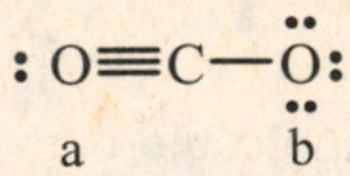

Which structure is more likely to be correct?

Solution

Let's establish formal charges for each atom.

Structure 1:

 C owns 4 electrons. Formal charge = 0
 O_a owns 6 electrons. Formal charge = 0
 O_b owns 6 electrons. Formal charge = 0

Structure 2:

 C owns 4 electrons. Formal charge = 0
 O_a owns 5 electrons. Formal charge = +1
 O_b owns 7 electrons. Formal charge = −1.

Structure 1 is more likely because all formal charges are zero.

Note: Structure 2 can be represented with its formal charges:

$$: O \equiv C - \ddot{O}:$$
$$\quad + \qquad\qquad -$$

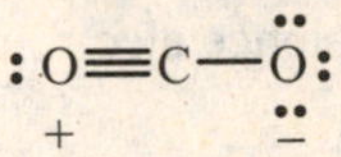

Example 13.12 B - Resonance

Draw all resonance structures, and select the most stable one for $(SCN)^-$.

Solution

First determine the number of bonds for the Lewis structures.

 There are **16 valence electrons** in the system.
 The total for happiness is **24 electrons**.
 There are $(24-16)/2 = $ **4 bonds** in this ion.

Now let's draw some possible resonance structures based on 4 bonds.

 a. $\left[:\ddot{S} - C \equiv N: \right]^-$ b. $\left[:S \equiv C - \ddot{N}: \right]^-$ c. $\left[\ddot{S} = C = \ddot{N} \right]^-$

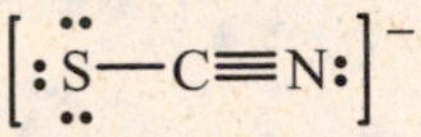

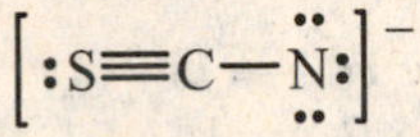

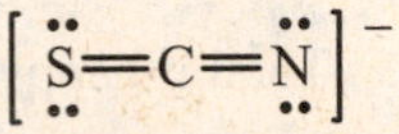

Now evaluate the formal charges on each atom.

 a. $\left[:\ddot{S} - C \equiv N: \right]^-$
 −1 0 0

 b. $\left[:S \equiv C - \ddot{N}: \right]^-$
 +1 0 −2

 c. $\left[\ddot{S} = C = \ddot{N} \right]^-$
 0 0 −1

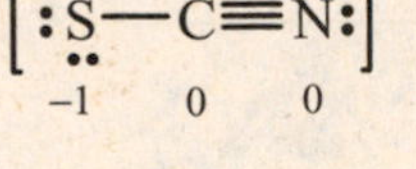

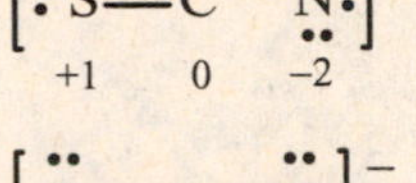

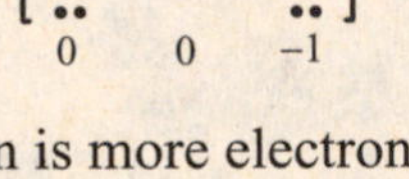

Nitrogen is more electronegative than sulfur. Therefore, structure "c" is the most likely.

Writing Lewis Structures for Exceptions to the Octet Rule

To determine if you have an exception to the rule, proceed as if your molecule obeys the octet rule. Let's use ICl_3 as an example.

1. Total of valence electrons = 7 for the I = 7
   7 per Cl × 3 = 21
   valence electrons = 28

2. Total if happy = 8 for the I = 8
   8 per Cl × 3 = 24
   electrons if happy = 32

3. # bonds = (32 − 28)/2 = **2 bonds**

CAUTION! We have 2 bonds for 3 chlorine atoms! **There are not enough bonds to account for all the atoms!** This is how you know that we have an exception to the octet rule.

To write the Lewis structure of exceptions, draw the structure with **one bond to each ligand, complete the octets, and add any extra electrons to the central atom**. With ICl_3,

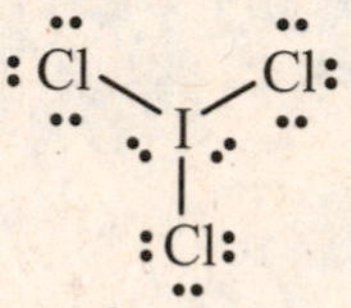

= 28 valence electrons (I has <u>10 electrons</u>)

Example 13.12 C - Exceptions To The Octet Rule

Write Lewis structures for the following molecules:

a. IF_2^{-1} b. CO_2 c. XeF_4

Solution

a. 1. Total of valence electrons = 7 for the I = 7
   7 per F × 2 = 14
   add 1 for "–" = 1
   valence electrons = 22

2. Total if happy = 8 for the I = 8
   8 per F × 2 = 16
   electrons if happy = 24

3. # bonds = (24 − 22)/2 = **1 bond** <u>NOT ENOUGH! EXCEPTION!</u>

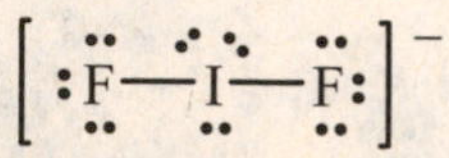

b. 1. Total valence electrons = **16**
   2. Total if happy = **24**
   3. # bonds = **4 bonds** Not an exception!

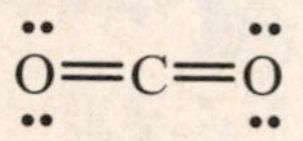

c. 1. Total of valence electrons = **36** (Xe = 8!)
   2. Total if happy = **40**
   3. # bonds = **2 bonds** <u>Exception!</u>

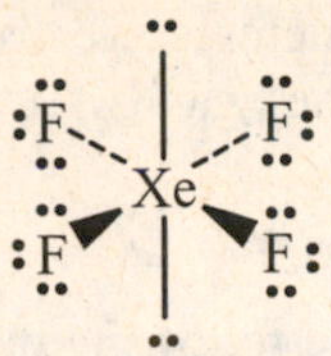

Your textbook notes (using SO_4^{2-} as an example) that formal charges do not necessarily dictate "the best" structure. The tendency of each period 3 central atom to obey the octet rule (see the discussion on hyperconjugation) could work against the notion of the best structure being that with the overall lowest formal charge.

13.13 Molecular Structure: The VSEPR Model

The Valence Shell Electron Pair Repulsion (VSEPR) model assumes that atoms will orient themselves so as to minimize electron pair repulsions around the central atom.

Memorize the information in Table 13.8 in your textbook. Each lone pair or bond around the central atom occupies a position in space. The effect of lone pairs around the central atom is to squeeze bonded pairs closer together (see your textbook's discussion regarding bond angles in CH_4, NH_3, and H_2O). Multiple bonds are counted as "one bonding pair" in the VSEPR model because the double bonds are constrained in space. Let's determine the VSEPR structure of formaldehyde, H_2CO, together.

Step 1: Determine the Lewis structure.

Total valence electrons = **12**
Total if happy = **20**
bonds = **4**

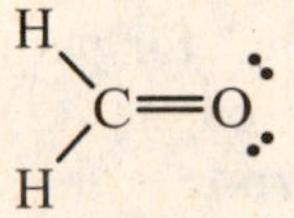

Step 2: Count the number of bonds and lone pairs on the central atom.

2 C—H + 1 C=O = **3 bonds**

Step 3: Determine the geometry based on Table 13.8 in your textbook.

3 "electron pairs" = **trigonal planar**

Now let's try **Ibr₂**.

Step 1:

Total valence electrons = 22
Total if happy = 24
bonds = 1 **Exception!**

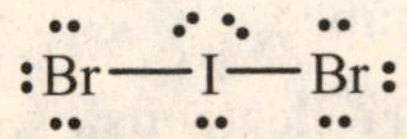

Step 2: There are **2 bonds** and **3 lone pairs** around the central atom. The total is **5 electron pairs**.

Step 3: The molecule will orient itself with a trigonal bipyramid geometry. The electron pairs will orient in the equatorial positions first (See discussion in the text), and the bonding pairs will make up the remaining positions:

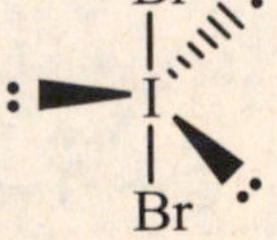

The molecule will be **linear.**

(Note that when we draw VSEPR structures, we are concerned only with the central atom. We often omit valence electrons around the ligands.)

Example 13.13 - VSEPR Structures

Determine the geometry of each of the following molecules or ions.

a. CO_2 b. SO_4^{2-} c. BrF_3 d. XeO_4 e. ICl_2^+

Solution

a. The Lewis structure for CO_2 (**16 valence electrons**) is

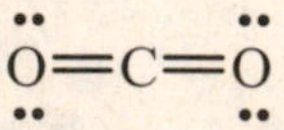

The double bonds each count in the VSEPR model as 1 restricted bond, so CO_2 acts as if it has **2 electron pairs** around it. Geometry = **linear.**

b. SO_4^{2-} = **32 valence electrons**

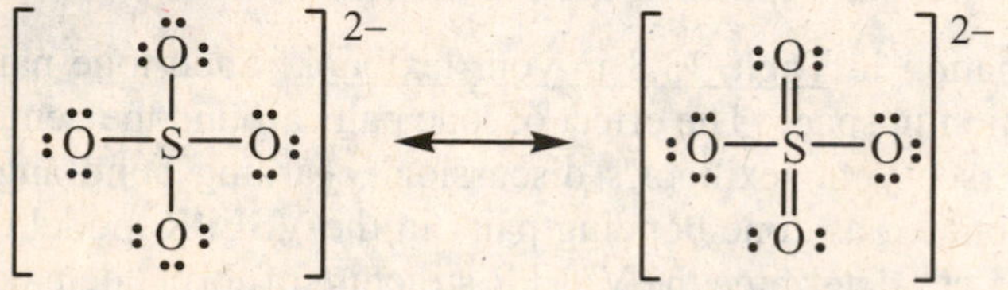

The double-bonded resonance structure seems to be better because of the lower formal charges on the oxygens, although even experts in inorganic chemistry still debate which is correct. It turns out that for purposes of the VSEPR model, both structures will give the same 4 electron pairs (2 double bonds "count" as 2 electron pairs). Geometry = **tetrahedron**.

c. BrF_3 = **28 valence electrons**

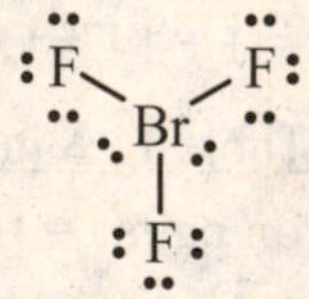

There are **2 lone pairs** and **3 bonding pairs** = **5 electron pairs**. It is a **trigonal pyramid** basis. The electrons take up two equatorial spots, leaving two F's axial and one equatorial. Geometry = **T-shaped.**

d. XeO_4 = **32 valence electrons**

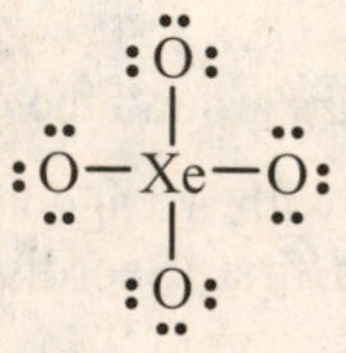

There are 4 bonds, thus **4 electron pairs**. Geometry = **tetrahedral.**

e. ICl_2^+ = **20 valence electrons**

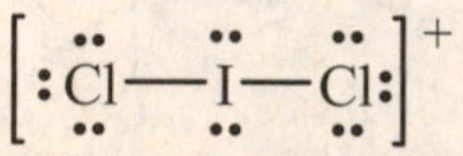

There are 2 bonding pairs and 2 lone pairs, thus 4 electron pairs. It is a **tetrahedral basis.** Geometry = **V-shaped.**

Exercises

Section 13.1

1. Indicate whether the bonds between the following would be primarily covalent, polar covalent, or ionic:

   a. O—H
   b. Cs—Cl
   c. H—Cl
   d. Br—Br

2. Calculate the energy of interaction for KCl if the internuclear distance is 0.314 nm.

3. Calculate the energy of interaction (in kJ/mole) between Ag^+ and Br^- if the internuclear distance of AgBr is 0.120 nm.

Section 13.2

4. Using a periodic table, order the following from lowest to highest electronegativity.

   a. Fr, Mg, Rb
   b. B, Al, C, N
   c. P, As, Ga, O
   d. Cl, S, P

5. Using the periodic table of elements, place the following in order from the lowest to the highest electronegativity:

   F, Nb, N, Si, Rb, Ca, Pt

6. Using Figure 13.3 in your textbook, calculate the difference in electronegativity (Δ) for each of the following bonds:

   a. Cl—Cl
   b. K—Br
   c. Fe—O
   d. H—O
   e. S—H

7. Place the following in order of increasing polarity:

   NaBr, I_2, H_2O, MnO_2, CN^-

8. Which of the following molecules contain polar covalent bonds? List in order of increasing bond polarity. (Use Figure 13.3 in your textbook.)

   O_3, P_8, NO, CO_2, CH_4, H_2S

9. How will the charge be distributed on each of the following molecules: HF, NO, CO, and HCl?

10. Why is it that BeF_2 is ionic, and $BeCl_2$ is covalent?

Section 13.3

11. Determine the orientation of the dipole of the following, if any.

    a. $AlCl_3$ (planar with aluminum atom at the center)
    b. CH_3F (tetrahedral with carbon at the center)
    c. N_2O (linear with N—N—O structure)
    d. $AgCl_4$ (planar molecule, silver atom at center, Ag—Cl bonds 90° apart)

12. Which of the molecules in Problem 11 contain one or more polar bonds?

13. Which of the following molecules would you expect to have a dipole moment of zero? Describe the dipole orientation of the other two molecules.

a. KI
b. CF_4 (tetrahedral structure)
c. H_2Se (bent structure)

Section 13.4

14. Determine the most stable ion for each of the following atoms, and indicate which element they would be isoelectronic with if they lost or gained electrons:

a. O
b. Be
c. I
d. Te
e. Na

15. List four ions that are isoelectronic with argon and have charges from −2 to +2. Arrange these in order of increasing ionic radius.

16. Determine the formula of the binary compound formed from the following sets of atoms.

a. Ca and O
b. K and Cl
c. Rb and S
d. Ba and P

17. Predict formulas for the following binary ionic compounds.

a. Mg and N
b. Na and F
c. Ca and S
d. Sr and Te

18. Using shorthand notation, list the core electron configurations for the ions in the compounds in Problem 17.

19. Place the following in order of increasing ionic size. Use the shorthand notation to list the core electron configuration for each of the ions.

a. Ba^{2+}, Te^{2-}, Cs^{+}, I^{-}
b. Rb^{+}, S^{2-}, O^{2-}, K^{+}

20. List four ions that are isoelectronic to Kr. Arrange these in order of increasing ionic radius.

21. Place the following in order of increasing ionic size. Use the shorthand notation to list the core electron configuration for each of the ions.

a. Cl^{-}, F^{-}, Sr^{2+}, Ca^{2+}
b. Na^{+}, Mg^{2+}, Li^{+}, Be^{2+}

Section 13.8

22. Using the bond energy values listed in Table 13.6 of your text, calculate the ΔH for the following reactions:

a. $2H_2(g) + O_2(g) \rightarrow 2H_2O$
b. $2C_2H_6(g) + 7O_2(g) \rightarrow 4CO_2(g) + 6H_2O(g)$
c. $HCN(g) + 2H_2(g) \rightarrow CH_3NH_2(g)$

23. Use bond energy values from Table 13.6 in your textbook to calculate ΔH for the following reactions:

a. $H—C\equiv C—H(g) + H_2(g) \rightarrow CH_2{=}CH_2(g)$
b. $2CH_4(g) + 3O_2(g) \rightarrow 2CO(g) + 4H_2O(g)$
c. $N_2(g) + 2H_2(g) \rightarrow N_2H_4(g)\ (NH_2—NH_2)$

24. Compare the values obtained in parts a and b of Problem 23 to ΔH values calculated from ΔH_f° data in Appendix 4 in your textbook.

25. Calculate the enthalpy of reaction, ΔH°, for the following reaction. Use the enthalpies of formation found in Appendix 4 of your textbook.

$$H_2(g) + C_2H_4(g) \rightarrow C_2H_6(g)$$

Section 13.10

26. Draw Lewis dot structures for the following atoms, ions, or molecules:

a. Sr
b. Br^-
c. ICN
d. Ga
e. $GaCl_4^-$
f. P^{3-}
g. NH_2^-
h. CSe_2

27. Draw Lewis structures for the following:

a. H^+
b. C
c. P
d. P^{5+}
e. Cl^-

28. Draw Lewis structures for the following:

a. AsF_3
b. O_3
c. H_3O^+
d. BH_4^-
e. NH_4^+
f. O_2

Section 13.11

29. Draw the remaining resonance forms for N_2O_4.

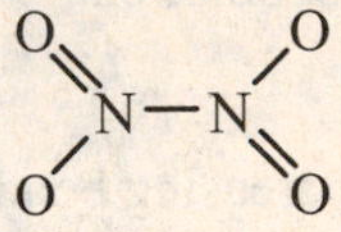

30. Draw a Lewis structure and any resonance forms of benzene, C_6H_6. (Benzene consists of a ring of six carbon atoms with one hydrogen bonded to each carbon.)

Section 13.12

31. Assign formal charges to each of the labeled atoms.

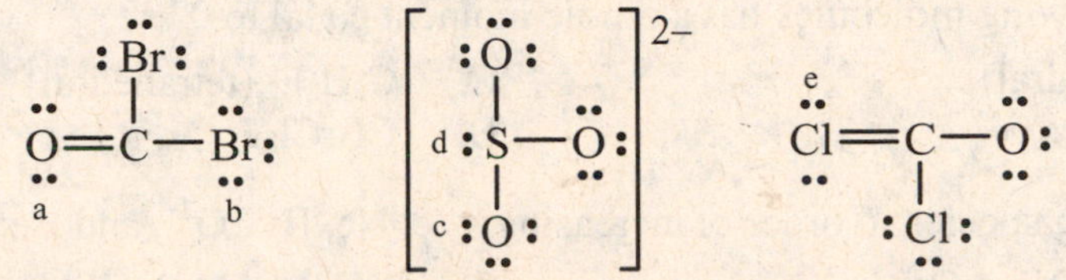

32. Draw Lewis dot structures for the following:

a. BCl_3
b. AsF_5
c. BrO_3^-
d. S_2F_{10} (contains an S—S bond)

33. Draw Lewis dot structures for the following:

a. $SbCl_3$
b. AlF_6^{3-}
c. PCl_5

Section 13.13

34. Predict the structure of each of the following molecules or ions:

a. SeF_6
b. N_2O
c. ClF_4^+
d. ClO^-
e. CF_3Cl (carbon is the central atom)

35. Using the VSEPR model, determine the molecular geometry for each of the following molecules:

a. SCl_4
b. H_2Se
c. IF_4^-
d. $SnCl_5^-$
e. $TlCl_2^+$

36. Which of the molecules or ions in Problem 34 contain polar covalent bonds? Are polar? For the polar species, which direction is the net dipole?

37. Discuss the nature of the bonding in each of the following. Indicate the number of sigma and pi bonds, the geometry, and the type of hybridization expected.

a. CO_2
b. $COCl_2$
c. H_2CO_2
d. HCN

Multiple-Choice Self-Test

1. The bond in RbF is:

A. covalent B. molecular C. polar covalent D. ionic

2. Which of the following bonds do you expect to be polar covalent?

A. H—N B. H—H C. Cs—F D. H—O

3. Which of the following bonds is the most polar one?

A. H—O B. Cs—Cl C. N—O D. C—H

4. Order the following bonds in order of decreasing bond polarity:

Ca—O, Ca—Cl, P—Cl, Fe—O, B—O, N—O

A. N—O > P—Cl > B—O > Fe—O > Ca—Cl > Ca—O
B. Ca—Cl > P—Cl > Ca—O > Fe—O > B—O > N—O
C. Ca—O > Ca—Cl > Fe—O > B—O > P—Cl > N—O
D. Fe—O > Ca—O > B—O > N—O > Ca—Cl> P—Cl

5. Which of the following molecules has a dipole moment equal to 0?

A. SiO_4 (tetrahedral)
B. H_2O (bent)
C. $C_2H_2F_2$ (tetrahedral)
D. $CBrCl_2F$

6. Place the following species in order of increasing size: Ne, B^{3+}, O^{2-}, and Be^{2+}.

A. $B^{3+} < Be^{2+} < Ne < O^{2-}$
B. $Ne < B^{3+} < Be^{2+} < O^{2-}$
C. $O^{2-} < Ne < Be^{2+} < B^{3+}$
D. $Ne < O^{2-} < B^{3+} < Be^{2+}$

7. Select the crystal that would have the largest lattice energy. Assume the internuclear distance is the same in all these crystals.

A. NaCl B. KCl C. K_2S D. CaO

8. The reaction of hydrogen with fluorine gas is highly exothermic (releases a high degree of energy). Calculate the F—F bond energy knowing that H—H = 432 kJ/mol, H—F = 565 kJ/mol, and $\Delta H = -543$ kJ.

$$H_2(g) + F_2(g) \rightarrow 2HF(g)$$

A. 155 kJ/mol B. 543 kJ/mol C. 698 kJ/mol D. 1019 kJ/mol

9. A truck uses propane (C_3H_8) to power its engine. Calculate how much heat will be released when 5 moles of propane are burned, knowing that the reaction of propane with oxygen gas produces carbon dioxide and water.

A. 7330 kJ B. 25 kJ C. 10,000 kJ D. 4784 kJ

10. Chlorine trifluoride is prepared by reacting chlorine gas with fluorine gas. The heat of the reaction is −803 kJ/mol of chlorine reacted. Calculate the Cl—Cl bond energy.

A. 1091 kJ/mol B. 155 kJ/mol C. 238 kJ/mol D. 50 kJ/mol

11. How many of the six valence electrons in oxygen are usually used in covalent bonding?

A. 4 B. 3 C. 6 D. 2

12. In the $POCl_3$ molecule, how many double bonds are there? How about single bonds?

A. 1 and 3 B. 4 and 1 C. 2 and 1 D. 1 and 2

13. Which one of the following molecules possesses a triple bond?

A. SF_4 B. PCl_5 C. C_2H_2 D. C_2H_6

14. Which one of the following molecules does not possess a double bond?

A. C_2F_4 B. $C_2H_4F_2$ C. OCH_2 D. HOCOCl

15. Which one of the following molecules contains a central atom that violates the octet rule?

A. SF_4 B. COF_2 C. $Si(OH)_4$ D. PBr_3

16. Calculate the formal charge on chlorine in $(ClO_4)^-$

A. −1 B. +3 C. +6 D. +4

Answers to Exercises

1. a. polar covalent
   b. ionic
   c. polar covalent
   d. covalent

2. -7.36×10^{-19} J

3. $E = -1160$ kJ/mole

4. a. Fr, Rb, Mg
   b. Al, B, C, N
   c. Ga, As, P, O
   d. P, S, Cl

5. Rb > Ca > Nb > Si > Pt > N > F

Rb	Ca	Nb	Si	Pt	N	F
0.8	1.0	1.6	1.8	2.2	3.0	4.0

6. a. 0 b. 2.0 c. 1.7 d. 1.4 e. 0.4

7. $I_2 < CN^- < H_2O < NaBr < MnO_2$
0 0.5 1.4 1.9 2.0

8. O_3, $P_8 < H_2S$, $CH_4 < NO < CO_2$

9. HF; NO; CO; HCl
+ − + − + − + −

10. The difference in the electronegativities between Be and F are far higher than Be and Cl.
$BeCl_2$: 3.0 − 1.5 = 1.5; covalent
BeF_2: 4.0 − 1.5 = 2.5; ionic

11. a. no dipole
b. negative toward F
c. negative toward O
d. no dipole

12. all

13. b. The opposing bond polarities in a tetrahedral structure cancel out. Thus CF_4 has no dipole moment. KI is a binary ionic compound that has a negative dipole toward I. Selenium will have a partial negative charge as its electronegativity is greater than that of hydrogen. Thus the resulting dipole moment of H_2Se would be orientated as shown:

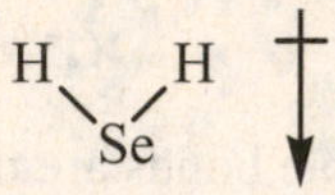

14. a. O^{2-}, isoelectronic with neon
b. Be^{2+}, isoelectronic with helium
c. I^-, isoelectronic with xenon
d. Te^{2-}, isoelectronic with xenon
e. Na^+, isoelectronic with neon

15. Ca^{2+}, K^+, Cl^-, S^{2-}

16. a. CaO
b. KCl
c. Rb_2S
d. Ba_3P_2

17. a. Mg_3N_2
b. NaF
c. CaS
d. SrTe

18. a. [Ne], [Ne]
b. [Ne], [Ne]
c. [Ar], [Ar]
d. [Kr], [Xe]

19. a. $Ba^{2+} < Cs^+ < I^- < Te^2$ all can be written as [Xe]
b. $K^+ < O^{2-} < Rb^+ < S^{2-}$ K^+ = [Ar], O^{2-} = [Ne], Rb^+ = [Kr], S^{2-} = [Ar]

20. Ne < Ar < Kr < Xe < Rn

21. a. $Ca^{2+} < Sr^{2+} < F^- < Cl^-$
Ca^{2+} = [Ar], Sr^{2+} = [Kr], F^- = [Ne], Cl^- = [Ar]
b. $Be^{2+} < Li^+ < Mg^{2+} < Na^+$
Be^{2+} = [He], Li^+ = [He], Mg^{2+} = [Ne], Na^+ = [Ne]

22. a. −509 kJ b. −2881 kJ c. −158 kJ

23. a. −169 kJ b. −1091 kJ c. +81 kJ

24. a. 6 kJ difference b. 7 kJ difference

25. $\Delta H_f^\circ = -136.7$ kJ/mole

26. a. · Sr ·

b. $\left[:\ddot{Br}:\right]^-$

c. :Ï—C≡N:

d. · Ga ·

e. $\left[\begin{array}{c} :\ddot{Cl}: \\ :\ddot{Cl}—Ga—\ddot{Cl}: \\ :\ddot{Cl}: \end{array}\right]^-$

f. $\left[:\ddot{P}:\right]^{3-}$

g. $\left[H—\ddot{N}—H\right]^-$

h. $\ddot{Se}=C=\ddot{Se}$

27. a. $[H]^+$

b. · Ç ·

c. : Ṗ ·

d. $[P]^{+5}$

e. $\left[:\ddot{Cl}:\right]^-$

28. a. $:\ddot{F}—\ddot{As}—\ddot{F}:$ with $:\ddot{F}:$ below

b. $\ddot{O}=\ddot{O}—\ddot{O}:$

c. $\left[\begin{array}{c} H—\ddot{O}—H \\ | \\ H \end{array}\right]^+$

d. $\left[\begin{array}{c} H—B—H \\ | \\ H \end{array}\right]^-$ with H above

e. $\left[\begin{array}{c} H \\ | \\ H—N—H \\ | \\ H \end{array}\right]^+$

f. $\ddot{O}=\ddot{O}$

29. [resonance structures of N₂O₄]

30. [benzene resonance structures]

31. a. 0
b. 0
c. −1
d. +1
e. +1
f. 0

32. a. $:\ddot{Cl}—B—\ddot{Cl}:$ with $:\ddot{Cl}:$ below

b. [AsF₅ structure]

c. $\left[:\ddot{O}—\ddot{Br}—\ddot{O}: \text{ with } :\ddot{O}: \text{ below}\right]^-$

d. [S₂F₁₀ structure]

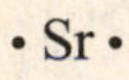

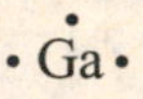

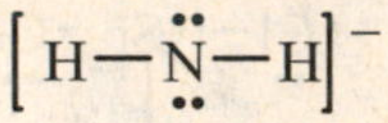

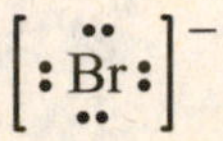

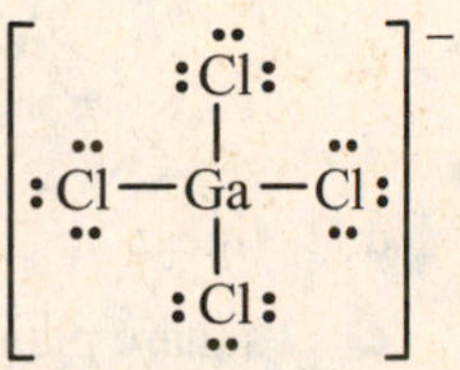

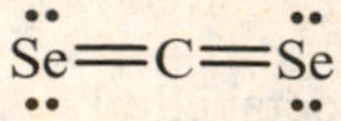

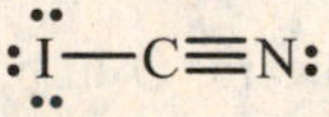

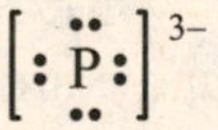

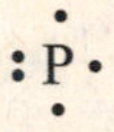

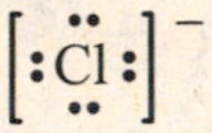

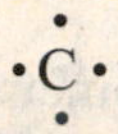

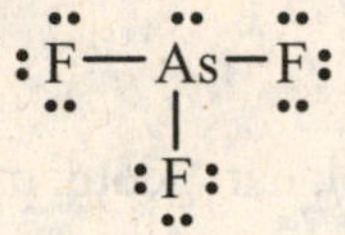

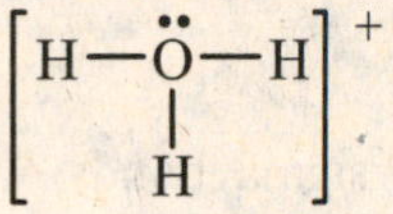

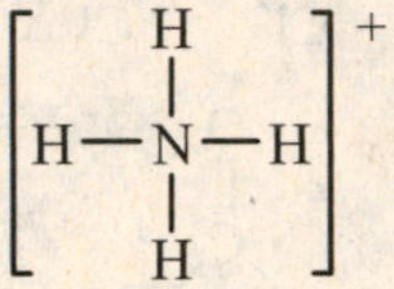

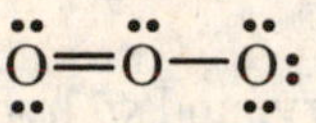

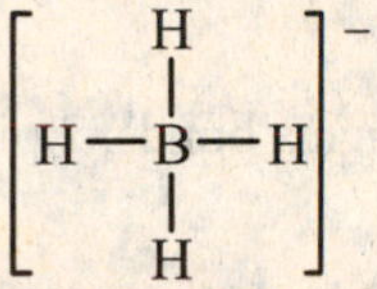

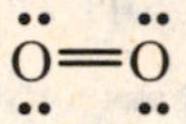

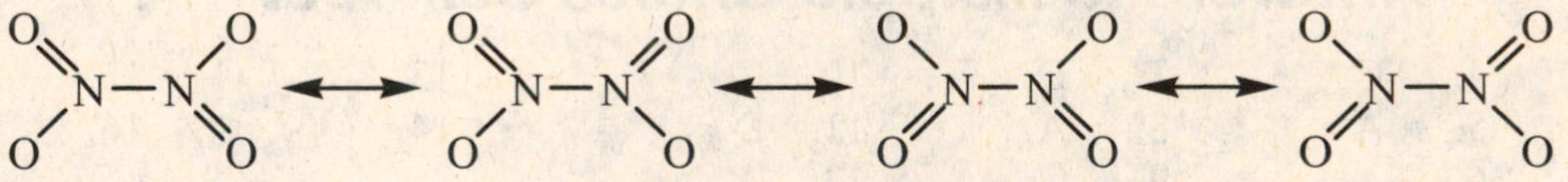

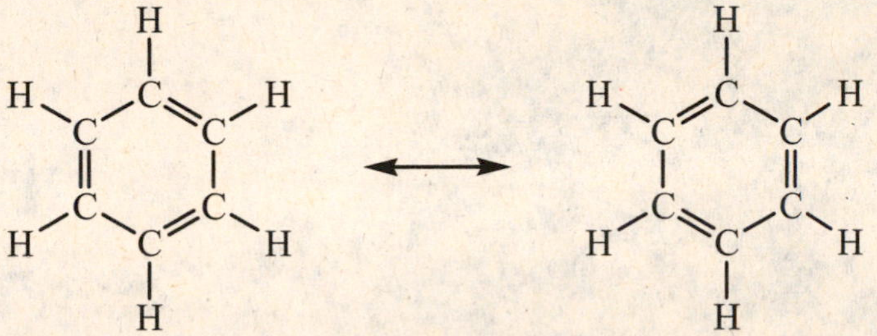

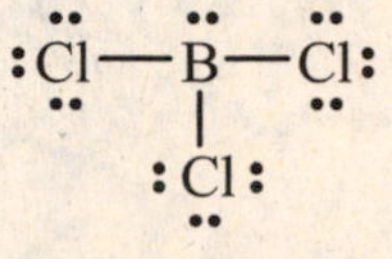

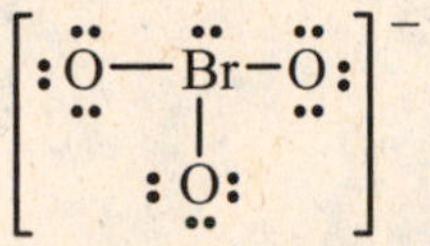

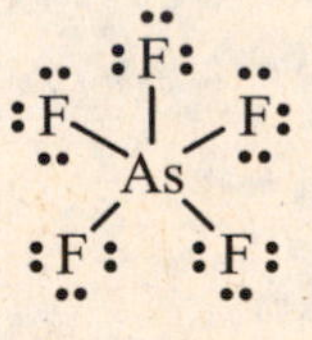

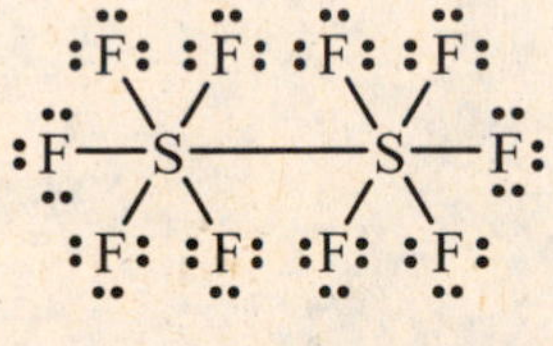

33. a. 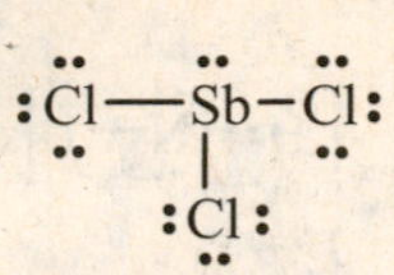 b. 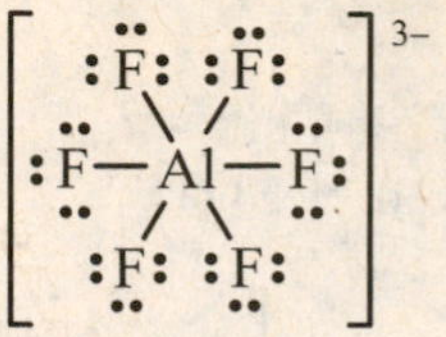 c. 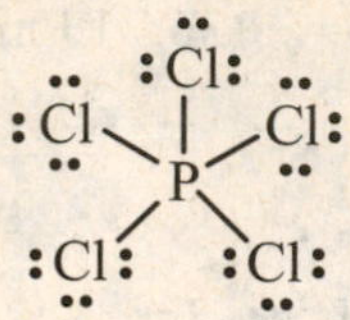

34. a. octahedral
    b. linear
    c. see-saw
    d. linear
    e. tetrahedral

35. a. see-saw
    b. bent
    c. square planar
    d. trigonal bipyramidal
    e. linear

36. All contain polar covalent bonds; b, c, d, and e are polar. Negative is toward: O in b; equatorial fluorine in c; O in d; fluorine in e.

37. a. O=C=O 2 sigma bonds, 2 pi bonds, linear geometry, and *sp* hybridization

    b. 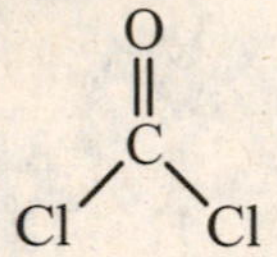 Carbon has 3 sigma bonds, 1 pi bond, trigonal planar geometry, and sp^2 hybridization

    c. 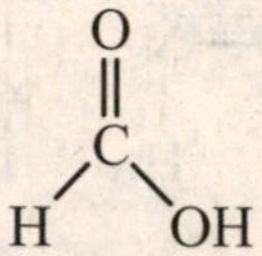 Carbon has 3 sigma bonds, 1 pi bond, trigonal planar geometry, and sp^2 hybridization

    d. H—C≡N Carbon has 2 sigma bonds, 2 pi bonds, linear geometry, and *sp* hybridization

Answers to Multiple-Choice Self-Test

1. D	4. C	7. D	10. C	13. C	15. A
2. D	5. A	8. A	11. D	14. B	16. B
3. B	6. A	9. C	12. A		

CHAPTER 14

Covalent Bonding: Orbitals

The Bottom Line: Chapter 14

Chapter 13 dealt with ways of representing structures on paper and with predicting the three-dimensional structure that covalent molecules should have. This chapter reviews the two most important models that attempt to explain covalent molecular and ionic structure and shape, the **Localized Electron (LE) Model** and the **Molecular Orbital (MO) Model**.

14.1 Hybridization and the Localized Electron Model

Your textbook points out the difficulty with assuming that the four C—H bonds in methane (CH_4) are made by interacting the 2*s* and 2*p* orbitals of carbon (valence orbitals) with the 1*s* orbitals of hydrogen. It says that the three 2*p*-1*s* C—H interactions should be located at 90° to one another. This does not agree with the observed bond angles of 109.5°. Recall that these bond angles **minimize electron-electron repulsions** in methane.

To work around this difficulty, the concept of **hybrid orbitals** is suggested. By combining the **one 2*s*** and the **three 2*p*** orbitals of carbon, **four hybrid** sp^3 orbitals are formed. These orbitals are equivalent, and that is consistent with the observed presence of four equivalent C—H bonds. Figure 14.6 in your textbook displays the interaction between the sp^3 hybrid orbitals of carbon and the 1*s* orbitals of the hydrogens.

The bottom line is this: When the VSEPR model indicates that you have a **tetrahedron** as the basis of your structure (4 effective electron pairs), the atom is sp^3 **hybridized**.

Example 14.1 A - sp³ Hybridization

Describe the bonding in the water molecule using the localized electron model.

Solution

We must establish the VSEPR structure. This is done by drawing the Lewis structure and determining the number of effective electron pairs around the central atom.

The Lewis structure of H_2O is (see Sections 13.10 and 13.13)

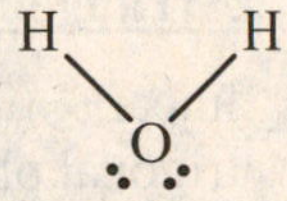

Four effective electron pairs around the oxygen atom indicate a **tetrahedral** basis leading to a V-shaped structure. Therefore the oxygen is sp^3 hybridized. Two sp^3 orbitals are occupied with 1*s* electrons from hydrogen. The other two sp^3 orbitals are occupied by lone pairs.

In molecules or ions with a **trigonal planar** configuration (3 effective electron pairs around an atom), sp^2 **hybridization** occurs. With this hybridization, an *s* orbital and two *p* orbitals are used. That leaves an **unhybridized** (unchanged) ***p* orbital perpendicular** to the sp^2 plane.

Look at Figures 14.11 and 14.12 in your textbook.

- Bonds formed from the overlap of orbitals in the plane between two atoms are called **sigma (σ) bonds**.
- Bonds formed by the overlap of unhybridized *p* orbitals (above and below the center plane) are called **pi (π) bonds**.
- A **single bond** is a **σ bond**.
- A **double bond** consists of **one σ and one π bond**.
- A **triple bond** consists of **one σ bond and two π bonds.**

Example 14.1 B - Sigma And Pi Bonds

How many σ bonds are there in the commercial insecticide Sevin® shown below? How many π bonds?

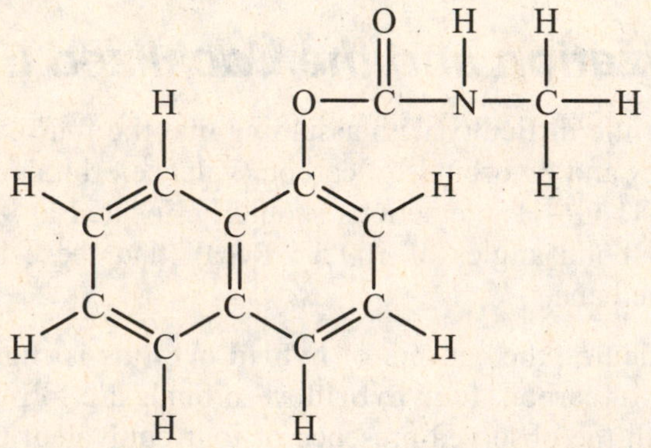

Solution

There are **27 σ bonds** (single bonds and one of the bonds in a double bond).
There are **6 π bonds** (the other bond in a double bond).

Your textbook goes over a variety of different hybridization schemes, all of which center on the idea that **the hybridization of the orbitals of an atom depends on the total number of effective electron pairs around it**. In order to determine the hybridization of an atom, it is essential that you can figure out its VSEPR structure.

The following table summarizes effective electron pairs around an atom with hybridization. Remember that, for purposes of the VSEPR model, **double and triple bonds count as only one effective electron pair**

Effective Electron Pairs Around an Atom	Arrangement	Hybridization
2	linear	sp
3	trigonal planar	sp^2
4	tetrahedral	sp^3
5	trigonal bipyramid	dsp^3
6	octahedral	d^2sp^3

Keep in mind that a ligand can have a hybridization different than that of the central atom. Each atom in a molecule must be considered separately based on the Lewis and VSEPR structures of that molecule.

Example 14.1 C - Practice With Hybrid Orbitals

Give the hybridization, and predict the geometry of each of the central atoms in the following molecules or ions.

a. IF_2^+

b. OSF_4 (sulfur is the central atom)

c. SiF_6^{2-}

d. HCCH (work with both carbons)

Solution

a. Lewis structure:

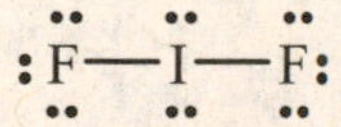

All three atoms have 4 effective electron pairs. They are **all sp^3 hybridized**. The VSEPR structure has a tetrahedral basis. Because the central atom (iodine) has two bonding pairs, it will take on a **V-shape**, but the bond angle will be smaller than 109.5°.

b. Lewis structure:

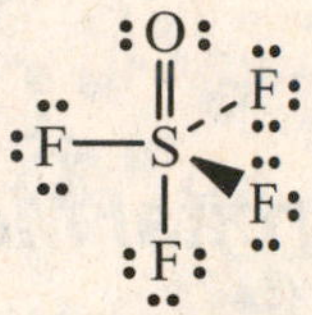

Sulfur has 5 effective electron pairs. The VSEPR structure is a trigonal bipyramid. **Sulfur is dsp^3 hybridized**. Each of the fluorines has 4 effective electron pairs.

c. Lewis structure:

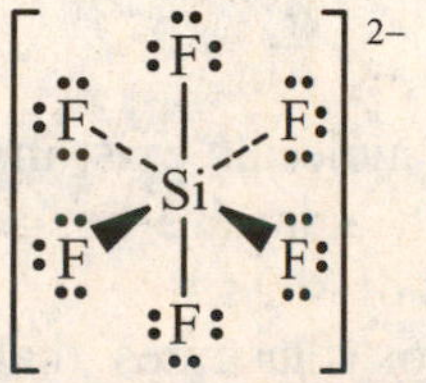

Silicon has 6 effective electron pairs. The VSEPR structure is octahedral. **Silicon is d^2sp^3 hybridized**. Each of the fluorines has 4 effective electron pairs.

d. Lewis structure: **H—C≡C—H**

Each carbon has 2 effective electron pairs. The VSEPR structure is linear. **Each carbon is *sp* hybridized**. Hydrogen atoms bond using 1*s* orbitals. The orbitals are unhybridized.

Example 14.1 D - Summing It All Up

Answer the following questions regarding aspartame (NutraSweet).

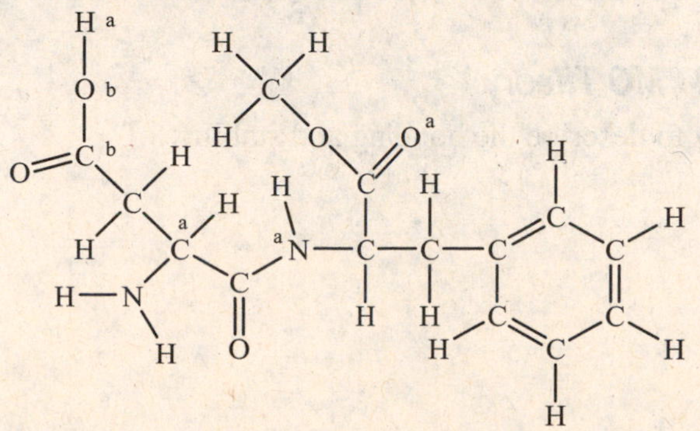

a. How many σ bonds are in the molecule?
b. How many π bonds?
c. What is the hybridization on carbon a? Carbon b?
d. What is the hybridization on nitrogen a? Oxygen a?
e. What is the C_b—O_b—H_a bond angle?

Solution

You must remember to complete the octets in this "shorthand" Lewis structure. Lone pairs are often "assumed," so always be on the lookout.

a. 39 σ bonds
b. 6 π bonds
c. sp^3, sp^2 (3 effective electron pairs)
d. sp^3 (remember the "assumed" lone pair to complete the octet), sp^2 (3 effective electron pairs, including the two to complete the octet)
e. O_b is sp^3 hybridized (Complete the octet!); therefore, the angle is based on a tetrahedron, with two lone pairs compressing the C—O—H bond angle to 104.5°.

14.2 The Molecular Orbital Model

The localized electron model does an excellent job of predicting and justifying molecular shapes. It does not deal with molecules with unpaired electrons. It also neglects bond energies. The molecular orbital (MO) model also gives a view of electrons in a molecule that allows us to get a clearer understanding of what we had called resonance.

Key Ideas of the MO Model

1. All the valence electrons in a molecule exist in a set of **molecular orbitals** of a given energy. The valence electrons of each atom are not acting independently, but rather act with other valence electrons to form a set of MO's.
2. Figure 14.27 in your textbook illustrates that there are **bonding** and **antibonding** MO's. Bonding results in **lower energy** than if no interaction occurred. Antibonding results in higher energy.
3. A molecule will be stable if there is more bonding than antibonding interaction.
4. **Bond order (BO)** is a measure of net bonding interactions.

$$BO = \frac{\#\text{ bonding electrons} - \#\text{ antibonding electrons}}{2}$$

5. BO must be greater than 0 for a stable molecule to form.
6. The higher the BO, the stronger the bond.

Your text goes over the bonding in H_2, H_2^-, and He_2. Let's try two more.

Example 14.2 - MO Theory

Use MO theory to describe the bonding and stability of

a. H_2^{2-}
b. H_2^+

Solution

We must fill in the molecular orbital energy-level diagram for each of the species.

a. H_2^{2-} has 4 electrons. They will fill pairwise, and with opposite spins, the σ_{1s} and σ_{1s}^* orbitals.

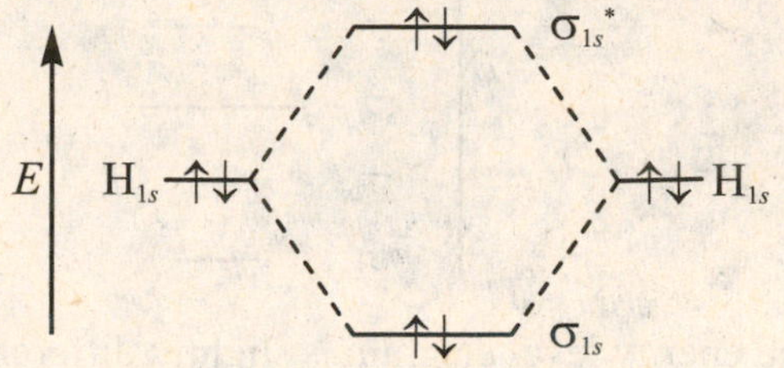

$$\textbf{Bond order} = \frac{2 \text{ bonding electrons} - 2 \text{ antibonding electrons}}{2} = \mathbf{0}$$

We would not expect H_2^{2-} to be a stable ion.

b. H_2^+ has 1 electron. We must again fill in our MO energy-level diagram.

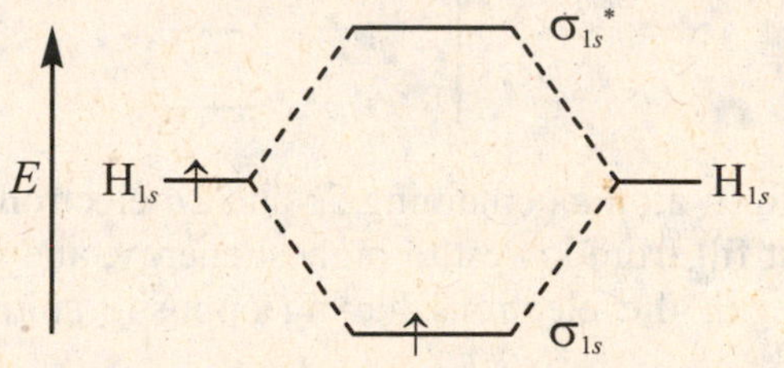

$$\textbf{Bonding order} = \frac{1 \text{ bonding electron} - 0 \text{ antibonding electrons}}{2} = ½$$

We would expect H_2^+ to form, but it would not be much more stable than two separate hydrogens.

14.3 Bonding in Homonuclear Diatomic Molecules

In the last section we saw how to use the MO model to assess the stability of diatomic molecules with 1*s* orbitals. In this section we will expand the discussion to diatoms containing 2*s* and 2*p* orbitals. Bonds can be formed when atomic orbitals overlap in space. This is not possible with 1*s* orbitals in period 2 elements because each of the orbitals is too close to its own nucleus. Only the 2*s* and 2*p* orbitals (containing the valence electrons) can participate in bond formation.

We know that σ bonds form between nuclear centers, and these MO's can be formed using 2*s* orbitals or $2p_x$ orbitals. In addition, π orbitals can be formed by the overlap of both **$2p_y$ or $2p_z$ orbitals**. Also σ and π antibonding orbitals exist. (See Figure 14.36 in your textbook.)

Your textbook uses **paramagnetism** and **diamagnetism** to prove the relative positions of MO energy levels. **Paramagnetism** indicates **unpaired electrons** in a substance. This causes the substance to be **attracted to** a magnetic field. **Diamagnetism** indicates **paired electrons** in a substance. This causes the substance to be **repelled from** the magnetic field. Paramagnetism is a much stronger effect than diamagnetism. It will dominate if both effects are present. (The substance will be attracted to a magnetic field.)

Your textbook describes how para- and diamagnetism help explain the order of molecular orbital energy-levels. This order for valence electrons in, for example, carbon and nitrogen is

$$
\begin{array}{ll}
\sigma_{2p}^{*} & \text{—} \\
\pi_{2p}^{*} & \text{— —} \\
\sigma_{2p} & \text{—} \\
\pi_{2p} & \text{— —} \\
\sigma_{2s}^{*} & \text{—} \\
\sigma_{2s} & \text{—}
\end{array}
$$

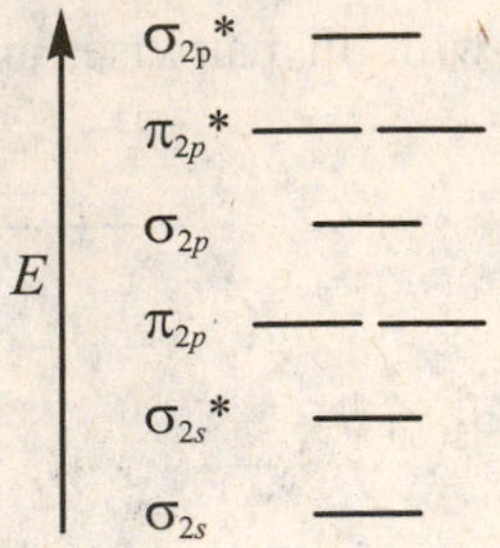

For oxygen and fluorine, the energy level diagram is slightly different as shown:

$$
\begin{array}{ll}
\sigma_{2p}^{*} & \text{—} \\
\pi_{2p}^{*} & \text{— —} \\
\pi_{2p} & \text{— —} \\
\sigma_{2p} & \text{—} \\
\sigma_{2s}^{*} & \text{—} \\
\sigma_{2s} & \text{—}
\end{array}
$$

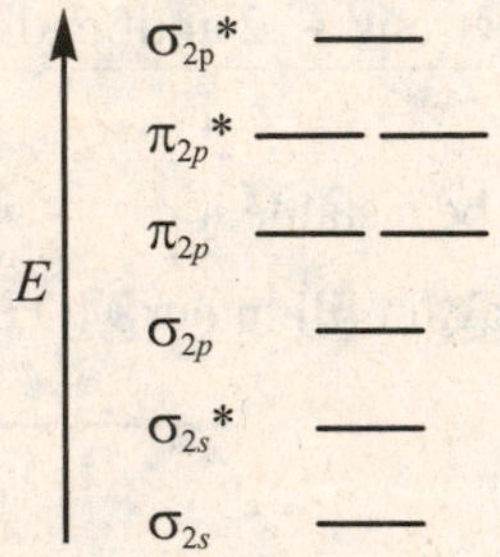

Filling molecular orbitals from atoms containing $2s$ and $2p$ electronic orbitals works the same as when filling from $1s$ orbitals. Just fill from lowest to highest energy, and remember to fill degenerate orbitals separately, then pairwise with the electrons having opposite spins. Let's try the following example together.

Example 14.3 A - σ And π Molecular Orbitals

Determine the following regarding F_2^-:

- electron configuration
- bond order
- para- or diamagnetism
- the bond energy relative to F_2

Solution

To get the **electron configuration,** we must fill in the molecular orbital energy-level diagram. We know that F_2^- has **15 valence electrons** (7 on F, 8 on F^-).

$$
\begin{array}{ll}
\sigma_{2p}^{*} & \uparrow \\
\pi_{2p}^{*} & \uparrow\downarrow \quad \uparrow\uparrow \\
\pi_{2p} & \uparrow\downarrow \quad \uparrow\uparrow \\
\sigma_{2p} & \uparrow\downarrow \\
\sigma_{2s}^{*} & \uparrow\downarrow \\
\sigma_{2s} & \uparrow\downarrow
\end{array}
$$

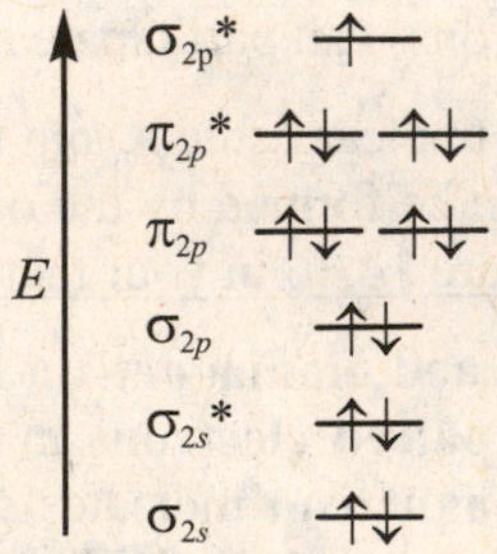

Electron Configuration (valence only) $F_2^- = (\sigma_{2s})^2(\sigma_{2s}*)^2(\sigma_{2p})^2(\pi_{2p})^4(\pi_{2p}*)^4(\sigma_{2p}*)^1$

$$\textbf{Bond Order} = \frac{8 \text{ bonding electrons} - 7 \text{ antibonding electrons}}{2} = \mathbf{1/2}$$

F_2^- has one unpaired electron. It would be expected to be paramagnetic. Molecular fluorine, F_2, has a bond order of one and is therefore more stable than F_2^-. Its bond energy would be expected to be higher.

Example 14.3 B - Practice With σ And π Orbitals

Determine the electron configuration and bond orders for S_2^{2-} and Cl_2^{2+}. If they can exist, discuss their magnetism.

Solution

S_2^{2-} has **14 valence electrons** (6 + 6 from the sulfur atoms and a −2 charge).
Cl_2^{2+}has **12 valence electrons** (7 +7 from the chlorine atoms and a +2 charge).

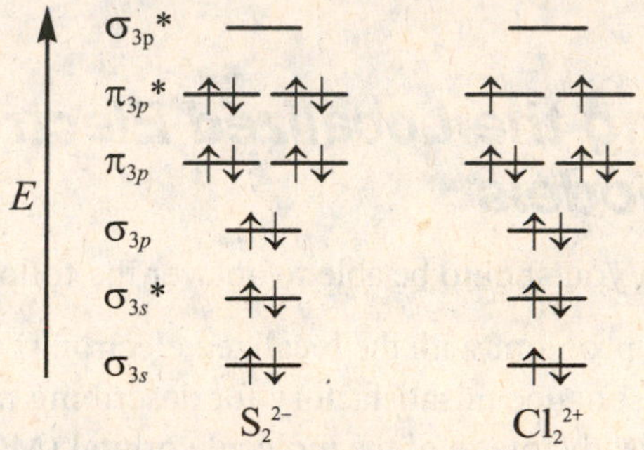

Electron Configuration of $S_2^{2-} = (\sigma_{3s})^2(\sigma_{3s}*)^2(\sigma_{3p})^2(\pi_{3p})^4(\pi_{3p}*)^4$
Electron Configuration of $Cl_2^{2+} = (\sigma_{3s})^2(\sigma_{3s}*)^2(\sigma_{3p})^2(\pi_{3p})^4(\pi_{3p}*)^2$

Bond Order of S_2^{2-} **= (8 − 6)/2 = 1**
Bond Order of Cl_2^{2+} **= (8 − 4)/2 = 2**

S_2^{2-} **would be diamagnetic.**
Cl_2^{2+} **would be paramagnetic.**

14.4 Bonding in Heteronuclear Diatomic Molecules

The MO model **for homonuclear molecules works well for describing the bonding in atoms** adjacent to one another **on the periodic table. The model breaks down with two** very different **atoms.**

Let's reinforce our understanding with practice on some more homonuclear species.

Example 14.4 - Practice With The MO Model

Using the MO model, describe the bonding, magnetism, and relative bond energies in the following species:

O_2, O_2^-, and O_2^{2-}

Solution

Valence Electrons

O_2 has 12

O_2^- has 13

O_2^{2-} has 14

Bond Orders

$O_2 = (8 - 4)/2 = 2$

$O_2^- = (8 - 5)/2 = 1.5$

$O_2^{2-} = (8 - 6)/2 = 1$

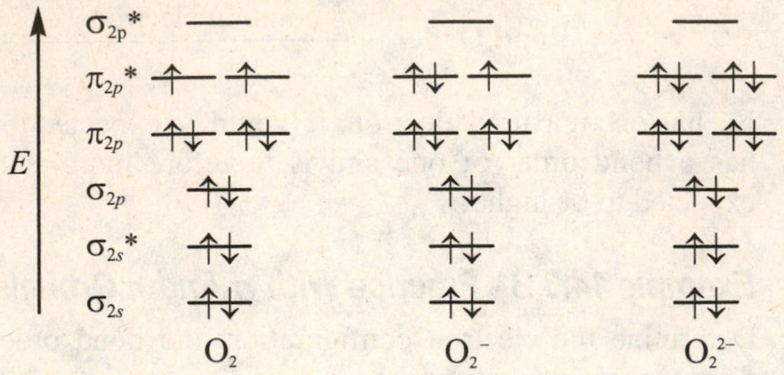

Both O_2 and O_2^- are expected to be paramagnetic. We expect O_2^{2-} to be diamagnetic. The bond energy of O_2 is expected to be the highest, followed by O_2^-, with O_2^{2-} having the lowest.

In the heteronuclear MO description of HF, your textbook notes that the MO model accounts for the polarity in that molecule. This is a great strength of the model.

14.5 Combining the Localized Electron and Molecular Orbital Models

When you finish this section you should be able to answer the following review questions:

- What is the primary problem with the localized electron (LE) model?
- Why is the use of resonance unsatisfactory for describing molecular bonding?
- What is the major disadvantage of the molecule orbital (MO) model?
- What bond is depicted as shifting in LE resonance structures? Why is that important in our combination model?
- In summary, what are the advantages and disadvantages of the LE model? The MO model?

14.6 Orbitals: Human Inventions

The key difference between your chemistry course and the conventional general chemistry is expressed in this section. In the "mainstream" course, students learn about orbitals as devices for understanding electron energies around nuclei. In this section we learn that orbitals are mathematical tools, rather than true physical representations, that help us understand atomic and molecular properties. Please read the section and try your hand at the following questions:

1. Why do we use models if they occasionally give us misleading information?
2. Do orbitals exist? Explain your answer.
3. Why doesn't your textbook present more sophisticated models for atomic and molecular properties?
4. Is science "value neutral"? Why or why not (with regard to orbital calculations)?

14.7 Molecular Spectroscopy: An Introduction

Over the next four chapters your textbook presents a general overview of molecular spectroscopy. Keep in mind as you read these sections and deal with the questions below that there are many different energy ranges involved in spectroscopic analysis, and each range involves an important atomic or molecular transition.

1. How do electronic, rotational, and vibrational transitions differ?
2. What information does each allow us to learn about molecular structure?

14.8 Electronic Spectroscopy

The following questions will help you review the material in this section:

1. Quick opener: In what regions of the electromagnetic spectrum do electronic transitions occur?
2. What is a "conjugated molecule," and can you give some examples **other than that** discussed in the section?
3. What does the particle-in-a-box model have to do with the wavelength of light absorbed in a conjugated molecule?
4. Based on the discussion and on Figure 14.57 in the text, explain why carrots are orange.

14.9 Vibrational Spectroscopy

The following questions will help you review the material in this section:

1. Quick opener: In what region of the electromagnetic spectrum do vibrational transitions occur?
2. Your textbook reminds us that molecular and large-body ("classical") physics lead us to important differences in our view of the world. What is the **key difference** regarding vibrational energy levels, and how can this difference be used in structural determinations?
3. How is IR spectroscopy used in structural determination? Give some examples.
4. Here is an IR spectrum. Can you identify some of the bonds in the molecule (see Table 14.1)?

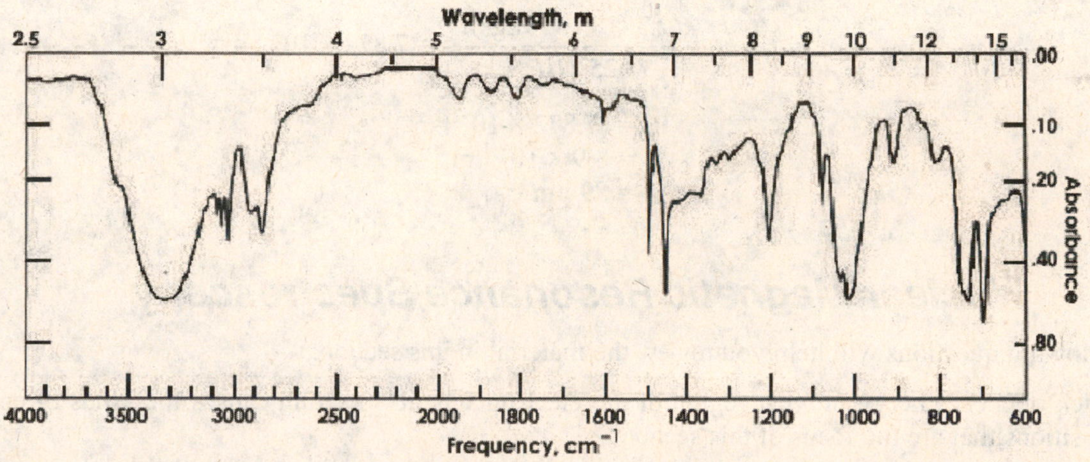

14.10 Rotational Spectroscopy

The following questions will help you review the material in this section:

1. Quick opener: In what region of the electromagnetic spectrum do rotational transitions occur?
2. How can the energy needed to promote a heteronuclear diatomic molecule to an excited rotational state be used to determine its average bond length?
3. How is the selection rule for rotational energy level changes different from that of, for example, electronic energy level changes? (You may need to revisit the quantum chemistry chapter to recall the selection rules for electronic transitions.)

Example 14.10 in your textbook goes through a calculation in which the bond length for $^{1}H^{35}Cl$ is determined. Let's run through the same type of calculation for $^{2}H^{19}F$.

Example 14.10 - Calculating Bond Length From Microwave Rotational Spectral Data

Below are the pertinent data from the NIST (National Institute of Standards and Technology) data files for the bond length calculation for $^{2}H^{19}F$.

J = 0 to 1
reduced mass, $\mu = 1.821$ amu
wavelength, $\lambda = 4.607 \times 10^{-4}$ m

Please calculate the bond length.

Solution

Following the rubric set forth in Example 14.10 in your textbook,

$$\frac{c}{\lambda} = 2B = \frac{2.998\times10^{8}\ \text{m s}^{-1}}{4.607\times10^{-4}\ \text{m}} = 6.507\times 10^{11}\ \text{s}^{-1}$$

$$B = 3.254\times10^{11}\ \text{s}^{-1}$$

$$I = \frac{h}{8\pi^2 B} = \frac{6.626\times10^{-34}\ \text{kg m}^2\ \text{s}^{-2}}{8\pi^2\,(3.254\times10^{11}\ \text{s}^{-1})} = 2.381\times10^{-47}\ \text{kg m}^2$$

$$\mu = 1.821\ \text{amu} \times \frac{1.661\times10^{-27}\ \text{kg}}{\text{amu}} = 3.025\times10^{-27}\ \text{kg}$$

(Remember that in the next step, 1 joule = 1 kg m^2 s^{-2})

$$R_e^{\,2} = \frac{I}{\mu} = \frac{2.381\times10^{-47}\ \text{kg m}^2}{3.025\times10^{-27}\ \text{kg}} = 7.872\times10^{-21}\ \text{m}^2$$

$$\mathbf{R_e} = (7.872\times10^{-21}\ \text{m}^2)^{1/2}$$
$$= 8.9\times10^{-11}\ \text{m}$$
$$= \mathbf{89\ pm}$$

14.11 Nuclear Magnetic Resonance Spectroscopy

The following questions will help you review the material in this section:

1. Quick opener: Energy in what region of the electromagnetic spectrum causes the kinds of energy transitions that are the focus of this section?
2. Your textbook notes that the nuclei don't actually spin in the traditional sense. Why do we call it nuclear spin?
3. Explain why the 3 hydrogen atoms on the methyl group in bromoethane (Figure 14.63 in your textbook) are said to be equivalent to each other, and what is the consequence of this equivalence for the NMR signal?
4. What causes a chemical shift in a signal?
5. How and why does the splitting pattern for the hydrogen atoms adjacent to a CH_3 group differ from the pattern adjacent to a CH_2 group?

6. Here is an NMR spectrum of diethyl ether. Can you identify which hydrogen atoms go with which peaks?

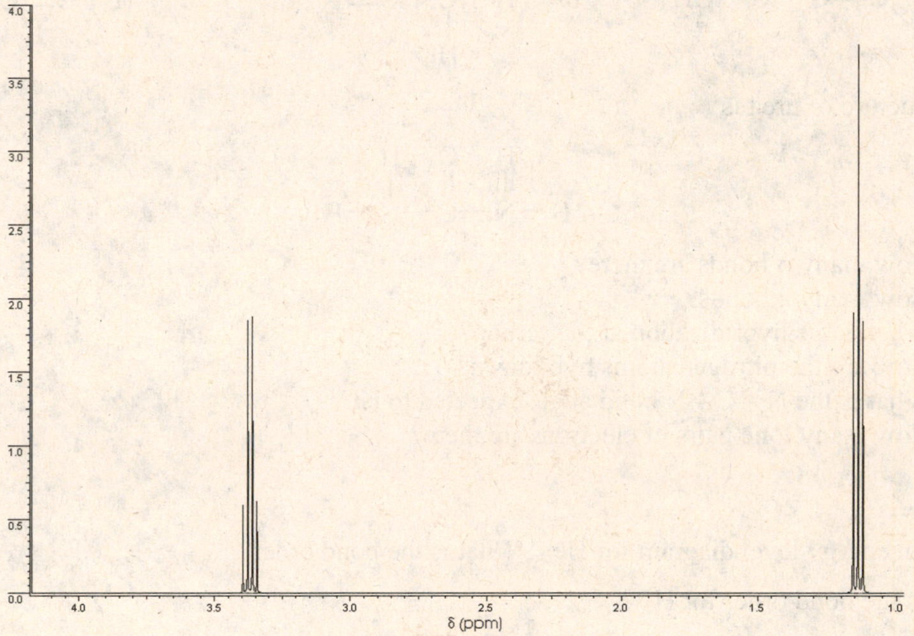

Exercises

Section 14.1

1. Predict the geometries of the following compounds:

a. SF_2
b. SF_4
c. XeF_2
d. XeF_4
e. IF_5
f. ClF_3

2. Predict the geometry about the indicated atom and identify the hybridization of each atom.

a. the two carbon atoms and the nitrogen atom of glycine

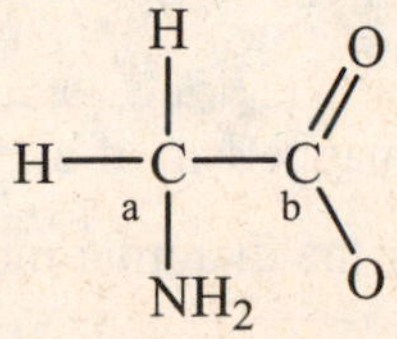

b. the carbon atom in CF_2Cl_2
c. the phosphorus atom in PCl_5
d. the nitrogen atom in NH_2^-

3. What geometry do the following hybrid bonds possess?

a. *sp*
b. sp^2
c. sp^3
d. dsp^3
e. d^2sp^3

4. Predict the type of hybrid orbital that the central atoms of each of the following compounds display.

a. SiH_4
b. H_3O^+
c. PCl_5
d. NCl_3
e. AsH_3
f. SiF_6
g. CH_3^+

5. Determine the number of sigma and pi bonds in each of the following:

a. $\begin{array}{c} H \\ \diagdown \\ C = C - C - OH \\ / \quad | \quad \| \\ H \quad H \quad O \end{array}$

b. $\begin{array}{c} H \\ | \\ H - C - C \equiv N \\ | \\ H \end{array}$

c. $\begin{array}{c} \qquad H \quad H \\ \qquad | \quad \diagup \\ H - C \equiv C - C = C \\ \qquad \diagdown \\ \qquad \qquad H \end{array}$

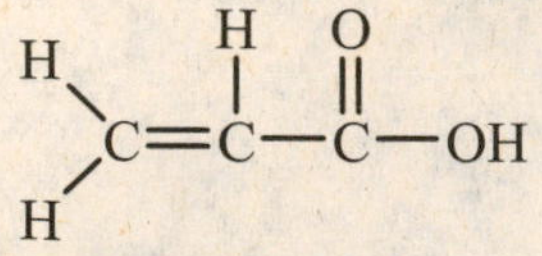

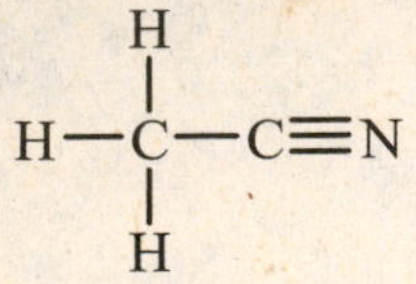

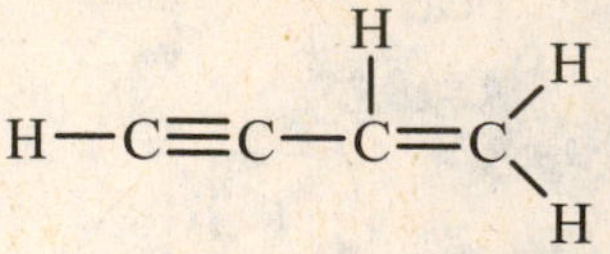

6. The structure of urea is

$$\begin{array}{c} H \quad O \quad H \\ | \quad \| \quad | \\ H - N - C - N - H \end{array}$$

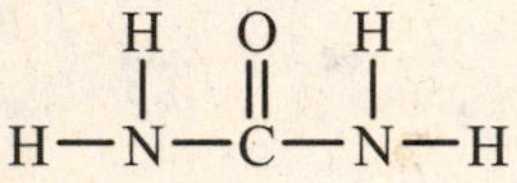

a. How many σ bonds are there?
b. How many π bonds?
c. What is the hybridization at the carbon?
d. How are the nitrogen atoms hybridized?
e. What is the N—C—N bond angle expected to be?
f. How many lone pairs of electrons are there?

Section 14.2

7. Draw an energy level diagram for He_2^+. What is the bond order?

8. What is the bond order for H_2^-?

9. The elements N, O, and F exist as diatomic molecules. Use MO theory to explain why Ne, the next element on the periodic table, does not exist as a diatomic molecule.

Section 14.3

10. Which of the following are paramagnetic species: Li_2, N_2, He_2^+, H_2, O_2, Be_2?

11. For N_2, N_2^+, N_2^-, and N_2^{2-}, use molecular orbitals to predict bond order and whether the species are diamagnetic or paramagnetic.

12. If liquid nitrogen were poured between the poles of a powerful magnet, would it "stick" to the poles or simply pour through the space between the poles? Why?

Section 14.4

13. What is the bond order and expected magnetism of carbon monoxide? Of CO^-?

14. Summarize the relative bond order for the diatomic molecules of the second period from boron to fluorine.

15. What is the bond order and expected magnetism of IBr?

16. Which of the following are diamagnetic: CN^-, CN, CN^+?

Multiple-Choice Self-Test

1. What hybridization describes square planar geometry?
 A. sp^3 B. dsp^3 C. d^2sp^3 D. sp^2

2. Formaldehyde is used as a preservative. In the presence of air, formaldehyde is oxidized to formic acid, HCOOH. What hybridization does the carbon atom have in formic acid?

   A. sp^2 B. sp^3 C. sp D. dsp

3. Phosphorus pentachloride is produced upon reaction of phosphorus trichloride with chlorine. What hybridization is present in the phosphorus atom of PCl_3 and PCl_5 molecules, respectively?

   A. dsp, d^2sp^3 B. sp^3, dsp^3 C. sp^3, d^2sp^3 D. d^2sp^3, dsp^3

4. How many σ, and how many π bonds, respectively, are there in the following molecule: $CH_3CH_2CH_2CH_2CHCCCH_2$? Remember that carbon needs to have four bonds to be satisfied.

   A. 19, 3 B. 16, 7 C. 16, 3 D. 20, 4

5. The following molecule, CH_3CH_2CHO, is reduced to $CH_3CH_2CH_2OH$. What orbital is most probably used in the reduction process?

   A. π orbital of one of the sp^3 carbons
   B. σ orbital of one of the sp^2 carbons
   C. σ orbital of one of the sp^3 carbons
   D. π orbital of the sp^2 carbon

6. What is the hybridization of phosphorus in PCl_6^-?

   A. d^2sp^3 B. dsp^3 C. sp^3 D. sp^2

7. How many π bonds are in the following molecule?

   $CH_3–CH_2–CH_2–CH_2CH=C=C=CH_2$

   A. 4 B. 3 C. 0 D. 1

8. Which of the following bond orders would show the greatest bond strength?

   A. ½ B. 1 C. 0 D. 0.3

9. According to the molecular orbital model, a bonding orbital:

   A. Is unstable.
   B. Is as stable as an antibonding orbital.
   C. Is more stable than an antibonding orbital.
   D. Is less stable than a non-bonding orbital.

10. How many bonding electrons does O_2^{2+} have? Calculate the bond order.

    A. 10, 3 B. 4, 3 C. 6, 1.5 D. 8, 1

11. According to molecular orbital theory, C_2^{2-} should be

    A. paramagnetic B. ferromagnetic C. diamagnetic D. antimagnetic

12. The bond order for C_2 is

    A. 0 B. 2 C. 2.5 D. 3

13. Which of the following species is paramagnetic?

    A. O_2 B. C_2 C. Be_2 D. N_2

14. Which of the following species has a bond order that differs from the others?

    A. O_2 B. NO C. BN D. NO^-

15. Which of the following species is paramagnetic?

    A. NO B. CO C. BN D. CN^-

Answers to Exercises

1. a. angular (like water)
   b. see-saw
   c. linear
   d. square planar
   e. square pyramidal
   f. T-shaped

2. a. carbon-sp^3, tetrahedral; sp^2, trigonal planar; nitrogen-sp^3, based on tetrahedron-trigonal pyramid
   b. sp^3-tetrahedral
   c. dsp^3-trigonal bipyramid
   d. sp^3-based on tetrahedron-bent

3. a. linear
   b. trigonal planar
   c. tetrahedral
   d. trigonal bipyramidal
   e. octahedral

4. a. sp^3
   b. sp^3
   c. dsp^3
   d. sp^3
   e. sp^3
   f. d^2sp^3
   g. sp^2

5. a. 8 sigma, 2 pi
   b. 5 sigma, 2 pi
   c. 7 sigma, 3 pi

6. a. 7
   b. 1
   c. sp^2
   d. sp^3
   e. 120°
   f. 4

7. ½

8. ½

9. Ne has an equal number of bonding and antibonding electrons.

10. He_2^+ and O_2

11. N_2: bond order = 3, diamagnetic
    N_2^+: bond order = 2.5, paramagnetic
    N_2^-: bond order = 2.5, paramagnetic
    N_2^{2-}: bond order = 2, paramagnetic

12. Liquid nitrogen would simply pour through the space between the poles because it is diamagnetic as opposed to oxygen, which would be held between the poles because it is paramagnetic.

13. CO: bond order = 3, diamagnetic; CO^-: bond order = 2.5, paramagnetic

14. $B_2 < C_2 < N_2 > O_2 > F_2$

15. IBr: bond order = 1, diamagnetic

16. CN^- and CN^+

Answers to Multiple-Choice Self-Test

1. C	4. A	7. B	10. A	12. B	14. B
2. A	5. D	8. B	11. C	13. A	15. A
3. B	6. A	9. C			

CHAPTER 15

Chemical Kinetics

The Bottom Line: Chapter 15

In the introductory remarks to this chapter, your textbook makes the *critical distinction between kinetics and thermodynamics. Thermodynamics* can be used to *predict if a reaction will occur.* It says nothing about how fast it occurs. *Kinetics* describes the *rate and mechanism of a reaction given that it occurs.* Thermodynamics says "if." Kinetics says "how" and "how fast."

15.1 Reaction Rates

Your textbook defines the **rate of a reaction** as

$$\text{rate} = \frac{\text{change in concentration of a substance}}{\text{change in time}}$$

Let's look at data for the reaction of hydrogen and oxygen to give water,

$$2H_2(g) + O_2(g) \rightarrow 2H_2O(l)$$

Time (s)	**$[H_2]$**	**$[O_2]$**	**$[H_2O]$**
0.0020	0.050	0.080	0
0.0040	0.025	0.0675	0.0250
0.0060	0.018	0.064	0.032
0.0080	0.0125	0.0612	0.0375

The rate of disappearance of H_2 must be twice the rate for O_2 and equal to the appearance of H_2O. This is because the *coefficients of the chemical equation tell you the relative rates.* That is,

$$\frac{-\Delta[H_2]}{\Delta t} = \frac{-2\Delta[O_2]}{\Delta t} = \frac{\Delta[H_2O]}{\Delta t}$$

where "[ ]" means **concentration in moles/liter.** This says that, for **every mole of O_2 that reacts** per unit time, 2 moles of H_2 react, and 2 moles of H_2O are produced. Note that the *rate of disappearance* of the reactants is given a negative ("–") sign. That implies that we are *losing reactants,* while gaining the products ("+" rate). Note also that your textbook uses the standard that all reaction rates should be *positive.*

Using the data from our reaction, the rate of reaction of hydrogen between 0.004 and 0.008 seconds is

$$\text{rate} = \frac{-\Delta[H_2]}{\Delta t} = \frac{-(0.0125 - 0.025)}{(0.008 - 0.004)} = \frac{-(-0.0125)}{0.004}$$

rate = 3.1 mol/L s

We are *losing* hydrogen at a rate of 3.1 mol/L s.

Example 15.1 - Reaction Rates

Answer the following questions using the hydrogen and oxygen rate data just presented.

a. Based on the *coefficients of the chemical equation alone*, what is the rate of oxygen reaction between 0.004 and 0.008 seconds?
b. What is the rate of water production during the same time period?
c. Use the concentration vs. time data to prove your answers.

Solution

a. One mole of oxygen reacts for every two moles of hydrogen. That is, *oxygen reacts at half the rate of hydrogen.* The rate of reaction of hydrogen is **3.1 mol/L s.** The **rate of reaction of oxygen** is therefore (3.1)/2 = **1.6 mol/L s** (rounded off to two significant figures).

b. The balanced chemical equation says that one mole of hydrogen reacts to produce one mole of water. Therefore, the *rate of consumption of hydrogen equals the rate of production of water.*

rate of water production = 3.1 mol/L s

c. For oxygen,

$$\frac{-\Delta[O_2]}{\Delta t} = \frac{-(0.0612 - 0.0675)}{(0.008 - 0.004)} = \mathbf{1.6\ mol/L\ s}$$

For water,

$$\frac{\Delta[H_2O]}{\Delta t} = \frac{(0.0375 - 0.0250)}{(0.008 - 0.004)} = \mathbf{3.1\ mol/L\ s}$$

Remember the order of operations of your calculator. After subtracting 0.0250 from 0.0375, press the "=" key before dividing by Δt; otherwise you will divide 0.0250 by Δt.

Look at Table 15.2 in your textbook. Notice that the *rate of NO_2 decomposition decreases with time* until equilibrium is reached. **The rate of a reaction is not constant.**

15.2 Rate Laws: An Introduction

Answer the following review questions after you have read the entire section.

1. What **simplifying assumption** do we make regarding the rates of forward and reverse reactions in our study of rate laws?
2. Define **rate law.**
3. What are **"*k*"** and **"*n*"**?
4. Why don't **products** appear in a rate law?
5. Why is it important to **specify the component** whose rate we are describing with the rate law?
6. What is a **differential rate law?**
7. What is an **integrated rate law?**
8. Give a couple of **practical examples** of the importance of determining rate laws.

15.3 Determining the Form of the Rate Law

Recall from the last section that the **differential rate law** deals with the **dependence of rate concentration.** The important question is, "How will the rate of reaction change if we change the concentration of our reactants?"

Let's take the hypothetical example of reactant "A" going to product "P."

$$\mathbf{A \rightarrow P}$$

A plot of the concentration of A, [A], vs. time is similar in shape to the one shown in Figure 15.2 in your textbook. Examine the following data where the **initial rate** = $\Delta[A]/\Delta t$, and is taken as close to $t = 0$ as possible.

$[A]_0$	Initial Rate of Reaction (mol/L s)
0.35	7.2×10^{-4}
0.70	2.90×10^{-3}
1.05	6.45×10^{-3}

("$[A]_0$" means the **initial concentration** of A. The subscript zero means initial.)

How does the rate of decomposition of A change as we change [A]?

As [A] doubles from 0.35 to 0.70, the rate goes from 7.20×10^{-4} mol/L s to 29.0×10^{-4} mol/L s. The rate has increased by a factor of 4. This 2:4, 3:9 relationship suggests that the rate depends on the square of the concentration or

$$\textbf{rate} = \frac{-\Delta[\mathbf{A}]}{\Delta t} = k[\mathbf{A}]^2$$

where k = the rate constant for the reaction.

Your textbook points out that such differential rate laws can be found only by experiments, not theoretically.

To solve for k, substitute for several values of rate with the corresponding values for [A], and average the resultant values for k. Using [A] = 0.70 M as an example,

$$\text{rate} = k[A]^2$$

$$2.90 \times 10^{-3} \text{ mol/L s} = k\,(0.70\ M)^2$$

$$k = \frac{2.90 \times 10^{-3} \text{ mol/L s}}{(0.70\ M)^2} = \mathbf{5.92 \times 10^{-3}\ L/mol\ s}$$

While the rate is not constant, *the rate constant is constant* for a particular reaction at a particular temperature.

Example 15.3 A - Differential Rate Law

Using the data given above, what would be the rate of reaction for A (rate = $-\Delta[A]/\Delta t$) if [A] = 0.16 M?

Solution

You know the rate law and the value for k (which you would check by doing further experiments). You can therefore solve for the rate by substitution.

$$\textbf{rate} = k[A]^2 = (5.92 \times 10^{-3} \text{ L/mol s})(0.16\ M)^2 = \mathbf{1.52 \times 10^{-4}\ mol/L\ s}$$

Note that the rate of reaction is *considerably slower* with the *lower concentration* than with [A] = 0.70 M. That makes sense because the rate equation says that *the rate will increase as the square of the concentration increases.* If the concentration decreases, the rate does as well.

Let's look at the more complex example of

$$\mathbf{A + B \rightarrow products}$$

We want to determine a rate law for the decomposition of A that takes into account both [A] and [B]. We can hypothesize that the rate of reaction is proportional to the concentrations of A and B. The more of each, the faster the reaction.

$$\text{rate} = \frac{-\Delta[\text{A}]}{\Delta t} = k[\text{A}]^n[\text{B}]^p$$

Gathering experimental data will allow us to determine the rate law. Our goal is to determine ***n*, *p*,** and ***k*.** The sum of the exponents, ***n* + *p*,** is called the reaction order. Let's say the data are:

Reaction	**$[A]_0$**	**$[B]_0$**	**Initial Rate of Reaction (mol/L s)**
1	0.100	0.100	1.53×10^{-4}
2	0.100	0.300	4.59×10^{-4}
3	0.200	0.100	6.12×10^{-4}
4	0.100	0.200	3.06×10^{-4}
5	0.300	0.600	8.26×10^{-3}

The key process here is to **determine how the rate varies with a varying concentration of one component while the concentration of the other is held constant.**

What happens to the rate when you vary [B] while holding $[A]_0$ constant?

In reactions 1, 2, and 4, $[A]_0$ is constant. Therefore any increase in rate must be related to the increase in $[B]_0$. Using data from reactions 1 and 2,

$$\text{increase in } [\text{B}]_0 = \frac{0.300\ M}{0.100\ M} = \textbf{factor of 3}$$

$$\text{increase in rate} = \frac{4.59\times10^{-4}\ \text{mol/L s}}{1.53\times10^{-4}\ \text{mol/L s}} = \textbf{factor of 3}$$

The rate increases *linearly* with $[B]_0$.

$$\text{rate} = k[\text{A}]^n[\text{B}]^1$$

What happens to the rate when you vary $[A]_0$ while holding $[B]_0$ constant?

In reactions 1 and 3, $[B]_0$ is constant. Therefore, any increase in rate must be related to the increase in $[A]_0$.

$$\text{increase in } [\text{A}]_0 = \frac{0.200\ M}{0.100\ M} = \textbf{factor of 2}$$

$$\text{increase in rate} = \frac{6.12\times10^{-4}\ \text{mol/L s}}{1.53\times10^{-4}\ \text{mol/L s}} = \textbf{factor of 4}$$

The rate increases with $[A]_0$ *squared.*

$$\textbf{rate} = k[A]^2[B] \text{ (If no exponential is given, 1 is assumed.)}$$

We now have a complete rate law. However, we need to **solve for *k*** and use other data to verify our solution. Using the data from reaction 4 (We really should average the values obtained from all five reactions.),

$$k = \frac{\text{rate}}{[A]^2[B]} = \frac{3.06 \times 10^{-4} \text{ mol/L s}}{(0.100\ M)^2(0.200\ M)} = \mathbf{0.153\ L^2/mol^2\ s}$$

(We will address the unusual units in the next section.)

As a **double check**, let's solve for the rate in reaction 5.

$$\textbf{rate} = (0.153 \text{ L}^2/\text{mol}^2 \text{ s})(0.300\ M)^2(0.600\ M) = \mathbf{8.26 \times 10^{-3}\ mol/L\ s}$$

Our rate law properly reflects the data.

Example 15.3 B - Rate Law Determination

Determine the rate law, and solve for the order and value of the rate constant for the reaction

$$C + D + E \rightarrow \text{Products}$$

given the following data:

Reaction #	$[C]_0$	$[D]_0$	$[E]_0$	Initial rate (mol/L s)
1	0.400	0.300	0.560	7.14×10^{-4}
2	0.100	0.500	0.200	4.55×10^{-5}
3	0.100	0.200	0.200	4.55×10^{-5}
4	0.400	0.300	0.750	1.28×10^{-3}
5	0.100	0.300	0.560	3.57×10^{-4}

Solution

$$\textbf{rate} = k[C]^n[D]^p[E]^q$$

We need to see how each component varies while the others remain constant.

1. ***Regarding dependence on* [C]**: $[D]_0$ and $[E]_0$ remain constant in reactions 1 and 5. In going from reaction 5 to 1, $[C]_0$ increases.

$$[C]_0 = \frac{0.400\ M}{0.100\ M} = \textbf{factor of 4}$$

$$\textbf{rate} = \frac{7.14 \times 10^{-4} \text{ mol/L s}}{3.57 \times 10^{-4} \text{ mol/L s}} = \textbf{factor of 2}$$

Therefore, rate increases as $[C]_0^{1/2}$!

$$\text{rate} = k[C]_0^{1/2}[D]_0^p[E]_0^q$$

2. ***Regarding dependence on* [D]**: $[C]_0$ and $[E]_0$ remain constant in reactions 2 and 3. In going from reaction 3 to 2, $[D]_0$ increases.

$$[D]_0 = \frac{0.500\ M}{0.200\ M} = \textbf{factor of 2.5}$$

$$\textbf{rate} = \frac{4.55 \times 10^{-5} \text{ mol/L s}}{4.55 \times 10^{-5} \text{ mol/L s}} = \textbf{NO CHANGE!}$$

Therefore the rate is **not dependent** on $[D]_0$.

$$\text{rate} = k[C]_0^{1/2}[D]_0^{0}[E]_0^{q}$$

3. ***Regarding dependence on* [E]**: $[C]_0$ and $[D]_0$ remain constant in reactions 1 and 4. In going from reaction 1 to 4, $[E]_0$ increases.

$$[E]_0 = \frac{0.750\ M}{0.560\ M} = \textbf{factor of 1.34}$$

$$\text{rate} = \frac{1.28\times10^{-3}\ \text{mol/L s}}{7.14\times10^{-4}\ \text{mol/L s}} = \textbf{factor of 1.79}$$

The square of 1.34 is 1.79, or $(1.34)^2 = 1.79$. Therefore rate increases as $[E]_0^2$.

$$\text{rate} = k[C]_0^{1/2}[E]_0^{2},\ \text{because}[D]_0^{0} = 1$$

4. ***Solving for the rate constant, k***: Pick any of the reactions (though it is best to average all five). Using reaction 3,

$$k = \frac{\text{rate}}{[C]_0^{1/2}[E]_0^{2}} = \frac{4.55\times10^{-5}\ \text{mol/L s}}{(0.100\ M)^{1/2}(0.200\ M)^2} = \mathbf{3.60 \times 10^{-3}\ L^{3/2}/mol^{3/2}\ s}$$

The order of k equals the sum of the exponents $n + p + q = 1/2 + 0 + 2 = 2^1/_2$ **order.**

Checking Your Work

Use **data from reaction 5,** and solve for the rate.

$$\text{rate} = 3.60 \times 10^{-3}\,(0.100)^{1/2}\,(0.560)^2 = \mathbf{3.57 \times 10^{-4}\ mol/L\ s}$$

This agrees with the observed rate.

15.4 The Integrated Rate Law

Units of k

In the last section, we reviewed the development of a differential rate law from initial rate data. The general form of the rate equation was

$$\text{rate} = \frac{-\Delta[A]}{\Delta t} = k[A]^n$$

where A is a reactant.

If $n = 1$ (a first-order reaction), then the units of k are

$$k = \frac{\text{rate}}{[A]} = \frac{\text{mol/L s}}{\text{mol/L}} = \mathbf{1/s}$$

If $n = 2$ (a second-order reaction), then the units of k are

$$k = \frac{\text{rate}}{[A]^2} = \frac{\text{mol/L s}}{\text{mol}^2/\text{L}^2} = \mathbf{L/mol\ s}$$

If $n = 3$ (a third-order reaction), then

$$k = \frac{\text{rate}}{[A]^3} = \frac{\text{mol/L s}}{\text{mol}^3/\text{L}^3} = \mathbf{L^2/mol^2\ s}$$

Do you see a pattern emerging? For an **nth-order reaction** the units of k will be $\mathbf{L^{(n-1)}/mol^{(n-1)}\ s}$. Thus, for our 2 1/2-order reaction in Example 15.3 B, k is in units of $M^{1-(5/2)}\ s^{-1} = M^{-3/2}\ s^{-1} \equiv \mathbf{L^{3/2}/mol^{3/2}\ s}$.

Example 15.4 - The Units Of The Rate Constant

Determine the units of k for the following differential rate law.

$$\text{rate} = k[\text{A}]^4[\text{B}][\text{C}]^2$$

Solution

The order of this reaction is 4 + 1 + 2 = 7. The units of k are $\text{mol}^{(1-7)}/\text{L}^{(1-7)}\ \text{s} = \mathbf{L^6/mol^6\ s}$.

First-Order Rate Law

Recall that the difference between a **differential rate law** and an **integrated rate law** is that the latter relates **concentration of a substance to reaction.**

Your textbook presents two *mathematically identical* forms of the first-order integrated rate law (rate = k[A]).

$$\ln[\text{A}] = -kt + \ln[\text{A}]_0$$

$$\ln\left(\frac{[\text{A}]_0}{[\text{A}]}\right) = kt$$

where **$[\text{A}]_0$** = initial concentration of A
[A] = concentration of A at time t
k = rate constant

I have added a third equation (by exponentiating both sides of the previous equations), which you may find useful.

$$[\text{A}] = [\text{A}]_0\, e^{-kt}$$

Your textbook makes several points regarding this concentration vs. time relationship. To reemphasize some of them,

- The equation can be expressed in the form $y = mx + b$, where $\mathbf{y = \ln[A]}$, $\mathbf{m = -k}$, $\mathbf{x = t}$, and $\mathbf{b = \ln[A]_0}$.
- As a result, if you plot ln[A] vs. time and get a straight line, your data is likely to be first-order. If you do not get a straight line, some other order may predominate.
- The form of the equation that you should use will depend upon what you are given. You generally will avoid equations where the unknown is part of a log term.

Example 15.4 B - Practice With First-Order Kinetics

All radioactive elements have nuclei that follow a first-order rate law when decaying. (We will learn more about this in Chapter 20.) Radon decays to polonium according to the following equation,

$$^{222}\text{Rn} \rightarrow {}^{218}\text{Po} + {}^{4}\text{He}$$

The first-order rate constant for decay is **0.181 days^{-1}**. If you begin with a 5.28-g sample of pure ^{222}Rn, how much will be left after **1.96 days**? **3.82 days**?

Solution

Your goal is to solve for [A]. You are given $[A]_0$, k, and t. Although any of the three forms of the integrated rate law is usable, the most useful one solves for [A] directly,

$$[A] = [A]_0 e^{-kt}$$

Given: $[A]_0 = 5.28$ g, $k = 0.181\ d^{-1}$, $t = 1.96$ d

$$[A] = 5.28e^{-(0.181 \times 1.96)} = 5.28 \times 0.701 = \mathbf{3.70\ g\ ^{222}Rn}$$

When we want to **increase** t to 3.82 days,

$$[A] = 5.28e^{-(0.181 \times 3.82)} = 5.28 \times 0.501 = \mathbf{2.64\ g\ ^{222}Rn}$$

Does the Answer Make Sense?

In each case, the amount we ended with was *less* than the amount we started with. That makes sense, as we are *decomposing* ^{222}Rn. The longer time period in the second part of the problem allowed even more to decompose. This answer should be less than in the first part. That too makes sense.

Note that in the problem you just finished **3.82 days** were required to **decompose ½ of the reactant** (5.28 g → 2.64 g). The time it takes to lose half of your reactant is called the **half-life,** or $t_{1/2}$. For ^{222}Rn, $t_{1/2} = 3.82$ days. For a first-order reaction (see your textbook for the proof),

$$t_{1/2} = \frac{0.693}{k}$$

In a first-order reaction, the half-life is *independent* of the initial concentration of the reactant.

Look at Figure 15.4 in your textbook. This says that you *lose 1/2 of the remaining substance for each half-life* that occurs. In effect,

$$[A] = \frac{[A]_0}{2^n}$$

where n = the number of half-lives that have occurred.

Example 15.4 C - First-Order Half-Life

Using the same ^{222}Rn nuclear decay data as in the previous example, how much ^{222}Rn would remain from a 5.82-g sample after **11.46 days? 21.01 days?**

(Solve for 21.01 days using the half-life method; then, for practice, by using the integrated rate law.)

Solution

The key is to calculate the number of half-lives (n) in the given time period. For $t = 11.46$ days,

$$n_{(11.46)} = \frac{1\text{ half-life}}{3.82\text{ days}} = 11.46\text{ days} = 3.0\text{ half-lives}$$

$$[A] = \frac{[A]_0}{2^n} = \frac{5.82\text{ g}}{2^3} = \frac{5.82\text{ g}}{8} = \mathbf{0.728\ g\ ^{222}Rn}$$

For $t = 21.01$ days (half-life method),

$$n_{(21.01)} = \frac{1 \text{ half-life}}{3.82 \text{ days}} = 21.01 \text{ days} = 5.50 \text{ half-lives}$$

$$[A] = \frac{5.82 \text{ g}}{2^{5.5}} = \frac{5.82 \text{ g}}{45.3} = \mathbf{0.129 \text{ g } ^{222}Rn}$$

For $t = 21.01$ days (integrated rate law, $k = 0.181 \text{ d}^{-1}$)

$$[A] = [A]_0 e^{-kt} = 5.82e^{-(0.181 \times 21.01)} = \mathbf{0.130 \text{ g } ^{222}Rn}^*$$

*The *slight difference* in the last two answers came as a result of rounding off when *converting* $t_{1/2}$ *to k.* Without round off, $k = 0.1814$, and $[A] = 0.129$ g.

Second-Order Rate Law

Your textbook gives the following equation relating to the **integrated rate law for second-order data.**

$$aA \rightarrow \text{Products}$$

$$\frac{\mathbf{1}}{\mathbf{[A]}} = \mathbf{kt} + \frac{\mathbf{1}}{\mathbf{[A]_0}}$$

The equation is in the form $\mathbf{y = mx + b}$ where $\mathbf{y = 1/[A]}$, $\mathbf{m = k}$, $\mathbf{x = t}$, and $\mathbf{b = 1/[A]_0}$. Thus, if your data fit a second-order model, a plot of 1/[A] vs. t should give a **straight line** with a **slope** = $\mathbf{k}$ and **intercept = 1/[A]₀**. (See Figure 15.5 in your textbook.)

The equation relating half-life to k is

$$\mathbf{t_{1/2}} = \frac{1}{k[A]_0}$$

Notice that, for a second-order system, the half-life depends on the initial concentration!

Example 15.4 D - Second-Order Rate Law

The differential rate law for the association of iodine atoms to molecular iodine,

$$2I(g) \rightarrow I_2(g)$$

is given by

$$\mathbf{rate} = \mathbf{k[I]^2}, \text{ where } \mathbf{k = 7.0 \times 10^9}$$

a. What are the proper units for k?
b. If $[I]_0 = 0.40\ M$, calculate [I] at $t = 2.5 \times 10^{-7}$ seconds.
c. If $[I]_0 = 0.40\ M$, calculate $t_{1/2}$.
d. If $[I]_0 = 0.80\ M$, how much time would it take for 75% of the iodine atoms to react?

Solution

a. The units for a second-order rate constant are L/mol s.

b. $$\frac{1}{[I]} = kt + \frac{1}{[I]_0}$$

$$= 7.0 \times 10^9 \text{ L/mol s } (2.5 \times 10^{-7} \text{ s}) + \frac{1}{0.40 \text{ mol/L}}$$

$$= 1.75 \times 10^3 \text{ L/mol}$$

$$\mathbf{[I]} = \mathbf{5.7 \times 10^{-4}}\ \boldsymbol{M}$$

c. $$t_{1/2} = \frac{1}{k[\mathrm{I}]_0}$$
$$= \frac{1}{7.0\times10^9\,(0.40)}$$
$$= \mathbf{3.6 \times 10^{-10}\ s}$$

Note that $t_{1/2}$ is so short that if we let the reaction proceed for as short as 2.5×10^{-7} seconds, most of the iodine atoms will have already reacted.

d. $[\mathrm{I}]_o = 0.80\ M$. After 75% has reacted, 25% is left. Therefore $[\mathrm{I}] = [\mathrm{I}]_0/4 = 0.20\ M$.

$$\frac{1}{0.20\ \text{mol/L}} = 7.0 \times 10^9\ \text{L/mol s}\ (t) + \frac{1}{0.80\ \text{mol/L}}$$
$$5 = 7.0 \times 10^9\ (t) + 1.25$$
$$3.75 = 7.0 \times 10^9\ (t)$$
$$\mathbf{t = 5.4 \times 10^{-10}\ s}$$

You might think that an alternative method is to recognize that 75% reacted is 2 half-lives. If $[\mathrm{I}]_o = 0.80\ M$

$$t_{1/2} = \frac{1}{7.0\times10^9\,(0.80)} = 1.8 \times 10^{-10}\ \text{s}$$

But this method is not valid here! The first half-lives would occur in 1.8×10^{-10} s. However, the **initial concentration** for the second half-life **is now 0.40 *M*** (½ of 0.80 *M*!). As a result, $t_{1/2}$ for the second half-life is **twice as long** (3.6×10^{-10} s). The sum of $t_{1/2}$ for both half-lives equals 1.8×10^{-10} s + 3.6×10^{-10} s = $\mathbf{5.4 \times 10^{-10}\ s}$**.** This agrees with our results, but is the difficult way of arriving at an answer.

The remainder of this section in your textbook discusses zero, multicomponent, and pseudo-first-order rate laws. The last one is especially important for you to read about because so many multicomponent kinetic analyses are done using a strategy in which one component is in much lower concentration than all of the others.

15.5 Rate Laws: A Summary

This section presents kinetics from two different standpoints. The first regards experimentation, which is dealt with in points 4 and 5 of the summary. In Table 15.6 in your textbook, a more mathematical approach is taken. Please pay special attention to the listing of the plot needed to give a straight line for each rate law.

15.6 Reaction Mechanisms

Your textbook begins this chapter by defining several terms. You should know the definitions of:

- **reaction mechanisms** - the series of steps by which a chemical reaction occurs.
- **intermediate** - a species that is neither a reactant nor product. It is formed and consumed in the reaction sequence.

- **molecularity** - the number of species that collide to produce the reaction indicated by an elementary step. Uni- and bimolecular are two important molecularities. (Why isn't a termolecular molecularity likely?)
- **rate-determining step** - the slowest step. The reaction can only go as fast as the rate-determining step.

One idea in this section is that a **chemical equation** tells us only **what** reactants become what products. It does not tell us **how** these changes occur. This is the goal of reaction mechanism studies.

Another key idea is that the differential rate law is determined by the **rate-determining step.** This step may be in one direction or reversible (both directions). Your goal is to be able to determine whether mechanisms are consistent with experimentally determined rate laws.

Review Example 15.6 in your textbook. Then try the following examples.

Example 15.6 A - Reaction Mechanisms

The balanced equation for the reaction of nitric oxide with hydrogen is

$$2NO + 2H_2 \rightarrow 2H_2O + N_2$$

The experimentally determined rate law is

$$\text{rate} = k[NO]^2[H_2]$$

The following mechanism has been proposed.

$$NO + H_2 \xrightarrow{k_1} N + H_2O \quad \text{(slow)}$$
$$N + NO \xrightarrow{k_2} N_2O \quad \text{(fast)}$$
$$N_2O + H_2 \xrightarrow{k_3} N_2 + H_2O \quad \text{(fast)}$$

Is this mechanism consistent with the observed rate law?

Solution

Your textbook says there are two criteria that must be met if a mechanism is to be considered acceptable.

1. **The sum of the steps must give the balanced equation.** If you add up the three steps, you will find this is, in fact, true here.
2. **The mechanism must agree with the observed rate law.** According to the proposed mechanism, the first step is rate-determining. This means the overall mechanism must be given by the first step.

$$\textbf{rate} = \boldsymbol{k_1[NO][H_2]}$$

This does not agree with the experimentally determined rate law; therefore **our observed mechanism is incorrect.**

Example 15.6 B - More Reaction Mechanisms

Consider the same balanced equation as in the previous example. Is the following mechanism consistent with the experimentally determined rate law?

$$NO + H_2 \underset{k_{-1}}{\overset{k_1}{\rightleftharpoons}} N + H_2O \quad \text{(fast, with equal rates)}$$
$$N + NO \xrightarrow{k_2} N_2O \quad \text{(slow)}$$
$$N_2O + H_2 \xrightarrow{k_3} N_2 + H_2O \quad \text{(fast)}$$

Solution

If you add up the individual steps, you will find they equal the balanced chemical equation. So far, so good. We must now **evaluate the mechanism.** The rate-determining step is the second one,

$$N + NO \xrightarrow{k_2} N_2O$$

based on this

$$\textbf{rate} = k_2[\mathbf{N}][\mathbf{NO}]$$

However, **N is an intermediate.** You can't control the initial amount of nitrogen you add because there is none present at the beginning. We want to always express rate laws in terms of reactants and/or products.

The **first step** allows us to solve for N in terms of reactants and products.

$$\text{forward rate} = k_1[NO][H_2]$$
$$\text{reverse rate} = k_{-1}[N][H_2O]$$

The rates are equal. Therefore,

$$k_1[NO][H_2] = k_{-1}[N][H_2O]$$

Solving for [N],

$$[\mathbf{N}] = \frac{k_1[NO][H_2]}{k_{-1}[H_2O]}$$

Substituting for [N] in our rate-determining equation,

$$\textbf{rate} = \frac{k_2k_1[NO][H_2][NO]}{k_{-1}[H_2O]} = \frac{\mathbf{k'[NO]^2[H_2]}}{\mathbf{[H_2O]}}$$

This is close, but **not quite there.**

Example 15.6 C - Even More Reaction Mechanisms

Same problem as before, with the same data. Test this mechanism:

$$NO + H_2 \xrightarrow{k_1} N + H_2O \qquad \text{(fast)}$$
$$N + NO \xrightarrow{k_2} N_2O \qquad \text{(fast)}$$
$$N_2O + H_2 \xrightarrow{k_3} N_2 + H_2O \qquad \text{(slow)}$$

Solution

Again, our mechanisms add up to give the balanced equation. The rate-determining step is the third.

$$\textbf{rate} = k_3[\mathbf{N_2O}][\mathbf{H_2}]$$

However, **N_2O is an intermediate.** There are no reverse reactions here. We can't explicitly substitute for $[N_2O]$ with rate equations. **However,** N_2O is formed **very quickly,** and irreversibly, from the first two reactions. Using **mole-ratios,**

- **1 mole of N_2O comes from the combination of 2 moles of NO** (2 nitrogens from 2 nitric oxides).

Therefore we can say

$$[N_2O] = 2[NO] \text{ (from stoichiometry)}$$

In the rate-determining step, we can then substitute

$$2NO + H_2 \rightarrow \text{products}$$

because $[N_2O]$ is directly related to [NO], and the formation of N_2O is fast. Our rate equation becomes

$$\text{rate} = k_3[NO]^2[H_2]$$

And *this agrees with the observed rate law.*

[Reference for 15.6 A, B, and C is T.W. Lippencott, A.B. Garret and F.H. Verhock, Chemistry, J.W. Wiley, New York, (1977)]

15.7 The Steady-State Approximation

If a rate-determining step cannot be ascertained, then the steady-state approximation can be used to find a rate expression. What assumptions are made in this approximation? Note in particular the summary listing at the end of the section.

15.8 A Model for Chemical Kinetics

Your textbook points out two empirical facts about the rate of chemical reactions. All other things being equal,

1. The more **concentrated** your reactants, the **faster** the reaction.
2. The **higher the temperature,** the **faster** the reaction.

These trends can be explained using **collision theory.** Among the key aspects of the theory are:

1. Molecules must hit with sufficiently high energy.
2. Molecules must hit with the proper orientation. (See Figure 15.13 in your textbook.)

The combination of these steps makes it unlikely that reaction will occur, even in the best of circumstances. However, the odds of collision are increased with higher concentrations. The odds of collision with a sufficiently high energy (kinetic energy is proportional to temperature) are increased with higher temperature.

The rate constant, ***k***, is a measure of the **fraction** of collisions with **sufficient energy** to produce a reaction,

$$k = Ae^{-E_a/RT}$$

where ***k*** is the **rate constant.**
A is the **frequency factor** (related to the collision frequency and orientation of collisions).
E_a is the **activation energy** (in J/mol).
R is the **universal gas constant** (8.3145 J/K mol).
T is the **temperature** (in Kelvins).

Your textbook puts the equation in "$y = mx + b$" form by taking the ln of both sides,

$$\underset{\substack{\uparrow \\ y}}{\ln(k)} = \underset{\substack{\uparrow \\ m}}{\frac{-E_a}{R}} \underset{\substack{\uparrow \\ x}}{\left(\frac{1}{T}\right)} + \underset{\substack{\uparrow \\ b}}{\ln(A)}$$

ln(k) can be plotted graphically vs. $1/T$ for a series of data to determine the slope and intercept or, for quick work, two sets of data can be used mathematically, yielding Equation 15.11 in your textbook.

$$\ln\left(\frac{k_2}{k_1}\right) = \frac{E_a}{R}\left(\frac{1}{T_1} - \frac{1}{T_2}\right)$$

Example 15.8 A - Activation Energy

Your textbook gives the activation energy for the reaction

$$H_2(g) + I_2(g) \rightarrow 2HI(g)$$

as being **167 kJ/mol**. It also gives the rate constant at 302°C as being **2.45×10^{-4} L/mol.** What is the rate constant for the reaction at 205°C?

Solution

You are given data at two temperatures. Your goal is to solve for k_2, the rate constant at the second temperature.

$k_1 = 2.45 \times 10^{-4}$ L/mol s $T_1 = 302 + 273.2 = 575.2$ K
$k_2 = ?$ $T_2 = 205 + 273.2 = 478.2$ K

$E_a = 167$ kJ/mol = 167,000 J/mol (the units (J) must match that in R)

$$\ln\left(\frac{k_2}{k_1}\right) = \frac{E_a}{R}\left(\frac{1}{T_1} - \frac{1}{T_2}\right)$$

$$\ln\left(\frac{k_2}{2.45\times10^{-4}\ \text{L/mol s}}\right) = \frac{1.67\times10^{5}\ \text{J/mol}}{8.3145\ \text{J/K mol}}\left(\frac{1}{575.2\ \text{K}} - \frac{1}{478.2\ \text{K}}\right)$$

$$\ln\left(\frac{k_2}{2.45\times10^{-4}\ \text{L/mol s}}\right) = -7.083$$

Taking the **inverse ln** of both sides,

$$\left(\frac{k_2}{2.45\times10^{-4}\ \text{L/mol s}}\right) = e^{-7.083}$$

$$\left(\frac{k_2}{2.45\times10^{-4}\ \text{L/mol s}}\right) = 8.394 \times 10^{-4}$$

$$\mathbf{k_2 = 2.06 \times 10^{-7}\ \text{L/mol s}}$$

Does the Answer Make Sense?

T_2 was lower than T_1. We would therefore expect k_2 to be lower than k_1. Also, for every 10°C, the rate should be cut in half. The temperature was reduced here by 100°C. The rate constant should be reduced substantially, as it was.

Example 15.8 B - Practice With Activation Energy

A second-order reaction has rate constants of **8.9×10^{-3} L/mol** and **7.1×10^{-2} L/mol** at 3°C and 35°C respectively. Calculate the value of the activation energy for the reaction.

Solution

$k_1 = 8.9 \times 10^{-3}$ L/mol s $T_1 = 276.2$ K
$k_2 = 7.1 \times 10^{-2}$ L/mol s $T_2 = 308.2$ K

Using Equation 15.11 from your textbook,

$$\ln\left(\frac{7.1\times10^{-2}\ \text{L/mol s}}{8.9\times10^{-3}\ \text{L/mol s}}\right) = \frac{E_a}{8.3145\ \text{J/K mol}}\left(\frac{1}{276.2\ \text{K}} - \frac{1}{308.2\ \text{K}}\right)$$

$$2.077 = (E_a)\, 4.52 \times 10^{-5}$$

$$E_a = 4.6 \times 10^{4}\ \text{J/mol} = \mathbf{46\ kJ/mol}$$

15.9 Catalysis

The following questions will help you review the material in this section.

1. Define **catalysis.**
2. What are **enzymes?**
3. Why are catalysts more useful for increasing the reaction rate than raising the temperature?
4. How do catalysts work? How does this reflect the collision-theory model?
5. What is a **heterogeneous catalyst?** Give some examples.
6. Differentiate between **absorption** and **adsorption.**
7. Describe the involvement of a heterogeneous catalyst in hydrogenation.
8. Discuss the role of catalysis in the conversion of sulfur dioxide to acid rain.
9. What is a **homogeneous catalyst?** Give an example.
10. How does nitric oxide act as a catalyst in the production of ozone?
11. How can chlorine catalyze the decomposition of ozone?
12. What is the mechanism for the hydrogen ion acting as a catalyst in chemical reactions?

Exercises

Section 15.1

1. At 40°C $H_2O_2(aq)$ will decompose according to the following reaction:

$$2H_2O_2(aq) \rightarrow 2H_2O(l) + O_2(g)$$

The following data were collected for the concentration of H_2O_2 at various times.

times(s)	**$[H_2O_2](M)$**
0	1.000
2.16×10^4	0.500
4.32×10^4	0.250

a. Calculate the average rate of decomposition of H_2O_2 between 0 and 2.16×10^4 s. Use this rate to calculate the rate of production of $O_2(g)$.

b. What are these rates for the time period 2.16×10^4 s to 4.32×10^4 s?

2. Given the following hypothetical equation and data,

$$fA \rightarrow gB + 3C$$

$$-\Delta[A]/t = 0.156\ \text{mol/L s}$$
$$+\Delta[B]/t = 0.026\ \text{mol/L s}$$
$$+\Delta[C]/t = 0.078\ \text{mol/L s}$$

find the coefficients f and g.

Section 15.3

3. Use the given data for the hypothetical reaction:

$$2A + B \rightarrow \text{products}$$

to determine the rate law and to evaluate the rate constant at 30°C.

Reaction #	[A]	[B]	Initial rate (mol/L s)
1	0.1	0.1	3×10^{-2}
2	0.1	0.3	3×10^{-2}
3	0.2	0.3	6×10^{-2}

4. Derive the rate law expression, and calculate the rate constant for the reaction:

$$A + B + 3C \rightarrow \text{products}$$

given the following data for 15°C.

Reaction #	[A]	[B]	[C]	Initial rate (mol/L s)
1	0.4	0.1	0.1	6.0×10^{-3}
2	0.4	0.2	0.1	6.0×10^{-3}
3	0.4	0.3	0.2	1.2×10^{-2}
4	1.2	0.4	0.2	0.11

5. Indicate the overall order of reaction for each of the following rate laws.

a. $R = k[NO_2][F_2]$

b. $R = k[I]^2[H_2]$

c. $R = k[H_2][Cl_2]^{1/2}$

6. The following hypothetical reaction was performed:

$$A + 2B + C + 4D \rightarrow \text{products}$$

Determine the rate law, and calculate the rate constant for the following data collected at 20°C.

Reaction #	[A]	[B]	[C]	[D]	Initial rate (mol/L s)
1	0.25	0.30	0.60	0.15	7.20×10^{-5}
2	0.75	0.30	0.60	0.15	2.17×10^{-4}
3	0.25	0.30	0.20	0.15	7.20×10^{-5}
4	0.75	0.30	0.60	0.45	6.51×10^{-4}
5	0.75	0.44	0.60	0.15	4.67×10^{-4}

Section 15.4

7. The decomposition of $H_2O_2(aq)$ into $H_2O(l)$ and $O_2(g)$ is first order. From the data in Exercise 1, determine the rate constant and the half-life.

8. Using the answer from Exercise 7, calculate the concentration of H_2O_2 remaining after 8 hours if the initial concentration was 5.0 *M*.

9. Determine the rate constant (k) and $t_{1/2}$ for the data pertaining to the decomposition of phosphine (PH_3).

$$4PH_3(g) \rightarrow P(g) + 6H_2O(g)$$

Reaction #	[PH_3]	Initial rate (mol/L s)
1	0.18	2.4×10^{-3}
2	0.54	7.2×10^{-3}
3	1.08	1.4×10^{-2}

10. Using the data and answers from the previous exercise, calculate the amount of PH_3 remaining after 2 minutes if the initial concentration of phosphine is 1.00 *M*. How much time would be needed for 90% of the phosphine to decompose?

11. Data for the decomposition of compound AB to give A and B is given below. Determine the rate expression, the rate constant, and $t_{1/2}$ for a 1 *M* solution.

Reaction #	[AB]	Initial rate (mol/L s)
1	0.2	3.2×10^{-3}
2	0.4	12.8×10^{-3}
3	0.6	28.8×10^{-3}

12. The rate law for the following second-order reaction at 10°C can be written as:

$$2NOBr(g) \rightarrow 2NO(g) + Br_2(g)$$

$$\text{rate} = k[NOBr]^2, \text{ where } k = 0.80 \text{ L/mol s}$$

a. Determine $t_{1/2}$ when $[NOBr]_0 = 0.650$ *M*.
b. Calculate [NOBr] at $t = 5.80 \times 10^{-3}$ s if $[NOBr]_0 = 0.650$ *M*.
c. If $[NOBr]_0 = 1.00$ *M*, how long would it take for 50% NOBr to react?

13. Using the information in Exercise 12, calculate $[NOBr]_0$ if it took 4.31 s for 75% to react.

14. The decomposition of NOCl is a second-order reaction with $k = 4.0 \times 10^{-8}$ L/mol s. Given an initial concentration of 0.50 *M*, what is the half-life? How much is left after 1×10^8 s? What is the half-life for an initial concentration of 0.25 *M*?

15. IBr(*g*) decomposes to form $I_2(g)$ and $Br_2(g)$. A plot of 1/[Br] v. time gave a straight line. Write the general rate law for the reaction.

16. The rate law for the reaction:

$$2NO(g) + Cl_2(g) \rightarrow 2NOCl(g)$$

is rate = $k[NO]^2[Cl]$. If an experiment was performed in which the partial pressure of NO(*g*) initially was 0.1 atm, and the initial partial pressure of $Cl_2(g)$ was 10 atm, what experimental data would give a straight-line plot?

Section 15.6

17. Write the rate laws for the following proposed mechanisms for the decomposition of IBr to I_2 and Br_2.

a. $IBr(g) \rightarrow I(g) + Br(g)$ (fast)
$IBr(g) + Br(g) \rightarrow I(g) + Br_2(g)$ (slow)
$I(g) + I(g) \rightarrow I_2(g)$ (fast)

b. $IBr(g) \rightarrow I(g) + Br(g)$ (slow)
$I(g) + IBr(g) \rightarrow I_2(g) + Br(g)$ (fast)
$Br(g) + Br(g) \rightarrow Br_2(g)$ (fast)

c. $IBr(g) + IBr(g) \rightarrow I_2Br^+(g) + Br^-(g)$ (fast)
$I_2Br^+(g) \rightarrow Br^+(g) + I_2(g)$ (slow)
$Br^-(g) + Br^+(g) \rightarrow Br_2(g)$ (fast)

d. $IBr(g) + IBr(g) \rightarrow I_2(g) + Br_2(g)$ (one step)

18. Which of the following mechanisms are consistent with the observed rate law,

$$\text{rate} = [H_2O_2][I^-][H_3O^+],$$

for the reaction

$$H_2O_2 + 3I^- + 2H^+ \rightarrow 2H_2O + I_3^-$$

If any are not, write a rate equation that is consistent with the mechanism.

a. $H_3O^+ + I^- \rightarrow HI + H_2O$ (fast)
$H_2O_2 + HI \rightarrow H_2O + HOI$ (slow)
$HOI + H_3O^+ + I^- \rightarrow 2H_2O + I_2$ (fast)
$I^- + I_2 \rightarrow I_3^-$ (fast)

b. $H_3O^+ + H_2O_2 \rightarrow H_3O_2^+ + H_2O$ (fast)
$H_3O_2^+ + I^- \rightarrow H_2O + HOI$ (slow)
$HOI + H_3O^+ + I^- \rightarrow 2H_2O + I_2$ (fast)
$I_2 + I^- \rightarrow I_3^-$ (fast)

c. $H_2O_2 + I^- \rightarrow H_2O + OH^-$ (slow)
$H_3O^+ + I^- \rightarrow H_2O + HI$ (fast)
$H_3O^+ + OI^- + HI \rightarrow 2H_2O + I_2$ (fast)
$I_2 + I^- \rightarrow I_3$ (fast)

19. Write the rate law for the following predicted mechanism for the production of nitrogen dioxide (NO_2):

$NO + O_2 \rightleftharpoons NO_3$ (fast)

$NO_3 + NO \rightarrow 2N_2O$ (slow)

20. Which of the following rate laws is consistent with the proposed mechanism for the reaction:

$$3ClO^- \rightarrow ClO_3^- + 2Cl^-$$

$ClO^- + ClO^- \rightarrow ClO_2^- + 2Cl^-$ (fast)
$ClO^- + ClO_2^- \rightarrow ClO_3^- + 2Cl^-$ (slow)

a. rate = $k[ClO][ClO_2^-]$
b. rate = $k[ClO^-]^3$
c. rate = $k[ClO^-]^2[ClO_2^-]$
d. rate = $k[ClO^-]^2$

Section 15.8

21. The activation energy for the decomposition of HI(g) to $H_2(g)$ and $I_2(g)$ is 186 kJ/mol. The rate constant at 555 K is 3.52×10^{-7} L/mol s. What is the rate constant at 645 K?

22. The rate constant for the decomposition of acetone to carbonic acid was determined to be 6.42×10^{-5} L/mol s at 10°C and 2.03×10^{-1} at 78°C. Calculate the activation energy.

23. The rate constant for the reaction:

$$C_4H_8 \rightarrow 2C_2H_4$$

at 325°C is 6.1×10^{-8} s^{-1}. At 525°C the rate constant is 3.16×10^{-2} s^{-1}. Calculate the activation energy.

24. The rate of a reaction increases 2.21 times as the temperature changes from 70°C to 80°C. Calculate the activation energy.

25. The activation energy for the hypothetical reaction $2A + B \rightarrow C$ is 8.9×10^4 J/mol. If the original rate of the reaction is tripled at a temperature of 325 K, at what temperature did the reaction begin?

Multiple-Choice Self-Test

1. The rate of decomposition of ammonia to hydrogen gas and nitrogen gas

$$2NH_3(g) \rightarrow N_2(g) + 3H_2(g)$$

is expressed as $-\Delta[NH_3]/\Delta t$. Express the rate of reaction in terms of $\Delta[H_2]/\Delta t$

A. rate = $^2/_3\Delta[H_2]/\Delta t$
B. rate = $\Delta[H_2]/\Delta t$
C. rate = $3\Delta[H_2]/\Delta t$
D. rate = $2\Delta[H_2]/\Delta t$

2. Given the following information, calculate the average rate, $\Delta[SO_2]/\Delta t$, between 10 and 40 minutes for the production of SO_2.

$$2SO_3(g) \rightarrow 2SO_2(g) + O_2(g)$$

min	$[SO_2]$
0.0	0.0
10.0	0.032
20.0	0.056
30.0	0.074
40.0	0.087
50.0	0.096

A. 1.8×10^{-3}
B. 1.5×10^{-3}
C. 3.0×10^{-3}
D. 3.2×10^{-3}

3. Given the following information, calculate the average rate, $-\Delta[NH_3]/\Delta t$, between 10 and 30 minutes for the production of NH_3.

$$2NH_3(g) \rightarrow N_2(g) + 3H_2(g)$$

min	$[NH_3]$
0.0	0.130
10.0	0.083
20.0	0.063
30.0	0.045
40.0	0.033
50.0	0.025

A. 8.0×10^{-3}
B. 1.9×10^{-3}
C. 1.5×10^{-2}
D. 2.3×10^{-3}

4. Based on the following data, determine the rate law of this reaction:

$$SO_2(g) + Cl_2(g) \rightarrow SOCl_2 + Cl_2O(g)$$

Experiment	$[SO_2]$	$[Cl_2]$	Initial rate (mol/L s)
1	0.400	0.400	0.2918
2	0.400	0.200	0.0730
3	0.400	0.800	1.1674
4	0.200	0.800	0.5837

A. rate = $k[SO_2]$
B. rate = $k[SO_2][Cl_2]$
C. rate = $k[Cl_2]$
D. rate = $k[SO_2][Cl_2]^2$

5. Based on the data given in multiple choice Question 4, determine the rate constant for that reaction.

A. 1.82 B. 9.12 C. 4.56 D. 0.0351

6. Based on the following data, determine the overall order of this reaction:

$$2PO(g) + Cl_2(g) \rightarrow 2POCl(g)$$

Experiment	[PO]	$[Cl_2]$	Initial rate (mol/L s)
1	0.20	0.20	0.40
2	0.20	0.40	0.80
3	0.60	0.20	3.2
4	0.60	0.60	9.6

A. 3 B. 2 C. 5/2 D. 4

7. The rate of decomposition of a substance is first-order. If $k = 2.46 \times 10^{-3}\ s^{-1}$, what concentration of this substance remains after 2 minutes, knowing that $[\text{Substance}]_0 = 0.550\ M$?

A. 0.409 *M* B. 0.547 *M* C. 0.553 *M* D. 0.739 *M*

8. A particular drug can be sold until 20% of the original drug has undergone change. Knowing that $k = 1.25 \times 10^{-2}$/day, and the change is first order, how long, in days, will it take before the drug no longer can be sold?

A. 17 days B. 18 days C. 25 days D. 35 days

9. A gas-phase reaction in which substance A reacts with substance B to produce AB is found to be second order on A. Knowing that $k = 0.0368\ M^{-1}\ hr^{-1}$, and that $[A]_0 = 2.25\ M$, what percent of A remains after 177 hours of reaction?

A. 3.31% B. 85.6% C. 14.4% D. 6.39%

10. The decomposition of N_2O gas obeys zero-order kinetics. Given a rate constant = $2.46 \times 10^{-3}\ M/s$ and $[N_2O]_{120} = 0.155\ M$, calculate $[N_2O]_0$.

A. 0.445 B. 0.450 C. 0.550 D. 0.225

11. Which of the following is not a factor determining the energy of activation according to the Arrhenius equation?

A. orientation of molecules
B. temperature
C. frequency factor
D. none of these choices

12. Calculate E_a when $k_1 = 2.00$ and $k_2 = 10.0$, when $T_1 = 318$ K and $T_2 = 371$ K.

A. 9.0 J B. 30.0 kJ C. 85.0 kJ D. 3.1 J

13. Calculate T_2 when $E_a = 30.0$ kJ, $T_1 = 285$ K, $k_1 = 3.00$, and $k_2 = 15.0$.

A. 327 K B. 253 K C. 158 K D. 2.53 K

14. In a "reaction progress" graph, reacting molecules are most unstable at:

A. their initial position
B. when they are about to collide
C. right after they collide
D. at the transition state

15. A catalyst
    A. is consumed during a reaction while effectively increasing the number of reacting molecules that can reach the energy of activation.
    B. changes an endothermic reaction into an exothermic reaction.
    C. increases the energy of the products.
    D. provides an alternate pathway to the reaction, effectively lowering E_a.

16. In which of the following examples is a heterogeneous catalyst NOT used?
    A. hydrogenation of fats
    B. oxidation of sulfur dioxide
    C. decomposition of ozone
    D. catalytic converters of automobile exhaust systems

Answers to Exercises

1. a. Average rate of decomposition of H_2O_2 = 2.31×10^{-5} mol/L s
   Rate of production of O_2 = 1.16×10^{-5} mol/L s
   b. Average rate of decomposition of H_2O_2 = 1.16×10^{-5} mol/L s
   Rate of production of O_2 = 5.79×10^{-6} mol/L s

2. $f = 6, g = 1$

3. rate = k[A], $k = 0.3\ s^{-1}$

4. rate = $k[A]^2[C]$, $k = 0.38\ L^2/mol^2\ s$

5. a. 2 b. 3 c. 1½

6. rate = $k[A][B]^2[D]$, $k = 2.14 \times 10^{-2}\ L^3/mol^3\ s$

7. $k = 3.21 \times 10^{-5}\ s^{-1}$, $t_{1/2} = 2.16 \times 10^4$ s

8. 2.0 M

9. $k = 1.3 \times 10^{-2}\ s^{-1}$, $t_{1/2} = 53$ s

10. $[PH_3] = 0.21\ M$, 177 s = 2 min. 57 sec.

11. rate = $k[AB]^2$, $k = 0.080$ L/mol s, $t_{1/2} = 12.5$ s

12. a. $t_{1/2} = 1.92$ s b. [NOBr] = 0.648 M c. t = 1.25 s

13. $[NOBr]_0 = 0.87\ M$

14. $t_{1/2}(0.50\ M) = 5.0 \times 10^7$ s, $[NOCl]_{10^8\ s} = 0.17\ M$, $t_{1/2}(0.25\ M) = 1 \times 10^8$ s

15. rate = $k[IBr]^2$

16. 1/[NO] vs. t (pseudo-second order)

17. a. rate = $k[IBr]^2$ b. rate = k[IBr] c. rate = $k[IBr]^2$ d. rate = $k[IBr]^2$

18. a and b; for c the rate law would be rate = $k[H_2O_2][I^-]$

19. rate = $k[NO]^2[O_2]$

20. d

21. 9.76×10^{-5} L/mol s

22. 98.0 kJ/mol

23. 2.61×10^5 J/mol = 261 kJ/mol

24. 79.9 kJ/mol

25. 315 K

Answers to Multiple-Choice Self-Test

1. A	4. D	7. A	10. B	13. A	15. D
2. A	5. C	8. B	11. D	14. D	16. C
3. B	6. D	9. D	12. B		

CHAPTER 16

Liquids and Solids

The Bottom Line: Chapter 16

Your textbook justifies grouping solids and liquids together, separate from gases, by examining common properties. The densities, compressibilities and heats of phase change all indicate that the forces that hold solids and liquids together are similar. Gases have no such forces. In this chapter, your textbook discusses the bonding models, structure and other properties of liquids and solids.

16.1 Intermolecular Forces

Your textbook differentiates between **intra**molecular and **inter**molecular forces.

- **Intra**molecular forces mean **forces within a molecule**. We dealt with these in Chapter 13.
- **Inter**molecular forces mean **forces between molecules**. These are the forces that hold molecules together as liquids and solids.

Three kinds of intermolecular forces are discussed in this section.

- **Dipole-dipole forces** result when the partial positive and negative charges of neighboring polar covalent molecules attract each other. These forces are about 1% as strong as intramolecular covalent bonds.
- **Hydrogen bonds** are a special case where dipoles in **small, highly electronegative atoms** (such as fluorine) form a **surprisingly strong** interaction with hydrogen, which has a highly positive charge per unit size (Figure 16.3 in your textbook). Hydrogen bonding leads to substances with unusually high intermolecular bond energies.
- **London dispersion forces** are caused by the instantaneous dipoles that arise in a molecule as a result of momentary imbalances in electron distribution. These are very weak forces that become more important as the size of the atom of interest increases. (See the discussion on **polarizability** in your textbook.)

Example 16.1 A - London Dispersion Forces

The boiling point of argon is −189.4°C.

a. Why is it so low?
b. How does this boiling point help prove that London dispersion forces exist?
c. The boiling point of xenon is −119.9°C. Why is it higher than that of Argon?

Solution

a. Argon does not interact with other substances because it is so small and has a complete octet of valence electrons. Argon must be made quite cool to allow liquefication via London dispersion forces.

b. If these forces did not exist, argon would never liquefy.

c. Xenon is bigger and has more electrons than argon. The likelihood of momentary dipoles is thus greater. (It has a greater polarizability than argon.)

Example 16.1 B - The Effect Of Intermolecular Forces

Put the following substances in order from lowest to highest boiling point.

$$C_2H_6, NH_3, F_2$$

Solution

$$\mathbf{F_2, C_2H_6, NH_3}$$

F_2 can exhibit only intermolecular London dispersion forces. $\mathbf{C_2H_6}$ is not especially polar, but it does have a very slight electronegativity difference between the carbons and the hydrogens. $\mathbf{NH_3}$ exhibits hydrogen bonding, thus giving it a relatively high boiling point.

16.2 The Liquid State

The following review questions will serve to test your understanding of the material in this section.

1. Why do liquids tend to bead up when on solid surfaces?
2. What are **cohesive forces**? **Adhesive forces**? What causes these forces?
3. What is surface tension? Why does it arise?
4. Why does water form a concave meniscus when in a thin tube? Why does mercury form a convex meniscus?
5. What is viscosity? What is a requirement for a liquid to be viscous?
6. Why do models of liquids tend to be more complex than those for either solids or gases?

Example 16.2 - Properties Of Liquids

Which would have a higher surface tension, H_2O or C_6H_{14}? Why? Would the shape of the H_2O meniscus in a glass tube be the same or different than C_6H_{14}?

Solution

Water, having a large dipole moment, has relatively **large cohesive forces**. **Hexane**, C_6H_{14}, is essentially **nonpolar**. It has low cohesive forces. **Water** would therefore have the **higher surface tension**.

The water meniscus is **concave** because the adhesive forces of water to polar constituents on the surface of the glass are stronger than its cohesive forces. **Hexane** would have a **convex meniscus**. It has very small adhesive forces, and the slightly larger cohesive forces would dominate.

16.3 An Introduction to Structures and Types of Solids

This section in your textbook begins by contrasting **crystalline solids** (highly regular atomic arrangements) and **amorphous solids** (disordered atomic arrangements). The focus of the section (and the bulk of the chapter) is on **crystalline solids**.

Example 16.3 A - Basic Terms

Answer the following questions regarding the terms introduced in this section of your textbook.

a. Define **lattice**.
b. Define **unit cell**.
c. List three types of **cubic unit cells**.
d. What is diffraction?

e. Why are x-rays used in diffraction analyses of solids?
f. What is the difference between **ionic**, **molecular**, and **atomic solids**?

Solution

a. **Lattice**: A three-dimensional system of points designating the centers of the constituent building units in a crystalline solid.

b. **Unit cell**: The smallest repeating unit of a lattice.

c. As shown in Figure 16.9 of your textbook, the three types of cubic unit cells are **simple cubic**, **body-centered cubic** and **face-centered cubic**.

d. **Diffraction** is the **scattering of light** from a **regular array** of points or lines. The spacings between the points or lines (in our case planes of atoms) are related to the wavelength of the light.

e. X-rays are used because their wavelengths are similar to the distances between atomic nuclei.

f. **Ionic solids** form electrolytes when dissolved in water. **Molecular solids** do not. An **atomic solid** contains atoms of only one element. These atoms are covalently bonded to each other.

Three types of atomic solids are described in your textbook: atomic, ionic, and molecular. Note the description and examples of these (as given in Figure 16.12 in your textbook). Review the meaning of each of the terms of the **Bragg equation** (Equation 16.3 in your textbook), which is used to determine the structures of crystalline solids. When you are comfortable with the terms, try the next example.

Example 16.3 B - The Bragg Equation

A topaz crystal has a lattice spacing (d) of 1.36 Å (1 Å = 1 × 10^{-10} m). Calculate the wavelength of x-ray that should be used if $\theta = 15.0°$ (assume $n = 1$).

Solution

The Bragg Equation is in the form

$$n\lambda = 2d \sin\theta$$

where **n** = the order of diffraction
λ = the wavelength of the "incident energy" (the beam hitting the sample)
d = the spacing between planes of atoms
$\sin\theta$ = the sine of the angle of reflection of the beam

sin(15.0°) = 0.259

$$\lambda = \frac{2d\sin\theta}{n} = \frac{2(1.36\text{Å})(0.259)}{1} = \mathbf{0.704\ Å = 70.4\ pm}$$

16.4 Structure and Bonding in Metals

This section of your textbook begins by listing properties of metals. These are

- high thermal and electrical conductivity,
- malleability (can form thin sheets), and
- ductility (can be pulled into a wire).

A model that describes the structure of metals must be able to explain these properties.

The proposed model assumes the atoms of metals are **uniform hard spheres** that are packed to best utilize available space **(closest packing)**. Figures 16.13 - 16.15 in your textbook give examples of packing in layers.

- **Hexagonal closest packing** (hcp) has every **other** layer spatially equivalent ("*ababab*..."). Examples include magnesium, zinc, cadmium, cobalt, and lithium.
- **Cubic closest packing** (ccp) or a **face-centered cubic** structure has every **third** layer spatially equivalent ("*abcabcabc*..."). Examples include silver, aluminum, nickel, lead, and platinum.

Each sphere in each structure has **12 equivalent nearest neighbors (the coordination number is 12)**: 6 in its layer, 3 in the layer above, and 3 in the layer below (see Figure 16.16 in your textbook). The exception to the above list is

- **Body-centered cubic structure** (bcc). In this structure, spheres are **not** closest packed. The **coordination number** for such a sphere is **8** (see Figure 16.9(b) in your textbook). Iron and alkali metals have bcc structures.

The calculation of properties such as **density** from crystal structures is possible if you know the type of unit cell that a metal (or ionic solid, as we shall see later) forms. Your textbook deals only with the **ccp** structure in this regard. The following example shows how density is calculated using the atomic radius and the ccp structure.

Example 16.4 A - Crystal Structure And Density

The radius of a nickel atom is 1.24 **Å** (1 Å = 1 × 10^{-8} cm). Nickel crystallizes with a cubic closest packed structure (face-centered cubic). Calculate the **density** of solid nickel.

Solution

The goal is to calculate density, which is mass/volume. The volume in this case can be represented by the unit cell. As shown in your textbook, the ccp unit cell contains 4 atoms of nickel.

Use your understanding of stoichiometry to calculate the mass of nickel in the unit cell.

The volume for the unit cell can be calculated using the Pythagorean Theorem, as illustrated in OWL Interactive Example 16.1 in your textbook. The volume of the unit cell, a cube, is

$$V = d^3$$

where d = the length on a side.

The Pythagorean theorem says the square of the hypotenuse is equal to the sum of the squares of the sides of a right triangle or, in our case,

$$h^2 = d^2 + d^2$$

$h = 4r$ (r = the radius of nickel), as illustrated in the sample exercise.

$$(4r)^2 = 2d^2$$
$$16r^2 = 2d^2$$
$$8r^2 = d^2$$
$$d = r\sqrt{8}$$

Therefore,

$$V = d^2 = r^3 8^{1.5} = 22.63r^3$$

We can now solve for density.

$$\textbf{mass} = 4 \text{ atoms} \times \frac{58.70 \text{ g Ni}}{6.022 \times 10^{23} \text{ atoms}} = \mathbf{3.900 \times 10^{-22} \text{ g}}$$

$$\textbf{volume} = 22.63 \times (1.24 \text{ Å} \times 1 \times 10^{-8} \text{ cm/Å})^3 = \mathbf{4.315 \times 10^{-23}}$$

$$\textbf{density} = 3.900 \times 10^{-22} \text{ g} / 4.315 \times 10^{-23} \text{ cm}^3 = \mathbf{9.04 \text{ g/cm}^3}$$

This is in good agreement with the actual value of 8.90 g/cm^3.

Your textbook justifies **conduction** of **electricity** and **heat** by metals by using the **band model**. The discussion says that, if you have a **few** molecular orbitals (MOs) when a couple of atoms covalently bond, then a covalently bonded system with **many** atoms should lead to MOs with many energies (a "continuum"). Electrons in filled MOs can be excited into empty MOs and thus travel throughout the metal. Notice how your knowledge of MO theory (Chapter 9) helps you understand bonding in metals.

The section ends with a discussion of **metal alloys**. When you think you are comfortable with the material, try the following example.

Example 16.4 B - Metal Alloys

Why are high-carbon steels **interstitial alloys** while brass is a **substitutional alloy**?

Solution

An **interstitial alloy** is formed when holes in the closest packed metal structure are occupied by **small atoms** (in high carbon steels the iron holes are filled by carbon).

A **substitutional alloy** contains **similar sized** atoms of more than one element (the holes are not occupied). An example is the combination of copper and zinc to form the brass alloy.

Note that the makeup of the alloy greatly affects its properties.

16.5 Carbon and Silicon: Network Atomic Solids

The questions listed below will help you review the material presented in this section.

1. Define **network** solid.
2. List the **properties** of network solids.
3. Why does diamond have carbons with sp^3 hybridizations while those in graphite are sp^2 hybridized?
4. Use MO theory to explain why graphite conducts electricity but diamond does not.
5. Why can carbon form π-bonds while silicon cannot?
6. What is the difference between **silica** and **silicate**?
7. What property does glass share with ceramics? How are they different?
8. What is the purpose of adding arsenic to a semiconductor?
9. What is a **p-type** semiconductor? How does it work?
10. Describe how a p-n junction acts as a rectifier.

16.6 Molecular Solids

1. Define **molecular solid**.
2. Give some **examples** of molecular solids.
3. Describe the relative bonding strength and bond distances **within** and **between** molecules of a molecular solid.
4. Why do some larger nonpolar molecular solids remain solid at 25°C?

16.7 Ionic Solids

Your textbook says the structure of most binary ionic solids can be explained by the **closest packing of spheres**. Anions, which are usually larger than the cations with which they combine (see Section 13.4 in your textbook), are packed in either an hcp or ccp arrangement. Cations fill the holes within the packed anions.

Key Idea: The packing arrangement is done in such a way as to minimize anion-anion and cation-cation repulsions.

The nature of the holes depends on the ratio of the anion to cation size. Trigonal holes are smallest, followed by tetrahedral, and octahedral is the largest.

Example 16.7 A - Packing In Ionic Structures

Would AlP have a closest packed structure which is more like NaCl or ZnS?

Ionic Radii are: $Al^{3+} = 0.50$ Å, $P^{3-} = 2.12$ Å
$Zn^{2+} = 0.74$ Å, $S^{2-} = 1.84$ Å
$Na^{+} = 0.95$ Å, $Cl^{-} = 1.81$ Å

Solution

We need to take the anion-to-cation radius ratio in each of the three cases.

$S^{2-} / Zn^{2+} = 2.49$ (tetrahedral holes)
$Cl^{-} / Na^{+} = 1.91$ (octahedral holes)
$P^{3-} / Al^{3+} = 4.24$ (?)

The aluminum ion is very small compared to the phosphorus ion. Therefore not much room is needed by the Al^{3+} cations. **Tetrahedral holes** are adequate. **AlP** is **more like ZnS** than NaCl.

Based on your overall knowledge of chemistry, try the next example, which deals with classifying solids.

Example 16.7 B - Types Of Solids

Based on their properties, classify each of the following substances as to the **type of solid** it forms.

a. Fe b. C_2H_6 c. $CaCl_2$ d. graphite e. F_2

Solution

a. Solid iron is an **atomic solid** with metallic properties.
b. Solid C_2H_6, ethane, contains nonpolar molecules and is a **molecular solid**.
c. Solid $CaCl_2$ contains Ca^{2+} and Cl^{-} ions and is an **ionic solid**.
d. Graphite is made up of nonpolar carbon atoms covalently bonded in directional planes. It is a **network solid**.
e. Solid fluorine is made up of nonpolar fluorine molecules. It is a **molecular solid**.

16.8 Structures of Actual Ionic Solids

Your textbook sets the stage for discussion of the structures of a number of ionic solids by noting that

- **stoichiometry** of the compound is determined by the **ion charges**. However,
- the **structure** of the compound is determined by the relative **sizes** of the ions.

Please test your understanding of the material in this section by answering the following questions:

1. Why are the octahedral holes in NaCl filled with Na^+ ions?
2. How and why does the CsF crystal structure differ from that of CsCl? Can you think of some other example compounds that might show the same differences?
3. How and why does ZnS differ structurally from NaCl?
4. Describe the structure CaF_2. How does it differ from the NaCl structure?

16.9 Lattice Defects

This section considers several common imperfections in crystals. The next series of questions will help you test your own understanding of these **lattice defects**.

1. Define the types of defects discussed in this section.
2. What also must be missing in a CsF crystal that has a Schottky defect with a missing Cs^+?
3. Why is a Frenkel defect useful in photography? Can you give another example where this type of defect might be helpful?
4. What does the empirical formula of wüstite tell us about its structure?

16.10 Vapor Pressure and Changes of State

Your textbook introduces some very useful terms in this section. You need to be able to define **vaporization, enthalpy of vaporization, condensation, sublimation, enthalpy of fusion, melting point,** and **boiling point.**

Dynamic equilibrium is a concept you will be using a great deal in the latter half of your chemistry studies. It means that *two opposing processes are occurring at the* ***same rate***. The net effect is *no* ***observable*** *change*. But the system is not static. In this section, **the equilibrium vapor pressure** means that evaporation and condensation by a liquid are occurring at the same rate. The net effect is to have a **constant** vapor pressure exerted by the liquid.

The vapor pressure of a liquid varies with the molecular weight of the liquid and other molecular properties such as polarity and hydrogen bonding.

A heavier substance will have a lower vapor pressure than a lighter substance, all other things being equal, because its atoms are more polarizable, leading to larger intermolecular forces. A substance with hydrogen-bonding interactions will have a lower vapor pressure (will be less volatile) than a nonpolar substance. Your textbook introduces the Clausius-Clapeyron equation (Equation 16.5 in your textbook), which interrelates the **vapor pressure, temperature**, and **enthalpy** of a liquid.

$$\ln\left(\frac{P_{vap}^{T_1}}{P_{vap}^{T_2}}\right) = \frac{\Delta H_{vap}}{R}\left(\frac{1}{T_2} - \frac{1}{T_1}\right)$$

Where T_1 and T_2 are **temperatures** in Kelvins,
ΔH_{vap} is the **enthalpy of vaporization** of the liquid,
and $P_{vap}^{T_1}$ and $P_{vap}^{T_2}$ **vapor pressures** of the liquid temperatures T_1 and T_2.

Example 16.10 A - Clausius-Clapeyron Equation

The vapor pressure of 1-propanol at 14.7°C is 10.0 torr. The heat of vaporization is 47.2 kJ/mol. Calculate the vapor pressure of 1-propanol at 52.8°C.

Solution

Let's list what we are given.

$\Delta H_{vap} = 47.2$ kJ/mol
$R = 8.314$ J/K mol $= 0.008314$ kJ/K mol
$T_1 = 14.7°C = 287.9$ K
$T_2 = 52.8°C = 326.0$ K
$P_{vap}^{T_1} = 10.0$ torr
$P_{vap}^{T_2} = x$

Substituting,

$$\ln\left(\frac{10.0}{x}\right) = \frac{47.2 \text{ kJ/mol}}{0.008314 \text{ kJ/K mol}}\left(\frac{1}{326.0 \text{ K}} - \frac{1}{287.9 \text{ K}}\right)$$
$$= 5677\ (-4.06 \times 10^{-4})$$
$$= -2.305$$

Taking the antilog of both sides,

$$x/10 = 10.02$$
$$\boldsymbol{x = P_{vap}^{T_1} = 100.2 = 100 \text{ torr}}$$

Look at Figure 16.50 in your textbook. This illustrates the **heating curve** for water. Note two important observations:

- The **temperature** of a substance **remains constant** during a **phase change**.
- The **temperature rises** when heat is input while a substance is in **one phase**.

You can find the amount of **energy required** to convert water from ice at T_1 to steam at T_2 by using the following information:

- heat capacity of ice **(2.1 J/g°C)**
- ΔH_{fusion} of water **(6.0 kJ/mol)**
- heat capacity of liquid water **(4.2 J/g°C)**
- ΔH_{vap} of water **(43.9 kJ/mol)**
- heat capacity of steam **(1.8 J/g°C)**

How much of the information will be used will depend on the problem you have to solve.

Example 16.10 B - Heating Curve

How much energy does it take to convert 130 g of ice to steam at 160°C?

Solution

There are **five steps** involved in this conversion from ice to steam.

1. Heating ice from −40°C to the melting point.
2. Melting ice to form liquid water.
3. Heating liquid water to its boiling point.
4. Boiling liquid water to form steam.
5. Heating to 160°C.

The **total energy** required is the **sum** of the energy required in each of the five steps. The appropriate constants for each step are given in the discussion preceding this problem. The units of heat capacity contain °C because the temperature is rising in each of these steps. The units "enthalpy of fusion and vaporization" do not because the temperature is constant at a phase change.

Energy used = sum of energies from individual steps. There are 7.22 mol of water in 130 g.

	Step 1 = 40°C × 130 g × 2.1 J/g°C	=	10.92	kJ
	Step 2 = 7.22 mol × 6.0 kJ/mol	=	43.3	kJ
	Step 3 = 100°C × 130 g × 4.2 J/g°C	=	54.6	kJ
	Step 4 = 7.22 mol × 43.9 kJ/mol	=	317.0	kJ
	Step 5 = 60°C × 130 g × 1.8 J/g°C	=	14.04	kJ
Total Energy		=	**440.**	**kJ**

16.11 Phase Diagrams

The beauty of this section is that it helps explain a large number of real-world phenomena. (See the Chemical Insight on diamonds in this chapter of your textbook.)

You should be able to define the following terms: **phase diagram, critical temperature, critical pressure, critical point,** and **triple point**. You also should be able to answer the following general questions regarding material presented in this section.

1. Why does the solid/liquid line in the phase diagram of water have a negative slope? Why is it positive for carbon dioxide?
2. Why does it take longer to cook an egg in the Rocky Mountains than at sea level?
3. Describe the change in thinking about why ice is slippery enough to facilitate ice skating. How does this relate to the discussion about phase diagrams?
5. How does the phase diagram for carbon dioxide help explain how a CO_2 fire extinquisher works?
6. Snow sometimes sublimes. How can this be so in spite of the phase diagram?

Example 16.11 - Phase Diagrams

What phase changes does water undergo (See Figure 16.55 in your textbook.) as the pressure changes while the temperature is held constant at −12°C?

Solution

At very low pressures, water exists as a gas at −12° C. As the pressure is increased, it turns into a solid. At very high pressures the water will liquefy.

16.12 Nanotechnology

Please read both the main and boxed-text parts of the section and be able to answer these questions.

1. How do we define nanotechnology?
2. Why do nano-size particles have such different chemical behavior than larger particles?
3. What are some applications of nano-sized particles?
4. What is a grain, and how is it different from the normal crystals we study in general chemistry?
5. What is the model presented for the good electrical conductivity of copper, and how does that relate to the discussion of nano-sized particles?
6. Describe the formation and uses of carbon nanotubes.
7. Describe the role of transport, including quantum dots, in medical applications.

Exercises

Section 16.1

1. Define the terms in your own words:
   a. dipole-dipole forces
   b. hydrogen bonding
   c. London dispersion forces

2. Define the following terms in your own words:
   a. crystalline solids
   b. amorphous solids
   c. lattice
   d. unit cell

3. Of HF, HCl, and HBr, which has the highest boiling point? Why? Which has the lowest boiling point?

4. Propane, C_3H_8, is a gas at room temperature; hexane, C_6H_{14}, is a liquid; and dodecane, $C_{12}H_{26}$, is a solid. Explain.

5. Which would you expect to have a lower melting point, C_3H_8 or CH_3OH? Why?

Section 16.2

6. Why does water bead up more on a car that is waxed than one that isn't?

7. Would mercury bead up more on a waxed or unwaxed car?

8. Which would have greater surface tension, $N_2(l)$ or $Br_2(l)$?

Section 16.3

9. X-rays of wavelength 0.1541 nm produced a reflection angle $\theta = 15.55°$. What is the spacing between crystal planes ($n = 1$)? What would be the angle of reflection for $n = 2$?

10. X-rays of wavelength 263 pm were used to analyze a crystal. A reflection was produced at $\theta = 13.9°$. Assuming $n = 1$, calculate the spacing between the planes of atoms producing this reflection.

11. The second-order reflection ($n = 2$) for a gold crystal is an angle of 22.20° for x-rays of 154 pm. What is the spacing between these crystal planes?

12. Classify the following as ionic, molecular, or atomic crystalline solids.
   a. dry ice, $CO_2(s)$
   b. graphite
   c. $CaF_2(s)$
   d. $MnO_2(s)$
   e. $C_{10}H_8(s)$ (naphthalene)

Section 16.4

13. Tungsten crystallizes in a body-centered cubic structure with a unit cell edge length of 315.83 pm. The density of tungsten metal is 19.3 g/cm^3, and its atomic weight is 183.85 g/mol. Calculate the value of Avogadro's number by this method.

14. Silver crystallizes in a face-centered cubic structure with a unit cell edge length of 407.76 pm. The density is determined to be 10.5 g/cm^3. Calculate the atomic weight of silver.

15. The density of gold is 19.3 g/cm^3. Gold crystallizes in a face-centered cubic unit cell. What is the radius of a gold atom?

16. Potassium crystallizes in a body-centered cubic structure. Calculate the atomic radius of potassium if the unit cell edge length is 533.3 pm.

17. Why is iron relatively soft, ductile, and malleable while high carbon steels are much harder, stronger, and less malleable?

Section 16.6

18. What accounts for the lubricating ability of graphite?
19. Which electrons account for the conductivity of graphite?
20. Explain why graphite conducts electricity parallel to the layers much better than it conducts electricity perpendicular to the layers.
21. How many bonds would each sulfur atom in an S_8 molecule have to make to have the nearest noble gas electron configuration?

Section 16.7

22. Cobalt fluoride crystallizes in a closest packed array of fluoride ions with cobalt ions filling half of the octahedral holes. What is the formula of this compound?
23. Explain why anions usually have larger radii than cations.
24. What is the formula for the compound that crystallizes with a closest packed array of sulfur ions and that contains zinc ions in 1/8 of the tetrahedral holes and aluminum ions in 1/2 of the octahedral holes?
25. Manganese and fluoride ions crystallize such that each cubic unit cell has manganese ions at the corners and fluoride ions at the center of each edge. What is the formula of this compound?
26. What is the maximum radius for a cation that will fit in an octahedral hole in a face-centered cubic arrangement of chloride ions if the chloride ions are to remain in contact? (Radius of Cl^- is 181 pm.)

Section 16.10

27. Using the following vapor pressure data for CCl_4, make a graph, and determine the normal boiling point of the liquid.

Temp (°C)	20.0	40.0	60.0	70.0	80.0	90.0	100.0
V.P. (torr)	91.0	213.0	444.3	617.4	836.0	1110	1459

28. The vapor pressure of water at 25°C is 23.8 torr. Confirm the value of 43.9 kJ/mol for the heat of vaporization of water. (Use data for the normal boiling point as well as the vapor pressure given.)
29. On top of one of the peaks in Rocky Mountain National Park the pressure of the atmosphere is 550 torr. Determine the boiling point of water at this location.
30. In Denver, Colorado, the pressure of the atmosphere is 697 torr. Determine the boiling point of water at this location.
31. Calculate the vapor pressure of water at 0°C (in kilopascals, kPa).
32. Isopropanol, C_3H_8O, also is known as rubbing alcohol. The heat of vaporization is 42.1 kJ/mol. How much heat is needed to evaporate 25 g of isopropanol?
33. How much isopropanol must evaporate to cool 1.00 kg of it from 25°C to 20°C? (specific heat of isopropanol is 2.59 J/g°C)
34. What quantity of heat is required to melt 1.0 kg of ice at its melting point?
35. What quantity of heat is required to vaporize (100°C) 1.0 kg of ice at 0°C?

36. What is the final temperature when 10.0 g of water at 0.0°C is added to 100.0 g of water at 75°C?

37. If 10 g of ice at 0°C comes into contact with 10 g of water at 50°C, calculate the final temperature reached by the system at equilibrium.

38. What is the final temperature when 10.0 g of ice at 0.0°C is added to 100.0 g of water at 75°C? (ΔH_f = 6.0 kJ/mol, heat capacity of water is 4.2 J/g°C)

39. If 10 g of ice at 0°C comes in contact with 50 g of water at 10°C, calculate the final temperature reached by the system at equilibrium.

Section 16.11

40. How is the change in density for a solid-to-liquid phase related to the slope of the liquid-solid line of a phase diagram?

Multiple-Choice Self-Test

1. Which of the following molecule pairs are not involved in hydrogen bonding?

   A. $HCOOH$, H_2O
   B. H_2O, NH_3
   C. CH_3OH, CH_3COOH
   D. H_2 and I_2

2. Knowing that solutes with a certain polarity (or absence of it) are best dissolved in solutions with similar polarity, which of the following solvents would be optimal for the solvation of CH_3COOH?

   A. CH_4 B. CH_3OH C. C_2H_6 D. C_6H_6

3. Which of the following molecules interact primarily through London dispersion forces?

   A. SO_2 B. CCl_4 C. CH_2Cl_2 D. H_2S

4. Which of the following has the highest boiling point?

   A. H_2O B. HF C. HI D. HBr

5. Which of the following liquids will be the most viscous?

   A. C_3H_8 B. C_6H_6 C. CH_4 D. C_2H_6

6. X-rays with a particular wavelength were used to analyze a crystalline solid. Assuming $n = 1$, and knowing that the distance between the planes of atoms, which produces a diffraction with an angle = 20.0°, is 200 pm, calculate the frequency of the x-rays.

   A. 365 B. 6.22×10^{15} C. 0.110 D. 2.19×10^{18}

7. The successive packing pattern for an hcp cell is which of the following?

   A. ABABAB B. ABAABA C. ABCABC D. ABCCBA

8. Conducting electrons in metals are situated in:

   A. localized orbitals B. *s*-orbitals C. metallic orbitals D. conduction bands

9. The slope of the solid-liquid equilibrium line for water is −99.4 atm/°C. Calculate the pressure, in atm, if the melting point is −4.30°C.

   A. 327 B. 428 C. 214 D. 99.4

10. Using the Clausius-Clapeyron equation, calculate P_1 in torr at 300 K for ether, knowing that T_2 (normal boiling point) = 319 K, and ΔH_{vap} = 29.69 kJ/mol.

A. 302 torr B. 403 torr C. 69.2 torr D. 579 torr

11. Calculate the normal boiling point (T_2), in K, of a liquid knowing that ΔH_{vap} = 29.09 kJ/mol, and P_1 = 69.2 torr at 300 K.

A. 305 K B. 378 K C. 250 K D. 325 K

12. Dichlorodifluoromethane (CCl_2F_2) is a liquid that cools by evaporating. How many kilograms of CCl_2F_2 must be evaporated to freeze a tray of water (1050 g of water) at 273.15 K to ice at the same temperature? Heat of evaporation of CCl_2F_2 is 17.4 kJ/mol.

A. 2.44 kg B. 1.22 kg C. 10.3 kg D. 12.2 kg

13. How much heat is necessary to melt 175.32 g of NaCl at 801°C? (heat of fussion NaCl = 28.16 kJ/mol)

A. 22.5 kJ B. 9.39 kJ C. 30.2 kJ D. 84.5 kJ

14. The heat of crystallization for substance A = −65.0 J/g. The heat of fusion of water is 335 J/g. If 2000.0 g of liquid A is added to an excess of ice, how many grams of ice will melt, assuming there is no temperature change?

A. 388 g B. 200.0 g C. 10.3 g D. 10300 g

15. Arrange the following liquids, A, B, C, with vapor pressures at room temperature of 88, 680, and 155, respectively, in order of decreasing boiling points.

A. B > C > A B. A > B > C C. A > C > B D. C > A > B

16. The melting point of ice will change in what direction as pressure decreases?

A. no change
B. decreases
C. increases
D. depends on the pressure

Answers to Exercises

1. student's own answers
2. student's own answers
3. HF has the highest boiling point because of hydrogen bonding. HCl has the lowest boiling point.
4. All other things being equal, the higher the molecular weight of nonpolar compounds, the greater the intermolecular London dispersion forces.
5. C_3H_8 has a lower melting point because CH_3OH exhibits hydrogen bonding.
6. There are fewer adhesive forces between the nonpolar wax and water molecules than among water molecules.
7. an unwaxed car
8. $Br_2(l)$
9. 0.2874 nm; $\theta_{n=2} = 32.42°$
10. 547 pm
11. 408 pm

12. a. molecular
    b. atomic
    c. ionic
    d. ionic
    e. molecular
13. 6.05×10^{23}
14. 107 g/mol
15. 144 pm
16. 230.9 pm
17. Iron forms new structures with carbon that exhibit stronger attractive forces involving directional bonding.
18. weak attractive forces between layers of bonded carbon atoms
19. the (delocalized) electrons in the π bonds
20. Parallel conduction is favored because π electrons are localized above and below the layers.
21. 2 bonds
22. CoF_2
23. Extra electrons repel each other and have a lower Z_{eff}.
24. $ZnAl_2S_4$
25. MnF_3
26. 75 pm
27. 77°C
28. The calculated value is 42.7 kJ/mol.
29. 91.6°C
30. 97.7°C
31. 0.61 kPa
32. 18 kJ
33. 18.5 g
34. 330 kJ
35. 3200 kJ
36. 68°C
37. 0°C
38. 61°C
39. 0°C
40. If solid is more dense, positive slope; if solid is less dense, negative slope.

Answers to Multiple-Choice Self-Test

1. D
2. B
3. B
4. A
5. B
6. D
7. A
8. D
9. B
10. B
11. B
12. A
13. D
14. A
15. C
16. C

CHAPTER 17

Properties of Solutions

The Bottom Line: Chapter 17

This chapter deals with **solutions—homogeneous mixtures** of solids, liquids, or gases—as shown in Table 17.1 of your textbook. Among other things, you will find in this chapter the thermodynamics of solution formation; factors that affect solubility, including temperature and pressure; and properties of solutions, called colligative properties, that are based on the concentration of solute.

17.1 Solution Composition

This section introduces you to three new concentration terms. **Mass percent, mole fraction, and molality** are useful because they are **temperature independent**. As the temperature of a solution changes, the **molarity changes slightly** because the **solution volume changes**. As your textbook points out, these new terms depend on only mass, which is temperature independent.

The formulas for the new concentration units are:

$$\textbf{mass percent} = \frac{\textbf{g solute}}{\textbf{g solution}} \times 100\%$$

$$\textbf{mole fraction of A } (\chi_A) = \frac{n_A}{n_{total}} = \frac{\textbf{number of moles of A}}{\textbf{total number of moles}}$$

$$\textbf{molality } (m) = \frac{n \textbf{ solute}}{\textbf{kg solvent}}$$

Review the definitions of solute and solvent (See the margin notes by this section.).

Regarding the **solute:**

1. If it and the solvent are present in the *same phase*, it is the one in *lesser amount*.
2. If it and the solvent are present in *different phases*, it is the one that *changes phase*.
3. Your book puts it in more general terms by saying that it is the one *being dissolved*.

Regarding the **solvent**:

1. If it and the solute are present in the *same phase*, it is the one in *greater amount*.
2. If it and the solute are present in *different phases*, it is the one that *retains its phase*.
3. Your book says that it is the *dissolving medium*.

Also, recall the definition of **molarity** (moles of solute/liter of solution) that we have used throughout the course.

Example 17.1 - Concentration Units

A solution is prepared by adding 5.84 g of formaldehyde, H_2CO, to 100.0 g of water. The final volume of the solution is 104.0 mL. Calculate the molarity, molality, mass percent, and mole fraction of the formaldehyde in the solution.

Solution

Molarity: You have $5.84\text{ g }H_2CO \times \frac{1\text{ mole }H_2CO}{30.03\text{ g }H_2CO} = 0.194\text{ mol }H_2CO$

$$M = \frac{0.194\text{ moles }H_2CO}{0.104\text{ L solution}} = \mathbf{1.87}\ \boldsymbol{M}\ \mathbf{H_2CO}$$

Molality:

$$m = \frac{0.194\text{ mol }H_2CO}{0.100\text{ kg }H_2O\text{ (solvent)}} = \mathbf{1.94}\ \boldsymbol{m}\ \mathbf{H_2CO}$$

Mass percent:

$$\frac{5.84\text{ g }H_2CO}{105.84\text{ g }(H_2CO + H_2O)} \times 100\% = \mathbf{5.52\%\ H_2CO}$$

(The mass percent of H_2O must be 94.48%. Can you see why?)

Mole fraction: We know there is 0.194 mole of H_2CO.

$$n_{H_2O} = 100.0\text{ g }H_2O \times \frac{1\text{ mol }H_2O}{18.02\text{ g }H_2O} = 55.5\text{ mol }H_2O$$

$$\chi_{H_2CO} = \frac{0.194\text{ mol }H_2CO}{(0.194\text{ mol }H_2CO + 5.55\text{ mol }H_2O)} = \mathbf{0.0338}$$

(The mole fraction of H_2O must be 0.9662. Can you see why?)

17.2 The Thermodynamics of Solution Formation

This section concerns **how**, and **under what conditions**, a solution will be formed from the interaction of two substances. Your textbook lists **three steps** that must occur for solution formation to occur.

1. Breaking up the solute into individual components (expanding the solute, endothermic, $\Delta H = +$).
2. Overcoming intermolecular forces in the solvent to make room for the solute (expanding the solvent, endothermic, $\Delta H = +$).
3. Allowing the solute and solvent to interact to form the solution (often exothermic, $\Delta H = -$).

In solution formation the enthalpy of solution (the sum of steps 1, 2, and 3) is often "–" (exothermic). However, this is not the determinant of whether a reaction will occur. It is just one outcome of solution formation. Although we can not yet determine explicitly if solution formation will occur, we can make the generalization that like dissolves like.

The key to predicting whether two substances will mix (are "miscible") is to establish the polarity of each. If they are similar, you can probably form a solution. If they are very different, a solution is not likely to form. The thermodynamic explanation is also important. The salient point here is that the **gain in entropy (+ΔS)** as a result of dispersion of substances in each other will **lower ΔG** enough to make the mixing spontaneous.

When trying to dissolve nonpolar solutes in polar solvents, the strong negative gain in enthalpy (ΔH_3) is not present. Therefore the lower ΔG is not achieved, and dissolution does not readily occur.

Example 17.2 - Predicting Solubility

The following substances are slowly added to a 250-mL graduated cylinder: 50 mL of carbon tetrachloride (CCl_4, density = 1.4 g/mL), 50 mL of water (density = 1.0 g/mL) and 50 mL of cyclohexane (C_6H_{12}, density = 0.8 g/mL). After cyclohexane has been added, **how would the liquids appear in the cylinder** (i.e., would there be solution formation)? If solid I_2 flakes are added to the system, **in which layers (if any) would they dissolve?**

Solution

CCl_4 has no dipole moment (the molecule is **nonpolar**).
H_2O is a **polar** molecule.
C_6H_{12} is essentially **nonpolar**.

Because of the different densities of the liquids, CCl_4 would be the bottom layer. Water would sit on top of CCl_4. C_6H_{12} would be the top layer.

Iodine is nonpolar. When solid I_2 flakes are added to the layers, it would dissolve in our nonpolar layers, CCl_4 and C_6H_{12}. There would ultimately be purple $CCl_4 + I_2$ and $C_6H_{12} + I_2$ layers surrounding a clear water layer.

The section ends with a discussion on why smaller, more highly charged salts (e.g., LiF) have a positive ΔS°_{soln} whereas larger salts with lower charge densities (e.g., KCl) have a $-\Delta S^\circ_{soln}$. The rationale for the insolubility of benzene in water is also presented.

17.3 Factors Affecting Solubility

In considering **solubility effects** it is important to **separate the behavior of solids and liquids from gases**. The solubility of gases is for the most part independent of structure. Solid and liquid solubility is highly structure dependent.

Reread the discussion on structure effects in your textbook as it relates to the solubilities of vitamins A and C; then work the next example.

Example 17.3 A - Factors Affecting Solubility

Determine whether or not each of the following compounds is likely to be **water soluble**.

a. H_3C——$CHNH_2$
 |
 OH

b. H
 |
 H_2C——N——C_6H_5
 ‖
 NC_6H_5

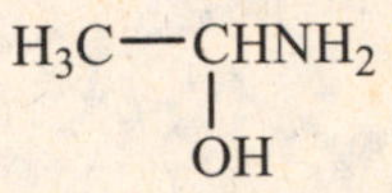

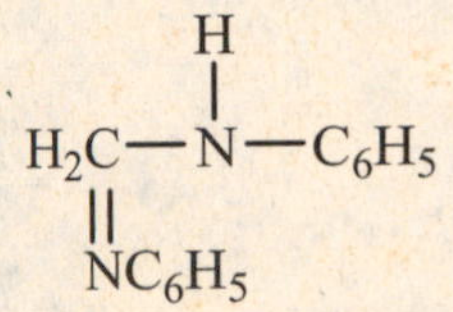

c. $C_4H_9CH{=}CH_2$

e. 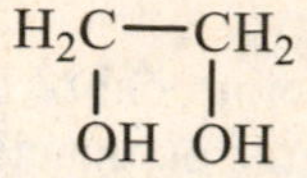

d. $CH_3(CH_2)_4CH_2NH_2$

Solution

An important factor in water solubility is the **ratio of polar to nonpolar groups** in the covalent molecule. The higher the ratio (with hydrogen-bonding groups being especially good), the more likely the covalent substance is to be water soluble, unless there is no dipole moment (such as in CCl_4).

However, there are no absolutes. There are other factors involved in solubility that you will learn more about when you are introduced to **organic chemistry** (Chapter 21 in your textbook).

a. water soluble (small molecule with 3 polar bonds)
b. insoluble (large molecule with only 1 polar bond)
c. insoluble (nonpolar molecule)
d. insoluble (large molecule)
e. water soluble (small molecule with 2 polar bonds)

Note how the answers here are consistent with Example 17.2 in this study guide.

Your textbook points out that **pressure has little effect on the solubilities of liquids and solids.** The solubility of gases is, for the most part, **independent of structure** and is given by **Henry's law,**

$$P = k_H\chi$$

where P = partial pressure of the gaseous solute above the solution (in atm)
χ = concentration of the dissolved gas (in mol/L)
k_H = a constant for a particular solution (in L atm/mol)

Keep in mind that Henry's law applies for **low gas concentrations** and gases that **do not react with the solution.**

Example 17.3 B - Henry's Law

The solubility of O_2 is 2.2×10^{-4} *M* at 0°C and 0.10 atm. Calculate the solubility of O_2 at 0°C and 0.35 atm.

Solution

Because the temperature is the same in both solutions, the Henry's law constant is the same. Therefore, if $k_H = P/\chi$ then

$$\frac{P_1}{\chi_1} = k_H = \frac{P_2}{\chi_2}$$

$P_1 = 0.10$ atm $\quad \chi_1 = 2.2 \times 10^{-4}\ M$
$P_2 = 0.35$ atm $\quad \chi_2 = ?$

$$\frac{0.10\text{ atm}}{2.2\times10^{-4}\ M} = \frac{0.35\text{ atm}}{\chi_2}$$

$$\chi_2 = 7.7 \times 10^{-4}\ M\ O_2$$

With regard to temperature effects, the solubility of most, but not all, solids increases with temperature. Your textbook points out that the solid dissolves **more rapidly** at higher temperatures. Also, review the discussion on **thermal pollution** and **boiler scale**.

17.4 The Vapor Pressures of Solutions

This section deals with the vapor pressure relationships of solutions containing both volatile and nonvolatile solutes. Let's deal with each of these separately.

Nonvolatile Solutes

This behavior is described by **Raoult's law**,

$$P_{soln} = \chi_{solvent} P^{o}_{solvent}$$

where P_{soln} = vapor pressure of the solution

$\chi_{solvent}$ = mole fraction of the solvent (See Section 5.5 in your textbook for a review of mole fractions.)

$P^{o}_{solvent}$ = vapor pressure of the pure solvent

This equation says the addition of a nonvolatile solute will cause the vapor pressure of the solution to fall in direct proportion to the mole fraction of the solute.

Let's extend the concept further using a **nonvolatile electrolyte** that **dissociates completely** in water. When solving the problem, we are interested in **moles of solute** actually present **in the solution** (after dissociation).

Example 17.4 A - Raoult's Law With Electrolytes

What is the vapor pressure of a solution made by adding 52.9 g of $CuCl_2$, a strong electrolyte, to 800.0 mL of water at 52.0°C? The vapor pressure of pure water at 52.0°C is 102.1 torr, and its density is 0.987 g/mL.

Solution

We know the vapor pressure of the solution will be lowered upon the addition of $CuCl_2$. Keep in mind however that $CuCl_2$ dissociates in aqueous solution:

$$CuCl_2(s) \rightarrow Cu^{2+}(aq) + 2Cl^{-}(aq)$$

As a result, we have **3 moles of solute formed for every mole we add.** To determine the new vapor pressure, we must calculate the **mole fraction** of water; then multiply by the vapor pressure of pure water.

$$\text{moles of } CuCl_2 = 52.9 \text{ g} \times \frac{1 \text{ mol}}{134.5 \text{ g}} = 0.393 \text{ mol}$$

$$\text{total moles of solute} = 0.393 \text{ mol} \times 3 = 1.18 \text{ mol solute}$$

$$\text{moles of water} = 800.0 \text{ mL} \times \frac{0.987 \text{ g}}{\text{mL}} \times \frac{1 \text{ mol}}{18.0 \text{ g}} = 43.8_6 \text{ mol } H_2O$$

$$\chi_{water} = \frac{n_{water}}{n_{water} + n_{solute}} = \frac{43.86}{43.86 + 1.18} = 0.974$$

$$P_{water} = \chi_{water} P^{o}_{water} = 0.974 \times 102.1 \text{ torr} = \mathbf{99.4 \text{ torr}}$$

Interactive Example 17.1 in your textbook shows how Raoult's law can be used to determine the molar mass of a nonvolatile solute. Review that example; then try this one.

Example 17.4 B - Molar Mass Via Raoult's Law

At 29.6°C pure water has a vapor pressure of 31.1 torr. A solution is prepared by adding 86.7 g of "Y," a nonvolatile nonelectrolyte, to 350.0 g of water. The vapor pressure of the resulting solution is 28.6 torr. Calculate the molar mass of Y.

Solution

In this problem, we know the vapor pressure drops due to the addition of Y.

If $P_{water} = \chi_{water} P^{o}_{water}$, then $\chi_{water} = \dfrac{P_{water}}{P^{o}_{water}}$. You can thus determine χ_{water}.

$$\chi_{water} = \frac{n_{water}}{n_Y + n_{water}}$$

You know n_{water} and χ_{water}. You can therefore solve for n_Y, which leads to the molar mass of Y.

$$\chi_{water} = \frac{28.6 \text{ torr}}{31.1 \text{ torr}} = 0.920$$

$$n_{water} = 350.0 \text{ g} \times \frac{1 \text{ mol}}{18.0 \text{ g}} = 19.4 \text{ mol}$$

$$\chi_{water} = \frac{n_{water}}{n_Y + n_{water}}$$

$$0.920 = \frac{19.4}{n_Y + 19.4}$$

$$\begin{aligned} 0.920(n_Y) + 17.8 &= 19.4 \\ 0.920(n_Y) &= 1.60 \\ n_Y &= 1.74 \text{ mol} \end{aligned}$$

$$\textbf{molar mass of Y} = \frac{86.7 \text{ g}}{1.74 \text{ mol}} = \mathbf{49.9 \text{ g/mol}}$$

We have thus far worked with nonvolatile solutes. **Volatile solutes** contribute to the vapor pressure such that

$$\mathbf{P_{total} = P_{solute} + P_{solvent}}$$

The vapor pressure of each component can be expressed by Raoult's law:

$$P_{solute} = \chi_{solute} P^{o}_{solute}$$

$$P_{solvent} = \chi_{solvent} P^{o}_{solvent}$$

therefore

$$P_{total} = \chi_{solute} P^{o}_{solute} + \chi_{solvent} P^{o}_{solvent}$$

By extension, for a solution that contains *n* components,

$$P_{total} = \chi_2 P_1^o + \chi_2 P_2^o + \ldots + \chi_n P_n^o$$

Let's apply this to the following example.

Example 17.4 C - Volatile Solutes

The vapor pressure of pure hexane (C_6H_{14}) at 60.0°C is 573 torr. That of pure benzene (C_6H_6) at 60.0°C is 391 torr. What is the expected vapor pressure of a solution prepared by mixing 58.9 g of hexane with 44.0 g of benzene (assuming an ideal situation)?

Solution

$$n_{hexane} = 58.9 \text{ g} \times \frac{1 \text{ mole}}{86.0 \text{ g}} = 0.685 \text{ mol}$$

$$n_{benzene} = 44.0 \text{ g} \times \frac{1 \text{ mole}}{78.0 \text{ g}} = 0.564 \text{ mol}$$

$$\chi_{hexane} = \frac{n_{hexane}}{n_{hexane} + n_{benzene}} = \frac{0.685 \text{ mol}}{0.685 \text{ mol} + 0.564 \text{ mol}} = \mathbf{0.548}$$

$$\chi_{benzene} = 1 - 0.548 = 0.452$$

$$P_{total} = \chi_{hexane} P^o_{hexane} + \chi_{benzene} P^o_{benzene} = 0.548\ (573) + 0.452\ (391) = \mathbf{491\ torr}$$

17.5 Boiling-Point Elevation and Freezing-Point Depression

This section in your textbook introduces us to **colligative properties**. These properties depend **only on the concentration of solute particles**; not on the nature of the particles.

Your textbook deals first with **boiling point elevation**, that is, **the increase in boiling point of a liquid due to the addition of a nonvolatile solute**. The boiling-point elevation, ΔT, is given by

$$\mathbf{\Delta T = K_b m_{solute}}$$

where $\mathbf{\Delta T}$ is the **boiling-point elevation** (in °C)
$\mathbf{K_b}$ is the **molal boiling-point elevation constant** (in °C kg/mol)
$\mathbf{m}$ is the **molality of the solute** (in mol solute/kg solvent)

Table 17.5 in your textbook gives values of K_b for some common solvents. One thing to keep in mind is that much useful information can be obtained from colligative property experiments, including molar mass of the solute, molality, and mass percent.

Example 17.5 A - Boiling-Point Elevation

A solution is prepared by adding 31.65 g of sodium chloride to 220.0 mL of water at 34.0°C (density = 0.994 g/mL; K_b for water is 0.51°C kg/mol). Calculate the boiling point of the solution, assuming complete dissociation of NaCl in solution (which is not true, strictly speaking—see Section 17.7).

Solution

In order to calculate ΔT, we need to calculate the **molality of NaCl**. In this case,

$$\text{molality} = \frac{\text{mol NaCl}}{\text{kg water}}$$

$$\text{mol NaCl} = 31.65\text{ g} \times \frac{1\text{ mol}}{58.4\text{ g}} = 0.542\text{ mol}$$

$$\text{kg water} = 220.0\text{ mL} \times \frac{0.994\text{ g}}{\text{mL}} \times \frac{1\text{ kg}}{1000\text{ g}} = 0.219\text{ kg}$$

$$m_{\text{NaCl}} = \frac{0.542\text{ mol}}{0.219\text{ kg}} = \mathbf{2.47_7}\ \boldsymbol{m}$$

But NaCl **dissociates** into Na^+ and Cl^-, giving 2 moles of solute per 1 mole dissolved. Therefore,

$$m_{\text{total}} = \mathbf{2.47_7 \times 2 = 4.95}\ \boldsymbol{m}$$

We can now calculate ΔT directly.

$$\Delta T = K_b m = 0.51°\text{C kg/mol} \times 4.95\ m = 2.5°\text{C}$$

The boiling point = 100 + 2.5 = **102.5°C**

As discussed in your textbook, the effect of adding a nonvolatile solute to a liquid is to lower the freezing point by an amount, ΔT, given by

$$\boldsymbol{\Delta T = K_f m_{\text{solute}}}$$

where **ΔT** is the **freezing-point depression** (in °C)
K_f is the **molal freezing-point depression** constant (in °C kg/mol)
m is the **molality of the solute** (in mol solute/kg solvent)

The overall effect of adding a nonvolatile solute to a liquid is to **extend the liquid range of the solvent**. The same information can be found from both freezing-point depression and boiling-point elevation problems.

Example 17.5 B - Freezing-Point Depression

How many grams of glycerin ($C_3H_8O_3$) must be added to 350.0 g to lower the freezing point to −3.84°C (K_f for water = 1.86°C kg/mol)?

Solution

$\Delta T = 3.84°\text{C}$
$K_f = 1.86°\text{C kg/mol}$

$$\text{molality} = \frac{\text{moles of glycerin}}{\text{kg of water}}$$

The bottom line, then, is to solve for the **molality of glycerin**, which leads to **moles** and finally **grams of glycerin.**

$$m_{\text{glycerin}} = \frac{\Delta T}{K_f} = \frac{3.84°C}{1.86°C\ \text{kg/mol}} = 2.06_4\ m$$

$$\text{moles of glycerin} = \frac{2.064\ \text{mol glycerin}}{\text{kg water}} \times 0.350\ \text{kg water} = 0.723\ \text{mol}$$

$$\textbf{g of glycerin} = 0.723\ \text{mol} \times 92.1\ \text{g/mol} = \textbf{66.5 g of glycerin}$$

17.6 Osmotic Pressure

This section in your textbook begins with the definition of **osmosis** as **the flow of solvent into a solution through a semipermeable membrane. Osmotic pressure** is **the pressure that just stops the osmosis**. Osmotic pressure is a **colligative property** because its value depends on the **concentration of the solute,** not its nature. Your textbook points out that osmotic pressure is especially useful for determining molar masses experimentally because **small solute concentrations produce large osmotic pressures.**

The relationship between osmotic pressure and solution concentration is given by

$$\pi = MRT$$

where π is the **osmotic pressure** (in atm)
M is the **molarity of the solute** (in mol/L)
R is the **universal gas constant** (0.08206 L/K mol)
T is the **temperature** (in K)

Let's try an example that uses osmotic pressure to determine the molar mass of a substance.

Example 17.6 - Osmotic Pressure

The osmotic pressure of a solution containing 26.5 mg of aspartame per liter is 1.70 torr at 30°C. Calculate the molar mass of aspartame.

Solution

$$\pi = MRT$$

We want to solve for molar mass. This can be done through **molarity**.

$\pi = 1.70\ \text{torr} \times \frac{1\ \text{atm}}{760\ \text{torr}} = 2.24 \times 10^{-3}\ \text{atm}$

$R = 0.08206\ \text{L atm/K mol}$

$T = 303\ \text{K}$

$$M = \frac{\pi}{RT} = \frac{2.24\times10^{-3}\ \text{atm}}{0.08206\ \text{L atm/K mol}\ (303\ \text{K})} = 9.01 \times 10^{-5}\ \text{mol/L}$$

$$\textbf{molar mass} = \frac{\text{g aspartame}}{\text{mol}} = \frac{1\ \text{L}}{9.01\times10^{-5}\ \text{mol}} \times \frac{0.0265\ \text{g}}{\text{L}} = \textbf{294 g/mol}$$

Several new terms are introduced in the discussion following Interactive Example 17.4 in your textbook. You should be able to define **dialysis, hypertonic solutions, isotonic solutions, crenation,** and **lysis.**

You should be able to describe the concept of **reverse osmosis** and how this can be used to **desalinate water.**

17.7 Colligative Properties of Electrolyte Solutions

Recall Example 17.5 A in this study guide, in which we calculated the boiling-point elevation of a solution to which NaCl had been added. We made an **assumption** that **NaCl completely dissociates**. Thus our 2.48 *m* solution dissociated into 2.48 *m* Na^+ and 2.48 *m* Cl^-.

This section of your textbook points out that this assumption is **not valid, especially at high solute concentrations**. The reason seems to be **ion pairing** in which some sodium and chloride ions encounter one another, pair, and are counted as a single particle.

The equation that takes experimentally observed dissociation into account is

$$\Delta T = imK$$

where ***i*** is the **van't Hoff factor** representing electrolyte dissociation in solution. Keep in mind that, for a given electrolyte, ***i*** is **concentration dependent.** Also, ***i* = 1 for all nonelectrolytes.** The expected and observed values for several electrolytes are given in Table 17.6 of your textbook.

Example 17.7 - Van't Hoff Factor

Use data from Table 17.6 in your textbook to calculate the freezing point and expected osmotic pressure of a 0.050 *m* $FeCl_3$ solution at 25.0°C. (Assume the density of the final solution equals 1.0 g/mL, and the liquid volume is unchanged by the addition of $FeCl_3$.)

Solution

a. ***Freezing point***

From Table 17.6 in your textbook, $i_{observed} = 3.4$ for this 0.050 *m* solution.

$\Delta T = imK_f = 3.4 \times 0.050\ m \times 1.86°C\ kg/mol = 0.32°C$

The **freezing point** would be **−0.32°C**.

b. ***Osmotic pressure***

Analogous to the formula for freezing-point depression, the formula for osmotic pressure (itself a colligative property) is

$$\pi = iMRT$$

We assumed that the density of the solution is 1.0 g/mL, and the volume of the liquid was unchanged by the addition of $FeCl_3$. In this **unusual** case, the molarity = 0.05 *M*.

$$\pi = \underset{i}{\underset{\uparrow}{3.4}} \times \underset{M}{\underset{\uparrow}{0.050\ M}} \times \underset{R}{\underset{\uparrow}{0.08206\ L\ atm/K\ mol}} \times \underset{T}{\underset{\uparrow}{298.2\ K}} = \mathbf{4.16\ atm}$$

Osmotic pressure (π) = 4.16 atm

17.8 Colloids

The following review questions will help guide your study of this section.

1. What is the **Tyndall effect?**
2. How can the Tyndall effect be used to distinguish between a suspension and a true solution?
3. Define **colloid.**
4. Give some examples of **colloids.**
5. How can we prove that **electrostatic repulsion** helps stabilize a colloid?
6. Define **coagulation.**

7. How does heating destroy a colloid?
8. How does the addition of an electrolyte destroy a colloid?
9. What is the relationship between proteins in fish and ice cream manufacture?

Exercises

Section 17.1

1. Rubbing alcohol contains 585 g of isopropanol (C_3H_7OH) per liter (aqueous solution). Calculate the molarity.

2. The density of a 10.0% (by mass) solution of NaOH is 1.109 g/cm^3. Calculate the concentration of this solution in molarity, molality, and mole fraction.

3. Toluene ($C_6H_5CH_3$), an organic compound often used as a solvent in paints, is mixed with a similar organic compound, benzene(C_6H_6). Calculate the molarity, molality, mass percent, and mole fraction of toluene in 200. mL of solution that contains 75.8 g of toluene and 95.6 g of benzene. The density of the solution is 0.857 g/cm^3.

4. A hydrochloric acid solution was made by adding 59.26 g HCl to 100. g H_2O. The density of the solution was 1.19 g/cm^3. Calculate the concentration of HCl in molarity, molality, mass percent, and mole fraction.

5. What mass of H_2SO_4 is required to prepare 250 mL of a 6.00 *M* solution?

6. In what circumstance is molality a more useful unit than molarity?

7. How many grams of NaOH and H_2O are required to prepare 100. g of a 28.7% NaOH by weight solution?

8. Find the concentration of NaOH in molarity, molality, and mole fraction for the solution prepared in Problem 7, assuming a solution density of 1.10 g/mL. Also determine the mole fraction of H_2O.

9. Seawater contains 1.94% chlorine (by mass). How many grams of chlorine are there in 400 mL of seawater if the density of seawater is 1.025 g/cm^3?

10. The molal concentration of a solution of salicylic acid ($C_7H_6O_3$), the primary organic molecule used in aspirin production, and ethanol ($C_2H_5O_3$) is 17.5 *m*. If 42.0 g of ethanol were used to make the solution, how many grams of salicylic acid were needed?

11. The concentration of glucose, $C_6H_{12}O_6$, in a biological fluid is 75 mg/100 g. What is the molal concentration?

Section 17.2

12. Rationalize the water solubilities for the gases listed below (units are g/100 cm^3).

Gas	Solubility
H_2	0.09191
CO_2	0.141
HCl	82.3
NH_3	89.9

What would you expect for relative solubilities in hexane (C_6H_{14})?

13. Would boric acid, $B(OH)_3$, be more soluble in ethanol, C_2H_5OH, or in benzene, C_6H_6?

14. Predict which substance would be more soluble in water.

 a. $NH_2CH_2CH_2OCH_3$ or NH_2CH_2COOH

 b. CH_3CH-CH_2 or $CH_3CH_2CHCH_2CH_3$

 | | |

 HO OH CH_2OH

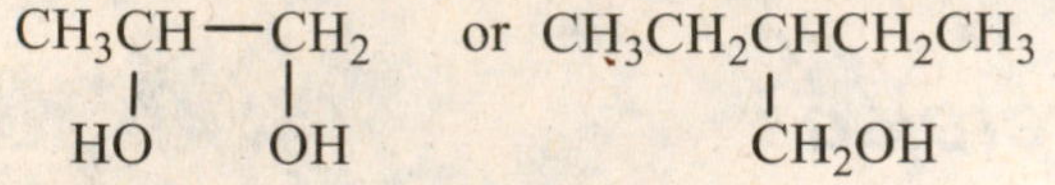

 c. C_4H_9OH or C_4H_9SH

Section 17.3

15. Ammonium salts can be used to make chemical cold packs. When these salts are dissolved in water the solution gets quite cold. What can we say about the heat of solution?

16. The solubility of oxygen in water at 0°C and 1 atm is 0.073 g per liter. What is k in Henry's law for this temperature?

17. The solubility of nitrogen at 0°C in water is 8.21×10^{-4} mol/L if the N_2 pressure above the water is 0.790 atm. What is the solubility at 1 atm of N_2 and 0°C?

18. What is the Henry's law k for nitrogen in Problem 17?

19. The solubility of nitrogen in blood at 37°C and 0.80 atm is 5.6×10^{-4} M. If a deep-sea diver breathes compressed air from a tank at a partial pressure of 3 atm, what would be the solubility of nitrogen in the diver's blood at 37°C?

20. Why doesn't Henry's law work for ammonia?

21. The solubility of CO_2 in water at 25°C and 1 atm is 0.034 M. What is the Henry's law constant (k)? What would the solubility of CO_2 in water be at 0.038 atm and 25°C?

Section 17.4

22. The vapor pressure of ethanol at 40°C is 135.3 torr. Calculate the vapor pressure of a solution of 53.6 g of glycerin, $C_3H_8O_3$, in 133.7 g of ethanol, C_2H_5OH, at 40°C.

23. Antifreeze solutions are mainly ethylene glycol, $C_2H_6O_2$, in water. Calculate the vapor pressure of a solution made by adding 101.6 g of ethylene glycol to 139.6 g of water at 50.0°C. At this temperature ethylene glycol is essentially nonvolatile, and the vapor pressure of water is 92.51 torr.

24. Sucrose, $C_{12}H_{22}O_{11}$ (a nonvolatile substance), is a sweetener. A solution was made with 35.2 g of sucrose and 78.0 g of water at 30°C. Calculate the vapor pressure of the solution if the vapor pressure of water at 30°C is 31.824 torr.

25. A solution was prepared by dissolving 16.8 g of camphor ($C_{10}H_{16}O$) in 86.1 g of benzene (C_6H_6). At 23°C, the vapor pressure of benzene is 86.0 torr. If the vapor pressure of the resulting solution was 78.2 torr, calculate the vapor pressure of camphor at 23°C. (Note: Solid camphor has a low volatility.)

26. How many grams of a nonvolatile compound B (molar mass = 97.8 g/mol) would need to be added to 250 g of water to produce a solution with a vapor pressure of 23.756 torr? The vapor pressure of water at this temperature is 42.362 torr.

27. Calculate the vapor pressure of a solution of 40.27 g $MgCl_2$ in 500.3 mL of water at 25.0°C. The density of water is 0.9971 g/mL, and the vapor pressure is 23.756 torr.

28. A 12.97-g sample of a metal chloride, general formula MCl, is added to 100.0 g of water at 30.0°C. The vapor pressure of the solution is 30.60 torr. The vapor pressure of pure water at 30.0°C is 31.824 torr. What is the formula of the metal chloride?

29. Calculate the vapor pressure of a solution made with 321 g of toluene (C_7H_8) and 398 g of benzene (C_6H_6). At 60°C, the vapor pressures of the two volatile components are 140. torr and 396 torr respectively.

30. What is the vapor pressure of a solution made by adding 26.93 g potassium sodium tartrate, $KNaC_4H_4O_6$, to 676.5 g of water at 25°C? The vapor pressure of pure water is 23.756 torr at 25°C. Assume that potassium sodium tartrate is a nonelectrolyte (a poor assumption).

Section 17.5

31. An aqueous solution of sucrose, $C_{12}H_{22}O_{11}$, boils at 112°C. What is the molality?

32. 1.51 g of white phosphorus (P_4) was dissolved in 18.0 g of carbon disulfide (CS_2). The boiling point elevation of the carbon disulfide solution was found to be 1.36°C. Calculate the molality of the white phosphorus. What is K_b for carbon disulfide?

33. A solution of carbon tetrachloride (CCl_4) has a molality of 1.35 *m*. The freezing-point depression constant for CCl_4 is 30°C kg/mol, and its freezing point is −22.99°C. At what temperature would the solution freeze?

34. What are the boiling and freezing points of a solution of 50.3 g of I_2 in 350 g of chloroform? (See Table 17.5 in your textbook for constants.)

35. A 4.367-g sample of an unknown hydrocarbon is dissolved in 21.35 g benzene. The freezing point of the solution is observed to be −0.51°C. Calculate the molar mass of the unknown.

36. During a Wisconsin winter, the temperature can reach −25°C (or colder!). How many grams of antifreeze (ethylene glycol, $C_2H_6O_2$) would you need to add to your radiator to keep 7.5 liters of water from freezing? Assume that the density is 1.0 g/mL (K_f for water = 1.86°C kg/mol).

Section 17.6

37. Calculate the osmotic pressure of a solution made by adding 13.65 g of sucrose, $C_{12}H_{22}O_{11}$, to enough water to make 250. mL of solution at 25°C.

38. What is the osmotic pressure of 1.38×10^{-2} *M* KBr at 25°C?

39. The osmotic pressure of blood is 7.65 atm at 37°C. How much glucose, $C_6H_{12}O_6$, should be used per liter for an intravenous injection having the same osmotic pressure as blood?

40. The osmotic pressure of an aqueous solution of a polypeptide solution was determined to be 4.80 torr at 25°C. How many grams of polypeptide would be in 2.00 L of solution? (The molar mass of the peptide is 269 g/mol.)

41. The osmotic pressure of a tryptophan solution containing 1.136 g per liter is 103.5 torr at 25°C. What is the molar mass of tryptophan?

42. What is the minimum pressure required to desalinate 1.0 *M* salt (NaCl) solution at 25°C?

43. Calculate the osmotic pressure (in atm) of a 0.225 *M* aqueous solution of urea that is isotonic with sea water at 10°C.

Section 17.7

44. Table 17.6 in your textbook lists values of observed van't Hoff factors for 0.05 *m* solutions. Would you expect the van't Hoff factors for 0.5 *m* solutions to be greater or less than the values for 0.05 *m*?

45. Using the van't Hoff factor from Table 17.6 in your textbook calculate the freezing point, boiling point, and osmotic pressure (at 25°C) of a 0.050 *m* $MgCl_2$ solution. (Assume 0.050 *m* = 0.050 *M*.)

46. Calculate the osmotic pressure of a 0.05 *M* solution of NaCl at 30°C. The van't Hoff factor for NaCl is 2.0.

Section 17.8

47. Indicate the type of colloid that each of the following represents (i.e., aerosol, foam, emulsion, sol, or gel):

    a. MOM (milk of magnesia)
    b. salad dressing
    c. meringue
    d. rain cloud

Multiple-Choice Self-Test

1. How many grams of nitric acid, 70.5% by weight, are present in 1500.0 mL of a solution with a density of 1.42 g/cm^3?

   A. 352 g B. 718 g C. 2130 g D. 1.50×10^3 g

2. The mole fraction of calcium chloride in water is 0.326. How many grams of $CaCl_2$ are present in 90.0 g of solution?

   A. 32.1 g B. 38.1 g C. 60.0 g D. 67.5 g

3. A solution contains 1300. g of solvent and 40.0 g of solute and is known to be 0.170 molal. What is the molar mass of the solute in g/mol?

   A. 90 g/mol B. 181 g/mol C. 133 g/mol D. 72.0 g/mol

4. Calculate the molarity of a solution prepared by mixing 50.0 cm^3 of 0.82 M NaCl with 30.0 cm^3 0.52 M NaCl solution, assuming volumes are additive.

   A. 0.71 *M* B. 1.02 *M* C. 0.80 *M* D. 0.66 *M*

5. What volume of solution is required to prepare a 0.014 *M* solution containing 1.40 g of $FeCl_2$?

   A. 0.100 L B. 1.80 L C. 0.79 L D. 0.010 L

6. How will increasing the pressure of a gas affect the solubility of the gas in a solvent?

   A. increase it C. no effect
   B. decrease it D. depends on pressure

7. Calculate the concentration of carbon dioxide in the atmosphere if the concentration of carbon dioxide gas in water, at equilibrium with the Earth's atmosphere at 25°C, is 1.1×10^{-4} molar. Assume the total pressure of the atmosphere to be 1.00 atm.

   A. 0.31% B. 0.62% C. 6.2% D. 2.7%

8. 4.3 g of Vitamin B_{12} is dissolved in 74 g of ether $[(C_2H_5)_2O]$ at 25°C. The change in vapor pressure amounts to 1.00 torr. Calculate the molar mass of this vitamin. The pressure of ether at 25°C is 315 torr.

   A. 422 g/mol B. 106 g/mol C. 675 g/mol D. 1.36×10^3 g/mol

9. 0.015 moles of a compound are dissolved in 18.0 fl. oz. of water. At what temperature will the solution begin to freeze? The density of water = 1.00 g/mL.

   A. −0.055°C B. −0.210°C C. −1.86°C D. −2.00°C

10. What is the boiling point, in degrees Celsius, of a solution that is prepared by mixing 500.0 mL of water with 5.844 g of sodium chloride?

    A. 100.1°C B. 102.0°C C. 99.80°C D. 100.05°C

11. Four distinct solvents, A, B, C, and D, have the following K_b values: 0.51, 4.59, 5.07, and 40.0°C/m, respectively. Their respective molar masses are: 60, 88, 98, and 152 g/mol. 3.0 g of a solute with a molar mass of 138.0 g/mol is dissolved in 200.0 g of one of the solvents. The boiling point is changed by 4.34°C. Which solvent is it?

    A. A B. B C. C D. D

12. What is the osmotic pressure created by 20.0 mg of insulin dissolved in 10.0 mL of solution at 27°C? The molar mass of insulin is 6.02×10^3 g/mol.

    A. 3.11 torr B. 6.22 torr C. 0.0493 torr D. 0.0818 torr

13. What molar glucose solution would be isotonic with blood? The osmotic pressure of blood at 37°C is 7.65 atm. The molar mass of glucose is 180 g/mol.

    A. 0.30 *M* B. 0.15 *M* C. 2.5 *M* D. 1.5 *M*

14. Calculate the van't Hoff factor for a solute that is a univalent-univalent weak electrolyte and is 25% ionized.

    A. 122 B. 1.25 C. 15.0 D. 1.75

15. What is the percent ionization of a univalent-univalent weak electrolyte solution if the van't Hoff factor is equal to 1.15?

    A. 30% B. 20% C. 15% D. 100%

16. How many grams of NaCl would be required to lower the freezing point of 2.00 kg of water to 0.0°F?

    A. 280 g B. 560 g C. 200 g D. 117 g

Answers to Exercises

1. 9.75 *M*
2. molarity = 2.77 *M*; molality = 2.78 *m*; mole fraction = 0.0476
3. molality = 8.62 *m* $C_6H_5CH_3$; molarity = 4.11 *M* $C_6H_5CH_3$; mass percent = 44.2%; mole fraction = 0.402
4. molarity = 12.1 *M*; molality = 16.3 *m*; mass percent = 37.2%; mole fraction = 0.226
5. 147 g
6. Molality does not change when temperature changes. The molarity will change.

7. 28.7 g NaOH; 71.3 g H_2O

8. molarity = 7.89 *M*; molality = 7.2 *m*; mole fraction of NaOH = 0.15; mole fraction of H_2O = 0.85

9. 7.95 g Cl

10. 101 g $C_7H_6O_3$

11. 4.2×10^{-3} *m*

12. H_2 and CO_2 are nonpolar, therefore not very soluble in water. HCl and NH_3 are polar, therefore more soluble in water. In hexane it is expected that H_2 and CO_2 will be more soluble than HCl and NH_3.

13. Ethanol. Ethanol is more polar than benzene. Because boric acid is also a polar substance, it would be more soluble in ethanol.

14. a. NH_2CH_2COOH b. 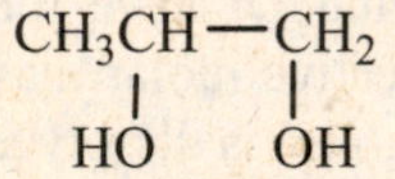 c. C_4H_9OH

15. It is an endothermic reaction; therefore $\Delta H_{sol} > 0$.

16. 2.3×10^{-3} mol/ L atm

17. 1.04×10^{-3} *M*

18. 962 L atm/mol

19. 2.1×10^{-3} mol/L

20. $NH_3 + H_2O \rightarrow NH_4^+ + OH^-$; ammonia reacts with water

21. k = 29 L atm/mol; 1.3×10^{-3} *M*

22. 113 torr

23. 76.39 torr

24. 31.1 torr

25. 0.05 torr

26. 1.063×10^3 g compound B

27. 23.40 torr

28. NaCl

29. 291 torr

30. 23.66 torr

31. 23.5 *m*

32. 6.77×10^{-1} *m*; 2.01°C kg/mol

33. −63.49°C

34. 63.3°C, −66.2°C

35. 174 g/mol

36. 6.3×10^3 g (or 6.3 kg)

37. 3.90 atm

38. 0.675 atm, 513 torr
39. 54.3 g/L
40. 0.138 g
41. 204 g/mol
42. 49 atm
43. 5.23 atm
44. 0.5 *m* factors would be less than 0.05 *m* factors. The more concentrated the solute the more chance for interactions.
45. Freezing pt. = −0.25°C; Boiling pt. = 100°C; osmotic pressure = 3.3 atm.
46. 2.49 atm
47. a. sol b. emulsion c. foam d. aerosol

Answers to Multiple-Choice Self-Test

1. D	4. A	7. A	10. A	13. A	15. C
2. D	5. C	8. D	11. D	14. B	16. B
3. B	6. A	9. A	12. B		

CHAPTER 18

The Representative Elements

The Bottom Line: Chapter 18

The chemistry of these elements can be explained by their electronic structures. As you go through the chapter, note how elements within a group show similar reactive properties.

18.1 A Survey of the Representative Elements

This section begins by reviewing the positions of various types of elements in the periodic table.

- **representative elements** are those in which ***s* and *p* orbitals are being filled**.
- **transition elements** are those in which ***d* orbitals are being filled**.
- **lanthanides and actinides** (sometimes called "inner-transition metals") are those in which the **4*f* or 5*f* orbitals are being filled**.

You should be able to determine the type of element by its position on the periodic table. (Review Sections 12.13-15 if you have forgotten!) You may also wish to review electron configurations to confirm your assignments. Try the following review examples.

Example 18.1 A - Electron Configuration Review

Write the electron configurations (either shorthand or longhand) for each of the following elements:

a. yttrium   b. silver   c. bismuth   d. iodine

Solution

We can use either our configuration triangle (Chapter 12) or the position in the periodic table to establish the electron configuration.

a. Yttrium has atomic number 39. It is in **Period 5** and **Group 3B**. It must therefore be $\mathbf{[Kr]5s^24d^1}$. Checking with the longhand form,

$$\mathbf{Y: 1s^22s^22p^63s^23p^64s^23d^{10}4p^65s^24d^1}$$

b. Silver has atomic number 47. It is in **Period 5** and **Group 1B**. It is $\mathbf{[Kr]5s^14d^{10}}$.

$$\mathbf{Ag: 1s^22s^22p^63s^23p^64s^23d^{10}4p^65s^14d^{10}}$$

c. Bismuth has atomic number 83. Elements with high atomic numbers get rather difficult to handle by using the configuration triangle. It is easier to use your knowledge of the periodic table. Bismuth is in **Period 6** and **Group 5A**. The shorthand form is $\mathbf{[Xe]6s^24f^{14}5d^{10}6p^3}$. The longhand form is

$$\mathbf{Bi: 1s^22s^22p^63s^23p^64s^23d^{10}4p^65s^24d^{10}5p^66s^24f^{14}5d^{10}6p^3}$$

d. Iodine has atomic number 53. It is in **Period 5** and **Group 7A**. It is $\mathbf{[Kr]5s^24d^{10}5p^5}$.

$$\mathbf{I: 1s^22s^22p^63s^23p^64s^23d^{10}4p^65s^24d^{10}5p^5}$$

Example 18.1 B - Types Of Elements

Classify each of the following elements as **representative** or **transition**:

a. iron b. sodium c. argon d. nickel

Solution

a. transition b. representative c. representative d. transition

The section focuses attention on chemical differences within groups caused by size differences in atoms. In particular, **the first member of a group tends to have different properties than the rest of the group because it is so much smaller than the rest of the atoms in the group.**

- Hydrogen is a nonmetal and forms covalent bonds. Lithium is a metal and forms ionic bonds.
- Beryllium oxide is amphoteric. Other Group 2A oxides are basic.
- Boron behaves as a nonmetal or semimetal. Other Group 3A elements are metal.
- Carbon can form π bonds due to effective overlap between 2*p* orbitals of carbon and carbon, carbon and nitrogen, and carbon and oxygen. Other Group 4A elements do not show this behavior.
- Nitrogen can form π bonds because of its size, just as carbon can. Other Group 5A elements normally do not. (A few phosphorus double-bonded compounds have been known to exist.)
- A similar situation exists with oxygen in Group 6A.
- Fluorine has a very weak F—F bond due to the close approach causing large electron-electron repulsions of F—F atoms resulting from the small size of the atoms.

18.2 The Group 1A Metals

The following questions and examples will help you review the material in this section.

1. (Review) Why do alkali metals react so well with nonmetals?
2. What electron configuration is common to each of the elements in this group?
3. Your textbook says that although lithium is a **stronger reducing agent**, it reacts **more slowly** with water than potassium or sodium. What does this illustrate?

Example 18.2 A - Preparation Of Alkali Metals

Write the reaction for the preparation of liquid sodium by electrolysis of molten sodium chloride

Solution

Chapter 11 deals with electrolysis. The reaction of interest (done at 600°C) is

$$2NaCl(l) \rightarrow 2Na(l) + Cl_2(g)$$

As a wrap-up to this section, study the material in Table 18.5 in your textbook, and try the following example.

Example 18.2 B - Practice With Reactions

Predict the products of the following reactions:

a. $2Rb(s) + Cl_2(g) \rightarrow ?$
b. $Rb(s) + O_2(g) \rightarrow ?$

c. $12K(s) + P_4(s) \rightarrow ?$
d. $2Na(s) + 2H_2O(l) \rightarrow ?$

Solution

a. $2Rb(s) + Cl_2(g) \rightarrow 2RbCl(s)$
b. $Rb(s) + O_2(g) \rightarrow RbO_2(s)$
c. $12K(s) + P_4(s) \rightarrow 4K_3P(s)$
d. $2Na(s) + 2H_2O(l) \rightarrow 2NaOH(aq) + H_2(g)$

Answers to review questions are at the end of this study chapter.

18.3 The Chemistry of Hydrogen

The following questions will help you review the material in this section.

1. (Review) Why does hydrogen act so differently from the metals in this group?
2. Why does hydrogen have such low boiling and melting points?
3. Write the combustion reaction between hydrogen and oxygen to form water.
4. Write the reaction between methane and water to produce hydrogen.
5. Why isn't electrolysis a very practical method of producing hydrogen?
6. List two commercial uses of hydrogen.
7. Define hydride, ionic hydride, covalent hydride, and interstitial hydride.
8. How is palladium used to help purify hydrogen?
9. Why might interstitial hydrides be used for storage of hydrogen gas?

Answers to review questions are at the end of this study chapter.

18.4 The Group 2A Elements

The following questions and examples will help you review the material in this section.

1. Why are the elements in this group called **alkaline earth** metals?
2. Why does beryllium display very different properties from the rest of the Group 2A elements?

Example 18.4 A - Reactions Of Group 2A Elements

Predict the products of the following reactions:

a. $Sr(s) + 2H_2O(l) \rightarrow ?$
b. $CaO(s) + H_2O(l) \rightarrow ?$
c. $Be(s) + 2H_2O(l) \rightarrow ?$

Solution

a. $Sr(s) + 2H_2O(l) \rightarrow Sr^{2+}(aq) + 2OH^-(aq) + H_2(g)$
b. $CaO(s) + H_2O(l) \rightarrow Ca^{2+}(aq) + 2OH^-(aq)$
c. $Be(s) + 2H_2O(l) \rightarrow$ No observable reaction!

3. Why are alkaline earth metals important?

Example 18.4 B - Practice With Group 2A Reactions

Based on the information in Table 18.8 in your textbook, predict the **reactants** given the following **products**.

a. ? + ? → $Ca_3N_2(s)$
b. ? + ? → $2RaO(s)$
c. ? + ? + ? → $Be(OH)_4^{2-}(aq) + H_2(g)$
d. ? + ? → $SrH_2(s)$

Solution

a. $\mathbf{3Ca(s) + N_2(g)} \rightarrow Ca_3N_2(s)$
b. $\mathbf{2Ra(s) + O_2(g)} \rightarrow 2RaO(s)$
c. $\mathbf{Be(s) + 2OH^-(aq) + 2H_2O(l)} \rightarrow Be(OH)_4^{2-}(aq) + H_2(g)$
d. $\mathbf{Sr(s) + H_2(g)} \rightarrow SrH_2(s)$

4. What is "hard" water?
5. Describe the **ion exchange method** for softening hard water.

Answers to review questions are at the end of this study chapter.

18.5 The Group 3A Elements

The following questions and examples will help you review the material in this section.

1. Why does boron exhibit different chemical behavior than the other elements in this group?
2. What is the simplest stable borane?
3. Describe the bonding in boranes.
4. Why are boranes highly reactive?
5. Why is aluminum amphoteric?
6. Why is gallium so useful in thermometers that measure a wide range of temperatures?

Example 18.5 - Reactions

Predict the products of each of the following reactions:

a. $2Al(s) + N_2(g) \rightarrow$?
b. $2Ga(s) + 2OH^-(aq) + 6H_2O(l) \rightarrow$?
c. $2In(s) + 6H^+(aq) \rightarrow$?
d. $4Tl(s) + 3O_2(g) \rightarrow$? (high temperatures)

Solution

a. $2Al(s) + N_2(g) \rightarrow 2AlN(s)$
b. $2Ga(s) + 2OH^-(aq) + 6H_2O(l) \rightarrow 2Ga(OH)_4^-(aq) + 3H_2(g)$
c. $2In(s) + 6H^+(aq) \rightarrow 2In^{3+}(aq) + 3H_2(g)$
d. $4Tl(s) + 3O_2(g) \rightarrow 2Tl_2O_3(s)$ (high temperatures)

Answers to review questions are at the end of this study chapter.

18.6 The Group 4A Elements

The following questions and examples will help you review the material in this section.

Example 18.6 A - Electron Configuration

Write shorthand electron configurations for each of the following Group 4A elements:

a. carbon b. germanium c. tin

Solution

a. carbon: $[He]2s^22p^2$
b. germanium: $[Ar]4s^23d^{10}4p^2$
c. tin: $[Kr]5s^24d^{10}5p^2$

Notice that Group 4A elements have $\boldsymbol{ns^2np^2}$ electron configurations.

1. What is the hybridization on the central atom in molecules such as $PbCl_4$ and $GeBr_4$?
2. Why can't carbon form a compound such as CCl_6^{2-} while tin can form $SnCl_6^{2-}$?
3. Why, with regard to bond energies, does the silicon to oxygen bond dominate silicon chemistry rather than a silicon to silicon bond?
4. Define **allotrope**.
5. What are the three allotropic forms of carbon?
6. What is the major industrial use of silicon? Germanium?
7. Under what conditions is each of the allotropes of tin stable?
8. What is **tin disease**?
9. What is the common source of lead?
10. Your textbook says that lead may have contributed to the demise of the Roman civilization. Why?
11. What has been a significant cause of lead poisoning in the twentieth century?

Example 18.6 B - Reactions Of Group 4A Elements

Predict the products of each of the following reactions:

a. $Sn(s) + 2H^+(aq) \rightarrow ?$
b. $Ge(s) + O_2(g) \rightarrow ?$
c. $Pb(s) + Cl_2(g) \rightarrow ?$

Solution

a. $Sn(s) + 2H^+(aq) \rightarrow Sn^{2+}(aq) + H_2(g)$
b. $Ge(s) + O_2(g) \rightarrow GeO_2(s)$
c. $Pb(s) + Cl_2(g) \rightarrow PbCl_2(s)$ (Note that Pb is the only Group 4A element that forms the dihalide. Others form the tetrahalide, MX_4.)

Answers to review questions are at the end of this study chapter.

18.7 The Group 5A Elements

The following questions and examples will help you review the chemistry of Group 5A elements.

Example 18.7 A - Electron Configurations

Give the shorthand configuration for each of the following elements:

a. nitrogen b. arsenic c. bismuth

Solution

a. nitrogen: **[He]$2s^2 2p^3$**
b. arsenic: **[Ar]$4s^2 3d^{10} 4p^3$**
c. bismuth: **[Xe]$6s^2 4f^{14} 5d^{10} 6p^3$**

Notice the ns^2np^3 configuration on each of these Group 5A elements.

1. Why are there no known ionic compounds containing bismuth(V) or antimony(V)?
2. Justify the fact that NH_3 and PH_3 act as Lewis bases.
3. What is the hybridization of As in $AsCl_3$?
4. Why can't nitrogen form compounds with 5 ligands?
5. Discuss the structure of compounds such as PCl_5.

Example 18.7 B - Reactions Of Group 5A Elements

Predict the products of the following reactions:

a. $P_4O_{10}(s) + 10C(s) \rightarrow ?$
b. $2Ca_3(PO_4)_2(s) + 6SiO_2(s) \rightarrow ?$ (decomposition)

Solution

a. $P_4O_{10}(s) + 10C(s) \rightarrow 4P(s) + 10CO(g)$
b. $2Ca_3(PO_4)_2(s) + 6SiO_2(s) \rightarrow 6CaSiO_3(s) + P_4O_{10}(s)$

Answers to review questions are at the end of this study chapter.

18.8 The Chemistry of Nitrogen

The following questions and examples will help you review the material in this section.

1. Why is N_2 such a stable molecule?
2. Why is nitrogen used as an atmosphere for reactants that normally react with water or oxygen?

Example 18.8 A - Thermodynamics Of Nitrogen Compounds

Using data from Appendix 4 in your textbook, calculate $\Delta S°$ and $\Delta G°$ for each of the following reactions.

a. $N_2O(g) \rightarrow N_2(g) + \frac{1}{2}O_2(g)$
b. $NO_2(g) \rightarrow \frac{1}{2}N_2(g) + O_2(g)$
c. $NH_3(g) \rightarrow \frac{1}{2}N_2(g) + \frac{3}{2}H_2(g)$

Solution

Recall that entropy and free energy are both **state functions**.

$$\Delta S^{o}_{\text{reaction}} = \sum S^{o}_{\text{products}} - \sum S^{o}_{\text{reactants}}$$
$$\Delta G^{o}_{\text{reaction}} = \sum G^{o}_{\text{f (products)}} - \sum G^{o}_{\text{f (reactants)}}$$

For entropy,

a. $\Delta S° = [S^{o}_{N_2} + \frac{1}{2} S^{o}_{O_2}] - S^{o}_{N_2O} = [192 \text{ J/K mol} + \frac{1}{2}(205 \text{ J/K mol})] - [220 \text{ J/K mol}]$

$\mathbf{\Delta S^{o}_{reaction} = 74.5\ J/K\ mol}$

b. $\Delta S° = [{}^1/_2 S°_{N_2} + S°_{O_2}] - S°_{NO_2} = [{}^1/_2(192 \text{ J/K mol}) + 205 \text{ J/K mol}] - [240 \text{ J/K mol}]$

$\mathbf{\Delta S°_{reaction} = 61 \text{ J/K mol}}$

c. $\Delta S° = [{}^1/_2 S°_{N_2} + {}^3/_2 S°_{H_2}] - S°_{NH_3} = [{}^1/_2(192 \text{ J/K mol}) + {}^3/_2(131 \text{ J/K mol})] - [193 \text{ J/K mol}]$

$\mathbf{\Delta S°_{reaction} = 100 \text{ J/K mol}}$

For free energy, recall that $\Delta G°$ of elements in their standard states = 0.

a. $\Delta G° = 0 - \Delta G°(N_2O) = \mathbf{-104 \text{ kJ/mol}}$
b. $\Delta G° = 0 - \Delta G°(NO_2) = \mathbf{-52 \text{ kJ/mol}}$
c. $\Delta G° = 0 - \Delta G°(NH_3) = \mathbf{+17 \text{ kJ/mol}}$

Notice that $\Delta G = \Delta H - T\Delta S$ will serve as a double check in each case!

3. Why are nitrogen-based explosives so effective (with regard to thermodynamics and gas volume)?
4. Why does your textbook say that the thermodynamics and kinetics of the Haber process are in opposition?
5. What are the conditions under which the Haber process is performed?
6. Define **nitrogen fixation**.
7. How does nitrogen fixation occur in an automobile?
8. List two natural mechanisms of nitrogen fixation.
9. Why is there such an interest in **nitrogen-fixing bacteria**?
10. What is **denitrification**?
11. What is the problem with accumulating excess nitrogen in soil and bodies of water?

Nitrogen Hydrides

12. Why does ammonia have a much lower boiling point than water?
13. What is the structure of hydrazine?
14. Why is hydrazine a useful chemical in the space program?

Example 18.8 B - Hydrazine As A Propellant

The reaction between the rocket fuel monomethylhydrazine and the oxidizer dinitrogen tetroxide is:

$$5N_2O_4(g) + 4N_2H_3(CH_3)(g) \rightarrow 12H_2O(g) + 9N_2(g) + 4CO_2(g)$$

How many liters of carbon dioxide are formed at 1.00 atm and 25°C from the reaction of 3.00 grams of N_2O_4 with 5.00 g of $N_2H_3(CH_3)$?

Solution

This problem combines a limiting reactant calculation with a gas law problem. The general strategy is

- Determine how much product is formed by each reactant.
- Determine how much product is formed by the limiting reactant.
- Convert moles of product to liters of product using the ideal gas law.

$$\text{mol } CO_2 \text{ from } N_2O_4 \;\|\; 3.00 \text{ g } N_2O_4 \times \frac{1 \text{ mol } N_2O_4}{92.0 \text{ g } N_2O_4} \times \frac{4 \text{ mol } CO_2}{5 \text{ mol } N_2O_4} = 0.0261 \text{ mol } CO_2$$

$$\text{mol } CO_2 \text{ from } N_2H_3(CH_3) \; \| \; 5.00 \text{ g } N_2H_3(CH_3) \times \frac{1 \text{ mol } N_2H_3(CH_3)}{46.0 \text{ g } N_2H_3(CH_3)} \times \frac{4 \text{ mol } CO_2}{4 \text{ mol } N_2H_3(CH_3)}$$

$$= 0.109 \text{ mol } CO_2$$

N_2O_4 is the limiting reactant, and 0.0261 mol CO_2 is formed.

$$V_{CO_2} = \frac{nRT}{P} = \frac{0.0261 \text{ mol}(0.08206 \text{ L atm/K mol})(298.2 \text{ K})}{1.00 \text{ atm}}$$

$$= \mathbf{0.639 \text{ L } CO_2 \text{ formed}}$$

15. List two uses of hydrazine in addition to that of a rocket propellant.

Nitrogen Oxides

16. List the formula, and determine the oxidation state of the nitrogen in each nitrogen oxide in Table 18.14 in your textbook.
17. List several uses of nitrous oxide.
18. What is the role of nitrous oxide in the Earth's climate control?
19. How is nitric oxide prepared?
20. Why does NO turn brown in air?
21. Why does NO^+ have a higher bond energy than NO?
22. What happens to NO_2 at low temperatures?

Oxyacids of Nitrogen

23. List some uses of nitric acid.
24. Give the reactions involved in the **Ostwald process**.
25. How can the concentration of nitric acid be increased from 68% to 95%?
26. Why does a nitric acid solution turn yellow upon constant exposure to sunlight?

Answers to review questions are at the end of this chapter.

18.9 The Chemistry of Phosphorus

The following questions and the example will help you review the material in this section.

1. List four reasons for the differences in chemical properties between nitrogen and phosphorus.
2. What are the three allotropic forms of phosphorus? How are they different from one another?
3. How is red phosphorus made?
4. How is black phosphorus obtained from either red or white phosphorus?
5. What is the oxidation state of phosphorus in phosphides? Give some examples of phosphides.
6. Why is phosphine viewed as an exception to the VSEPR model?
7. Your textbook says that H_3PO_3 and H_3PO_2 are diprotic and monoprotic, respectively. Why aren't they triprotic?
8. Why isn't naturally occurring phosphorus in soil easily usable by plants?
9. What is the oxidation state of phosphorus in PF_3? PF_5?

Example 18.9 - Reactions Of Phosphorus

Give the products of each of the following reactions (discussed in this section of your textbook):

a. $2Na_3P(s) + 6H_2O(l) \rightarrow ?$
b. $P_4O_{10}(s) + 6H_2O(l) \rightarrow ?$
c. $P_4(s) + 3O_2(g;\text{ limited}) \rightarrow ?$
d. $P_4(s) + 5O_2(g;\text{ excess}) \rightarrow ?$
e. $P_4O_6(s) + 6H_2O(l) \rightarrow ?$

Solution

a. $2Na_3P(s) + 6H_2O(l) \rightarrow 2PH_3(g) + 6Na^+(aq) + 6OH^-(aq)$
b. $P_4O_{10}(s) + 6H_2O(l) \rightarrow 4H_3PO_4(aq)$
c. $P_4(s) + 3O_2(g;\text{ limited}) \rightarrow P_4O_6(s)$
d. $P_4(s) + 5O_2(g;\text{ excess}) \rightarrow P_4O_{10}(s)$
e. $P_4O_6(s) + 6H_2O(l) \rightarrow 4H_3PO_3(aq)$

Answers to review questions are at the end of this chapter.

18.10, 18.11, and 18.12 The Chemistry of the Group 6A Elements Focusing on Oxygen and Sulfur

The following questions and examples will help you review the material in these sections.

Example 18.10 A - Electron Configurations

Write the shorthand configurations for each of the following elements:

a. oxygen b. selenium c. polonium

Solution

a. oxygen: **[He]$2s^22p^4$**
b. selenium: **[Ar]$4s^23d^{10}4p^4$**
c. polonium: **[Xe]$6s^24f^{14}5d^{10}6p^4$**

Note that all Group 6A elements have the configuration ns^2np^4.

1. What is the most common oxidation state of Group 6A elements in ionic compounds?
2. Give some examples of Group 6A covalent hydrides.
3. Why do Group 6A elements (other than oxygen) form compounds with more than eight electrons around the central atom?

Example 18.10 B - VSEPR Review

Predict the geometry of TeI_4.

Solution

We learned about the VSEPR model in Chapter 8. Our first step is to draw the Lewis structure.

1. # valence electrons in the system = 6 for Te + 4 × 7 for I = 34 electrons
2. # electrons if happy = 8 for Te + 4 × 8 for I = 40 electrons

3. # bonds = (40−34)/2 = 3 bonds

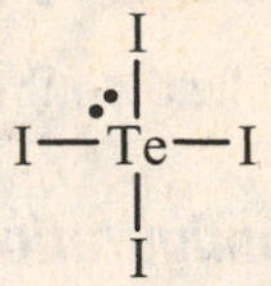

This is an exception! We must therefore draw the Lewis structure and put extra electron pairs around the central atom. (We have omitted electrons around the iodine atoms for clarity.)

There are 5 effective electron pairs around the central atom. The structure is **based on a trigonal bipyramid**. The lone pair will be in the equatorial plane. This is a **seesaw** structure.

4. Why don't Group 6A elements form +6 cations?
5. Why has there been growing interest in the chemistry of selenium?
6. Why might polonium be related to cancer in smokers?
7. What are some of the important uses of oxygen in our world?
8. What percent of the Earth's atmosphere is oxygen?
9. How is oxygen isolated from air?
10. How can we demonstrate the paramagnetism of oxygen?
11. What is the VSEPR geometry of ozone?
12. Why is the bond angle in ozone less than 120°?
13. How is ozone prepared in the laboratory?
14. Why might ozone be useful in municipal water purification?
15. What is the importance of the ozone layer of the Earth's atmosphere?
16. Name some minerals that contain sulfur.
17. Describe the **Frasch Process**.
18. Why does elemental oxygen exist as O_2 whereas sulfur exists as S_8?
19. What is the structural difference between rhombic and monoclinic sulfur?
20. Why is O_2 more stable than SO?
21. What role might dust play in the conversion of SO_2 to SO_3?

Example 18.12 - Reactions Of Sulfur

Based on the material in this section, predict the products of each of the following reactions:

a. $C_{12}H_{22}O_{11}(s) + 11H_2SO_4(conc) \rightarrow ?$
b. $SO_2(aq) + H_2O(l) \rightarrow ?$
c. $2SO_2(g) + O_2(g) \rightarrow ?$

Solution

a. $C_{12}H_{22}O_{11}(s) + 11H_2SO_4(conc) \rightarrow 12C(s) + 11H_2SO_4{\bullet}H_2O(l)$
b. $SO_2(aq) + H_2O(l) \rightarrow H_2SO_3(aq)$
c. $2SO_2(g) + O_2(g) \rightarrow 2SO_3(g)$

Answers to review questions are at the end of this study chapter.

18.13 and 18.14 The Group 7A and 8A Elements

The following questions and examples will help you review the material in these sections.

Example 18.13 A - Electron Configurations

Write the shorthand configurations for each of the following elements:

a. bromine b. iodine

Solution

a. bromine: **[Ar]$4s^2 3d^{10} 4p^5$**
b. iodine: **[Kr]$5s^2 4d^{10} 5p^5$**

Note that the halogens have the configuration ns^2np^5.

Example 18.13 B - The Decomposition Of Astatine

Astatine has been used for cancer therapy because of its short half-life. The longest-lived isotope, ^{210}At, has $t_{1/2}$ = 8.3 hours. If you have a 1.000-g sample of ^{210}At, how much will remain after 24.0 hours? (The radioactive decomposition of astatine obeys a **first-order** kinetic rate law.)

Solution

For a first-order decay, the rate constant

$$k = 0.693/t_{1/2} = 0.693/8.3 \text{ hr} = \mathbf{8.35 \times 10^{-2}\ hr^{-1}}$$

$$\ln\left(\frac{A_t}{A_0}\right) = -kt$$

where $A_0 = 1.000$ g
$A_t = ?$
$k = 8.35 \times 10^{-2}$ hr^{-1}
$t = 24.0$ hr

$$\ln\left(\frac{A_t}{1.000}\right) = -8.35 \times 10^{-2} \text{ hr}^{-1} (24.0 \text{ hr})$$

$$\ln\left(\frac{A_t}{1.000}\right) = -2.00$$

Taking the antilog of both sides,

$$\frac{A_t}{1.000} = 0.135$$

$$\mathbf{A_t = 0.135 \text{ g of astatine remaining at time } t}$$

You can see that, after just one day, the majority of a sample of astatine will decompose.

1. What kind of bonds do halogens tend to form with nonmetals? Metals in lower oxidation states?
2. What about bonding with metals in higher oxidation states?
3. Why does HF have such a high boiling point relative to other hydrogen halides?
4. Why can't the relative strength of hydrogen halides as acids be assessed in water?

5. What is the order of acid strength of hydrogen halides in acetic acid?
6. What is the rationale for HF being the least acidic hydrogen halide?
7. Why is hydrochloric acid such an important industrial chemical?
8. Why must $HOClO_3$ be handled so carefully?
9. Define **"disproportionation reaction."**
10. List some uses of chlorate salts.
11. Why is OF_2 called oxygen difluoride rather than difluorine oxide?
12. What are the pertinent reactions that make the darkening of sunglasses reversible?
13. What is the importance of copper ion in the "automatic sunglasses" process?

Example 18.13 C - Reactions Of Halogens

Predict the products of the following reactions:

a. $H_2(g) + Cl_2(g) \xrightarrow{\text{U.V. light}}$
b. $SiO_2(s) + 4HF(aq) \rightarrow ?$
c. $4F_2(g) + 3H_2O(l) \xrightarrow{\text{NaOH}} ?$
d. $Cl_2(aq) + H_2O(l) \rightleftharpoons ?$

Solution

a. $H_2(g) + Cl_2(g) \xrightarrow{\text{U.V. light}} 2HCl(g)$
b. $SiO_2(s) + 4HF(aq) \rightarrow SiF_4(g) + 2H_2O(l)$
c. $4F_2(g) + 3H_2O(l) \xrightarrow{\text{NaOH}} 6HF(aq) + OF_2(g) + O_2(g)$
d. $Cl_2(aq) + H_2O(l) \rightleftharpoons HOCl(aq) + H^+(aq) + Cl^-(aq)$

14. What is the major source of helium on Earth?
15. What are the major uses of helium, neon, and argon?
16. What are some xenon and krypton compounds that have been prepared?

Answers to review questions are at the end of this study chapter.

Answers to Review Questions

Section 18.2

1. They react so well together because they lose electrons very readily to form M^+ ions.
2. The ns^1 configuration is common to each of the elements in this group.
3. This illustrated that reaction rate and thermodynamics are not directly related.

Section 18.3

1. Hydrogen is a nonmetal that is very small and can either gain or lose an electron, but does so covalently rather than ionically.
2. Hydrogen has low boiling and melting points because of its low molecular weight and nonpolarity.
3. $2H_2(g) + O_2(g) \rightarrow 2H_2O(l)$
4. $CH_4(g) + H_2O(g) \rightarrow CO(g) + 3H_2(g)$
5. It isn't very practical because of the high cost of electricity.
6. Production of ammonia and hydrogenating unsaturated vegetable oil

7. **hydride:** Binary compound containing hydrogen.
   ionic hydride: Hydrogen combined with Group 1A or 2A metals.
   covalent hydride: Hydrogen combined with nonmetals.
   interstitial hydride: Hydrogen atoms occupy holes (interstices) in a metal's crystal structure.
8. Hydrogen diffuses through the palladium metal wall (see "interstitial hydride"), leaving impurities behind.
9. Interstitial hydrides lose absorbed (stored) hydrogen when heated.

Section 18.4

1. They are called **alkaline earth** metals because their oxides form bases in water.
2. Beryllium displays different properties because of its relatively small size and high electronegativity.
3. Calcium is an essential element in bones and teeth. Magnesium is vital in metabolism and muscle functions. Magnesium is also useful in structural materials when it is alloyed with aluminum.
4. Hard water is that which contains large concentrations of Ca^{2+} and Mg^{2+} ions.
5. In a nutshell, hard water cations become bound to sites on a polymer resin that releases sodium ions as a result, thus softening the water.

Section 18.5

1. Boron is very small and has a relatively high electronegativity, thus forming mostly covalent compounds.
2. B_2H_6 is the simplest stable borane.
3. Boranes have one or more three-center two-electron bonds similar to that in solid BeH_2. The remaining bonds are normal covalent bonds.
4. As with beryllium, boranes are highly electron-deficient, which accounts for their reactivity.
5. Covalent bonding is responsible for the amphoteric nature of aluminum.
6. Gallium is useful in thermometers because, of all the elements, it has the widest temperature range in which it is a liquid.

Section 18.6

1. The central atom is sp^3 hybridized.
2. There are no *d* electrons available in carbon to allow it to exceed an octet.
3. The bond energy of the Si—O bond is substantially higher than that of the Si—Si bond.
4. An allotrope has the same element but different structures (i.e., S, S_2 and S_8, or O_2 and O_3).
5. Graphite, buckminsterfullerene and diamond are allotropic forms of carbon.
6. The major use of silicon is in semiconductors. The major use of germanium is also for semiconductors.
7. White tin is stable at ambient temperatures. Gray tin is stable below 13.2°C. Brittle tin is stable above 161°C.
8. Tin disease is the conversion of white tin to powdery gray tin in cold temperatures.
9. Galena (PbS) is the common source of lead.
10. High concentrations of lead are toxic. Significant levels of lead have been found in the bones of the Roman era.
11. Lead in gasoline, lead-based paints, and lead containing pottery and crystal are significant contributors to lead poisoning.

Section 18.7

1. Too much energy is required to remove all five electrons from the neutral atom.
2. NH_3 and PH_3 have lone pairs they can donate.

3. The hybridization is sp^3.
4. Nitrogen cannot because of its small size.
5. Bonding in PCl_5 is an exception to the octet rule. It is trigonal bipyramidal hybridization.

Section 18.8

1. N_2 is so stable because it has a very high bond strength.
2. Nitrogen is used as an "inert atmosphere" because it is so unreactive.
3. Nitrogen-based explosives are effective because the decomposition of the molecules to N_2 (gas) is a thermodynamically favorable process and it releases large volumes of gas and heat.
4. In the Haber process, the equilibrium lies in the direction of NH_3 at room temperature. Because the reaction is exothermic, increasing the temperature to increase the reaction rate decreases the equilibrium constant.
5. High pressure and moderately high temperature.
6. The process of transforming N_2 to other nitrogen-containing compounds.
7. Nitrogen from the air reacts with oxygen to form NO and then NO_2.
8. They are lightning and nitrogen-fixing bacteria.
9. Such bacteria produce ammonia in the soil at standard conditions.
10. Denitrification is the return of nitrogen-containing compounds to the atmosphere as N_2 gas.
11. Algae and other undesirable organisms may accumulate because of the presence of too much fixed nitrogen.
12. Water contains two polar bonds and two lone pairs. It can exhibit much more substantial hydrogen bonding than ammonia.
13. See Figure 18.14 in your textbook.
14. Hydrazine is an excellent reducing agent that produces a great deal of energy when reacted with oxygen.
15. It is used as a blowing agent in plastics and is also used in the production of agricultural pesticides.
16. See Table 18.14 in your textbook.
17. It is used by dentists as an anesthetic. It is also used as a propellant in aerosol cans of whipped cream.
18. It strongly absorbs infrared radiation, thus helping with climate control.
19. NO is prepared by the reaction of copper with dilute (6 *M*) nitric acid.
20. NO is oxidized in air to brown NO_2.
21. Because an antibonding electron is removed the resulting ion has a stronger bond. The M.O. bond order for NO^+ is 3, and for NO is 2.5. Therefore, NO^+ has a higher bond energy than NO.
22. NO_2 can dimerize to form N_2O_4.
23. It is used in the manufacture of fertilizer and explosives, among other things.
24. a. $4NH_3(g) + 5O_2(g) \rightarrow 4NO(g) + 6H_2O(g)$
    b. $2NO(g) + O_2(g) \rightarrow 2NO_2(g)$
    c. $3NO_2(g) + H_2O(l) \rightarrow 2HNO_3(aq) + NO(g)$
25. The concentration is increased by treatment with sulfuric acid, a dehydrating agent.
26. HNO_3 decomposes in sunlight to give $NO_2(g)$, a brown gas that dissolves in the solution and colors it yellow.

Section 18.9

1. a. Nitrogen can form stronger bonds.
   b. Nitrogen is more electronegative.
   c. Phosphorus is larger.
   d. Phosphorus has available *d* orbitals.
2. White, red and black phosphorus differ in structure as outlined in Figure 18.18 in your textbook.

3. Red phosphorus is formed by heating white phosphorus in the absence of air.
4. Black phosphorus is made by heating white or red phosphorus at high pressures.
5. The oxidation state is −3. Examples are Na_3P and Ca_3P_2.
6. The bond angle is far smaller (94°) than expected for a tetrahedral structure with one lone pair.
7. They are not triprotic because hydrogens attached to phosphorus are not acidic.
8. Phosphorus in soil is often present in insoluble (and thus unusable) minerals.
9. The oxidation state is +3 in PF_3 and +5 in PF_5.

Section 18.10, 18.11, and 18.12

1. The most common oxidation state is −2.
2. Examples are H_2S and H_2O.
3. Other than oxygen, all Group 6A elements have available *d* orbitals.
4. It takes far too much energy to remove six electrons from these elements.
5. It is possible that selenium has a role in cancer protection.
6. Polonium is found in tobacco and is an α emitter.
7. In general, we need oxygen for respiration and for combustion.
8. Oxygen makes up 21% of the Earth's atmosphere.
9. Nitrogen distillation isolates oxygen.
10. We can pour liquid oxygen between poles of a magnet.
11. The VSEPR structure is trigonal planar.
12. The lone pair requires more room than bonded pairs.
13. Ozone is prepared by passing an electric charge through pure oxygen gas.
14. Ozone is a strong oxidant.
15. Ozone absorbs ultraviolet radiation.
16. Galena, cinnabar, pyrite, and gypsum, among others.
17. The Frasch process is used to obtain sulfur from underground deposits. Superheated water is pumped in to melt sulfur, which is recovered by air pressure.
18. Oxygen can form π-bonds, thus stabilizing O_2. Sulfur can form only σ-bonds, thus stabilizing the large form S_8.
19. The difference has to do with the way the rings are stacked.
20. O_2 has stronger bonding than SO.
21. Dust acts as a catalyst for the conversion.

Section 18.13 and 18.14

1. Halogens form polar covalent bonds with nonmetals and ionic compounds with metals in lower oxidation states.
2. Bonds with metals in higher oxidation states tend to be polar covalent.
3. HF can hydrogen-bond (due to the small size and high electronegativity of fluorine).
4. Hydrogen halides (except HF) completely dissociate in water, so they appear equally strong.
5. HI > HBr > HCl >> HF
6. The H—F bond strength is higher than the other H—X acids. Note the extensive discussion about the relationship between bond strength and entropy in which he concludes "the deciding factor appears to be entropy."
7. HCl is used for, among other things, cleaning steel.
8. $HOClO_3$ is an incredibly strong oxidizing agent that reacts explosively with many organic compounds.
9. A disproportionation reaction occurs when an element is both oxidized and reduced in the same reaction.
10. Among the uses are as weed killers and as oxidizers in fireworks.

11. Fluorine is the anion because it has higher electronegativity. IUPAC nomenclature rules dictate that it gets the -ide ending.
12. $Ag^+ + Cl^- \xrightarrow{h\nu} Ag + Cl$ (electron transfer)
    $Cl + Cu^+ \rightarrow Cu^{2+} + Cl^-$
    $Cu^{2+} + Ag \rightarrow Cu^+ + Ag^+$
13. The copper ion allows for the *reversibility* of the process (as opposed to the irreversibility of the photography process).
14. The major sources are natural gas deposits.
15. Helium is used as a coolant, among other things; neon and argon are used in lighting.
16. $XeFPtF_6$, XeF_4, XeO_3, KrF_4, and KrF_2 among others.

CHAPTER 19

Transition Metals and Coordination Chemistry

The Bottom Line: Chapter 19

Transition metals show a wonderful and remarkable variety of bonding properties that make them useful to our bodies and our world. This chapter explores how and why they act as they do.

19.1 The Transition Metals: A Survey

This section begins by noting that transition metal chemistry shows **much greater consistency** than that of the representative elements as we go across the periodic table. This is because the filling of transition metals involves *d*- or *f*-block electrons, which are core electrons. These do not participate as easily in bonding as *s* and *p* electrons. Read the information on General Properties in your textbook, and then try the following example.

Example 19.1 A - Properties Of Transition Metals

Answer the following questions regarding properties of first-row transition metals.

1. List the properties they have in common.
2. What is the trend with regard to
   a. density?
   b. electrical conductivity?
   c. atomic radius?
   d. common oxidation states?

Solution

1. Properties in common: metallic luster, high electrical conductivity, high thermal conductivity.
2. a. The density shows a steady increase with the exception of zinc.
   b. In general, electrical conductivity increases.
   c. Atomic radius decreases, then increases after iron. The key argument here is the balance between increased nuclear charge and electron-electron repulsion. In any case the radii are not drastically different across the period.
   d. Vanadium, chromium, and manganese have many common oxidation states due to the availability of *s* and *d* electrons. Atoms such as nickel, copper, and zinc have available only *s* electrons (it would require too much energy to strip *d* electrons away), so these elements have fewer oxidation states available.

Your textbook points out that the reasons for electron configurations of the transition metals are quite complex—we note the configurations and establish patterns where possible.

Example 19.1 B - Electron Configurations Of Ions

Write the shorthand electron configurations for each of the following:

a. V b. V^{2+} c. V^{3+} d. V^{5+}

Solution

a. V: **[Ar]4s^2 3d^3**
b. V^{2+}: **[Ar]3d^3**
c. V^{3+}: **[Ar]3d^2**
d. V^{5+}: **[Ar]**

The discussion that closes out this section deals with the 4*d* and 5*d* transition series. The **lanthanide contraction** is the slight shrinkage in atomic size that occurs across these periods due to extra nuclear-electron charge attraction as 4*f* or 5*f* electrons are added. The lanthanide contraction also accounts for the similarity in properties among 4*d* and 5*d* elements in a particular group.

19.2 The First-Row Transition Metals

The following questions will help you review your knowledge of the materials in this section.

1. What elements are being discussed? What do they have in common?
2. What is the most common oxidation state for scandium in its compounds?
3. Why are scandium compounds generally colorless and diamagnetic?
4. How is scandium prepared?
5. What is the major industrial use of scandium?
6. What properties make titanium a useful structural material?
7. Why is titanium used for reaction vessels in the chemical industry?
8. List some uses of titanium(IV) oxide.
9. What are the main natural sources of titanium?
10. Write the reactions for the purification of TiO_2 from its ores.
11. What are the major industrial uses of vanadium?
12. How is pure vanadium prepared?
13. What is vanadium steel?
14. Why do V^{5+} and V^{4+} exist in solution as VO_2^+ and VO^{2+}?

Example 19.2 - Bit Of Detective Work

An aqueous solution of ammonium vanadate (NH_4VO_3) was yellow. It was then added to a "Jones Reductor," which is a Zn-Hg amalgam. The solution was swirled over the reductor.

1. An aliquot of the solution was removed. It was **blue**.
2. The remaining solution was swirled over the reductor.
3. Another aliquot was removed. It was **green**.
4. The remaining solution was swirled over the reductor.
5. A **final** aliquot was removed. It was **violet**.
6. Upon standing in air, this **final** aliquot turned **green**.

Describe, in terms of the oxidation states of vanadium, the chemistry that occurred.

Solution

a. The original solution (yellow) is vanadium(V) (as VO_2^+)
b. When swirled, the vanadium was reduced to V^{4+} (as VO^{2+}), which is blue.

c. When further swirled, VO^{2+} was reduced to V^{3+}, which is green.
d. A final mixing reduced V^{3+} to V^{2+} (violet).
e. Upon sitting, V^{2+} was air-oxidized to V^{3+} (green).

15. How is ferrochrome prepared?
16. What are the common oxidation states of chromium?
17. Write the half-reaction for the reduction of the dichromate ion in acidic solution.
18. What are the structural differences between chromium(VI) in acidic and in basic solutions?
19. What is "cleaning solution" composed of? What is it used for?
20. What are the most common uses of manganese?
21. What are "manganese nodules"?
22. What are the common oxidation states of manganese?
23. What are the important oxidation states of iron?
24. Why are iron(II) solutions light green?
25. Why are iron(III) solutions yellow?
26. Why does $Fe(H_2O)_6^{3+}$ behave as an acid in aqueous solution?
27. What are some commercial uses of cobalt?
28. What species causes the rose color of cobalt in aqueous solution?
29. Why is nickel used for plating active metals?
30. Why is copper so useful?
31. What is the major use of copper?
32. What species causes the characteristic blue color of aqueous copper solutions?
33. What is the main industrial use of zinc?

19.3 Coordination Compounds

This section begins with several important definitions:

1. **coordination compound**: It consists of a **complex ion** (a transition metal ion with attached ligands) and **counter ions** (anions or cations as needed to balance the charge of the compound).
2. **ligand**: a neutral molecule or ion **having a lone pair** that can be used to form a bond to a metal ion.
3. **monodentate ligand**: Can form one bond to a metal ion.
4. **bidentate ligand**: Can form two bonds to a metal ion.
5. **polydentate ligand**: Can form more than two bonds to a metal ion.

Table 19.13 in your textbook gives examples of each kind of ligand.

Nomenclature of coordination compounds can get sticky, so let's do a variety of problems together. **Memorize** the rules given before Interactive Example 19.1 in your textbook. Let's try to name $[Cr(H_2O)_4Cl_2]Cl$.

The coordination compound is composed of a **complex ion, $[Cr(H_2O)_4Cl_2]^+$**, which is a cation, and a **counter anion, Cl^-**. According to the rules of nomenclature, the cation (complex ion in this case) will be named first.

Naming the cation: The ligands are named before the metal ion. The two different ligands are **water** (aqua) and **chlorine** (chloro). The ligands are placed **alphabetically, so aqua goes before chloro**. This is done before adding any prefixes to the ligands.

aqua chloro __________ __________

There are **four** waters (tetra) and **two** chlorines (di) so we have

tetraaqua dichloro __________ __________

We must determine the oxidation state of the chromium ion. Water is a neutral base, and each chlorine has a −1 charge. The chromium is therefore in the +3 oxidation state.

tetraaqua dichloro chromium(III) ___________

The complex ion is treated as one word.

tetraaquadichlorochromium(III) ___________

Naming the anion: The anion is chloride. This finishes the naming of the entire compound. Remember, the cation and anion are named separately.

tetraaquadichlorochromium(III) chloride

Example 19.3 A - Naming Coordination Compounds

Name each of the following compounds:

a. $K_2[Ni(CN)_4]$
b. $K_4[Fe(CN)_6]$
c. $(NH_4)_2[Fe(H_2O)Cl_5]$
d. $[Co(NH_3)_2(en)_2]Cl_2$
e. $[Ag(NH_3)_2]Cl$

Solution

a. The complex ion is an **anion**. You know this because potassium, the counter ion, must be a cation. Therefore potassium is listed first. We **don't** say **di**potassium because we can explicitly determine the number of potassiums necessary based on the complex ion charge.

potassium ___________ ___________

The complex ion has four cyanide ions (tetracyano).

potassium tetracyano ___________

The nickel is in the **+2 oxidation state** because 4 CN = −4, and 2 K = +2. We put -ate on the end of a complex anion.

potassium tetracyanonickelate(II)

b. The complex ion is again an anion. The cation is named first.

potassium ___________ ___________

There are 6 cyanides in the complex cation (hexacyano).

potassium hexacyano ___________

The iron (ferrate) has an oxidation state of +2. (Can you see why?) The compound name is

potassium hexacyanoferrate(II)

c. The counter ion, ammonium, is a cation. It is named first.

ammonium ______________________

The complex ion has one water (aqua) and five chlorines (pentachloro). The iron (ferrate) is in the +3 oxidation state ($NH_4 = +1 \times 2 = +2$ and $Cl_5 = -5$).

ammonium aquapentachloroferrate(III)

d. The counter ion, chloride, is an anion. It is named last.

_____________________ **chloride**

The complex ion has 2 ammonia ligands (diammine) and 2 ethylenediamines [bis(ethylenediamine)]. The cobalt is in the +2 oxidation state (en and NH_3 are neutral).

diamminebis(ethylenediamine)cobalt(II) chloride

e. Again, the counter ion is an anion, chloride.

_____________________ **chloride**

The complex ion has two NH_3 (diammine) and silver in the +1 oxidation state.

diamminesilver(I) chloride

Let's try the reverse procedure.

Example 19.3 B - Writing Coordination Compound Formulas

Write the formulas for each of the following compounds.

a. potassium pentacyanocobaltate(II)
b. tris(ethylenediamine)nickel(II) sulfate
c. potassium dicarbonyltricyanocobaltate(I)

Solution

a. pentacyano = $\mathbf{(CN)_5}$ (−1 per CN × 5 = −5)
cobaltate(II) = **Co** (oxidation state = +2)

The complex anion is $Co(CN)_5$. The total charge = −3. We must therefore have 3 potassium ions to balance electrically.

$\mathbf{K_3[Co(CN)_5]}$

b. tris(ethylenediamine) = $\mathbf{(en)_3}$ (neutral Lewis base)
nickel(II) = **Ni** (oxidation state = +2)
sulfate = $\mathbf{SO_4}$ (−2 charge)

$\mathbf{[Ni(en)_3]SO_4}$

c. dicarbony l= $\mathbf{(CO)_2}$ (neutral Lewis base)
tricyano = $\mathbf{(CN)_3}$ (−1 per CN × 3 = −3)
cobaltate = **Co** (oxidation state = +1)

The complex anion is $\mathbf{Co(CN)_3(CO)_2}$. The total charge = −2. We must therefore have 2 potassium ions to balance electrically.

$\mathbf{K_2[Co(CN)_3(CO)_2]}$

19.4 Isomerism

There are several new definitions introduced in this section:

1. **Isomerism:** same chemical formula, different compounds.
2. **Structural isomerism:** same atoms, different bond arrangement.
   a. **Coordination isomerism:** complex ion and counter ion interchange members.
   b. **Linkage isomerism:** point of attachment of ligand to metal is different.

3. **Stereoisomerism:** same bond arrangement, different spatial arrangement.
   a. **Geometrical isomers**: atoms or groups of atoms assume different positions around the central atom (*cis-* or *trans-* isomerism).
   b. **Optical isomers**: rotates plane polarized light in different directions. Such molecules have chiral centers.

Your textbook gives some examples of each kind of isomerism. After reading the section, try the following example.

Example 19.4 - Recognizing Isomers

Identify the type of isomerism exhibited by each pair of substances.

a. *trans*-$[RuCl_2(H_2O)_4]^+$ and *cis*-$[RuCl_2(H_2O)_4]^+$
b. $[Co(en)_2(NH_3)Cl]^{2+}$ exists in a form that can rotate plane polarized light to the left and one that can rotate it to the right.
c. $[Co(NH_3)_5(NO_2)]^{2+}$ and $[Co(NH_3)_5(ONO)]^{2+}$

Solution

a. geometrical isomers
b. optical isomers
c. linkage isomers

19.5 Bonding in Complex Ions: The Localized Electron Model

Your textbook points out that the VSEPR model fails when considering a coordination number of 4, which can mean a tetrahedral or square planar arrangement. The localized electron model can be used to rationalize general bonding schemes, but does not predict key properties of complex ions.

19.6 The Crystal Field Model

The purpose of the crystal field model is to attempt to **explain the magnetism and colors of complex ions**. The underlying assumptions of the model are:

1. That ligands are **negative point charges.**
2. That the bonding between ligands and the central atom is **totally ionic**. That is, electrostatic interactions keep ligands bonded to the central atom. (There are no electrons shared or donated by the ligands.)

Regarding **octahedral complex ions**, your textbook discusses two types of field splitting between t_{2g} and e_g states for the *d* orbitals on the metal. Δ is the splitting energy.

low-spin = strong-field = large energy difference (Δ) = **minimum** number of unpaired electrons. (See Figure 19.23 in your textbook.)

high-spin = weak-field = small Δ = **maximum** number of unpaired electrons.

In octahedral complex ions, two things determine the type of spin state (low or high).

- The spectrochemical series (given after Interactive Example 19.4 in your textbook).
- The **charge on the metal ion**. (A higher charge causes stronger field-splitting.)

Example 19.6 A - Crystal Field Splitting

The complex ion $Mn(H_2O)_6^{3+}$ is observed to have a high-spin state. How many unpaired electrons are in the complex?

Solution

Mn has an $[Ar]4s^2 3d^5$ configuration. The Mn^{3+} is $\mathbf{4s^0 3d^4}$. Having a **high-spin state** means that water induces **weak-field** splitting in the complex ion. That implies a small Δ.

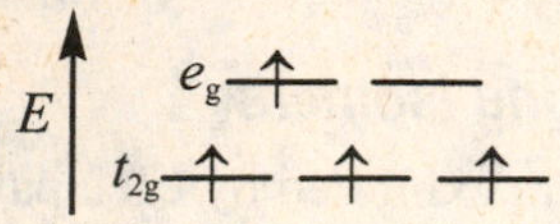

There are 4 unpaired electrons in the complex.

Example 19.6 B - Determining Splitting From Unpaired Electrons

The complex ion $Co(NH_3)_6^{3+}$ is observed to have no unpaired electrons. Is it a low- or high-spin complex?

Solution

Cobalt(III) is a $\mathbf{3d^6}$ ion. The two options are

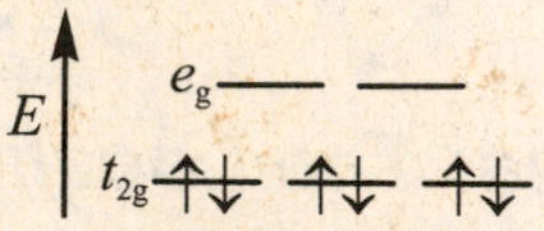

large E, low-spin

OR

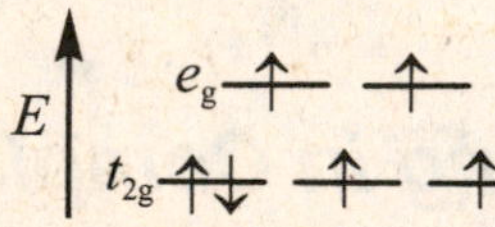

small E, high-spin

The high-spin complex would result in four unpaired electrons. The low-spin complex would result in none. This complex is therefore a **low-spin complex.**

Color of the Complex

Your textbook points out that it takes energy to promote electrons between the t_{2g} and e_g states. The **higher the energy**, the **shorter the wavelength** of absorbed color. That is, higher energy means blue and violet will be absorbed. (You will see a reddish compound.) Lower energy means red will be absorbed. (You will see a bluish compound.)

If you look at Table 19.17 in your textbook, you can see that, as the number of NH_3 groups diminishes, the color goes toward violet, indicating more red is being absorbed (a lower Δ value).

Example 19.6 C - Colors And Field Splitting

A solution of $[Cu(en)_2]^{2+}$ is green. The color of a $[CuBr_4]^{2-}$ solution is violet. What does this tell you about the relative crystal field splitting energies? Which ligand causes the greater splitting, **en** or $\mathbf{Br^-}$?

Solution

Because the $[CuBr_4]^{2-}$ solution is violet, it means lower energy radiation is being absorbed (so that violet is emitted). This indicates a relatively small splitting energy. The $[Cu(en)_2]^{2+}$ solution transmits green, indicating that it absorbs blue, a relatively high energy (along with red and orange). This means the **splitting energy is relatively large.**

The ethylenediamine ligand **causes greater splitting than the bromide ligand** (all other things being equal).

19.7 The Molecular Orbital Model

The following questions will help you review the material in this section. Remember, crystal field theory is too good to be true, and molecular orbital theory is too true to be good.

1. What are the weaknesses of the crystal field model?
2. What are the advantages of the molecular orbital (M.O.) model in describing the bonding and properties of complex ions? What are the disadvantages?
3. Discuss the two important considerations in predicting the extent of orbital overlap.
4. What orbitals combine to give σ_s, σ_p, and σ_d orbitals in the octahedral M.O. in Figure 19.31 of your textbook?
5. Why are e_g^* orbitals antibonding?
6. What is the effect on M.O. splitting of having an electronegative ligand?
7. If the M.O. model accounts for the properties of complex ions, why did your textbook bother to introduce the crystal field model at all?

19.8 The Biological Importance of Coordination Complexes

The following questions will help you review the material in this section.

1. What are some biological uses of coordination complexes?
2. Why are coordination complexes ideal for biological applications?
3. What are the principal sources of energy in mammals?
4. What is the **respiratory chain**?
5. What are **cytochromes**? What are their two main parts?
6. What is the structure of chlorophyll?
7. How can Fe^{2+} have room to attach to myoglobin?
8. What is the mechanism for the oxidation of Fe^{2+} in heme?
9. Describe the transport of O_2 in blood.
10. What biochemical mechanism leads to sickle cell anemia?
11. What does Le Châtelier's principle have to do with altitude sickness?
12. Describe how CO and CN^- interact with iron to cause death.
13. Define **respiratory inhibitor**.

Exercises

Section 19.1

1. Write electron configurations for each of the following metals.

   a. Mn b. Pd c. Zr d. Zn e. Rh

2. Write electron configurations for each of the following ions.

   a. Cr^{6+} b. Cr^{2+} c. Fe^{6+} d. Fe^{3+} e. Mn^{2+}

Section 19.3

3. What is the coordination number on the metal in each of the following complex ions?

   a. $[CuF_4]^{2-}$ c. $[Fe(en)(H_2O)_4]^{2+}$
   b. $[CuF_6]^{3-}$ d. $[Fe(en)_3]^{2+}$

4. Gadolinium has the electron configuration $[Xe]6s^25d^14f^7$. Why is this configuration more energetically favorable than $[Xe]6s^25d^04f^8$?

5. The final step in the preparation of cobalt is the reduction of Co_3O_4 with aluminum:

$$3Co_3O_4(s) + 8Al(s) \rightarrow 9Co(s) + 4Al_2O_3(s)$$

How much cobalt can be obtained from 125.0 g of Co_3O_4?

6. Iron in aqueous solution can undergo a series of reactions with ethylenediamine (en).

$$[Fe(H_2O)_6]^{2+} + en \rightarrow [Fe(en)(H_2O)_4]^{2+} + 2H_2O$$
$$[Fe(en)(H_2O)_4]^{2+} + en \rightarrow [Fe(en)_2(H_2O)_2]^{2+} + 2H_2O$$
$$[Fe(en)_2(H_2O)_2]^{2+} + en \rightarrow [Fe(en)_3]^{2+} + 2H_2O$$

a. Name each complex ion.
b. What is the overall geometry of the complexes?
c. What is the coordination number of the iron?
d. Will the crystal field splitting increase or decrease as these reactions proceed?

7. What is the oxidation state of the metal ion in each of the following complex ions?

a. $[Mn(CO)_5]^+$ b. $Pt(CO)Cl_2$ c. $[Co(CN)_5OH]^{3-}$

8. Name each of the complex ions in Problem 7.

9. Give the formula for each of the following coordination compounds.

a. Potassium tetrahydroxynickelate(II)
b. tetraaquamanganese(II) sulfate
c. tris(ethylenediamine)cobalt(III) chloride

Section 19.6

10. How many unpaired electrons will be present in each of the following complex ions?

a. $[Fe(NH_3)_6]^{2+}$ (high-spin) b. $[Co(NH_3)_6]^{3+}$ (low-spin)

11. Determine the spin of $[Ru(H_2O)_6]^{3+}$ if it has one unpaired electron.

12. Based on the spectrochemical series, predict whether each of the following will be high or low-spin.

a. $[Fe(CN)_6]^{4-}$ b. $[CoF_6]^{3-}$ c. $[MnCl_6]^{4-}$

13. How many unpaired electrons will be present in each of the complex ions in Problem 12?

14. Arrange the following colors in order of increasing wavelength.

red, violet, green, blue, yellow

15. Arrange the colors in Problem 14 in terms of increasing energy.

16. Two compounds are synthesized. One is red. The other is green. Which compound has the larger value for Δ? Why?

17. An electron is excited from the t_{2g} to the e_g state by radiation of wavelength 620 nm.

a. What color light did the electron absorb?
b. What is the energy of the absorbed light?

18. The complex ion $[Cu(H_2O)_6]^{2+}$ has an absorption maximum at around 800 nm. When four ammines replace water, $[Cu(NH_3)_4(H_2O)_2]^{2+}$, the absorption maximum shifts to around 600 nm. What do these results mean in terms of the relative field splittings of NH_3 and H_2O?

Multiple-Choice Self-Test

1. Which one of the following statements is not true about the lanthanide series?
   A. Going from left to right, the atomic size decreases.
   B. Going from right to left, the atomic size decreases.
   C. Going from left to right, the lanthanide elements are filling their 4*f* orbitals.
   D. Due to the lanthanide elements, the 4*d* and 5*d* elements in a vertical group have similar properties.

2. Which one of the following metals is the best reducing agent?
   A. Ni   B. Co   C. Ti   D. Cr

3. Which one of the following statements about scandium is not true?
   A. The most common oxidation state of scandium is +3.
   B. Its chemistry strongly resembles that of lanthanides.
   C. Its electron configuration is $[Ar]4s^23d^1$.
   D. Scandium is prepared by electrolysis of molten $ScCl_5$.

4. Which one of the following statements about vanadium is not true?
   A. Its electron configuration is $[Ar]4s^23d^2$.
   B. It can be used to produce steel-gray, which is hard and corrosion resistant.
   C. Its principal oxidation state is +5.
   D. Its higher oxidation states do not exist as hydrated ions of the type V^{n+}. They cause the attached waters to become very acidic.

5. Which one of the following statements about chromium is not true?
   A. Chromite, produced by reacting carbon with ferrochrome, is added to iron in the steelmaking process.
   B. Its most common oxidation state is +2.
   C. Chromium(VI) species are excellent oxidizing agents.
   D. The lower the pH, the higher the oxidizing abilities of chromium(VI).

6. Which one of the following statements about copper is not true?
   A. Copper is second best (after silver) in conducting heat and electricity.
   B. It is very prone to corrosion, especially when oxidized.
   C. It is used in many alloys, such as sterling silver and brass.
   D. Aqueous solutions of copper(II) are typically blue in color due to the presence of the $Cu(H_2O)_6^{2+}$ ion.

7. Which one the following ligands is unable to act in a bidentate manner?
   A. S^{2-} B. CN^- C. Br^- D. NH_3

8. The ligands in a complex ion act as:
   A. Lewis acids   B. Lewis bases   C. Arrhenius bases   D. oxidizing agents

9. If a ligand acts in an octadentate manner, how many electrons does it donate?
   A. 4   B. 6   C. 8   D. 16

10. A complex ion is found to contain 31.56% zinc and 68.44% chlorine. Which one of the following geometries is consistent with these data?

A. linear B. square planar C. tetrahedral D. octahedral

11. What is the formula for the following compound, tetrachlorodiaquaferrate(III) ion?

A. $[Fe(H_2O)_2Cl_2]^-$ B. $[Fe(H_2O)_2Cl_4]^-$ C. $[Fe(H_2O)_2Cl_4]^{2-}$ D. $[FeH_2OCl_4]^{2+}$

12. What is the name of the nonpolar complex ion, $[Cr(H_2O)_4I_2]^+$?

A. *trans*-tetraaquadiiodochromium(III) ion
B. *cis*-tetraaquadiiodochromium(III) ion
C. *trans*-tetraaquadiiodochromium ion
D. *cis*-tetraaquadiiodochromium(I) ion

13. What is the name of the nonpolar complex ion, $[CuCl_2F_2]^{2-}$?

A. *trans*-dichlorodifluorocopper(II) ion
B. *trans*-dichlorodifluorocuprate(II) ion
C. *cis*-dichlorodifluorocuprate(II) ion
D. *cis*-dichlorodifluorocopper(II) ion

14. Which one of the following pairs are coordination isomers?

A. *trans*-dichlorodifluorocuprate(II) ion and *cis*-dichlorodifluorocuprate(II) ion
B. $[Cr(H_2O)_4I_2]Cl$ and $[Cr(H_2O)_4Cl_2]I$
C. *trans*-dichlorodifluorocuprate(II) ion and *trans*-dichlorodifluorocopper(II) ion
D. $[Cu(H_2O)_6]^{2+}$ and $[Cu(H_2O)_6]I_2$

15. Which one of the following metal ions should exhibit the largest crystal field splitting energy?

A. Fe^{3+} B. Ir^{3+} C. Mn^{3+} D. Co^{3+}

16. How many unpaired electrons are there in the elevated *d*-orbitals of $[FeCl_4]^-$?

A. 0 B. 1 C. 2 D. 3

17. If a complex ion is square planar, which *d*-orbital is highest in energy?

A. $d_{x^2-y^2}$ B. d_{x^2} C. d_{xy} D. d_{zz}

18. In myoglobin:

A. Fe^{2+} is oxidized to Fe^{3+} when it transports oxygen.
B. Fe^{3+} is bound to a porphyrin ring containing bromine as a ligand.
C. Fe^{2+} is not oxidized to Fe^{3+} when it transports oxygen.
D. The iron-containing heme is bound to a protein.

19. In the process of roasting, sulfur can be separated from zinc by:

A. oxidizing sulfur and removing elemental zinc
B. oxidizing both sulfur and zinc
C. oxidizing zinc and removing elemental sulfur
D. oxidizing zinc and removing hydrated sulfur

Answers to Exercises

1. a. Mn: $[Ar]4s^23d^5$
b. Pd: $[Kr]5s^04d^{10}$
c. Zr: $[Kr]5s^24d^2$
d. Zn: $[Ar]4s^23d^{10}$
e. Rh: $[Kr]5s^14d^8$

2. a. Cr^{6+}: [Ar]
b. Cr^{2+}: $[Ar]3d^4$
c. Fe^{6+}: $[Ar]3d^2$
d. Fe^{3+}: $[Ar]3d^5$
e. Mn^{2+}: $[Ar]3d^5$

3. a. 4 b. 6 c. 6 d. 6

4. Although electron configurations are hard to nail down unequivocally this deep in the periodic table, it would seem that the 4*f* and 5*d* energy levels are close enough together to allow electron crossover. The half-filled 4*f* helps minimize the electron-electron repulsion that would occur if the configuration were $5d^0 4f^8$.

5. 91.78 g cobalt

6. a. $[Fe(H_2O)_6]^{2+}$ = hexaaquairon(II) ion
   $[Fe(en)(H_2O)_4]^{2+}$ = tetraaquaethylenediamineiron(II) ion
   $[Fe(en)_2(H_2O)_2]^{2+}$ = diaquabis(ethylenediamine)iron(II) ion
   $[Fe(en)_3]^{2+}$ = tris(ethylenediamine)iron(II) ion
   b. All complexes are octahedral.
   c. The iron has a coordination number of 6.
   d. According to the spectrochemical series, the splitting will increase (larger Δ).

7. a. +1 b. +2 c. +3

8. a. pentacarbonylmanganese(I) ion
   b. carbonyldichloroplatinum(II)
   c. pentacyanohydroxycobaltate(III) ion

9. a. $K_2[Ni(OH)_4]$ b. $[Mn(H_2O)_4]SO_4$ c. $[Co(en)_3]Cl_3$

10. a. 4 b. 0

11. Ru^{3+} is $4d^5$. One unpaired electron corresponds to **low-spin** (strong field case).

12. a. low-spin b. high-spin c. high-spin

13. a. 0 b. 4 c. 5

14. (lowest) violet, blue, green, yellow, red (highest)

15. (lowest) red, yellow, green, blue, violet (highest)

16. The red compound has the higher Δ value because it absorbs blue light. The green compound absorbs red and yellow light, which is of lower energy.

17. a. red light b. $E = hc/\lambda = 3.2 \times 10^{-19}$ J/photon

18. An absorption maximum shift to 600 nm means Δ is larger (600 nm represents greater radiation than 800 nm). Therefore a stronger field splitting is occurring. This is consistent with the spectrochemical series.

Answers to Multiple-Choice Self-Test

1. B	5. A	8. B	11. B	14. B	17. A
2. C	6. B	9. C	12. A	15. B	18. D
3. D	7. D	10. C	13. B	16. D	19. B
4. A					

CHAPTER 20

The Nucleus: A Chemist's View

The Bottom Line: Chapter 20

In our study of chemistry we have been dealing so far with the movement of electrons within and between atoms. We have been relatively unconcerned about nuclear structure—until now. This chapter deals with **nuclear transformations**. An important goal is to relate **radioactive decay** to nuclear energy and to the age of terrestrial objects.

20.1 Nuclear Stability and Radioactive Decay

Radioactive decay is the process by which a nucleus decomposes to form a different nucleus along with additional particles.

The key to determining nuclear decay products (from a general chemistry point of view) is to remember that *the sum of the protons after the decay equals that before the decay. The same is true with the number of neutrons.* For example, neptunium undergoes decay as follows:

$$^{237}_{93}\text{Np} \rightarrow {}^{4}_{2}\text{He} + {}^{233}_{91}\text{Pa}$$

Notice that the **sum** of protons and neutrons is the same on both sides of the equation as is the number of protons.

Your textbook discusses **six types of radioactive processes**. Each involves isotopes of various atoms. Note that, for *individual atoms*, we prefer the term **nuclide.** These are, briefly,

- **α-particle production** results in the release of ${}^{4}_{2}\text{He}$.

  example: ${}^{240}_{94}\text{Pu} \rightarrow {}^{4}_{2}\text{He} + {}^{236}_{92}\text{U}$

- **β-particle production** results in the release of ${}^{0}_{-1}\text{e}$

  example: ${}^{194}_{76}\text{Os} \rightarrow {}^{194}_{77}\text{Ir} + {}^{0}_{-1}\text{e}$

- **γ-ray production** (High-energy radiation released as other nuclear transformations occur.)
- **positron production** results in the release of ${}^{0}_{1}\text{e}$.

  example: ${}^{17}_{9}\text{F} \rightarrow {}^{17}_{8}\text{O} + {}^{0}_{1}\text{e}$

- **electron capture** is accomplished when the nucleus captures an inner orbital electron.

  example: ${}^{229}_{92}\text{U} + {}^{0}_{-1}\text{e} \rightarrow {}^{229}_{91}\text{Pa}$

- **spontaneous fission** is the decomposition of the nucleus into two relatively large fractions.

The following example tests your knowledge of the various nuclear transformation processes.

Example 20.1 A - Nuclear Decay

Write equations for each of the following processes:

a. $^{241}_{96}Cf$ undergoes electron capture.

b. $^{241}_{95}Am$ produces an α particle.

c. $^{121}_{54}Xe$ produces a positron.

d. $^{138}_{53}I$ produces a β particle.

Solution

Remember that the **sum** of protons and neutrons, as well as the **number** of protons is **consistent** on both sides of the equation. The key, then, is to calculate these values and then determine the element that has the values associated with it.

a. $^{241}_{96}Cf + ^{0}_{-1}e \rightarrow ^{241}_{95}Am$

b. $^{241}_{95}Am \rightarrow ^{237}_{93}Np + ^{4}_{2}He$

c. $^{121}_{54}Xe \rightarrow ^{121}_{53}I + ^{0}_{1}e$

d. $^{138}_{53}I \rightarrow ^{138}_{54}Xe + ^{0}_{-1}e$

There are a limited number of ways a nucleus can decay. It may release many particles before finally becoming stable.

A **decay series is a number of sequential decays** by an unstable nuclide. The series continues until a stable nuclide is formed. For example,

$$^{241}_{95}Am \rightarrow ^{237}_{93}Np + ^{4}_{2}He \rightarrow ^{233}_{91}Pa + ^{4}_{2}He$$

Each step is consistent with the previous one in terms of proton and neutron totals. Keep in mind that when a particle (such as α) is released, it is lost to the system, so it should not enter into the calculation.

Example 20.1 B - Decay Series

$^{247}_{97}Bk$ undergoes decay to $^{208}_{82}Pb$ in the following order:

α, α, β, α, α, β, α, β, α, α, α, α, β, β, α

Write equations for the first **six** steps.

Solution

1. $^{247}_{97}Bk \rightarrow ^{243}_{95}Am + ^{4}_{2}He$
2. $^{243}_{95}Am \rightarrow ^{239}_{93}Np + ^{4}_{2}He$

3 $^{239}_{93}Np \rightarrow ^{239}_{94}Pu + ^{0}_{-1}e$

4. $^{239}_{94}Pu \rightarrow ^{235}_{92}U + ^{4}_{2}He$
5. $^{235}_{92}U \rightarrow ^{231}_{90}Th + ^{4}_{2}He$
6. $^{231}_{90}Th \rightarrow ^{231}_{91}Pa + ^{0}_{-1}e$

20.2 The Kinetics of Radioactive Decay

The decay of nuclides follows a **first-order rate law.** That means that the kinetics is governed by **the same equations** that were introduced for first-order kinetics in Section 15.4. These are

$$\ln\left(\frac{N}{N_0}\right) = -kt$$

where N_0 = the original number (mass) of nuclides
N = the number remaining at time t
k = the first-order rate constant
t = the time

Also,

$$t_{1/2} = \frac{0.693}{k}$$

where $t_{1/2}$ is the half-life of the nuclide.

The following examples are intended to show how you can apply these equations directly to radioactive decay data.

Example 20.2 - Half-Life And Concentration

The half-life of $^{239}_{94}Pu$ is 2.411×10^4 years. How many years will elapse before 99.9% of a given sample decomposes?

Solution

We have no specific amounts. However, we do know that (from a **fractional** point of view) **0.999 of our original 1.000** decomposes, leaving **0.001** remaining. We can thus establish the **ratio** N/N_0 **as 0.001/1.000**. We can find k from $t_{1/2}$.

$$k = \frac{0.693}{k} = \frac{0.693}{2.411\times10^4\ \text{y}} = 2.87 \times 10^{-5}/\text{y}$$

$$\ln\left(\frac{N}{N_0}\right) = -kt$$

$$\ln(0.001) = 2.87 \times 10^{-5}/\text{y}\ (t)$$

$$\mathbf{t = 2.4 \times 10^5\ y}$$

Does the Answer Make Sense?

Our total time, 2.4×10^5 y, is about **10 half-lives**. This means we should have about $1/2^{10}$, (or 0.001) of our original material remaining. This is in fact the case, so the answer makes sense.

20.3 Nuclear Transformations

The following problem is intended to give you practice with the particles involved in nuclear transformations. Remember that in this case nuclides are not spontaneously decaying, but are instead being bombarded with particles (either neutrons or other nuclides) that **cause a heavier element to form.**

Example 20.3 - Nuclear Transformations

Fill in the missing particles in each of the following nuclear transformations:

a. $? + {}^{4}_{2}\text{He} \rightarrow {}^{243}_{97}\text{Bk} + 2\,{}^{1}_{0}\text{n}$

b. ${}^{253}_{99}\text{Es} + {}^{4}_{2}\text{He} \rightarrow ? + {}^{1}_{0}\text{n}$

c. ${}^{250}_{98}\text{Cf} + {}^{11}_{5}\text{B} \rightarrow {}^{257}_{103}\text{Lr} + ?$

Solution

a. ${}^{241}_{95}\textbf{Am} + {}^{4}_{2}\text{He} \rightarrow {}^{243}_{97}\text{Bk} + 2\,{}^{1}_{0}\text{n}$

b. ${}^{253}_{99}\text{Es} + {}^{4}_{2}\text{He} \rightarrow {}^{256}_{101}\textbf{Md} + {}^{1}_{0}\text{n}$

c. ${}^{250}_{98}\text{Cf} + {}^{11}_{5}\text{B} \rightarrow {}^{257}_{103}\text{Lr} + \mathbf{4\,{}^{1}_{0}n}$

20.4 Detection and Uses of Radioactivity

The **key assumptions** when using radioactivity to date objects using **radiocarbon dating** are:

- The ratio of ${}^{14}C/{}^{12}C$ has been constant over the years, and
- living systems have that **same constant ratio** until they die.

The mathematical basis of the technique is that once the living thing dies, the **ratio of ${}^{14}C/{}^{12}C$ diminishes** with a **half-life of 5730 years**. This means that after 17,190 years (3 half-lives), the ${}^{14}C/{}^{12}C$ ratio (be it amount, or disintegrations per second or whatever) will be $1/2^3 = 1/8$ what it was before.

Example 20.4 A - Radiocarbon Dating

A sample of bone taken from an archeological dig was determined by radiocarbon dating to be 12,000 years old. If we assume that a constant atmospheric **${}^{14}C/{}^{12}C$ ratio has 13.6 disintegrations per minute per gram of carbon, how many disintegrations per minute per gram** did our sample give off ($t_{1/2}$ for ${}^{14}C$ = 5730 years).

Solution

We know the age of our solution, and we know the half-life. We can therefore determine, by using our first-order decay formula, the ratio N/N_0, which in this case represents the ratios of disintegrations per minute per gram.

$$\begin{aligned} k &= 0.693/t_{1/2} \\ &= 0.693/5730\text{ y} \\ &= 1.21 \times 10^{-4}/\text{y} \\ \ln(N/N_0) &= -kt \\ &= -1.21 \times 10^{-4}/\text{y}\ (12000\text{ y}) \\ &= -1.45 \\ N/N_0 &= e^{-1.45} = \mathbf{0.234} \end{aligned}$$

N_0 = 13.6 disintegrations; therefore $N = 0.234(13.6)$

N = 3.2 disintegrations per minute per gram

In determining the age of exceptionally old objects (billions of years), your textbook discusses the nuclear transformation

$${}^{238}_{92}\text{U} \rightarrow {}^{206}_{82}\text{Pb} \qquad t_{1/2} = 4.5 \times 10^9\text{ y}$$

Example 20.2 in your textbook deals with the age of a rock based on comparing **atoms** of $^{238}_{92}\mathbf{U}$ and $^{206}_{82}\mathbf{Pb}$. There is a 1:1 stoichiometry in this relationship. Let's take this one step further and compare **grams** of U and Pb. The relationship in this case is **not** 1:1, but rather **238 g U/206 g Pb.**

Example 20.4 B - Dating Via Uranium

A rock contains 0.141 g of $^{206}_{82}$Pb for every 1.000 g $^{238}_{92}$U. How old is the rock ($t_{1/2}$ $^{238}_{92}$U = 4.5 × 10^9 y, and you are to assume that the intermediate nuclides decay instantaneously)?

Solution

The goal really is to find the ratio N/N_0 for uranium decay. The amount of lead found is a **direct reflection** of the amount of uranium that has decayed.

$$^{238}_{92}\text{U decayed} = 0.141 \text{ g Pb} \times \frac{238 \text{ g U}}{206 \text{ g Pb}} = \mathbf{0.163 \text{ g } ^{238}_{92}U}$$

The original amount of $^{238}_{92}\mathbf{U}$ = g found + g decayed = 1.000 + 0.163 = **1.163 g.**

$$k = \frac{0.693}{t_{1/2}} = \frac{0.693}{4.5 \times 10^9 \text{ y}} = \mathbf{1.54 \times 10^{-10}/y}$$

$$\ln(N/N_0) = -kt$$
$$\ln(1.000/1.163) = 1.54 \times 10^{-10}/\text{y } (t)$$
$$t = \mathbf{9.8 \times 10^8 \text{ years old}}$$

20.5 Thermodynamic Stability of the Nucleus

A nucleus is a bundle of protons and neutrons bound together. The nucleus is more energetically stable than the individual array of protons and neutrons. This is known because the **sum of the masses** of protons and neutrons in a nucleus is **less than** the sum of the masses of individual nucleons. The difference in the mass is called the **mass defect (Δ*m*)**. The mass has been converted to energy, as predicted by

$$E = mc^2, \text{ or } \Delta E = \Delta mc^2$$

ΔE is called the binding energy. For example,

$$^{187}_{77}\mathbf{Ir} = 187.958830 \text{ g/mol}$$

1 mole of 77 (protons + electrons) + 110 neutrons =

77 × 1.007825 g per proton or electron	=	77.602525 g
110 × 1.008665 g per neutron	=	110.95315 g
		188.555675 g

The **mass defect Δ*m*** = 188.555675 g − 187.958830 g = **0.596845 g/mol**. That is the mass that has been converted to energy. To find out how much, remember that

1 Joule = 1 kg m^2/s^2
speed of light = c = 3.00 × 10^8 m/s

$$\Delta E = \Delta mc^2 = \frac{-0.596845 \text{ g}}{\text{mol}} \times \frac{1 \text{ kg}}{1000 \text{ g}} \times \left(\frac{3.00 \times 10^8 \text{ m}}{\text{s}}\right)^2 = 5.37 \times 10^{13} \text{ kg m}^2/\text{s}^2 \text{ mol}$$

$\mathbf{\Delta E = -5.37 \times 10^{13}}$ **J/mol** (The "−" indicates energy is released (exothermic process) when the nucleus is formed.)

In order to calculate ΔE per nucleus, we must divide by Avogadro's number.

$$\Delta E = \frac{-5.37\times10^{13}\ \text{J}}{\text{mol}} \times \frac{1\ \text{mol}}{6.02\times10^{23}\ \text{nuclei}} = \mathbf{-8.92 \times 10^{-11}\ J/nucleus}$$

In order to convert to J/nucleon, divide by 187 nucleons in the $^{187}_{77}\text{Ir}$ nucleus.

$$\Delta E = \frac{-8.92\times10^{-11}\ \text{J}}{\text{nucleus}} \times \frac{1\ \text{nucleus}}{187\ \text{nucleons}} = \mathbf{-4.77 \times 10^{-13}\ J/nucleon}$$

Finally, the binding energy per nucleon in million electron volts (MeV),

$$\mathbf{\Delta E} = \frac{-4.77\times10^{-13}\ \text{J}}{\text{nucleon}} \times \frac{1\ \text{MeV}}{1.60\times10^{-13}\ \text{J}} = \mathbf{-2.98\ MeV/nucleon}$$

This means that 2.98 MeV **per nucleon** would be released if the $^{187}_{77}\text{Ir}$ nucleus were formed from individual protons and neutrons (with electrons surrounding the nucleus).

Example 20.5 - Mass Defect And Binding Energy

Determine the binding energy in **J/mol** and **MeV/nucleon** for $^{101}_{46}\text{Pd}$ (atomic mass = 100.908287 g/mol).

Solution

Mass of individual nucleons.

46 × 1.007825 g per proton or electron	=	46.35995 g
55 × 1.008665 g per neutron	=	55.476575 g
		101.836525 g

mass defect (Δm) = 101.836525 g − 100.908287 g = **0.928238 g**

$$\Delta E = \Delta mc^2 = -9.28238 \times 10^{-4}\ \text{kg}\ (3.00 \times 10^8\ \text{m/s})^2$$

(minus sign because mass is <u>lost</u> in forming the nuclide)

$$\mathbf{\Delta E = -8.35 \times 10^{13}\ J/mol}$$

$$\frac{\text{MeV}}{\text{nucleon}} \parallel \frac{-8.35\times10^{13}\ \text{J}}{\text{mol}} \times \frac{1\ \text{MeV}}{1.60\times10^{-13}\ \text{J}} \times \frac{1\ \text{mol}}{6.02\times10^{23}\ \text{nuclides}} \times \frac{1\ \text{nuclide}}{101\ \text{nucleons}}$$

$$\mathbf{\Delta E = -8.59\ MeV/nucleon}$$

20.6 Nuclear Fission and Nuclear Fusion

The following questions will help you review the material in this section of your textbook.

1. Define **fusion.**
2. Define **fission.**
3. Is $^{235}_{92}\text{U} + ^{1}_{0}\text{n} \rightarrow ^{137}_{52}\text{Te} + ^{97}_{40}\text{Zr} + 2^{1}_{0}\text{n}$ an example of fission or fusion?
4. How does a **chain reaction** work?
5. Describe the process of fission in a nuclear warhead (also define **subcritical, critical,** and **supercritical).**
6. Outline how nuclear reactors work. Focus on the functions of the **reactor core, moderator, control rods, and cooling systems.**
7. What are the consequences of failure of each of the systems in a nuclear reactor?

8. What is the principle behind a **breeder reactor?**
9. Why is fusion a process worth exploring?
10. What is the difficulty with doing fusion reactions on Earth?

20.7 Effects of Radiation

The following questions will help you review the material in this section of your textbook.

1. Why is any energy source potentially dangerous?
2. Why does exposure to radioactivity seem less hazardous than it really can be? (Discuss the exposure per event.)
3. Define **somatic damage** and **genetic damage.**
4. What is a **free radical,** and what is its role in the biological effects of radiation?
5. What variables affect the degree of damage radiation can cause? Discuss each one.
6. Your textbook says our exposure to natural radiation is much greater than to man-made sources. Why are we so worried about a relatively small exposure level?

Exercises

Section 20.1

1. Write equations for each of the following processes:

   a. $^{73}_{31}Ga$ produces a β particle.

   b. $^{68}_{31}Ga$ undergoes electron capture.

   c. $^{192}_{78}Pt$ produces an α particle.

2. Write equations for each of the following processes:

   a. $^{207}_{87}Fr$ produces an α particle.

   b. $^{234}_{90}Th$ produces a β particle.

   c. $^{62}_{29}Cu$ produces a positron.

3. Fill in the missing particle in each of the following equations:

   a. $^{129}_{51}Sb \rightarrow \_\_\_\_ + ^{0}_{-1}e$

   b. $\_\_\_\_ + ^{0}_{-1}e \rightarrow ^{7}_{3}Li$

   c. $^{205}_{83}Bi \rightarrow ^{205}_{82}Pb + \_\_\_\_$

   d. $^{206}_{87}Fr \rightarrow \_\_\_\_ + ^{4}_{2}He$

4. Fill in the missing particle in each of the following equations:

   a. $\_\_\_\_ + ^{0}_{-1}e \rightarrow ^{212}_{86}Rn$

   b. $\_\_\_\_ \rightarrow ^{208}_{85}At + ^{4}_{2}He$

   c. $^{226}_{90}Th \rightarrow \_\_\_\_ + ^{4}_{2}He$

   d. $^{186}_{77}Ir \rightarrow \_\_\_\_ + ^{0}_{1}e$

5. The radioactive isotope $^{242}_{96}Cm$ undergoes decay to $^{206}_{82}Pb$ in the following order:

$$\alpha, \alpha, \alpha, \alpha, \alpha, \alpha, \alpha, \beta, \beta, \alpha, \beta, \beta, \alpha.$$

Write equations for the first six steps.

6. The sixth step in the previous problem should leave you with $^{218}_{84}Po$. Continue writing equations in the series until you reach $^{206}_{82}Pb$.

Section 20.2

7. Calculate the rate constant for each of the following values of half-life:

   a. $^{182}_{72}Hf$, $t_{1/2} = 9 \times 10^6$ y

   b. $^{228}_{91}Pa$, $t_{1/2} = 26$ hr

   c. $^{225}_{88}Ra$, $t_{1/2} = 14.8$ d

   d. $^{181}_{78}Pt$, $t_{1/2} = 51$ s

8. The half-life of $^{161}_{65}Tb$ is 6.9 days. How many grams of an original 3.000-g sample will remain after two weeks?

9. The half-life of $^{129}_{53}I$ is 1.7×10^7 y. How many grams of an original 50.00-g sample will remain after 5.0×10^8 y?

10. How long will it take for 98.6% of a sample of $^{188}_{79}Au$ to decompose ($t_{1/2} = 8.8$ min)?

11. How long will it take for 99.99% of a sample of $^{199}_{84}Po$ to decompose ($t_{1/2} = 5.2$ min)?

12. How many half-lives have passed if 87.5% of a substance has decomposed? How many if 99.999% has decomposed?

Section 20.4

13. Let us assume a constant $^{14}C/^{12}C$ ratio of 13.6 disintegrations per minute per gram of living matter. A sample of a petrified tree was found to give 1.2 disintegrations per minute per gram. How old is the tree? ($t_{1/2} = {}^{14}C = 5730$ years)

14. The radioactive isotope $^{237}_{90}Th$ has a rate constant of $k = 4.91 \times 10^{-11}$ y^{-1}. Is this nuclide useful for determining the age of bone samples? Why or why not?

15. A rock contains 0.688 g of $^{206}_{82}Pb$ for every 1.00 g of $^{238}_{92}U$. How old is the rock? ($t_{1/2}$ $^{238}_{92}U$ = 4.5×10^9 y. Assume intermediate nuclides decay instantaneously to the stable Pb nuclide.)

Section 20.5

16. Calculate the mass defect in grams for one mole of each of the following:

   a. $^{235}_{92}U$, atomic mass = 235.0439 g/mol

   b. $^{127}_{53}I$, atomic mass = 126.9004 g/mol

   c. $^{75}_{33}As$, atomic mass = 74.9216 g/mol

17. Determine the binding energy in J/mol and MeV/nucleon for $^{66}_{30}Zn$ (atomic mass = 65.9260 g/mol).

18. Determine the binding energy in J/mol and MeV/nucleon for $^{150}_{60}Nd$ (atomic mass = 149.9207 g/mol).

Multiple-Choice Self-Test

1. A ^{238}U nucleus decays by alpha emission. What is the product of this reaction?
   A. ^{234}U B. ^{234}Th C. ^{242}Pu D ^{242}Ra

2. What is the final product of the following decay series of ^{230}Th?

   $$\alpha, \alpha, \alpha, \beta, \alpha, \beta$$

   A. ^{218}Po B. ^{214}Pb C. ^{210}Po D. ^{214}Po

3. Positron production results in:
   A. Higher proton/neutron ratio
   B. Same proton/neutron ratio
   C. Smaller proton/neutron ratio
   D. Smaller neutron/proton ratio

4. Which one of the following decay series would change ^{210}Pb to ^{206}Pb?
   A. α, α, α, β, α, γ B. β, β, α C. β, β, α, α D. α, β, α, γ

5. How long, in years, will it take for the ^{208}Po activity to be reduced by 90.0%? The half-life of ^{208}Po is 2.83 years.
   A. 9.40 B. 26.0 C. 28.3 D. 3.14

6. How much of a 2.00-g ^{239}Pu sample would have decayed after 1.6×10^4 years? The half-life for ^{239}Pu is 2.4×10^4 years.
   A. 1.98 g B. 0.74 g C. 0.2 g D. 1.28 g

7. It took 109.8 years for 300.0 mg of a sample of an unknown radioactive material to completely disintegrate. Calculate the mass of the element (in grams/mole), assuming that the disintegrations per second are constant over the lifetime of the sample, and the sample is labeled 100.0 Ci (Curie = 3.7×10^{10} disintegrations per second).
   A. 14.0 B. 90.0 C. 209 D. 238

8. How long, in years, will it take for ^{208}Po activity to be reduced down to 0.01% of the original activity? The half life of ^{208}Po is 2.83 years.
   A. 283 B. 18.8 C. 2.82 D. 37.6

9. The number of disintegrations per second of ^{14}C in a living organism is 31 for every two grams of carbon. How long, in years, will it take for the level of activity to reach 29 disintegrations per second per gram of carbon? The half-life of ^{14}C is 5730 years.
   A. 11.0 B. 551 C. 110 D. 5.36×10^3

10. If the number of disintegrations of ^{14}C in a living organism is 930 disintegrations per second per gram of carbon, how old is the content of a clay amphora that displays 9.22 disintegrations per minute per gram of carbon?
    A. 2100 years B. 1200 years C. 2150 years D. 4300 years

11. A certain map that is claimed to be the map of the lost continent of Atlantis displays a ^{14}C disintegration rate of 12.0 disintegrations per gram of carbon per minute. The continent is supposed to have disappeared approximately 2500 years ago, and the map is claimed to be original. If the ^{14}C rate of decay in living organisms is 15.5 decays per gram of carbon per minute, how old is the map, and could it be original?
    A. No, approximately 2100 years old
    B. Yes, approximately 3800 years old
    C. Yes, dates from approximately 1800 BC
    D. No, approximately 630 years old

12. Calculate the total binding energy, in joules, in one atom of ^{32}S (31.97207 amu). m_p = 1.00728 amu, and m_n = 1.00866 amu.
    A. 6.82×10^{-11} B. 4.10×10^{-5} C. 4.10×10^{-11} D. 1.36×10^{-12}

13. Calculate the mass defect in 2 moles of ^{32}S (31.97207 amu). (m_p = 1.00728 amu, and m_n = 1.00866 amu.)
    A. 0.58351 g B. 0.84891 g C. 1.2203 g D. 0.62355 g

14. Calculate the binding energy per nucleon, in MeV per mole, for ^{33}S nucleus (32.97146 amu). m_p = 1.00728 amu and m_n = 1.00866 amu.
    A. 8.52 B. 8.25 C. 9.01 D. 30.2

15. In nuclear fission, a critical mass means that
    A. less than one neutron per fission is available.
    B. the process is capable of sustaining itself at a constant rate of fission.
    C. the process is capable of expanding its rate of fission.
    D. more than one neutron per fission are available.

16. The role of the moderator in a nuclear reactor is to
    A. absorb neutrons; therefore to regulate the rate of reaction.
    B. speed up neutrons released by the reactor core so the control rods can capture them more efficiently.
    C. slow down neutrons so the uranium fuel can capture them more efficiently.
    D. cool down the reactor core and prevent possible explosions from heat.

17. According to the linear model of exposure
    A. the higher the dose of radiation, the higher the danger.
    B. radiation is dangerous only above a certain threshold dose.
    C. the higher the dose, and the shorter the time of exposure, the smaller the danger.
    D. radiation is dangerous at only one certain dose.

18. Which one of the following types of radiation cause the greatest damage due to ionization ability?
    A. alpha particles B. beta particles C. gamma rays D. positrons

Answers to Exercises

1. a. $^{73}_{31}Ga \rightarrow {}^{73}_{32}Ge + {}^{0}_{-1}e$
   b. $^{68}_{31}Ga + {}^{0}_{-1}e \rightarrow {}^{68}_{30}Zn$
   c. $^{192}_{78}Pt \rightarrow {}^{188}_{76}Os + {}^{4}_{2}He$

2. a. $^{207}_{87}Fr \rightarrow {}^{203}_{85}At + {}^{4}_{2}He$
   b. $^{234}_{90}Th \rightarrow {}^{234}_{91}Pa + {}^{0}_{-1}e$
   c. $^{62}_{29}Cu \rightarrow {}^{62}_{28}Ni + {}^{0}_{1}e$

3. a. $^{129}_{51}Sb \rightarrow {}^{129}_{52}Te + {}^{0}_{-1}e$
   b. $^{7}_{4}Be + {}^{0}_{-1}e \rightarrow {}^{7}_{3}Li$
   c. $^{205}_{83}Bi \rightarrow {}^{205}_{82}Pb + {}^{0}_{1}e$
   d. $^{206}_{87}Fr \rightarrow {}^{202}_{85}At + {}^{4}_{2}He$

4. a. ${}^{212}_{87}\text{Fr} + {}^{0}_{-1}\text{e} \rightarrow {}^{212}_{86}\text{Rn}$
   b. ${}^{212}_{87}\text{Fr} \rightarrow {}^{208}_{85}\text{At} + {}^{4}_{2}\text{He}$
   c. ${}^{226}_{90}\text{Th} \rightarrow {}^{222}_{88}\text{Ra} + {}^{4}_{2}\text{He}$
   d. ${}^{186}_{77}\text{Ir} \rightarrow {}^{186}_{76}\text{Os} + {}^{0}_{1}\text{e}$

5. a. ${}^{242}_{96}\text{Cm} \rightarrow {}^{238}_{94}\text{Pu} + {}^{4}_{2}\text{He}$
   b. ${}^{238}_{94}\text{Pu} \rightarrow {}^{234}_{92}\text{U} + {}^{4}_{2}\text{He}$
   c. ${}^{234}_{92}\text{U} \rightarrow {}^{230}_{90}\text{Th} + {}^{4}_{2}\text{He}$
   d. ${}^{230}_{90}\text{Th} \rightarrow {}^{226}_{88}\text{Ra} + {}^{4}_{2}\text{He}$
   e. ${}^{226}_{88}\text{Ra} \rightarrow {}^{222}_{86}\text{Rn} + {}^{4}_{2}\text{He}$
   f. ${}^{222}_{86}\text{Rn} \rightarrow {}^{218}_{84}\text{Po} + {}^{4}_{2}\text{He}$

6. a. ${}^{218}_{84}\text{Po} \rightarrow {}^{214}_{82}\text{Pb} + {}^{4}_{2}\text{He}$
   b. ${}^{214}_{82}\text{Pb} \rightarrow {}^{214}_{83}\text{Bi} + {}^{0}_{-1}\text{e}$
   c. ${}^{214}_{83}\text{Bi} \rightarrow {}^{214}_{84}\text{Po} + {}^{0}_{-1}\text{e}$
   d. ${}^{214}_{84}\text{Po} \rightarrow {}^{210}_{82}\text{Pb} + {}^{4}_{2}\text{He}$
   e. ${}^{210}_{82}\text{Pb} \rightarrow {}^{210}_{83}\text{Bi} + {}^{0}_{-1}\text{e}$
   f. ${}^{210}_{83}\text{Bi} \rightarrow {}^{210}_{84}\text{Po} + {}^{0}_{-1}\text{e}$
   g. ${}^{210}_{84}\text{Po} \rightarrow {}^{206}_{82}\text{Pb} + {}^{4}_{2}\text{He}$

7. a. $k = 7.7 \times 10^{-8}$/y b. $k = 0.0267$/h c. $k = 0.04683$/d d. $k = 0.0136$/s

8. 0.74 g will remain.

9. 7.0×10^{-8} g will remain.

10. 54 min

11. 69 min

12. a. 3 half-lives. b. 19.93 half-lives

13. The tree is 20,000 years old.

14. No, it is not. If we assume that bones may be as old as 2×10^6 y, this is far less than one half-life, and therefore probably undetectable.

15. The rock is 3.8×10^9 y old.

16. a. mass defect = 1.915095 g
    b. mass defect = 1.55535 g
    c. mass defect = 0.700555 g

17. ΔE = binding energy = -5.59×10^{13} J/mol and −8.79 MeV/nucleon

18. $\Delta E = -1.20 \times 10^{14}$ J/mol and −8.31 MeV/nucleon.

Answers to Multiple-Choice Self-Test

1. B	4. B	7. A	10. D	13. A	16. C
2. D	5. A	8. D	11. A	14. A	17. A
3. C	6. B	9. B	12. C	15. B	18. A

CHAPTER 21

Organic and Biochemical Molecules

The Bottom Line: Chapter 21

Organic chemistry is the study of compounds that contain carbon. Such compounds are ubiquitous because carbon **forms strong bonds** to hydrogen, oxygen, and nitrogen, among others. It also has the **unique** ability to form **chains and rings** with other carbon atoms. This chapter serves as a simple introduction to the tens of thousands of known organic compounds.

21.1 Alkanes: Saturated Hydrocarbons

Alkanes are a group of **saturated hydrocarbons. Saturated** means the carbon is bound to **four atoms** (each by a single bond). Each carbon is sp^3-hybridized. Alkanes have the general formula C_nH_{2n+2}. Table 21.1 in your textbook gives names, formulas, and some properties of straight-chain (n) alkanes.

Example 21.1 A - Alkanes

Give the formula and name for the straight-chain alkanes with $n = 6$ and $n = 8$. Compare their boiling and melting points. Justify the difference.

Solution

$$\text{For } n = 6, C_nH_{2n+2} = \textbf{C}_6\textbf{H}_{14} = \textbf{hexane}$$
$$\text{For } n = 8, C_nH_{2n+2} = \textbf{C}_8\textbf{H}_{18} = \textbf{octane}$$

The boiling point of octane is 58° higher than hexane. The melting point is 38° higher. Octane is considerably heavier and longer. The London dispersion forces are more extensive, thus more energy is required to break intermolecular bonds.

Isomers are compounds with the **same formula** but **different structures**. Let's draw some of the isomers for heptane, C_7H_{16}. The key is to make sure that if the main chain lengths are the same, **the type** or **location** of groups on the chain are **unique**.

Given the straight chain,

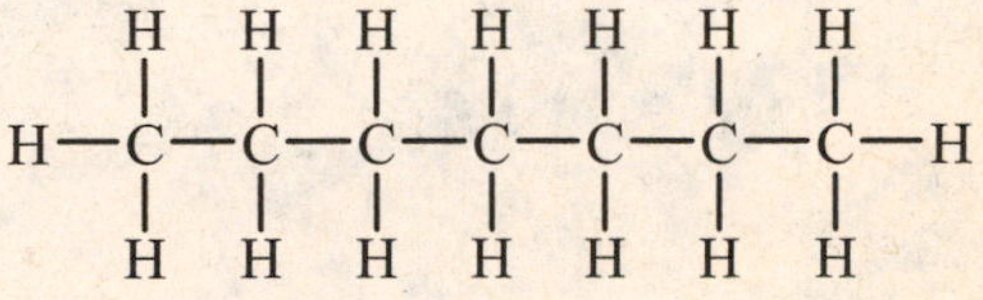

$$H-\underset{\underset{H}{|}}{\overset{\overset{H}{|}}{C}}-\underset{\underset{H}{|}}{\overset{\overset{H}{|}}{C}}-\underset{\underset{H}{|}}{\overset{\overset{H}{|}}{C}}-\underset{\underset{H}{|}}{\overset{\overset{H}{|}}{C}}-\underset{\underset{H}{|}}{\overset{\overset{H}{|}}{C}}-\underset{\underset{H}{|}}{\overset{\overset{H}{|}}{C}}-\underset{\underset{H}{|}}{\overset{\overset{H}{|}}{C}}-H \qquad H_3C-CH_2-CH_2-CH_2-CH_2-CH_2-CH_3 \qquad C_7H_{16}$$

an example of an isomer is

$$CH_3\underset{\underset{\textbf{CH}_3}{|}}{CH}CH_2CH_2CH_2CH_3$$

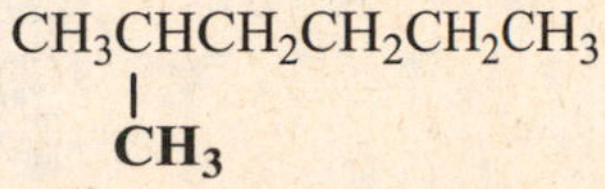

- The formula is still **C_7H_{16}.**
- The longest chain has **6 carbons, a** hexane.

- The numbering of carbons puts the functional group closest to the #1 carbon. Therefore the **methyl** group is attached to the **#2 carbon.**
- The name of this isomer of heptane is **2-methylhexane.**

Carefully examine the nomenclature rules listed after Example 21.1 in your textbook.

Another isomer is

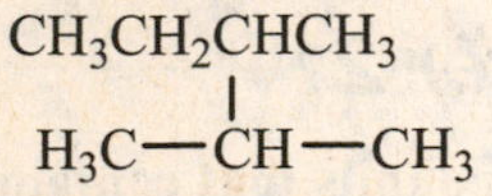

The longest chain looks like 4 carbons. It is actually 5 (eliminating hydrogens for clarity):

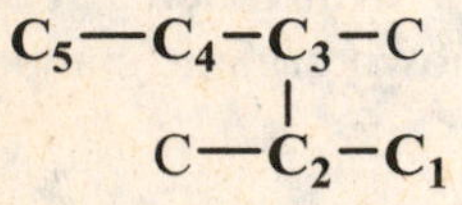

This is 2,3-dimethylpentane

Remember what you **draw on paper** is only a **shorthand representation** of a three-dimensional structure. Look very hard for the longest chain.

Example 21.1 B - Isomers

There are 9 isomers of heptane, C_7H_{16}. We have drawn and named 3. **Draw and name the other 6.** (You may eliminate hydrogens for clarity if you wish.)

Solution

Structure — Name

4. 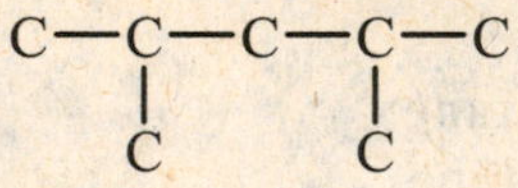 2-4-dimethylpentane

5. 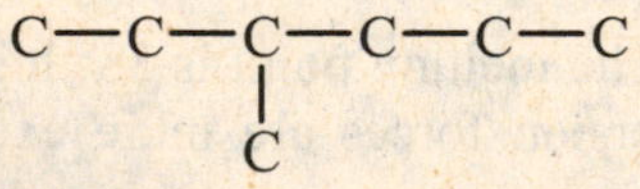 3-methylhexane

6. 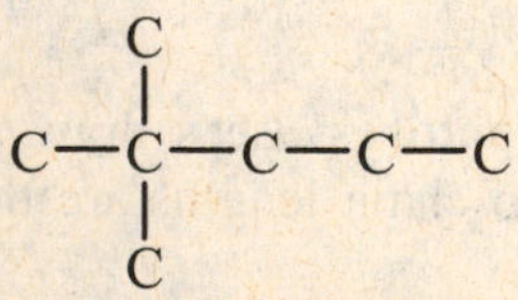 2,2-dimethylpentane

7. 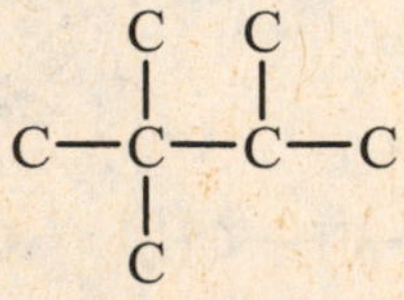 2,2,3-trimethylbutane

8. 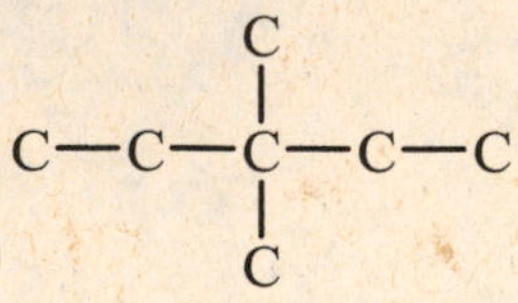 3,3-dimethylpentane

9. 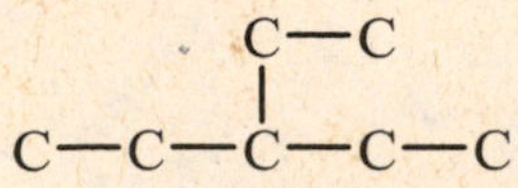 3-ethylpentane

Let's try naming some compounds and drawing some structures.

Example 21.1 C - Nomenclature

Give IUPAC names for the following structures:

a. $H_3C-CH_2-CH\cdot CH_2\cdot CH\cdot CH_2\cdot CH_3$
 | |
 CH_3 $H_2C-CH_2-CH_3$

b. $H_3C-CH-CH-CH_2-CH_3$
 | |
 Cl Cl

 CH_3
 |
c. $H_3C-CH_2-CH-CH_2\cdot CH-CH-CH_3$
 | |
 CH_3 $H_2C-CH_2-CH_3$

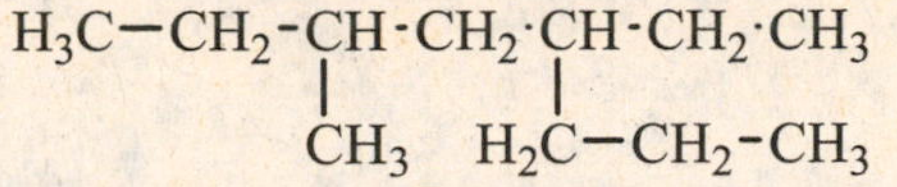

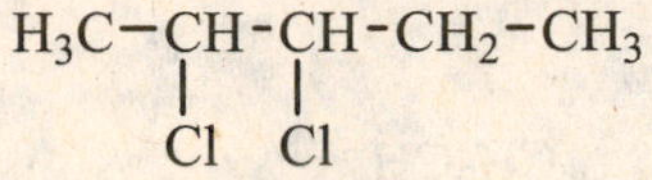

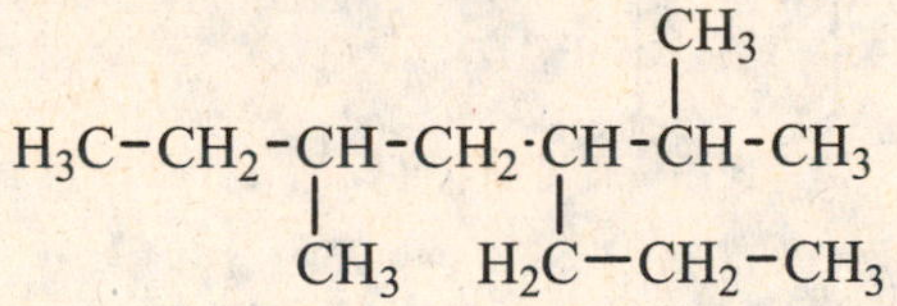

Solution

a. The longest chain has 8 members, an octane.

$$C_1-C-C-C-C-C-C$$
$$\quad\quad\quad | \quad\quad\quad\quad |$$
$$\quad\quad\quad C \quad\quad C-C-C_8$$

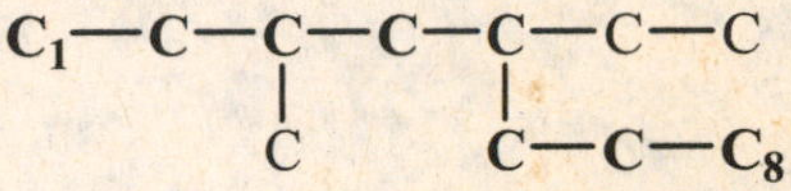

 The number would start on the **left-hand carbon** because the methyl group is closest to that side (in the #3 position). The name of the structure is **3-methyl-5-ethyloctane.**

b. **2,3-dichloropentane**

c. The longest chain has 8 members, an octane.

$$\quad\quad\quad\quad\quad\quad\quad\quad C$$
$$\quad\quad\quad\quad\quad\quad\quad\quad |$$
$$C_1-C-C-C-C-C-C$$
$$\quad\quad\quad | \quad\quad\quad\quad |$$
$$\quad\quad\quad C \quad\quad C-C-C_8$$

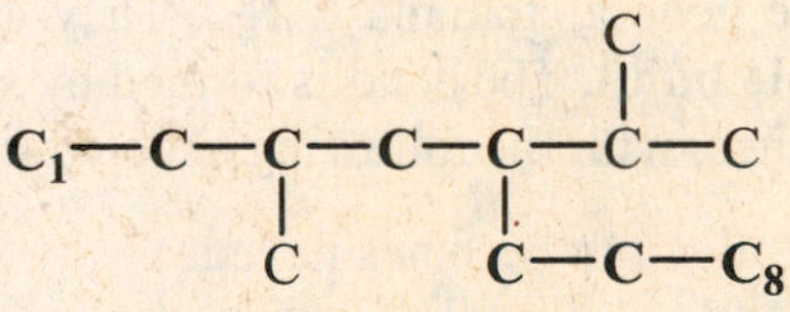

 The closest group to the #1 carbon is the methyl group in the #3 position. The name of the structure is **3-methyl-5-isopropyloctane**.

Example 21.1 D - Drawing Structures

Draw structures for the following compounds:

 a. 3,4-dimethyl-4-ethylnonane
 b. 1-chloro-2-bromobutane

Solution

 CH_3
 |
a. $H_3C-CH_2-CH-C-CH_2-CH_2-CH_2-CH_2-CH_3$
 | |
 H_3C CH_2-CH_3

b. $H_2C-CH-CH_2-CH_3$
 | |
 Cl Br

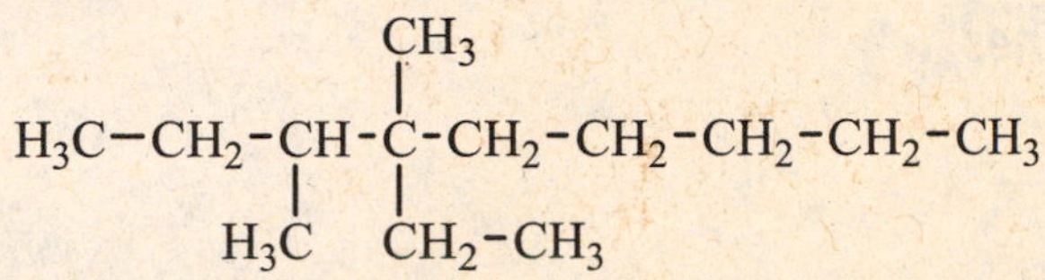

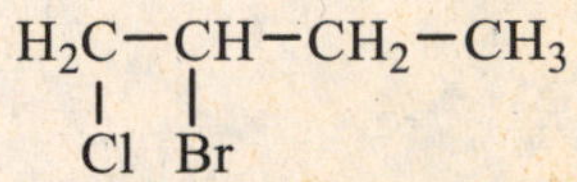

Cyclic alkanes have the general formula C_nH_{2n} (for example, C_6H_{12} is cyclohexane). Note the use of the **short-hand notation** for each of the cyclic compounds in your textbook. Notice also that as with straight-chain alkanes, numbering is done so that **substituents are attached to carbons with the lowest possible numbers.**

Example 21.1 E - Cyclic Alkanes

Name the following compounds:

a. 

c. 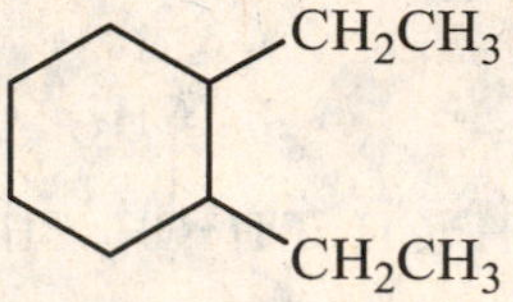

b. 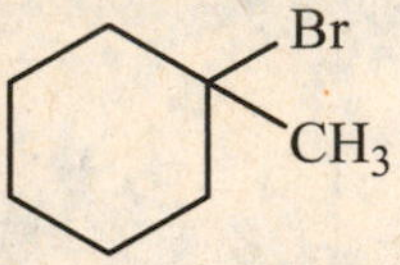

Solution

a. chlorocyclopropane
b. 1-bromo-1-methylcyclohexane
c. 1,2-diethylcyclohexane

21.2 Alkenes and Alkynes

Straight-chain **alkenes** have the general formula C_nH_{2n}. They are characterized by the presence of **at least one carbon-carbon double bond.** The bond is formed by sharing *p*-orbitals. The rules for naming alkenes are the same as for alkanes, with the following exceptions:

- **-ane** is changed to **-ene** (i.e., hex**ane** becomes hex**ene**).
- The position of the **double bond** has **highest priority** in terms of nomenclature. For example

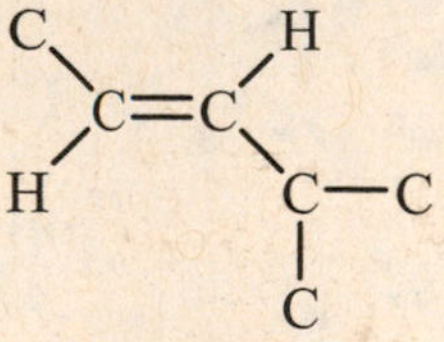

is **4-methyl-*trans*-2-pentene** NOT 2-methyl-*trans*-4-pentene.

- ***cis*** and ***trans*** isomers exist because rotation around a carbon-carbon double bond is **restricted** due to *p-p* orbital interaction between carbons.

Example 21.2 A - Naming Alkenes

Name the following alkenes:

a. 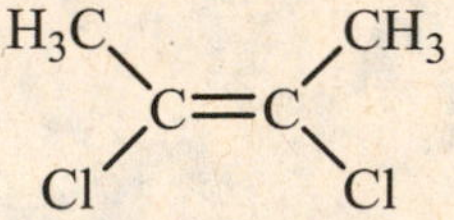

b.
$$\begin{array}{c} \text{Cl} \qquad\qquad \text{CH}_3 \\ \diagdown \qquad\diagup \\ \text{C}=\text{C} \\ \diagup \qquad\diagdown \\ \text{H}_3\text{C} \qquad\qquad \text{Cl} \end{array}$$

c. $\text{H}_3\text{C-CH}_2\text{-CH}=\text{C}\begin{array}{l}\nearrow \text{CH}_2\text{CH}_2\text{CH}_3 \\ \searrow \text{CH}_2\text{CH}_2\text{CH}_3\end{array}$

d. $\text{H}_3\text{C-CH}_2\text{-CH}_2\text{-CH}_2\text{-CH}_2\text{-C}=\text{CH}_2$
$$\qquad\qquad\qquad\qquad\qquad\qquad\quad | \\ \qquad\qquad\qquad\qquad\qquad\qquad\;\text{CH}_3$$

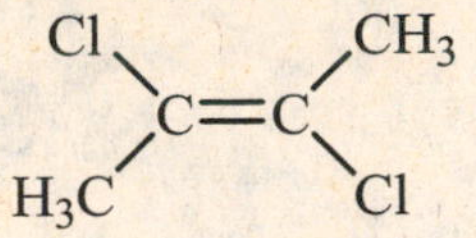

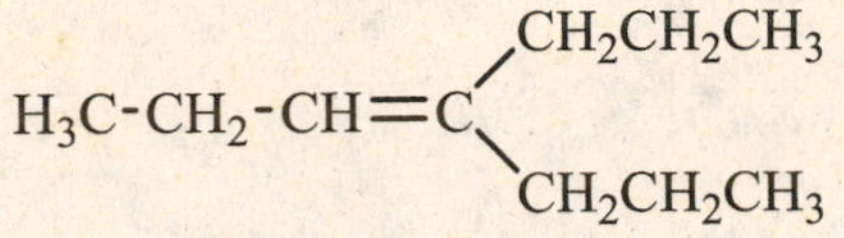

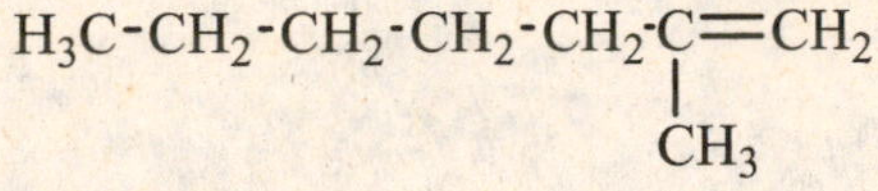

Solution

a. The longest straight chain is a butene. In this case it is **2-butene.** The chlorines are *cis* to each other and in the **#2 and #3 positions.**

$$\textbf{2,3-dichloro-}\textit{cis}\textbf{-2-butene}$$

b. Everything is the same except the chlorines are *trans* to one another.

$$\textbf{2,3-dichloro-}\textit{trans}\textbf{-2-butene}$$

c. The longest straight chain is a 7-member, or **heptene,** chain. The double bond is in the #3 position. The attached group is a **propyl** group.

$$\textbf{4-propyl-3-heptene}$$

d. The longest chain here is a **heptene.** The double bond is in the #1 position. The methyl group is in the #2 position.

$$\textbf{2-methyl-1-heptene}$$

Example 21.2 B - Drawing Alkenes

Draw the following alkenes:

 a. 3-methyl-1-hexene
 b. 1-chloro-4-ethyl-3-hexene
 c. 4-methyl-*cis*-2-pentene

Solution

a. $\text{H}_2\text{C}=\text{CHCHCH}_2\text{CH}_2\text{CH}_3$
$$\qquad\qquad\quad | \\ \qquad\qquad\;\text{CH}_3$$

b. $\text{CH}_2\text{CH}_2\text{CH}=\text{CCH}_2\text{CH}_3$
$$\quad | \qquad\qquad\qquad | \\ \;\text{Cl} \qquad\qquad\qquad \text{C}_2\text{H}_5$$

c.
$$\begin{array}{c} \text{H} \qquad\qquad \text{H} \\ \diagdown \qquad\diagup \\ \text{C}=\text{C} \\ \diagup \qquad\diagdown \\ \text{H}_3\text{C} \qquad\qquad \text{CHCH}_3 \\ \qquad\qquad\qquad | \\ \qquad\qquad\quad \text{CH}_3 \end{array}$$

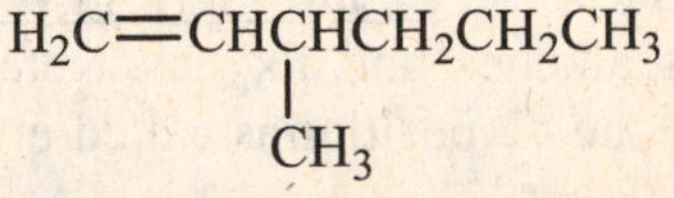

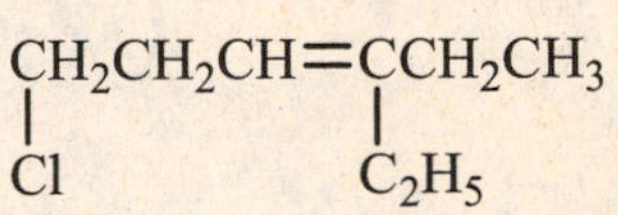

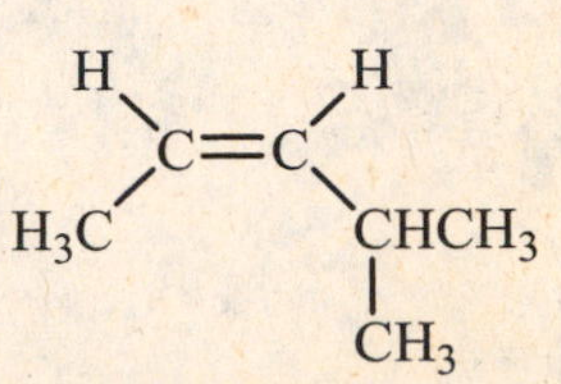

Alkynes are molecules that contain triple bonds (1s, 2 bonds involving 2 carbons). The nomenclature follows the same strategy as always, except the compound ends in **-yne.**

Example 21.2 C - Naming Alkynes

Name the following:

a. $HC{\equiv}C(CH_2)_6CH_3$

b. $HC{\equiv}C-\underset{\underset{CH_3}{|}}{\overset{\overset{CH_3}{|}}{C}}-CH_3$

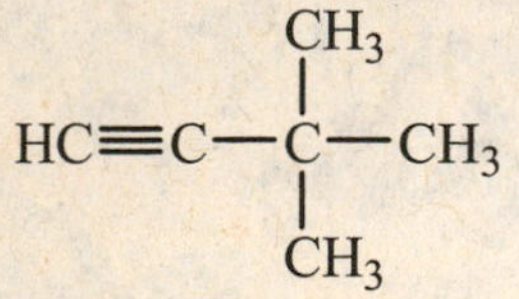

Solution

a. 1-nonyne
b. 3,3-dimethyl-1-butyne

Example 21.2 D - Just For Fun

Name this compound:

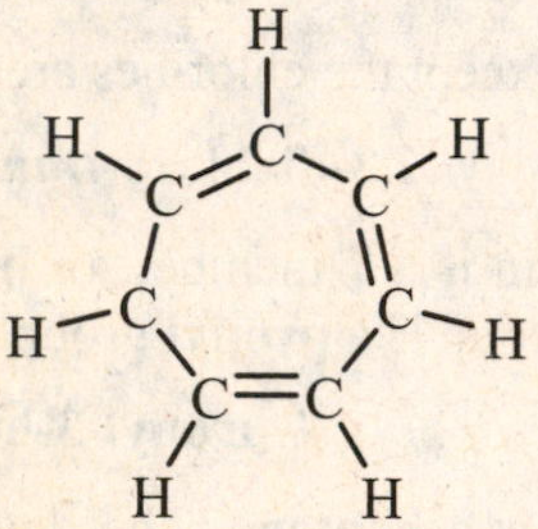

Solution

The three double bonds are in the **1, 3, and 5** positions. This is a **7-member ring**.

1,3,5-cycloheptatriene (also called **tropilidene**)

21.3 Aromatic Hydrocarbons

The main idea in Section 21.3 is that **electrons** in benzene and benzene-related compounds are **delocalized** (can move freely around the molecule). This makes benzene family compounds unreactive to addition, but reactive instead to **substitution** (where hydrogen atoms are replaced by other atoms).

Examples of benzene-related compounds with their names are shown in <u>Figure 21.12 in your textbook.</u> Notice that the numbering system is similar to that with alkanes, alkenes, and alkynes. Note as well that when there is a substituent in the #1 position, the substituent in the #2 position is called ortho-. The #3 position is meta- and the #4 position is para-.

Example 21.3 A - Naming Aromatic Compounds

Name the following compounds (NO_2 = nitro):

a.

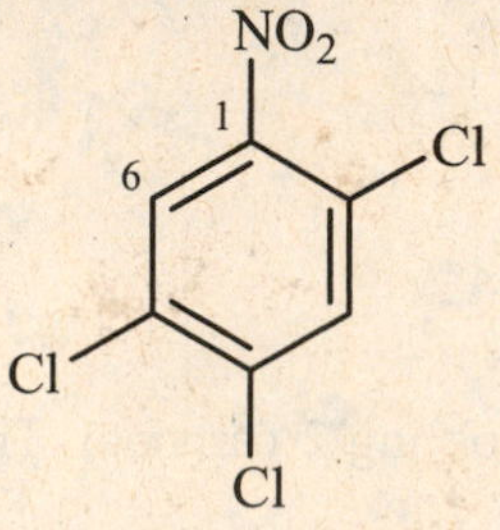

b.

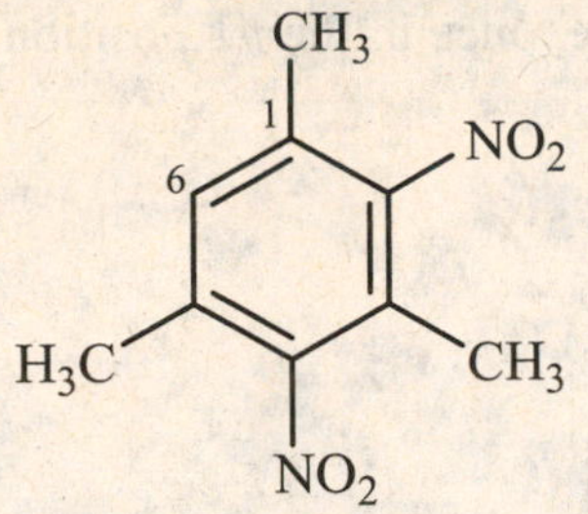

c.

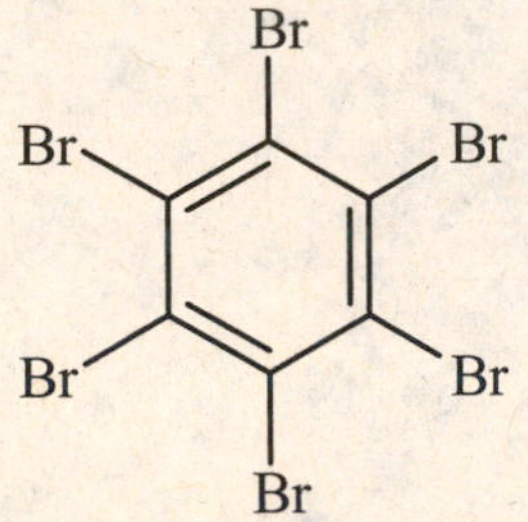

Solution

a. 1-nitro-2,4,5-trichlorobenzene
b. 1,3,5-trimethyl-2,4-dinitrobenzene
c. hexabromobenzene

Example 21.3 B - Drawing Aromatic Compounds

Draw the following compounds:

a. 1-nitro-2,3,6-triiodobenzene
b. 1-ethyl-2-methylbenzene ("2-ethyltoluene")
c. 1,4-bis(dibromomethyl)benzene

Solution

a.

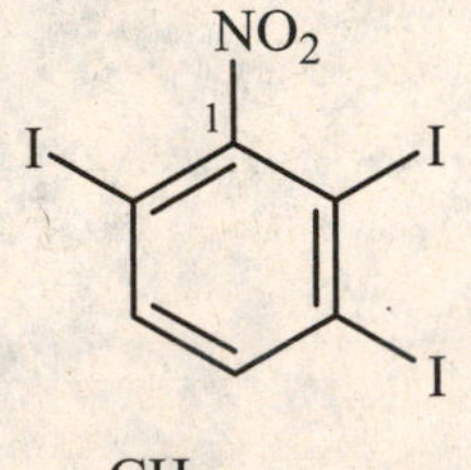

b.

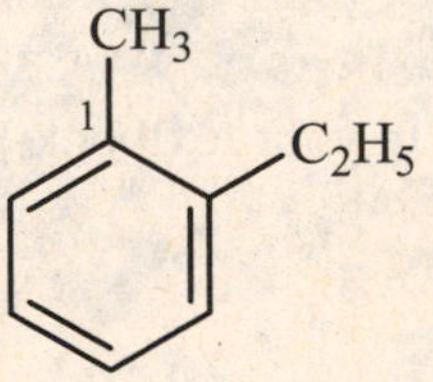

c. **bis** implies that the dibromomethyl **appears twice**, once in the #1 position and once in the #4 position.

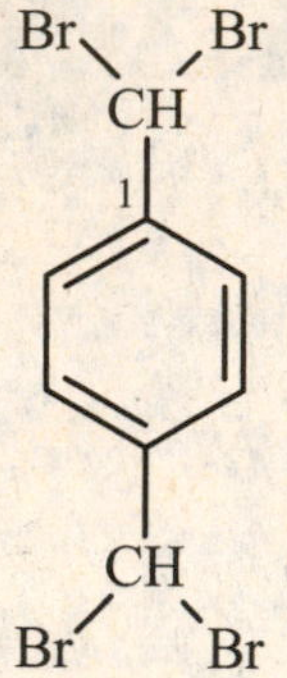

21.4 *Hydrocarbon Derivatives*

Hydrocarbon derivatives are molecules that have substituents (**functional groups**) that contain some atoms that are **not carbon or hydrogen.** Your textbook discusses the properties of several functional groups in this section. You should know the properties of

- alcohols
- aldehydes and ketones
- carboxylic acids and esters
- amines

The functional groups are summarized in <u>Table 21.4 in your textbook</u>. Let's try some naming and drawing exercises (remembering that when **benzene** is an attached group rather than the main focus of the molecule, it is called a **phenyl** group).

Example 21.4 A - Functional Groups

Name the following compounds:

a.

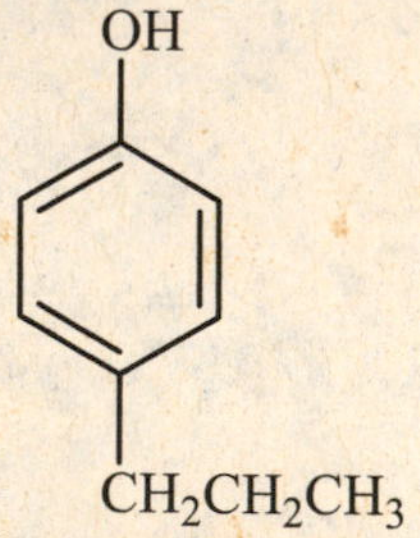

c.

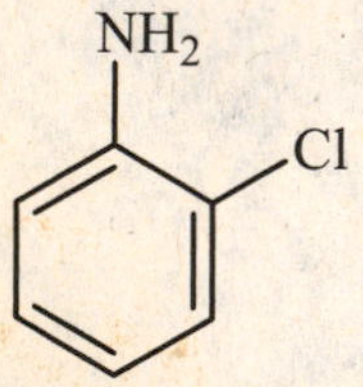

(See Table 21.6 in your textbook.)

b.

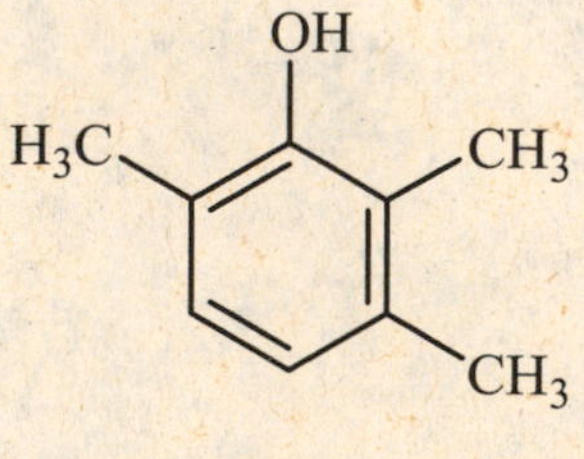

d. $CH_2CH_2\overset{\displaystyle O}{\overset{\displaystyle \|}{C}}CH_3$
 |
 Cl

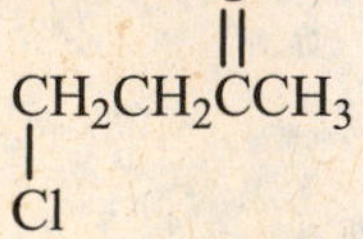

Solution

a. 4-propylphenol
b. 2,3,6-trimethylphenol

c. 2-chloroaniline (or *o*-chloroaniline)

d. 4-chloro-2-butanone

Example 21.4 B - More Functional Groups

Draw the following compounds:

a. 1,2-pentanediol

b. 3-fluorobenzoic acid

c. 1,2-cyclopentanedicarboxylic acid

Solution

a. $H_3C-CH_2-CH_2-CH-CH_2$
 | |
 OH OH

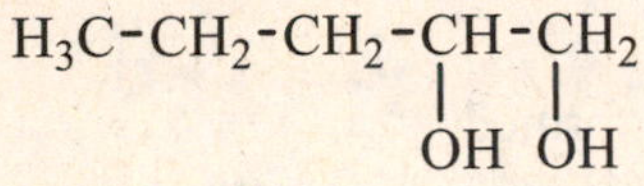

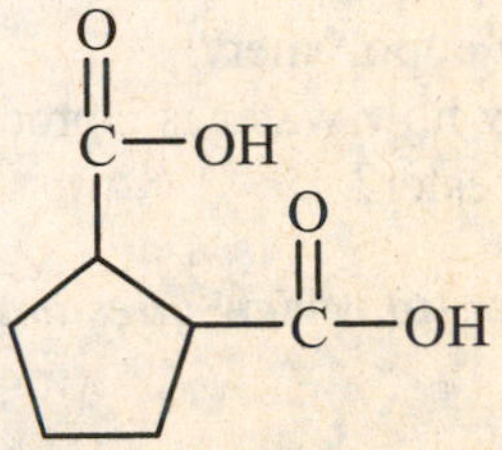

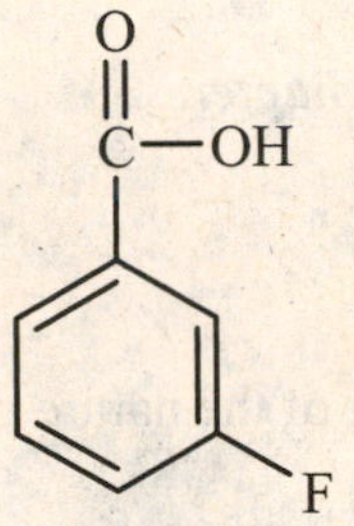

Example 21.4 C - Revenge Of The Functional Groups

List all the functional groups in each of the following molecules:

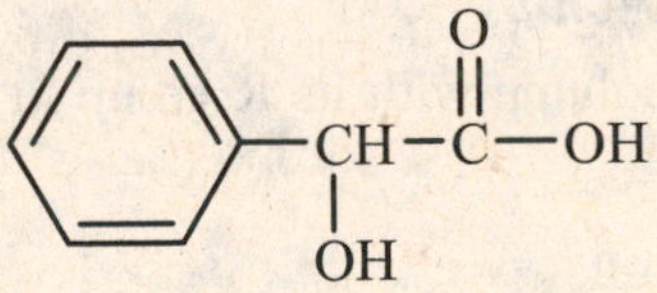

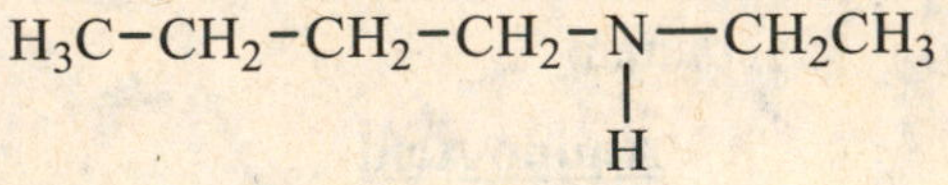

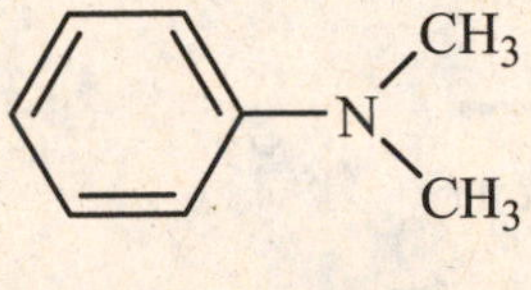

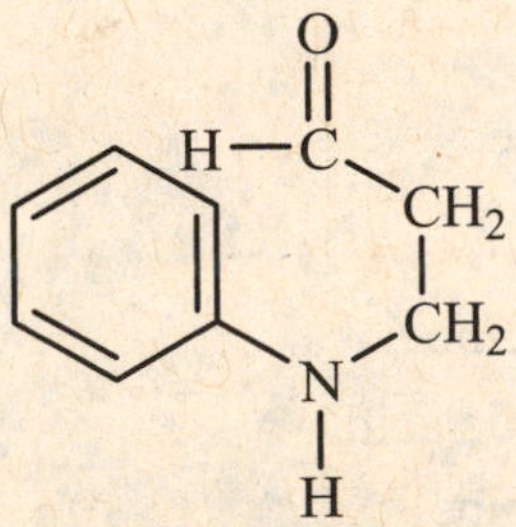

Solution

a. alcohol, carboxylic acid

b. tertiary amine (3 hydrogens have been substituted)

c. secondary amine (2 hydrogens have been substituted)

d. secondary amine, aldehyde

21.5 Polymers

The following questions will help you test your knowledge of the material in this section.

1. Define **polymer**.
2. What are polymers made from?
3. How can you vary properties when making polymers?
4. What is Teflon made of?
5. Why is Teflon so widely used?
6. Define **addition polymerization**.
7. What is a free radical?
8. Define **condensation polymerization**.
9. Why is nylon called a copolymer?
10. Use equations to show how water is a product when nylon is formed.
11. Why is Dacron a polyester?

Answers to review questions are at the end of this study chapter.

21.6 Natural Polymers

The following questions and exercises will help to review your understanding of the nature of proteins.

1. What are **proteins**?
2. What range of molar masses can proteins have?
3. Why are the acids that comprise proteins called **α-amino** acids? That is, what does α mean? What does amino mean?

Example 21.6 A - The Common Protein Amino Acids

Match the name of the amino acid in the left-hand column with its R group in the right-hand column.

Amino Acid	R group
1. glycine	a. $-CH_2-CH_2-\overset{\overset{\large O}{\|\|}}{C}-NH_2$
2. methionine	b. $-CH_2-CH_2-\overset{\overset{\large O}{\|\|}}{C}-OH$
3. glutamic acid	c. $-CH_2-(C_6H_5)$
4. glutamine	d. $-(CH_2)_4-NH_2$
5. lysine	e. $-CH_2-CH_2-S-CH_3$
6. phenylalanine	f. $-H$

Solution

Amino Acid	R group
1. glycine	f. $-H$
2. methionine	e. $-CH_2-CH_2-S-CH_3$
3. glutamic acid	b. $-CH_2-CH_2-\overset{\overset{\large O}{\|\|}}{C}-OH$

4. glutamine — a. $-CH_2-CH_2-C(=O)-NH_2$
5. lysine — d. $-(CH_2)_4-NH_2$
6. phenylalanine — c. $-CH_2-(C_6H_5)$

4. What is a peptide linkage?
5. Define **dipeptide.**

Example 21.6 B - Peptide Linkage

Draw the structure of the dipeptide formed from the condensation reaction between **leucine** and **phenylalanine**.

Solution

As pointed out in your textbook, the peptide linkage occurs between the **carboxylic acid** end of one molecule and the **amine** end of the other. Also, as standard procedure the terminal amine is always put on the left, and the terminal carboxylic acid is put on the right.

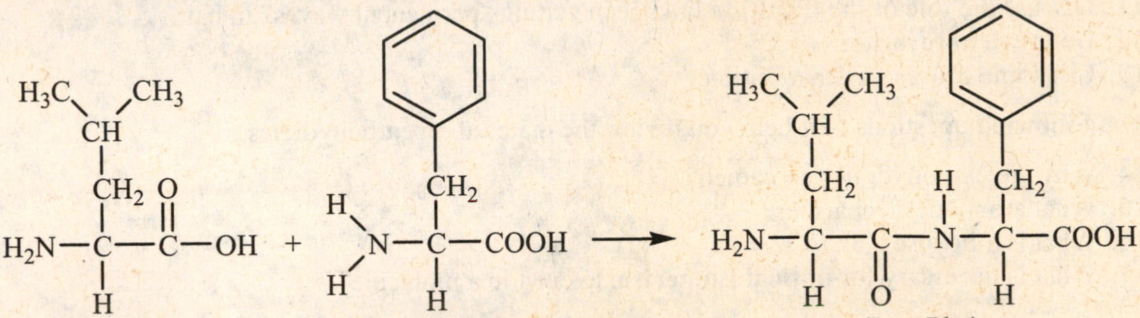

Leucine (Leu) — Phenylalanine (Phe) — (Leu-Phe)

Example 21.6 C - Practice With Peptide Linkages

Draw the structure of the polypeptide with the sequence **Trp-Ser-Asp**.

Solution

This is formed as a result of 2 peptide linkages, one between **Trp and Ser** and one between **Ser and Asp**.

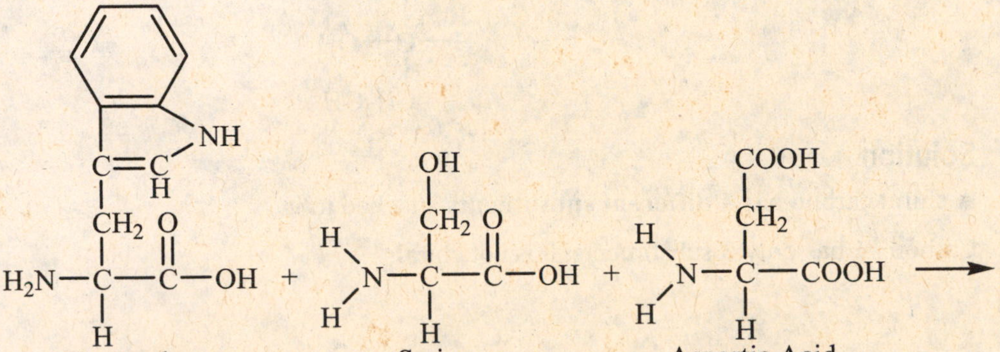

Tryptophan (Trp) — Serine (Ser) — Aspartic Acid (Asp)

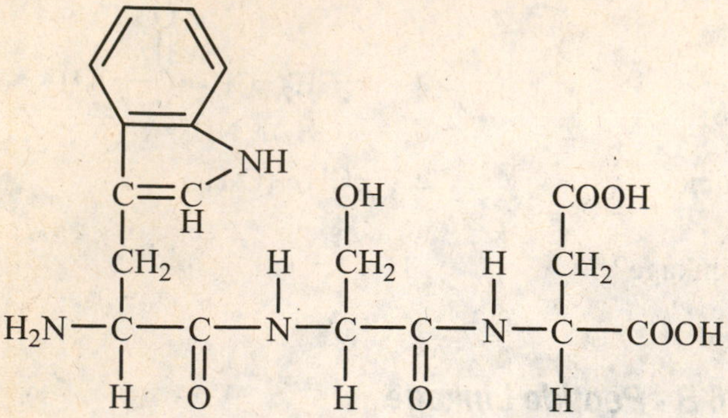

(Trp-Ser-Asp)

6. How many sequences can possibly exist for a polypeptide chosen from 10 unique amino acids?
7. What are the three levels of structure in proteins?
8. What kinds of bonding interactions are responsible for each level of structures?
9. What type of structure is an α-helix?
10. Give some practical examples of primary, secondary, and tertiary structures.
11. Discuss the role of the **disulfide** linkage in getting "permanent waves" in hair.
12. Define **denaturation**.
13. List some causes of denaturation.

The following questions will help you review the material on carbohydrates.

14. Why are carbohydrates so named?
15. What are monosaccharides?
16. What is a **hexose**?
17. What is necessary for **optical isomerism** to exist in a molecule?

Example 21.6 D - Chiral Carbons

How many chiral carbons are there in D-Glucose?

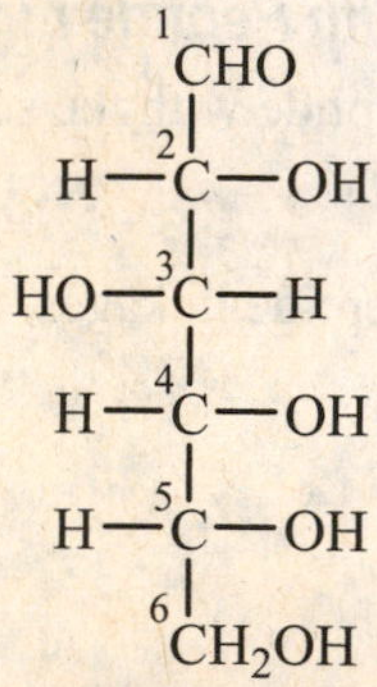

Solution

A chiral carbon has **4 different substituents** attached to it.

Carbon #1 has only 3 substituents. It is not chiral.

Carbon **#2 is chiral**.

```
        CHO
        |
  H ─── C ─── OH
        |
HO ─── C ─── H
        |
  H ─── C ─── OH
        |
  H ─── C ─── OH
        |
       CH₂OH
```

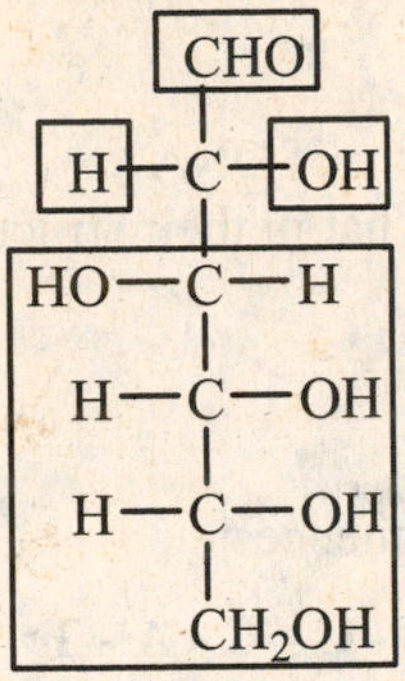

Carbon **#3 is chiral**.

```
        CHO
        |
  H ─── C ─── OH
        |
HO ─── C ─── H
        |
  H ─── C ─── OH
        |
  H ─── C ─── OH
        |
       CH₂OH
```

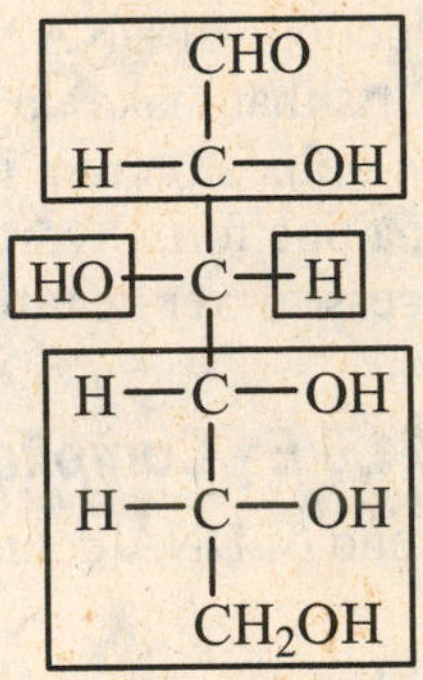

Carbon **#4 is chiral**.

```
        CHO
        |
  H ─── C ─── OH
        |
HO ─── C ─── H
        |
  H ─── C ─── OH
        |
  H ─── C ─── OH
        |
       CH₂OH
```

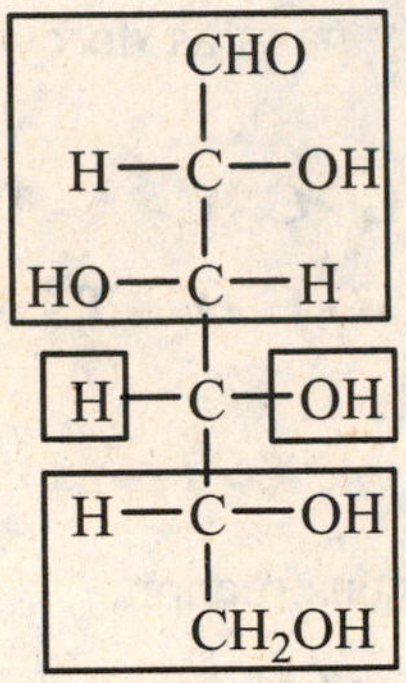

Carbon **#5 is chiral**.

```
        CHO
        |
  H ─── C ─── OH
        |
HO ─── C ─── H
        |
  H ─── C ─── OH
        |
  H ─── C ─── OH
        |
       CH₂OH
```

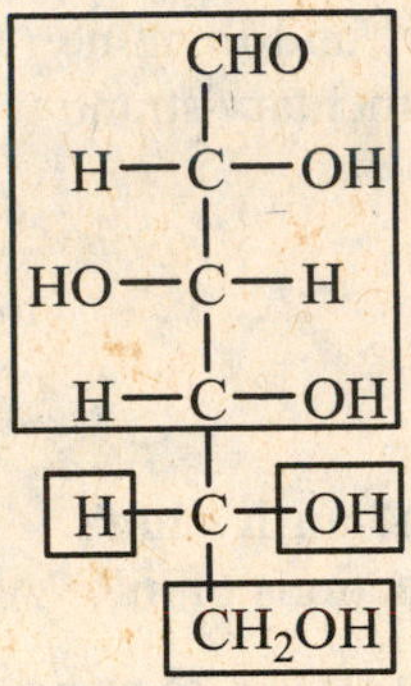

Carbon **#6** has two identical groups. It **is not chiral**.

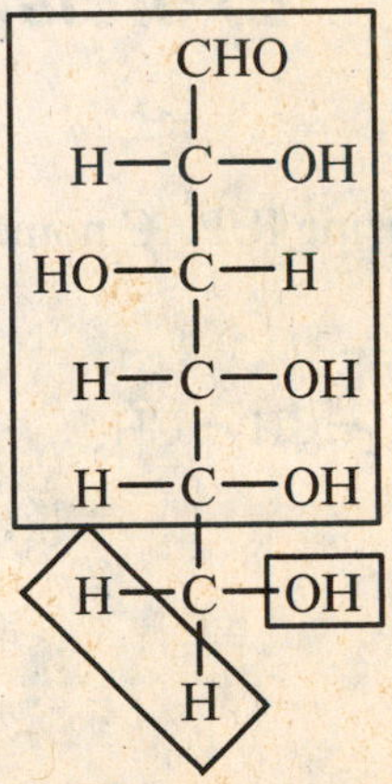

In summary, **carbons 2, 3, 4, and 5 are chiral**. Hexoses have $2^4 = 16$ optical isomers.

18. What are the bonds involved in cyclizing pentoses? hexoses?
19. Define **disaccharide**.
20. What is a **glycoside linkage**?
21. Describe the structure of **starch**.
22. Why is it advantageous for "fuel" storage to have starch as one long molecule instead of many small ones?
23. Why is cellulose **indigestible** by humans, but can be digested by certain animals?

The following questions will help you review the material on nucleic acids.

24. What are the functions of DNA?
25. List the basic parts that make up nucleotides.
26. Why is a double-helix structure important to the function of DNA?
27. Cytosine and guanine form hydrogen-bonding pairs. What in their structure makes this possible?
28. Outline the process for replication of DNA.

Example 21.6 E - Complimentary Sequences

A single strand of DNA contains the nucleotide sequence

$$A - A - G - T - T - G - C - C - A - T$$

List its complimentary strand.

Solution

Adenine (A) and thymine (T) form complimentary pairs as do cytosine (C) and guanine (G). The complimentary strand would be

$$old \; A - A - G - T - T - G - C - C - A - T$$

$$new \; T - T - C - A - A - C - G - G - T - A$$

29. Define **gene**.
30. What is a **codon**? **anticodon**?
31. Describe the functions of mRNA and tRNA to protein construction.

Answers to review questions are at the end of this study chapter.

Exercises

Section 21.1

1. Name the following compounds using IUPAC nomenclature.

 CH_3
 |
a. $H_3C - C - CH_2 - CH_2 - CH_2 - CH_2 - CH_3$
 |
 CH_3

 CH_3
 |
 CH_2
 |
b. $H_3C - CH_2 - CH - CH_2 - CH - CH_3$
 |
 CH_2
 |
 CH_2
 |
 CH_3

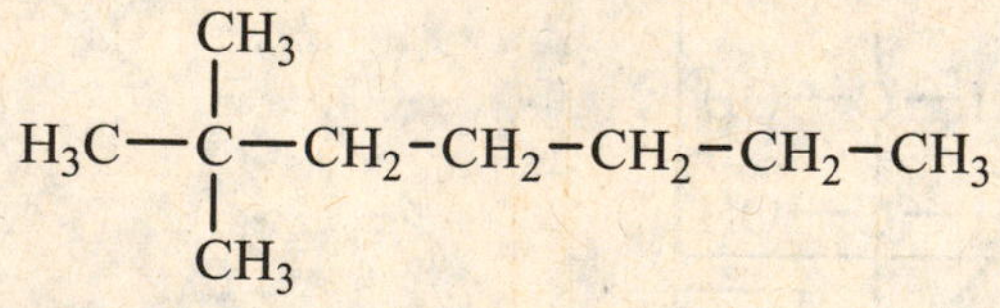

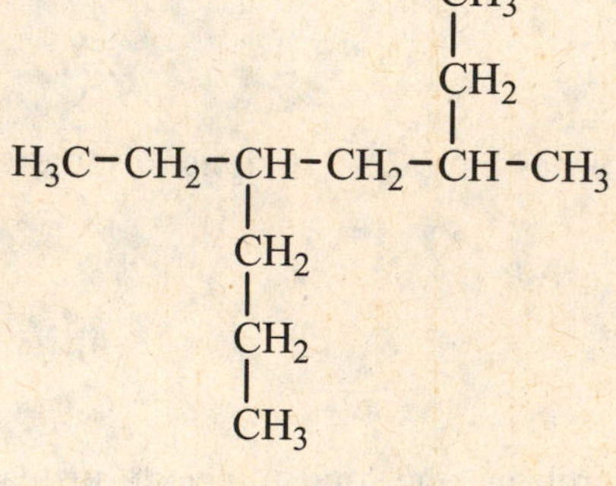

$$H_3C-CH-CH_3$$

c. $H_3C-C-CH_3$
 $|$
 CH_3

 CH_3
 $|$
d. $H_3C-C-CH_3$

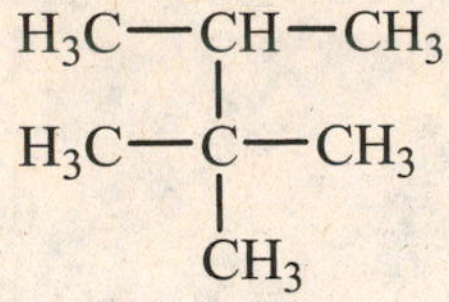

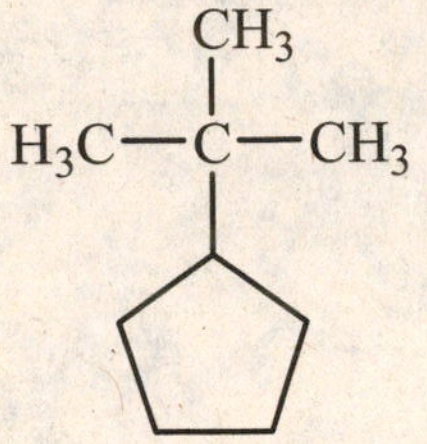

2. Write structures for the following systematic names.

 a. 1-ethyl-3-propylcyclohexane
 b. 1,1,2-trichloroethane

3. Are saturated hydrocarbons (alkanes) soluble in water?

Section 21.2

4. Match the following structures with their correct systematic name.

 a. $H_2C=CHCHCH_3$ b. $H_3C-CH-C\equiv CH$ c. $CH_3CH_2CCH_3$
 $|$ $|$ $||$
 CH_3 CH_3 CH_2

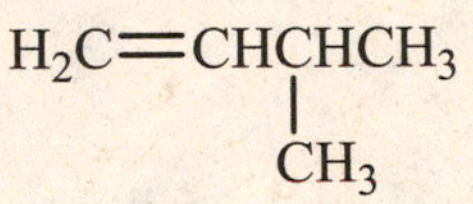

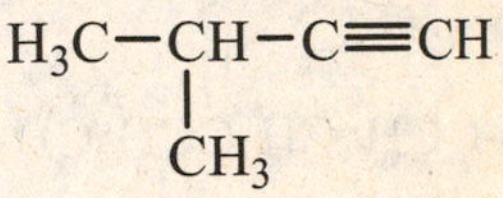

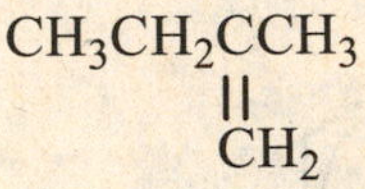

 1. 3-methyl-1-butyne 2. 3-methyl-1-butene 3. 2-methyl-1-butene

5. What products would form after hydrogenation of 2-methyl-1-butene? After halogenation (Cl_2)?

6. Why, in general, are alkenes more reactive in addition reactions than alkanes?

Section 21.3

7. What are the products of the following reactions?

 a. + $Br_2/FeBr_2$ →

 b. + Br_2/CCl_4 →

 O
 $||$
 c. + $H_3C-C-Cl /AlCl_3$ →

 d. + H_2SO_4 (fuming) →

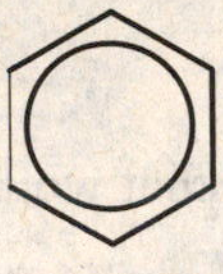

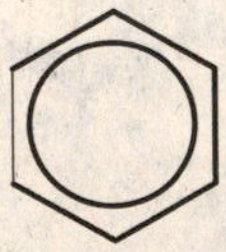

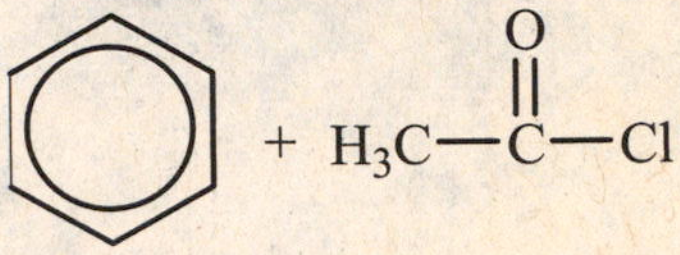

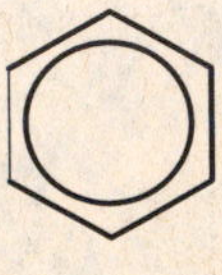

Section 21.4

8. Name the functional group(s) in each of the following compounds:

 a. $H_3C-O-CH_2-CH_2-OH$
 1 2

 b. $\underset{\underset{1}{\underset{\displaystyle O}{\|}}}{CH_3CH}$

 c. $\underset{\underset{1}{\underset{\displaystyle O}{\|}}}{CH_3CH_2C}-O-\underset{\underset{2}{\underset{\displaystyle NH_2}{|}}}{CH_2CHCH_3}$

 d. $H_3C-\underset{\underset{\displaystyle CH_3}{|}}{\overset{\overset{\displaystyle CH_3}{|}}{C}}-\underset{1}{\overset{\overset{\displaystyle O}{\|}}{C}}-OH$

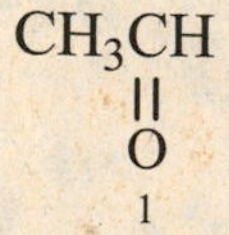

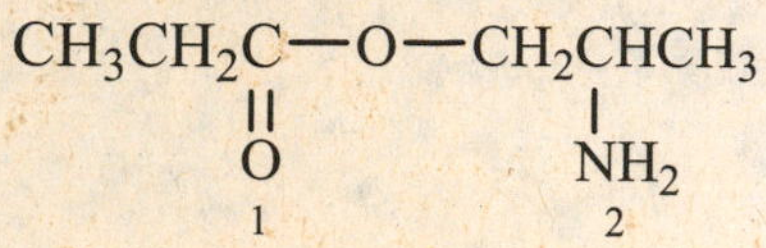

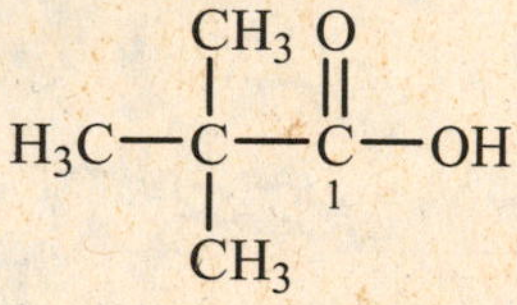

9. Name the following compounds or give their structure:

 a. (phenol structure, benzene ring with OH)

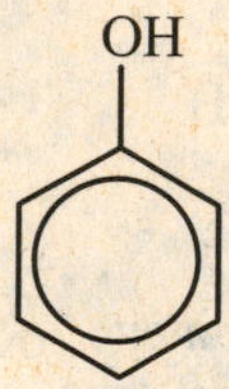

 b. 3-chlorbenzaldehyde

 c. $CH_3CH_2\overset{\overset{\displaystyle O}{\|}}{C}CH_2CH_2CH_3$

 d. $CH_3CH_2\underset{\underset{\displaystyle OH}{|}}{CH}CH_2\underset{\underset{\displaystyle Cl}{|}}{CH}CH_3$

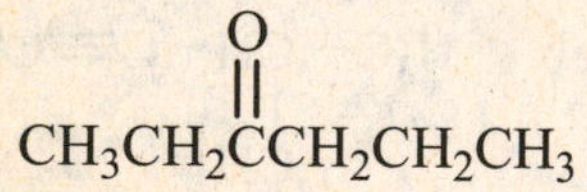

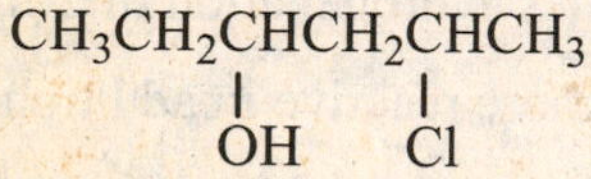

10. Arrange the molecules in order from lowest to highest boiling point:

 $CH_3CH_2CH_2OH$ $CH_3CH_2\overset{\overset{\displaystyle O}{\|}}{CH}$ $CH_3\overset{\overset{\displaystyle O}{\|}}{C}CH_3$

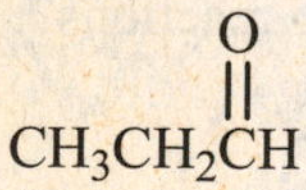

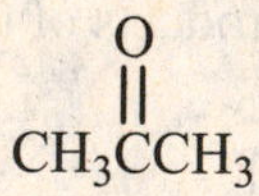

11. Name the reactants in each equation below. Give the structure of the products that would form.

 a. $CH_3\underset{\underset{\displaystyle CH_3}{|}}{CH}-\overset{\overset{\displaystyle O}{\|}}{C}-OH$ + (benzene ring with CH_2OH) \rightarrow

 b. $CH_2ClCOOH + H_3C-\underset{\underset{\displaystyle OH}{|}}{CH}-CH_3 \rightarrow$

 c. $H_3C-\underset{\underset{\displaystyle CH_3}{|}}{\overset{\overset{\displaystyle CH_3}{|}}{C}}-CH_2CH_2OH \xrightarrow{KMnO_4(aq)}$

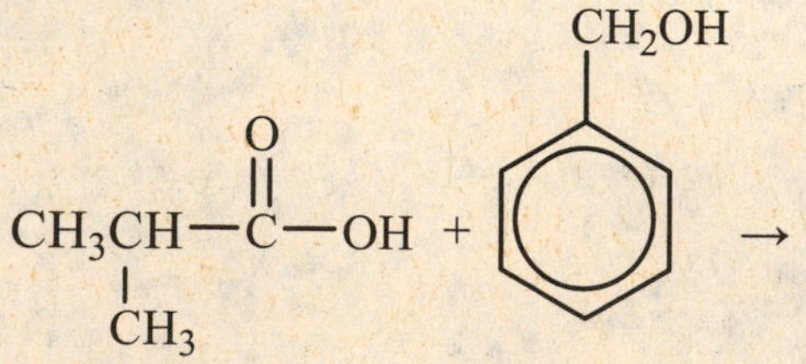

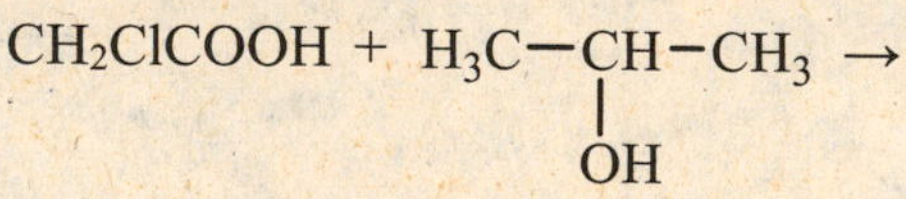

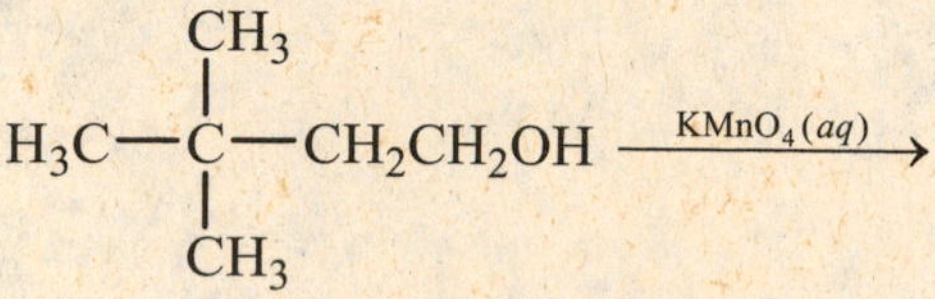

12. Label the following amines as 1°, 2°, or 3°.

 a. $N(CH_2CH_3)_3$

 CH_3
 |
 b. $(CH_3)_2N-CH$
 |
 CH_3

 c. $CH_3CH_2NHCH_3$

 CH_3
 |
 d. NH_2CH_2CH
 |
 CH_3

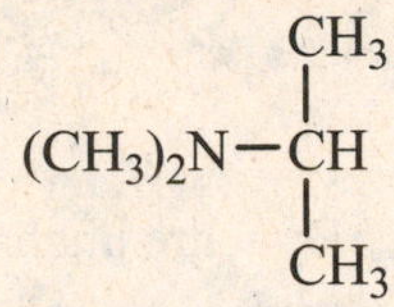

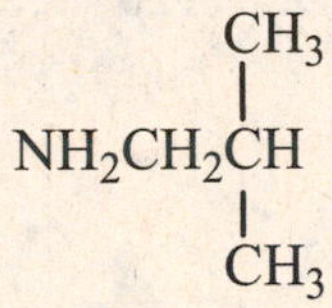

Section 21.5

13. Define or explain the following terms:

 a. dimer c. copolymer e. polymer
 b. free radical d. homopolymer

14. Distinguish between addition polymerization and condensation polymerization.

15. Write the *cis*- and *trans*- chair conformations of 1,2-dichlorocyclohexane.

16. Arrange the following alkenes from most stable to least stable. (Hint: Stability is directly related to the substitution of the double bond.)

 H H H $CH_2CH_2CH_3$
 | | | |
 a. $H-C{=}C-H$ c. $H_3CH_2C-C{=}C-H$

 CH_3 H CH_3 CH_2CH_3
 | | | |
 b. $H-C{=}C-H$ d. $H_3C-C{=}C-CH_3$

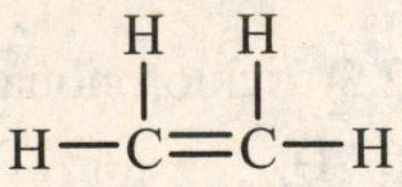

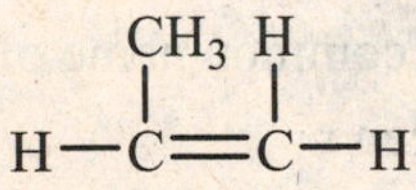

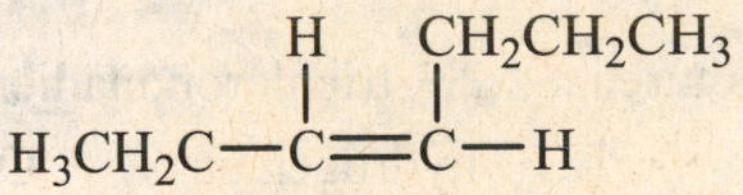

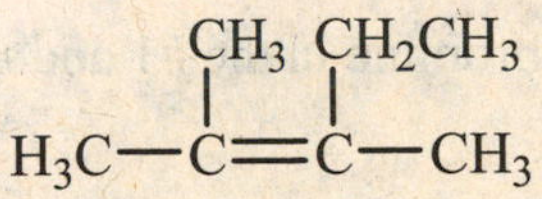

17. Alcohols are capable of forming strong hydrogen bonds to each other, making them polar. Why is ethyl alcohol greatly soluble in water while heptyl alcohol is almost insoluble in water?

18. Arrange the following amines from the highest to the lowest boiling point. Give an explanation of your answer.

 a. $(CH_3-CH_2)_2N-H$ b. $CH_3-CH_2-N(CH_3)_2$ c. $CH_3(CH_2)_3-NH_2$

Multiple-Choice Self-Test

1. What is the number of possible isomers for C_4H_8?

 A. 6 B. 3 C. 5 D. 2

2. 1,1,2-trimethylcyclopentane is an isomer of which one of the following compounds?

 A. nonane C. 2-isopropyl-pentane
 B. isoheptane D. isohexane

3. Which one of the following compounds can react with chlorine gas to produce 1,2-dichlorocyclohexane?

A. hexane
B. cyclohexene
C. 3-methylcyclohexane
D. 2-methylhexane

4. When ethane is converted to ethylene (CH_2CH_2), the carbon atoms

A. are oxidized B. are reduced C. act as oxidizers D. are unchanged

5. What is the bond angle between H—C—C in acetylene?

A. 180° B. 90° C. 109° D. 120°

6. What is the proper name of the following compound?

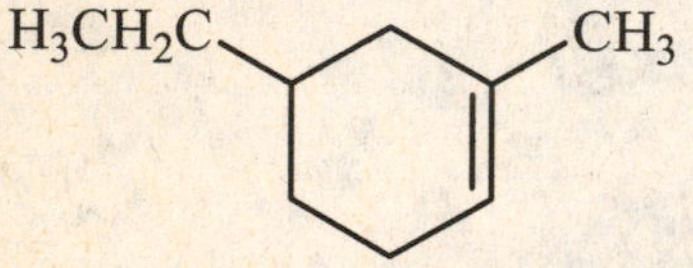

A. 4-ethyl-2-methylcyclohexene
B. 5-ethyl-1-methyl-cyclohexene
C. 4-ethyl-2-methylcyclohex-1-ene
D. 2-methyl-5-ethylcyclohex-1-ene

7. What is the proper name of the following compound?

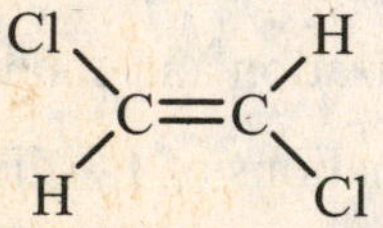

A. *cis*-1,2-dichlorobutene
B. *trans*-1,2-dichlorobutane
C. *trans*-1,2-dichloroethene
D. *cis*-1,2-dichloroethene

8. With what would you react 2,2,3-trichlorononadiene to convert it to 2,2,3-trichlorononane?

A. oxygen B. hydrogen gas C. chlorine gas D. water

9. A benzene compound with bromine in the #1 and #3 positions has the common name of:

A. *o*-dibromobenzene
B. *p*-dibromobenzene
C. *m*-dibromobenzene
D. dibromobenzene

10. The process by which hexane is converted into methylcyclopentane is known as:

A. esterification
B. pyrolysis
C. catalytic reforming
D. isomerization

11. Which one of the following processes is not used to increase octane rating?

A. polymerization B. alkylation C. isomerization D. esterification

12. Which one of the following alcohols would you expect to have the highest boiling point?

A. methanol B. propanol C. decanol D. hexanol

13. Oxidation of which one of the following compounds would lead to an aldehyde?

A. cyclohexanol B. 2-butanol C. methanol D. phenol

14. What functional group(s) are present in the compound $CH_3CHOHCOOH$?

A. acid B. alcohol, acid C. ketone, acid D. ether, acid

15. The following compound can be prepared by reacting which one of the following pairs of reagents?

$$CH_3CH_2CH_2COOCH_2CH_3$$

A. butyric acid with ethanol
B. butyraldehyde with ethanoic acid
C. ethanoic acid with butyrate
D. 2-butanone with acetaldehyde

16. Which one of the following amines is a primary amine?

A. diethylamine B. 1-aminohexane C. trimethylamine D. diphenylamine

Answers to Exercises

1. a. 2,2-dimethylheptane
   b. 5-ethyl-3-methyloctane
   c. 2,2,3-trimethylbutane
   d. t-butylcyclopentane

2. a.

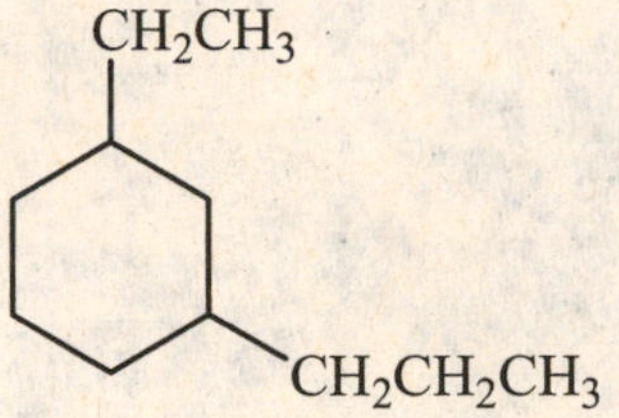

   b. $Cl{-}CH(Cl){-}CH_2(Cl)$

3. Alkanes are almost totally insoluble in water. This is due to their nonpolar nature (water is polar) and their inability to form hydrogen bonds.

4. a. (2) b. (1) c. (3)

5. Hydrogenation:

$$H_2C{=}C(CH_3){-}CH_2{-}CH_3 + H_2 \rightarrow H_3C{-}CH(CH_3){-}CH_2{-}CH_3 \text{ (2-methylbutane)}$$

Halogenation:

$$H_2C{=}C(CH_3){-}CH_2{-}CH_3 + Cl_2 \rightarrow H_2C(Cl){-}C(Cl)(CH_3){-}CH_2{-}CH_3 \text{ (1,2-dichloro-2-methylbutane)}$$

6. Alkenes have a carbon-carbon double bond consisting of a C—C σ bond and C—C π. Alkanes consist of C—C σ bonds. Thus the presence of the π bond and its exposed electrons make alkenes more susceptible to addition reaction than alkanes.

7. a. 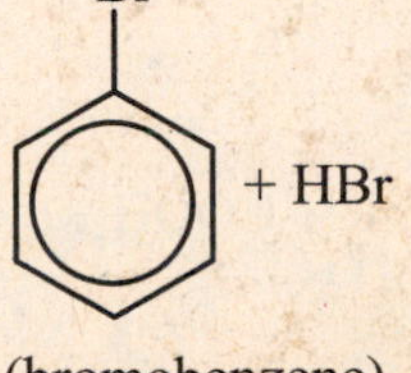 + HBr

(bromobenzene)

   b. No Reaction. (No Lewis acid catalyst)

7. c.

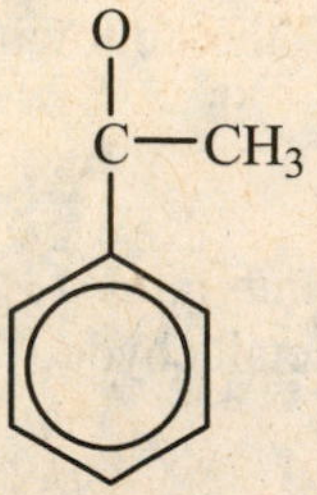

$$\overset{\displaystyle O}{\overset{\displaystyle \|}{C}}\text{—CH}_3$$

[benzene ring] + HCl

(methylphenylketone)

d.

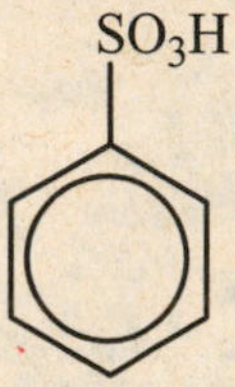

SO_3H

[benzene ring] + H₂O

(benzenesulfonic acid)

8. a. (1) ether, (2) 1° alcohol c. (1) ester, (2) 1° amine
 b. (1) aldehyde d. (1) carboxylic acid

9. a. phenol

 b. $\text{H}-\overset{\displaystyle O}{\overset{\displaystyle \|}{C}}$

 [benzene ring with Cl]

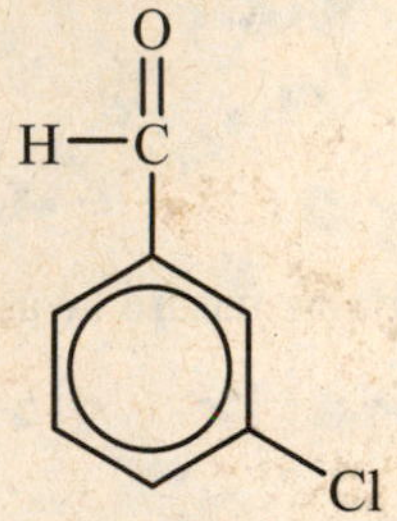

 c. 3-hexanone
 d. 2-chloro-4-hexanol

10. $\underset{}{\text{CH}_3\text{CH}_2\overset{\displaystyle O}{\overset{\displaystyle \|}{C}}\text{H}} \quad < \quad \text{CH}_3\overset{\displaystyle O}{\overset{\displaystyle \|}{C}}\text{CH}_3 \quad < \quad \text{CH}_3\text{CH}_2\text{CH}_2\text{OH}$

11. a. 2-methylpropanoic acid + benzyl alcohol → $\text{H}_3\text{C}-\underset{\displaystyle \underset{}{\text{CH}_3}}{\text{CH}}-\overset{\displaystyle O}{\overset{\displaystyle \|}{C}}-\text{O}-\text{CH}_2$ [benzene ring]

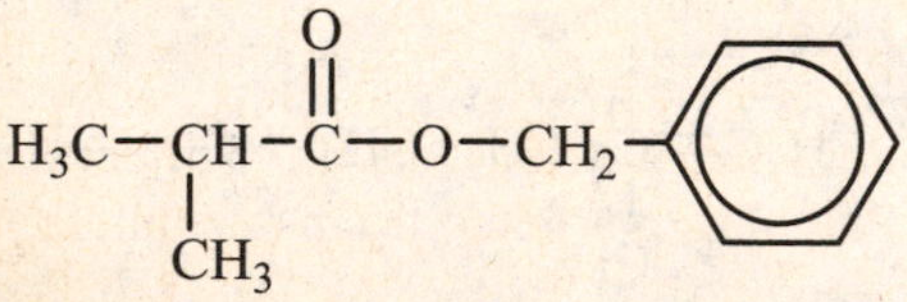

 b. chloroethanoic acid + 2-propanol or isopropyl alcohol → $\text{H}_2\text{CClC}\overset{\displaystyle O}{\overset{\displaystyle \|}{}}-\text{O}-\underset{\displaystyle \underset{}{\text{CH}_3}}{\text{CH}}-\text{CH}_3 + \text{H}_2\text{O}$

 c. 3,3-dimethyl-1-butanol KMnO₄(aq) $\xrightarrow{\text{KMnO}_4(aq)}$ $\text{H}_3\text{C}-\underset{\displaystyle \underset{}{\text{CH}_3}}{\overset{\displaystyle \overset{}{\text{CH}_3}}{\text{C}}}-\text{CH}_2\text{COOH}$

12. a. 3° b. 3° c. 2° d. 1°

13. a. two identical monomers joining together to form a molecule
 b. a species with an unpaired electron
 c. more than one type of monomer combining to form the chain in a polymer
 d. identical monomers combining to form the chain in a polymer
 e. large 1, 2, or 3-dimensional molecules consisting of large numbers of repeating monomer units

14. **Addition polymerization** involves the formation of the polymer by a free-radical mechanism. The only product formed is the polymer, and the polymerization stops when two radicals react to form a bond without producing any other radicals.

 Condensation polymerization produces a product other than the polymer itself. The side products are most commonly small molecules such as water or alcohols.

15.

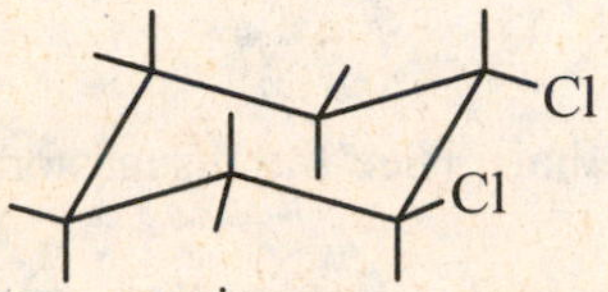

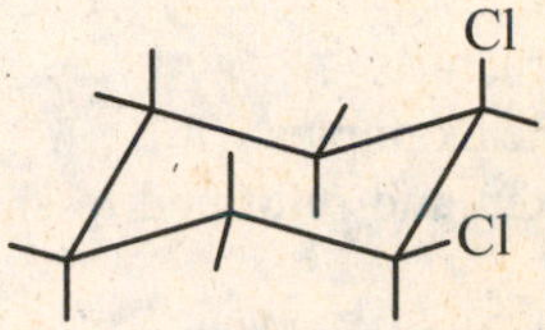

16. d > c > b > a

17. Heptyl alcohol has a seven-carbon chain while ethyl alcohol has a two-carbon chain. Due to this fact, the nonpolar characteristics of the seven-carbon chain dominate over the polar characteristics of the two-carbon chain and make it insoluble in water.

18. c > a > b, due to hydrogen bonding. A 1° amine has a greater amount of hydrogen bonding when compared to a 2° amine. A 3° amine has little hydrogen bonding when compared to a 1° and 2° amine.

Answers to Multiple-Choice Self-Test

1.	A	4.	A	7.	C	10.	C	13.	C	15.	A
2.	C	5.	A	8.	B	11.	D	14.	B	16.	B
3.	B	6.	B	9.	C	12.	C				

Answers to Section 21.5

1. Polymers are large chainlike molecules that are built from small molecules.
2. Polymers are made from monomers. See Table 21.7 in your textbook for a list of some common monomers.
3. Properties can be varied by using substituted monomers.
4. Teflon is made of tetrafluoroethylene monomers.
5. Because the C—F bonds resist chemical attack, Teflon is inert, tough, and nonflammable.
6. A process in which monomers combine together to form a polymer. The process is initiated by a free radical.
7. A free radical is a species with an unpaired electron.
8. A process in which a small molecule, such as water, is formed for each extension of the polymer chain.
9. Because two different types of monomers combine to form the polymer.
10. See the reaction after the definition of homopolymer in your textbook.
11. The two monomers combine to form an ester group.

Answers to Section 21.6

1. Proteins are large amino acid-based "natural polymers" in our bodies that perform a variety of biological functions.
2. The molecular weights can range from 6,000 to over 1,000,000.
3. They are called "α-amino acids" because the amino group is attached to the α-carbon. Amino means an NH_2 group.
4. A peptide linkage occurs when the carboxyl group of a carboxylic acid interacts with a hydrogen from an amine group.
5. A dipeptide involves 2 amino acids in a peptide linkage.
6. $10! = 10 \times 9 \times 8 \times 7 \times 6 \times 5 \times 4 \times 3 \times 2 \times 1 = 3.6 \times 10^6$ sequences.
7. The levels are primary, secondary, and tertiary.
8. Primary = peptide linkages, secondary = hydrogen bonding, tertiary = a variety of interactions including hydrogen bonding, ionic bonds, and covalent bonds, among others, quaternary = bonding between individual subunits.
9. An α-helix represents a secondary structure.
10. See the discussion in the text for some examples.
11. The disulfide linkage in hair is broken and reformed to shape the hair. (See the discussion right before the carbohydrates section.)
12. Denaturation involves breaking down the three-dimensional structure of a protein, thus rendering it inactive.
13. Heat and intense radiation are two causes.
14. Historically, they were thought to be hydrates of carbon. For example, $C_{12}H_{22}O_{11}$ was thought to be $C_{12} \cdot 11H_2O$.
15. Monosaccharides are simple sugars.
16. A hexose is a sugar with 6 carbon atoms.
17. Optical isomerism requires a chiral carbon.
18. The oxygen of the terminal OH group combines with the carbon of the ketone group.
19. A disaccharide is a combination of 2 simple sugars.
20. A C—O—C linkage between rings of glucose and fructose is a glycoside linkage.
21. Starch is a polymer of α-glucose. (See Figure 21.33 in your textbook.)
22. There is less stress on the plant's internal structure. (See the discussion on osmotic pressure in Chapter 17.)
23. We do not have the necessary enzymes, β-glycosidases, to break down cellulose.
24. DNA stores and transmits genetic information.
25. a. A five-carbon sugar.
    b. A nitrogen-containing organic base.
    c. A phosphoric acid molecule.
26. The double helix structure allows DNA to produce complementary strands.
27. Polar C═O and N—H bonds lead to hydrogen bonding which leads to the formation of the double helix.
28. See Figure 21.39 in your textbook.
29. A gene is a segment of DNA that contains the code for a specific protein.
30. A codon consists of 3 bases and codes for a specific amino acid. An anticodon is a part of tRNA that decodes the "genetic message" from mRNA.
31. mRNA migrates from DNA to the cell cytoplasm where protein synthesis occurs. tRNA decodes the genetic message from mRNA.